Contemporary Electroanalytical Chemistry

Contemporary Electroanalytical Chemistry

Edited by

Ari Ivaska, Andrzej Lewenstam,
and Rolf Sara
Åbo Akademi University
Turku-Åbo, Finland

Plenum Press • New York and London

Library of Congress Cataloging-in-Publication Data

ElectroFinnAnalysis International Conference on Electroanalytical
  Chemistry (1988 : Turku, Finland)
    Contemporary electroanalytical chemistry / edited by Ari Ivaska,
  Andrzej Lewenstam, and Rolf Sara.
        p.   cm.
    Proceedings of the ElectroFinnAnalysis International
  Conference on Electroanalytical Chemistry, held June 6-9, 1988, in
  Turku-Åbo, Finland.
    Includes bibliographical references and index.
    ISBN 0-306-43818-6 (hard)
    1. Electrochemical analysis--Congresses.   I. Ivaska, Ari.
  II. Lewenstam, Andrzej.   III. Sara, Rolf.   IV. Title.
  QD115.E527   1988
  543'.0871--dc20   90-27542
                                                            CIP

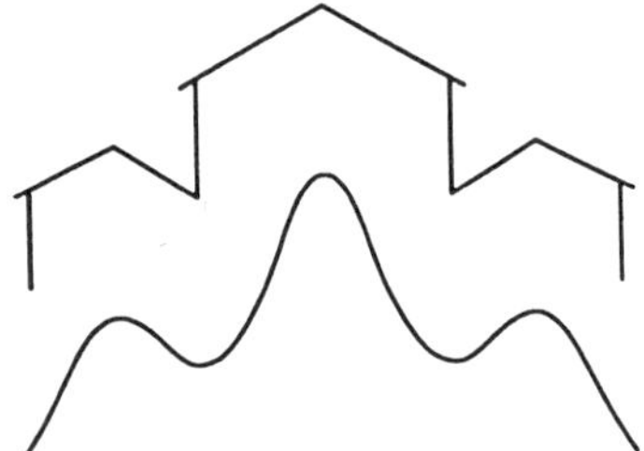

ElectroFinnAnalysis

An International Conference on
Electroanalytical Chemistry

Proceedings of the ElectroFinnAnalysis International Conference on
Electroanalytical Chemistry, held June 6–9, 1988,
in Turku-Åbo, Finland

ISBN 0-306-43818-6

© 1990 Plenum Press, New York
A Division of Plenum Publishing Corporation
233 Spring Street, New York, N.Y. 10013

Printed in the United States of America

PREFACE

This volume is based on the presentations given at the ElectroFinnAnalysis conference held on June 6-9, 1988 in Turku-Åbo, Finland. This event was the second in a series of electroanalytical conferences. The first was held in Ireland 1986 and the next will be held in Spain 1990.

The aim of these conferences is to bring together scientists who use electroanalytical methods in their research. This is also reflected in the disposition of this volume where instrumentation and applications from the different fields have their own chapters.

The editors are grateful to Mr. Johan Nyman, Mr. Kent Westerholm and Mr. Markku Lehto for their technical assistance during the editorial work of this volume.

Ari Ivaska
Andrzej Lewenstam
Rolf Sara

CONTENTS

Introduction 1
Ari Ivaska

ELECTROCHEMICAL INSTRUMENTATION AND METHODS

**New Instrumental Approaches to Fast Electro-Chemistry at
Ultramicroelectrodes** 5
Larry R. Faulkner, Michael R. Walsh and Chuanjing Xu

Photoelectroanalytical Chemistry — Methods and Instrumentation 15
Jouko J. Kankare

**Experiences of an On-Line Fourier Transform Faradaic Admittance
Measurement (FT-FAM) System Based on Digital Signal
Processors** 21
Sten O. Engblom, Mikael Wasberg, Johan Bobacka and
Ari Ivaska

Processor-Controlled Fast Potentiostat 31
J. Kankare and J. Lukkari

Smoothing of AC Polarographic Data by FFT Filtering 37
Johan Bobacka and Ari Ivaska

**Reverse Pulse Voltammetry at Microelectrodes. New Possibilities
in Analytical Chemistry** 47
Zbigniew Stojek

Multiple Sensor Arrays: Advantages and Implications 51
Dermot Diamond

Simultaneous ESR-Electrochemical Investigations at Solid Electrodes 59
Lothar Dunsch

INDUSTRIAL APPLICATIONS

**Industrial Applications and Perspectives of Electro-Analytical
Methods** 71
E. Pungor, Zs. Niegreisz and L. Pólos

**Examples of Electrometric Methods Applied to Process and
Product Control in the Explosives Industry** 85
Jan Asplund

**Applied Polarography and Voltammetry in Day-to-Day
Environmental Analysis, Possibilities and Limitations** 109
Pierre M. Bersier and Jacques Bersier

The Use of Electro-Analytical Techniques in Biotechnology 139
Bauke te Nijenhuis

Potentiometric Determination of Copper in Various Plating Baths 145
Adam Hulanicki, Tomasz Sokalski and Andrzej Lewenstam

Controlled-Growth Mercury Drop Electrode and Perspectives in
Process Monitoring Application 149
Zygmunt Kowalski and Jan Migdalski

ELECTROCHEMICAL SENSORS

Solid State Potentiometric Sensors 159
Jiří Janata

Biosensing Based on Gas Sensitive Semiconductor Devices 173
I. Lundström and F. Winquist

Chemically Modified Electrodes for the Electrocatalytic
Oxidation of NADH 183
L. Gorton, B. Persson, M. Polasek and G. Johansson

Solid Polymer Electrolytes for Gas Sensing Electrodes 191
Lionel S. Goldring

Carbon Fiber Microelectrodes 199
Karin Potje-Kamloth, Petr Janata and Mira Josowicz

Voltammetric Determination of Organic Compounds Using
Clay Modified Carbon Paste Electrodes 205
Lucas Hernandez, Pedro Hernandez and Encarna Lorenzo

The Role of Surface Processes in Signal Formation with Solid-State
Ion-Selective Electrodes – Chloride Interference on Copper
Ion-Selective Electrode 213
A. Lewenstam, A. Hulanicki and E. Ghali

FTIR-ATR and Ion Chromatographic Investigations of the Ion
Transport through Ion-Selective PVC-Membranes 223
R. Kellner, E. Zippel, E. Pungor, K. Toth and E. Lindner

On the Electrochemical Approach to Solid-State Ion Selective
Membrane Preparation 231
M. Neshkova

Sensor Technique for Monitoring Changes in pH, Ca^{2+}, p_{O_2}, p_{CO_2}
and Electrical Conductivity in Milk During Fermentation............ 237
Pekka O. Lehtonen, Hanna Laitinen, Tuomo Tupasela and
Matti Antila

ELECTROCHEMICAL FLOW ANALYSIS

Exploitation of Electrochemical Techniques by Flow Injection
Analysis 245
Elo Harald Hansen

Potentiometric Detection in High-Performance Ion-Chromatography 255
Marek Trojanowicz

Stable Modified Electrodes for Use in Amperometric Detectors
in Flow Systems 267
James A. Cox and Thomas J. Gray

Flow Stream Detectors Based on the Electrocatalytic Oxidation
of Polyhydroxy Compounds at Silver Oxide Electrodes 275
Terrence P. Tougas, Edwin G.E. Jahngen and
Michael Swartz

Application of Modified Electrodes for Analysis in Flowing Solutions 283
Gordon G. Wallace, Mary Meaney and Malcolm R. Smyth

Studies of the Modulated Flow Technique for Flow Potentiometric
Stripping Analysis 289
Gerhard Schulze, Raymond Sänger and Edzard Han

Electroluminescence Detector for Flow Analysis 293
K. Haapakka, J. Kankare and K. Lipiäinen

Determination of Polyamines (Spermine, Spermidine, Putrescine)
in Biological Samples 299
R. Rips and C. Guette

CLINICAL APPLICATIONS

Opportunities and Limitations in the Use of ISEs in Clinical
Chemistry: Assays Without Calibration? 305
Gudrun G. Rumpf, Lucas F.J. Dürselen, Hans W. Bühler
and Wilhelm Simon

Proposed IFCC Recommendations for Electrolyte Measurements
with ISEs in Clinical Chemistry 311
Anton H.J. Maas and Ron Sprokholt

Influence of Some Drugs on ISE Measurements of Serum Electrolytes 317
Ryszard Lewandowski, Tomasz Sokalski and Adam Hulanicki

Buffer System for the Simultaneous Standardization of pH and
Electrolytes by ISE Determination in Whole Blood 323
Angelo Manzoni and Mario Belluati

An Analytical Approach to the Determination of Some Mixtures of
Selected Pteridines by Adsorptive Stripping Voltammetry 329
P. Tuñón Blanco, J.M. Fernández Alvarez and
A. Costa Garcia

Electrochemical Behaviour of Metronidazole 339
P. Siva Sankar and S.J. Reddy

PHARMACEUTICAL APPLICATIONS

The Determination of Selected Antibiotics, Antibacterials and
Anticonvulsants by Voltammetric and Liquid Chromatographic
Techniques 349
W.F. Smyth

Applications and Potentiality of Electroanalytical Methods for
Inorganic Trace Analysis in the Pharmaceutical Industry 359
Juerg B. Reust

Direct Electrochemical Immunoassays Involving Adsorbed or
Immobilised Species 367
Malcolm R. Smyth, Eileen Buckley, Juana Rodriguez Flores,
Richard John and Gordon G. Wallace

Electrochemical Behaviour at Solid Electrodes and Metabolic
Fate of Drugs 373
J-M. Kauffmann, J-C. Vire, O. Chastel and G.J. Patriarche

Adsorptive Stripping Square-Wave Voltammetry of Pharmaceutical
Quinonic Derivatives 379
J-C. Vire, G.J. Patriarche, H. Zhang, B. Gallo and R. Alonso

The Determination of Timolol in Biological Fluids by Adsorptive
Stripping Voltammetry 387
R.J. Barrio Diez-Caballero and J.F. Arranz Valentin

Pulse Voltammetric Determination of Sulphur Containing Organic Compounds 395
Zenon J. Karpinski

GENERAL

Adsorptive Stripping Voltammetry in Trace Analysis 403
Robert Kalvoda

Adsorption Effects Used in Electroanalysis 407
Hendrik Emons and Gerhard Werner

Microanalytical Application of a Kissinger Type Thin Layer Cell Polarized by LSV, NPV and DPV Excitation 413
G. Farsang and T. Dankházi

Differential Pulse Polarographic Determination of TeO_4^{2-} and VO_3^- in Presence of some Oxyhalide Anions 419
M.M. Kamal, Z.A. Ahmed and Y.M. Temerk

Differential Pulse Polarographic Determination of some Substituted Benzylidene Acethydrazones 423
M.M. Kamal, Y.M. Temerk, Z.A. Ahmed and G. Abd-El-Wahab

Simultaneous Determination of Copper, Cadmium, Lead and Zinc in Indian Snuff Samples by Differential Pulse Anodic Stripping Voltammetry (DPASV) 431
M.S. Nayak and S.S. Dhaktode

Electrochemical Behaviour of EPN 437
T.N. Reddy and S. Jayarama Reddy

Electroanalysis of the Acrylic Monomer/Polymer System 443
Lothar Dunsch

Contributors 453

Index 455

INTRODUCTION

Ari Ivaska

Laboratory of Analytical Chemistry
Åbo Akademi University
SF-20500 Turku-Åbo, Finland

Electroanalysis is perhaps the branch of electrochemistry with the greatest number of practical applications of all the different areas of electrochemistry.

One of the main advantages of electroanalytical methods is that they allow both quantitative and qualitative determinations. The electrical signal of an electrochemical system can easily be used in electronic instrumentation and computer processing. Improvements in instrumentation are of central importance in developing new instrumental analytical methods. Instruments are like vehicles which scientists use to take them over difficult terrain into new areas. Electroanalytical chemistry is rather instrument oriented and good instrumentation is of great importance in obtaining high quality data. The strong impact of research on instrumentation can clearly be seen in the instrumentation section of these Proceedings. Automation and instrumentation always go together. The development of flow methods with electrochemical detection is a good example of how combination of the two approaches can give good results.

Progress and current activity in the field of chemically modified electrodes is rapid and remarkable. By coating an inert electrode with redox polymers and polymers with suitable mediators the electrochemical signal can be transferred through insulating layers. The use of conductive polymers to modify electrodes is also increasing. Specific compounds can also be incorporated into a polymeric matrix to increase the selectivity and specificity of a given method. Examples of this are the use of enzymes and antigen-antibody systems for specific determination of single compounds. Combining biochemical systems with electroanalytical techniques is certainly a field where a lot of progress will be made in the future. Another way of increasing specificity is to use a chromatographic method to separate electrochemically similar compounds and then measure them by a less specific method which is selective for this family of compounds. In this case the specificity is obtained by chromatographic separation.

Electrode reactions are heterogeneous reactions taking place at the surface of the electrode. Sensitivity in detection of some compounds has been increased by several orders of magnitude by taking advantage of the adsorption phenomena at the electrode surface. In adsorptive stripping methods accumulation of compounds can be improved by proper selection of the electrode potential.

The importance of electroanalytical methods in industrial applications is clearly evident from the papers in that section of the Proceedings. There are cases where a whole chemical plant producing explosive materials is mainly controlled by electroanalytical methods.

Clinical and pharmaceutical applications of electroanalytical methods are also numerous and they have been accorded an established status in these laboratories.

One method is never enough when analysing samples with complicated matrices and a combination of electroanalytical techniques with other techniques like spectroscopy and chromatography offers new possibilities in analytical work. These combined methods form another line of development in electroanalysis and progress on these lines will certainly be seen in the future.

ELECTROCHEMICAL INSTRUMENTATION
AND
METHODS

NEW INSTRUMENTAL APPROACHES TO FAST ELECTRO-CHEMISTRY AT ULTRAMICROELECTRODES

Larry R. Faulkner, Michael R. Walsh and Chuanjing Xu

University of Illinois, Department of Chemistry
1209 W. California Street
Urbana, Illinois 61801, U.S.A.

The advent of ultramicroelectrodes has opened up the possibility of performing electrochemical kinetic measurements conveniently on timescales of nanoseconds. Thus, it should become possible to resolve fast kinetic events in electrode processes that could only be surmised on the basis of conventional measurements in the millisecond or high microsecond domains. The fast kinetic methods used in homogeneous chemistry, such as pulse radiolysis and flash photolysis, have already demonstrated that many chemical steps proceed in the picosecond, nanosecond, and low microsecond regimes. Among them are examples of certain importance to electrochemistry, including electron transfers, proton transfers, ejections of leaving groups, ligand exchanges, and isomerizations. By and large, electrochemistry has been blind to chemistry on these timescales and has suffered from a relative inability to define the mechanisms of electrode processes as a consequence.

This paper will cover a number of strategic issues related to fast electrochemical measurements at ultramicroelectrodes. For several reasons, it is not possible to implement measurements on the fastest accessible timescales by a simple extension of conventional methods based on instrumentation constructed from operational amplifiers. Therefore, we are exploring new approaches to the measurement of current.

The use of ultramicroelectrodes for fast measurements has been recognized and developed by others and has been discussed in the literature.[1-5] For the most part, the existing reports have concentrated on cyclic voltammetry. We believe that potential step methods are more appropriate to the realization of the promise of high speed kinetic measurements at ultramicroelectrodes, so we have concentrated on them. A preliminary report of our activity has appeared.[6]

Basic Considerations

Every electrochemical measurement has a lower limit of timescale that is imposed by the cell time constant, which is the product of the uncompensated resistance and the double-layer capacitance at the working electrode. The potential of the electrode cannot be changed appreciably on timescales shorter than the cell time constant, because the double layer cannot be charged except through the cell resistance. Ultramicroelectrodes offer dramatically reduced cell time constants because the electrode area is so small. In general, we can write the double-layer capacitance at a disk-shaped ultramicroelectrode as $C_d = \pi r^2 C_o$, where r is the radius of the electrode and C_o is the specific capacitance of the electrode, e.g. in microfarads per cm^2. The small size of the electrode imposes a price in increased uncompensated resistance, essentially because the entire cell current is required to pass through a region of solution with a physical size comparable to the size of the electrode. A detailed analysis shows that the uncompensated resistance is approximately $R_u = 1/4\kappa r$, where κ is the conductivity of the solution.[7] Thus, the cell time constant becomes,

$$R_u C_d = \pi r C_o / 4\kappa \tag{1}$$

Contemporary Electroanalytical Chemistry, Edited by A. Ivaska *et al.*
Plenum Press, New York, 1990

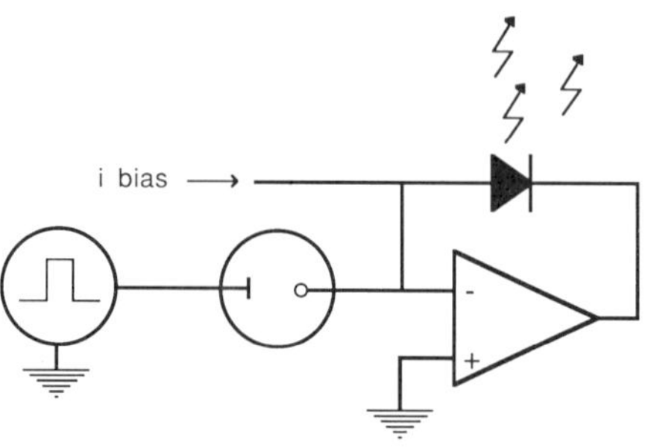

Fig.1. Circuit for current-to-photon conversion. Working electrode is connected to the inverting input of the amplifier. Circle at far left indicates function generator.

which is proportional to the radius r. Apparently, arbitrarily small time constants can be achieved by shrinking the radius sufficiently. We will return below to limitations on this idea, but it certainly is true that cell time constants in the nanosecond regime can be achieved with disk-shaped ultramicroelectrodes having radii of a few microns.

One of the prices of using the small electrodes to obtain appropriate cell time constants is that the currents also become small. It is well known that in experiments at conventional timescales, these currents are usually in the nanoampere and picoampere ranges. They are readily measured with fairly simple circuitry based on operational amplifiers when the experimental timescale is milliseconds or longer, but problems develop when one wishes to measure even microampere-level currents with submicrosecond resolution. This is true because conventional approaches to the measurement of the currents call for the current to be dropped across a resistor large enough to produce a conveniently recordable voltage. With small currents, the required resistances are large, usually in the range of 100 kΩ to 10 MΩ. When these resistances couple with the inevitable capacitances of the measurement circuitry, one develops measurement system time constants that are unacceptably long. For example, a capacitance of just 10 pF coupled with a 100 kΩ resistor yields a 1 μs system time constant. In practice, we observe time constants that are even larger because the stray capacitances from resistors, wire, and connectors typically exceed 10 pF. An alternative is to use a smaller feedback resistor and several amplification stages, but then one is faced with a set of propagation delays through the successive stages of amplification.

A Measurement Strategy Based on Current-to-Photon Conversion

The problem seems to call for a new approach to the measurement of the small currents. We have carried out research on a scheme designed to take advantage of well-developed methods for the precise measurement of fast transients in luminescence spectroscopy. By transducing the electrochemical cell current to a photon flux, instead of a voltage, one transforms an electrochemical measurement with a difficult electronic solution into an optoelectronic problem with several known solutions.

In our approach, the current-to-photon conversion is achieved by using the circuit shown in Fig.1, which is basically a logarithmic amplifier circuit, but with a light emitting diode (LED) in the feedback loop. The cell current is passed through the diode and produces a photon flux that reflects the magnitude of the cell current. The operational amplifier is used in order to hold the working electrode at virtual ground, so that potential control is simplified. We use only a two-electrode configuration, so that we can minimize the number of operational amplifiers involved in the overall control and measurement problem. It is difficult to find single amplifiers that perform in the frequency regime needed for the work discussed here. It is probably impossible to develop a network of even two amplifiers that will perform satisfactorily. Fortunately, the cell currents are sufficiently small that they do not polarize

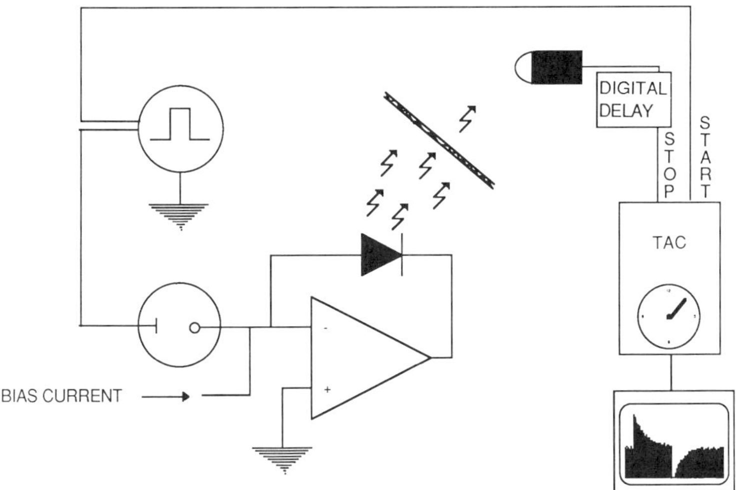

Fig.2. Current measurement by current to photon conversion.
Schematic diagram of the time-correlated single photon counting apparatus. Light passes through a neutral density filter to reduce the flux at the photomultiplier. Box at lower right is a multichannel analyzer showing a histogram of photon counts vs. delay.

the reference electrode, which serves simultaneously as the counter electrode; thus the output of the function generator can be applied directly to the reference electrode, and the working electrode can be expected to follow that output (with the opposite polarity) vs. the reference. This is a standard trick in the field.

Several high-speed photometric methods can be used to measure the photon flux. Our approach is to employ time-correlated single-photon counting (TCSPC), which is an established method for obtaining transient responses with very high precision. The scheme is shown in Fig.2. The working electrode is subjected to a potential step, and on the rising edge of the step, a time-to-amplitude converter (TAC, a voltage ramp generator) is started. The current transient passes through the circuit of Fig.1 and produces photons. When the first of these is detected, the TAC is stopped; hence its final output is a measure of the time delay between the start of the step and the arrival of the first detected photon. The output of the TAC is presented to a multichannel analyzer, which increments the channel corresponding to this delay. The step experiment is then repeated many times. A histogram of detection events vs. delay is built up in the multichannel analyzer. If the experiment is carried out in a manner that provides for a detected photon in only a few percent of the step experiments, then the probability of detection is the same for all photons, whether they come early or late, and the shape of the histogram is the same as the shape of the photon transient.

The advantages of this method are that its time resolution is controlled by the timing jitter of the electronics, rather than the transfer function of the whole photometric apparatus, and that one can achieve practically any desired level of precision by executing enough pulses. The drawback is that the method makes very inefficient use of the available data. With only one photon detected per 50-100 pulses, it requires 10-500 million pulses to acquire one averaged transient at appropriate precision. In photoluminescence, this is often not a limiting factor, because each cycle is begun with a low-intensity flash that does not degrade the sample, and repetition rates of 10 kHz to 100 MHz are feasible. In electrochemical practice, there are some serious consequences, which will be discussed later in this paper.

An issue that requires attention, in any scheme based on current-to-photon conversion,

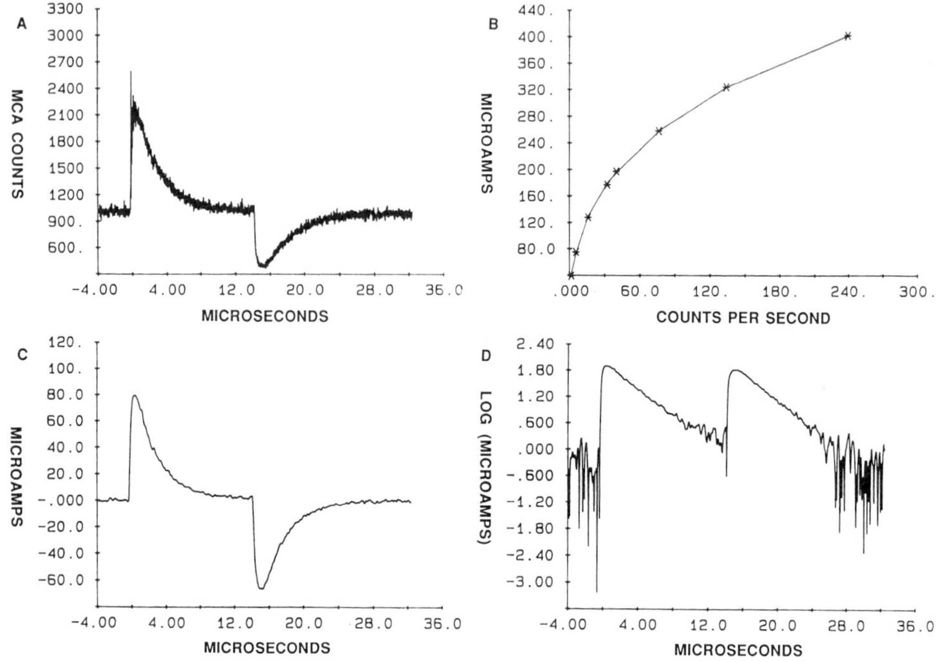

Fig.3. Different stages in processing of light intensity data. Plot A is the raw data (multichannel analyzer counts vs. time). Plot B is the calibration data. The stars show measured calibration points and the line shows the calibration equation obtained by fitting the starred data. Plot C is the current vs. time data after completing the calibration process. Plot D is the log of the absolute value of the current from plot C. The slopes correspond to a 2.7 μs time constant for the forward step and a 3.0 μs time constant for the reverse step. The nominal RC time constant was 3.0 μs.

is the establishment of a means for measuring bipolar currents. Most electrochemical kinetic measurements are based on reversal methods, in which both anodic and cathodic currents flow. It is essential that one be able to measure currents in both directions in a single experiment; yet the LED will produce a photon flux only when it is forward- biased. We circumvent the conflict by inserting a bias current into the summing junction of the operational amplifier, as shown in Fig.1. This current is chosen such that its sum with the cell current always maintains the LED in a forward-biased condition. In practice, it is usually advantageous to choose a positive bias current just larger than the greatest negative excursion.

The TCSPC experiment produces a histogram of photon counts vs. time, but to be useful for electrochemical purposes, this curve must be translated into a current-time plot. We achieve the mapping in the manner shown in Fig.3. First, we establish the relationship between photon count rate and current by using a series of known bias currents and leaving the working electrode disconnected. Because the quantum yield of the diode depends on the current, this curve is nonlinear, as shown in Fig.3B. Using the working curve, one can translate each point in the raw histogram of photon counts (Fig.3A), into a corresponding current level (Fig.3C). The bias current used in the experiment produces a constant, intermediate photon flux corresponding to zero cell current, and positive and negative photon fluxes measured with respect to this level can be mapped into positive and negative cell currents.

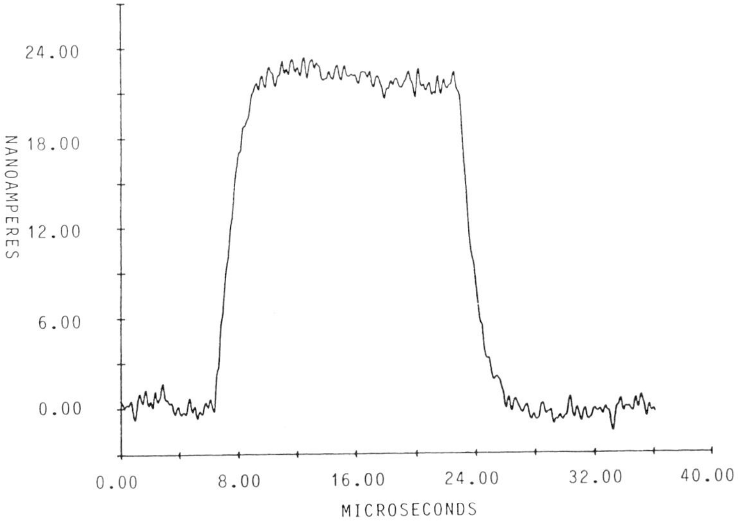

Fig.4. Data plot for a 20 nA current pulse. The pulse width was 16.5 μs. The circuit used for this data consisted of a single transistor in a common emitter configuration with a light emitting diode as the load element. This circuit was used instead of the operational amplifier circuit in Fig.1 because it provides improved response at low current levels.

Some Initial Results and Problems

The methodology and apparatus discussed above has been tested extensively on RC networks, and has performed very satisfactorily. Some results of this type are shown in Fig.3C and 3D. We have been able to measure RC time constants near 500 ns with 30 ns accuracy. These and other tests suggest that the instrument response function has a characteristic time that can be as short as about 30 ns, depending on the level of the current. Capacitive effects in the LED's junction region cause the response time of the diode to lengthen at reduced current levels, but a quite usable measurement system time constant was recorded even for a 20 nA current pulse, as shown in Fig.4. We have not been able to record current transients of this magnitude by any other means with a time resolution even close to that established in Fig.4.

We have also carried out tests of the methodology in several different chemical studies. They will not be discussed in detail here, because we choose to concentrate on issues of measurement strategy. In general, we have had success in studying the rate of reductive hydrogen adsorption on clean platinum and in examining the rate at which an oxidized platinum surface reorganizes itself to produced oxygen atoms in sub-surface locations.[8]

Some significant difficulties have also presented themselves. Two are related to the inefficiency in data usage in TCSPC. First, one has to allow time for the chemical system to restore itself after each pulse, and this requires a time at least ten times longer than the pulse width. Repetition rates are therefore generally less than 10 kHz, and the collection of enough photons to provide a histogram with good precision requires times as long as 10 h. Moreover, the electrochemical systems change with time during these trains of pulses, so that the results are compromised. This has been particularly serious for studies of species that produce radical intermediates, such as halogenated aromatics, which eject the halogen upon reduction. In these cases, the electrode seems to develop a film during the course of data acquisition by TCSPC.

Our studies have made it amply clear that fast experiments will be much easier to carry out on the dynamics of reactions involving species bound to the electrode surface, as opposed to dissolved species. In the time regime of interest here, the diffusion layer is so small (only a few hundred Å at 100 ns to a few thousand Å at 10 μs) that it does not contain enough molecules to provide good signal-to-background ratios unless double-layer charging currents are fully eliminated. We have attempted to cancel charging currents by subtraction of the current through an RC network approximating the characteristics of the double layer in systems of interest, and we have found this approach to be functional, but we have concentrated mostly on surface reactions because of the comparative ease of doing them.

The degradation of chemical systems under the repeated pulsing required by TCSPC implies that one must look for ways to use the cyclic pulse sequence to reinitialize the chemical system in each observation cycle. We have done this in our studies of platinum electrodes to restore a clean, reduced surface in each cycle. To expand the scope of the chemistry that can be studied reliably, we have begun to consider other means that can allow similar restoration of desired conditions on short timescales. Mercury electrodes are attractive, because they can be partially stripped and plated *in situ* and because one can often expel adsorbates by simply polarizing the electrode to sufficiently extreme potentials.

A Well-Behaved Mercury Ultramicroelectrode

Fast kinetic measurements on adsorbates at mercury/water interfaces seem the most appropriate entry point for chemical studies in this field just now, not only for the reasons mentioned above, but also because there is a wealth of descriptive and thermodynamic knowledge about these kinds of systems. We have therefore sought to establish mercury ultramicroelectrodes that show adsorptive behavior and faradaic background properties comparable to those of conventional dropping mercury electrodes and hanging drops. There is literature suggesting that mercury ultramicroelectrodes can be established on carbon fibers,[9-11] but we have not been able to achieve satisfactory performance with such systems. Our criterion has been the voltammetric behavior of adsorbed 2,6-anthraquinone disulfonic acid (2,6-AQDS) in 0.1 M nitric acid solution, which has a nearly perfect thin-layer characteristic unless the highest packing densities are approached. On mercury-plated carbon fibers we observe grossly distorted voltammetry for the adsorbate, and we observe rather high background currents.

A good solution to this problem has been achieved by the approach depicted in Fig.5. We begin with a platinum ultramicroelectrode disk sealed in glass. It is etched in aqua regia for 1-3 hr with the intention of developing a channel in the glass that can later be filled with mercury by electrodeposition. The glass is then silanized with 5 % dichlorodimethylsilane in carbon tetrachloride, and finally mercury is deposited from an aqueous solution of mercurous nitrate. The background characteristics and the adsorptive properties of electrodes prepared in this manner on 25 μm Pt disks (Fig.6) are very similar to those of conventional hanging mercury drops. We have characterized their areas by exposing the electrode to a solution of 2,6-AQDS in 0.1 M nitric acid and measuring the reductive charge by chronocoulometry. From the packing density measured at an HMDE, we obtain an estimate of the area. A parallel estimate is obtained from the interfacial capacitance, and good agreement is found. Microscopic examination suggests that the electrode is approximately hemispherical, and the measurements of area suggest that the radius is approximately equal to that of the underlying Pt disk.

Studies of the kinetics of the redox processes involving adsorbed 2,6-AQDS have begun, but we are not yet prepared to interpret the results.

How Small a Cell Time Constant Could Be Achieved?

In work of this kind, one naturally would like to be able to define the lower bounds of timescale. Equation (1) suggests that one could obtain arbitrarily small cell time constants by relying on sufficiently small electrodes, but eventually there will be a breakdown of equation (1) because of stray capacitance in the cell. In other words, when the interfacial capacitance falls to sufficiently small values, as a consequence of the reduction of area, other components of capacitance will begin to dominate. These will arise from capacitive coupling between leads

10

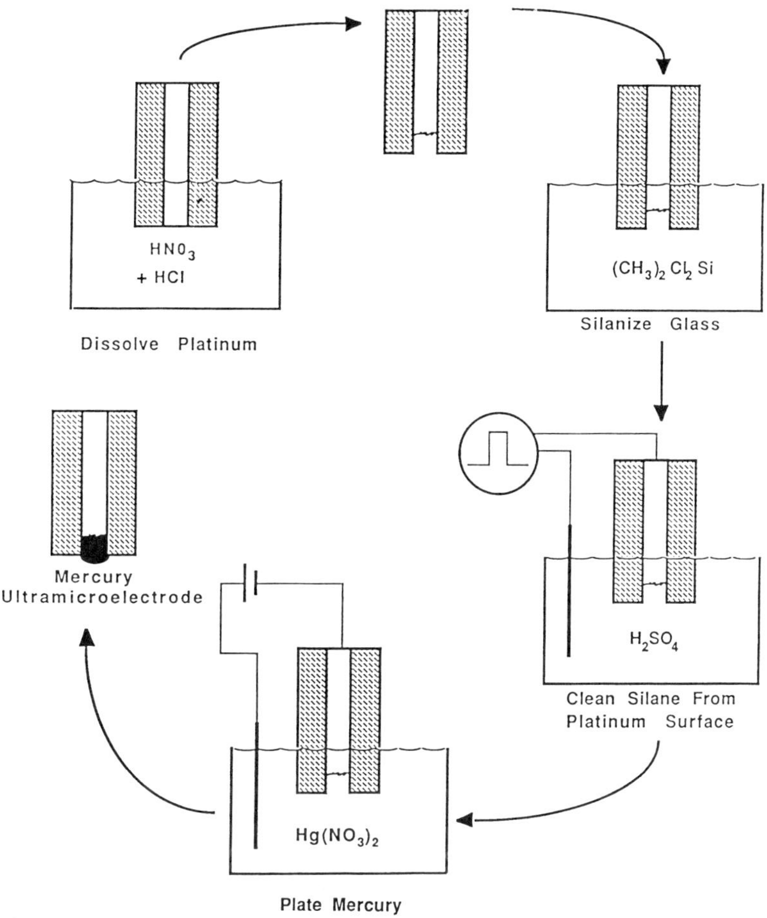

Fig.5. Procedure schematic for preparing well behaved mercury ultra-microelectrodes for fast electrochemical experiments.

or from the required small degree of interfacial charging on surfaces of conductors not facing the solution. In our experience, it is difficult to reduce stray capacitance below 20 pF, so we have considered the time constants that could be obtained with a 20 pF stray capacitance in parallel with that of the double layer. The results are shown in Fig.7. The striking effect is that there is a predicted minimum in the cell time constant. It occurs because the uncompensated resistance continues to rise as the disk radius falls, but the capacitance levels off at the value of the stray capacitance. The plot in Fig.7 is intended to show approximately the best case, because the conductivity of the solution was taken as that of a concentrated acid in water. The results suggest that time constants significantly less than 50 ns will be difficult to achieve in any medium, at least in the usual geometry involving a large counter electrode at infinite separation.

Fig.8 shows the present state of the art. The current pulses are for double layer charging at a platinum disk in 5 M perchloric acid. The time constants for the forward and reverse phases are both about 60 ns for the 5 μm diameter platinum electrode and about 120 ns for the 10 μm diameter electrode.

A consideration of parallel plate geometry suggests that generally shorter time constants

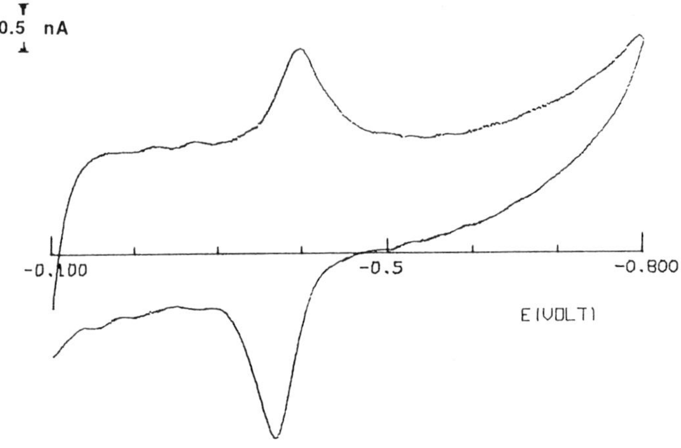

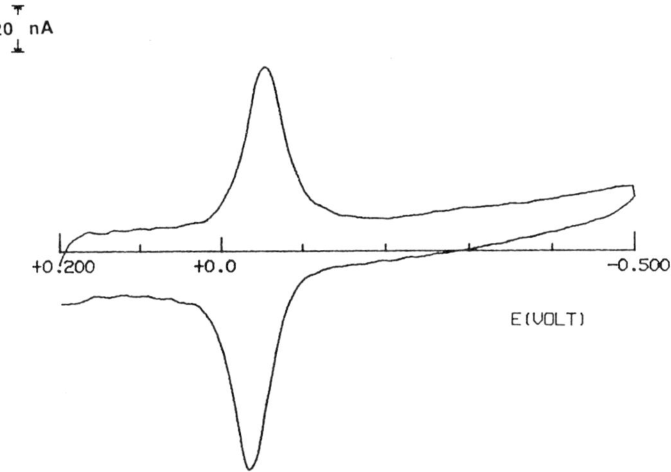

Fig.6. Cyclic voltammetry of 2,6-anthraquinone disulfonic acid at mercury ultramicroelectrodes. Scan rate = 10240 mV/s. The top voltammogram was obtained with a platinum electrode overcoated with mercury. The bottom voltammogram was obtained using a platinum electrode that was etched and silanized before coating with mercury as shown in Fig.5.

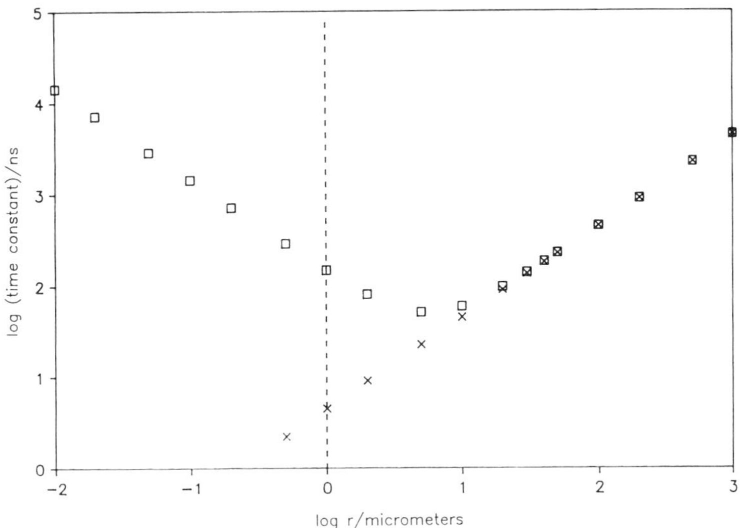

Fig.7. Plot of cell time constant vs. electrode radius in 1 M acid with a 20 pF stray capacitance.

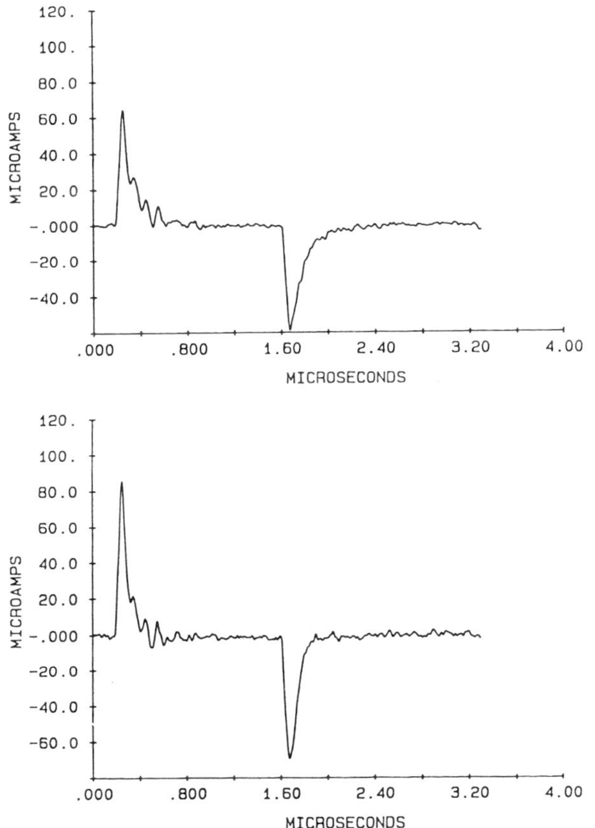

Fig.8. Double layer charging current for a pulse experiment in 5 M perchloric acid solution. The top plot is for a 10 μm diameter platinum disk and the bottom plot is for a 5 μm diameter disk. $V_{step} = 0.25$ V to 0.00 V vs. Ag/AgCl, 1.43 μs pulse width.

might be reached if the counter electrode could be spaced from the working electrode by a distance smaller than the radius of the working electrode. This is probably feasible, but it would introduce other problems. We believe that a good deal of interesting science can be done on the timescales that are accessible now.

Acknowledgements

We are grateful to the U. S. National Science Foundation for supporting this work under Grant CHE-86-07984, to the Phillips Petroleum Foundation for a fellowship to M. R. Walsh, and to the State Education Commission of the People's Republic of China for financial support of Chuanjing Xu during his term as a visiting scholar at the University of Illinois. In addition, we thank Neste, Ltd. for supporting the participation of L. R. Faulkner in ElectroFinnAnalysis.

References

1. J.O. Howell and R.M. Wightman, Anal. Chem., **56** (1984) 524.
2. J.O. Howell and R.M. Wightman, J. Phys. Chem., **88** (1984) 3915.
3. J.O. Howell, W.G. Kuhr, R.E. Ensman and R.M. Wightman, J. Electroanal. Chem., **209** (1986) 77.
4. C.A. Amatore, A. Jutand and F. Pfluger, J. Electroanal. Chem., **218** (1987) 361.
5. D.O. Wipf, E.W. Kristensen, M.R. Deakin and R.M. Wightman, Anal. Chem., **60** (1988) 306.
6. L.R. Faulkner, P. He, D. Ingersoll, H-J. Huang and M.R. Walsh, in: "Ultramicroelectrodes", M. Fleismann, B.S. Pona, D.R. Rolison, and P.P. Schmidt (Eds), Datatech, Morgonton, N.C., U.S.A., 1987, pp. 225-239.
7. J. Newman, J. Electrochem. Soc., **113** (1968) 501.
8. H. Angerstein-Kozlowska, B.E. Conway and W.B.A. Sharp, J. Electroanal. Chem., **43** (1973) 9.
9. J. Golas and J. Osteryoung, Anal. Chim. Acta, **181** (1986) 211.
10. G. Schulze and W. Frenzel, Anal. Chim. Acta, **159** (1984) 95.
11. M.R. Cushman, G.B. Bennet and W.C. Anderson, Anal. Chim. Acta, **130** (1981) 323.

PHOTOELECTROANALYTICAL CHEMISTRY — METHODS AND INSTRUMENTATION

Jouko J. Kankare

Department of Chemistry, University of Turku
SF-20500 Turku, Finland

Introduction

We are talking about "spectroelectrochemistry" when measuring absorption spectra on the surface of electrodes, "electrogenerated luminescence" or "electroluminescence" when talking about photons emanating from an electrode surface as the result of an electrode process, "photoelectrochemistry" usually when light impinging on the surface of an electrode generates electric current, etc. The common feature of all these methods is that information is retrieved from the electrode surface or electrode process by using photons. The word "photoelectroanalytical chemistry" could be coined for those methods in which information from an electrode surface or electrode process and the species participating in it is gained by using any method applying photons.

Spectroelectrochemistry is one of the many facets of photoelectroanalytical chemistry. It can be used for numerous purposes in solving the mechanisms of electrochemical processes but especially with electrically conducting polymers it shows its main advantages.[1] The original drive to study conductive polymers arose from the applications anticipated in the energy storage, but these polymers are also interesting from the analytical point of view as potential sensor materials.

There is already a large number of different conductive polymers. A typical monomer is 3-methylthiophene, which can be electrically polymerized to a polymer coupled by the 2- and 5-positions of the monomer. In the oxidized form, usually called "doped", the chains contain positive charges at about every fourth monomer unit. In order to keep the polymer layer electrically neutral, also counter anions should be present in the polymer matrix. It is analytically interesting that the diffusion rate of these counter anions controls the rate of oxidation and reduction of the polymer, and the diffusion rate depends on the size, degree of solvation etc. of the anion. Hence, by a suitable choice of the polymer, it should be possible, at least in principle, to tailor-make sensors for different anions. In addition, it has been shown,[2] that electrically neutral polymers can be incorporated from the solution into the polymer matrix during the polymerization process. This of course extends enormously the possibilities for developing selective sensors without undue efforts to synthesize new electrically polymerizable monomers.

Experimental

The rate of anion diffusion can be measured in various ways. The conventional way is to use classical electrochemical methods, e.g., chronoamperometry or chronocoulometry. The measurement of electrochemical impedance is also sometimes used. However, the electronically conducting polymers have a special property, potential-dependent absorption spectrum, which can be advantageously used to monitor the oxidation state of the polymer. In addition to the neglect of capacitive current, monitoring of the spectral change gives additional information. For instance, the presence of an isosbestic point shows that most likely

Contemporary Electroanalytical Chemistry, Edited by A. Ivaska *et al.*
Plenum Press, New York, 1990

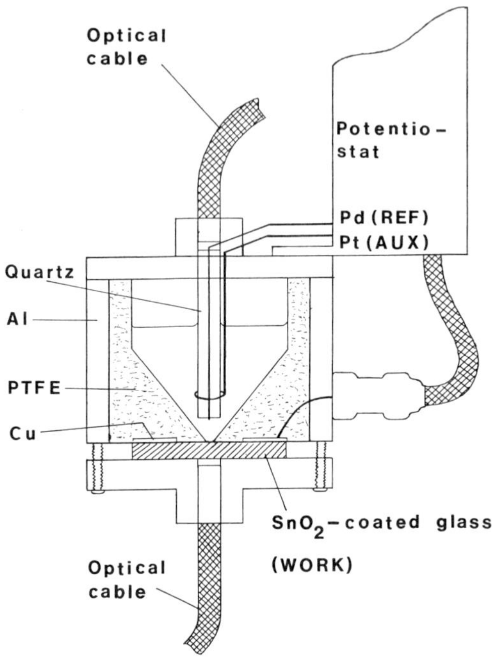

Fig.1. Cell for *in situ* spectroelectrochemical measurements of conductive polymers.

only two different oxidation states are present in the system. The more profound method of principal component analysis can be used for the evaluation of the number of species with different spectra in the system. To follow the fast changes in the composition of the polymer membrane, the main requirement is the ability of measuring successive spectra fast enough after applying a voltage transient. The recording time for a single spectrum should be so short that the spectrum does not appreciably change during the scan. This can be only done by using an optical multichannel analyzer (OMA), e.g., a diode array detector with a pulsed light source or a gatable OMA with a continuous source. The system used in our laboratory for fast recording of absorption spectra at the wavelength range from 350 nm to 850 nm is based on the optical multichannel analyzer OMA 1461 made by Princeton Applied Research combined with a Jarrel-Ash 0.25 m spectrograph. The light source is either a Xe flash lamp with ca. 1 s pulse width or a conventional tungsten filament lamp. The detector is an intensified gated diode array detector PAR OMA 1421. At the full length of the diode array, 1024 diodes, ca. 50 spectra can be recorded in 1 s. The cell which allows the *in situ* recording of absorption spectra is shown in Fig.1. The cell body is machined from PTFE, the working electrode made of tin dioxide coated glass forms the bottom of the conical cell interior. Light is conducted to the working electrode using a fiber optic cable and a quartz capillary the end of which is quite close to the bottom. Palladium wire which is fitted into the capillary serves as a pseudoreference electrode. A loop of platinum wire around the capillary serves as a counter electrode. A simple potentiostat wired from a few operational amplifiers is constructed onto a printed circuit board and mounted on the lid of the cell. Hence all the wiring is made as short as possible and also the electrodes are positioned as symmetrically and close to each other as possible in order to improve the high-frequency response of the system.

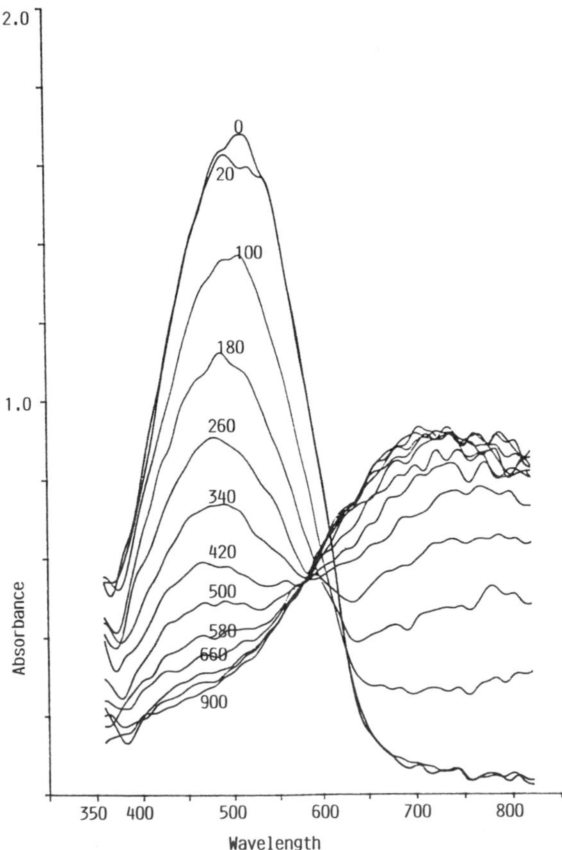

Fig.2. *In situ* absorption curves for poly-3-methylthiophene as a function of time after a voltage step from -0.6 V to +0.6 V. The numbers in curves refer to the time in ms elapsed from the step.

Results and Discussion

Conductive Polymers

Conductive polymers are usually prepared by electrolytic deposition at constant current on a suitable conducting substrate. Thickness is generally controlled by the consumed charge. Another and more direct way is to control the spectrum of the deposit during the electrolysis. The absorbance vs. charge curve is not always linear, and seems to depend on the previous history of the substrate. A likely reason is the variation in the number of nucleation centres. Hence already in the preparation of the polymer film it is advantageous to be able to monitor the absorbance at a single wavelength or even better, the whole spectrum.

As an example, the change of the spectrum of poly-3-methylthiophene by applying a step voltage from -0.6 V to +0.6 V to the working electrode and recording absorption spectra as a function of time is shown in Fig.2. An interesting feature is the lack of a clear isosbestic point which may indicate either a third component in the redox system or a medium effect which shifts the spectra in the wavelength scale when going from the nonpolar reduced form to the highly polar oxidized form.

Another interesting photoelectroanalytical method for the characterization of polymer films is a method which might be called photoimpedance spectrum. A small-amplitude sine-wave signal is applied to the working electrode and the resulting absorbance response is recorded at different frequencies. Alternatively, several frequencies are applied simultaneously and the response analyzed by using Fourier transform. The main advantage compared with the conventional electrical impedance measurements is naturally that only faradaic current

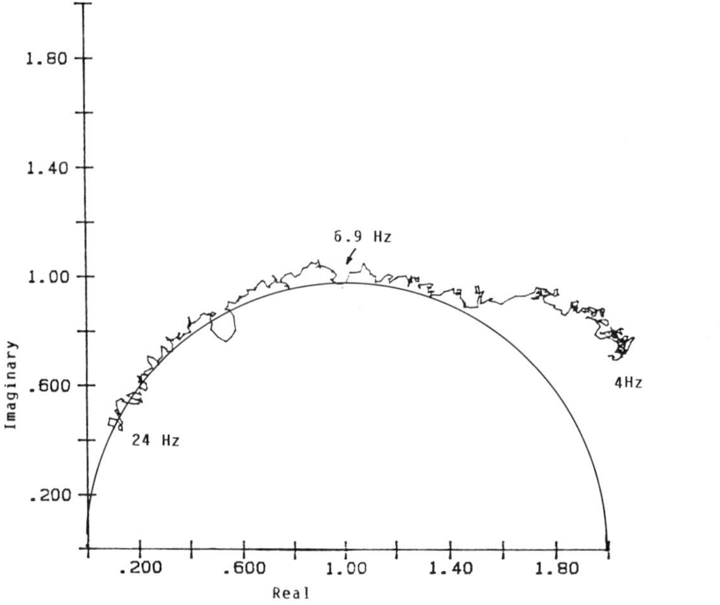

Fig.3. Complex-plane photoimpedance spectrum of
poly-3-methylthiophene in acetonitrile (0.1 M tetrabutylammonium
perchlorate). Absorbance response is recorded at 510 nm and 0.800 V
with 0.1 V excitation amplitude.

has any influence and even then only that part which induces changes in the doping level of the polymer film. The information obtained is directly related to the diffusion rates of coanions in the membrane. Fig.3 shows the Nyquist plot of the absorbance response for poly-3-methylthiophene. The equivalent circuit of the membrane is a capacitor and resistor in series, and the characteristic frequency is 6.9 Hz. The details of this method will be published later.[3]

The main difficulty when working with thin conducting polymer membranes is the lack of quantitative theory of ion diffusion within the membrane. Various theoretical schemes and approximations have been suggested, but the most difficult problem seems to be in the analytical solution or even approximation for the boundary problem of the combined Nernst-Planck and Poisson equations. The latter equation comes from the fact that electroneutrality cannot be assumed to prevail inside the thin membrane. Doblhofer et al.[4] have made an attempt to solve the problem numerically , but even then certain initial approximations were made. Also the "brute force" method of finite differences does not allow to see clearly the influence of different parameters.

Electroluminescence

Spectroelectrochemistry is one facet of photoelectroanalytical chemistry. Another facet is luminescence from the electrode induced by some electrolytic process. Electrogenerated luminescence is actively studied by a few groups, but the number of analytical applications is still quite small, the most important of which seem to be in the fields of chromatographic detectors and immunoassay. One difficulty in the applications may be the small number of possible electroluminescent compounds in aqueous solutions using conventional electrodes. We have found that by using an oxide-covered electrode like aluminium or tantalum, and a suitable oxidizing agent like peroxydisulfate, luminescence of nearly any conventionally fluorescent compound can be generated electrically. This should have a positive impact on the analytical use of the phenomenon.

Electroluminescence on oxide-covered electrodes is best generated by using pulsed wave-form. The shape of the pulse train is of primary importance. It is usually composed of

18

sequential positive and negative pulses with intermittent zero levels. The positive pulses are important for regenerating the oxide layer. Light is generated during the negative pulse, and zero levels are used for providing the baseline level for gated light measurement.[5] The spectrum of luminescence and the shape of light pulses are important factors for solving the mechanisms of these phenomena. For the analytical purposes the highest sensitivity is obtained by recording the light intensity directly either without any wavelength-selective device or through a simple optical filter. The main requirement is a gated integrator and a pulse generator which can be programmed to generate various pulse trains with variable amplitude.

Electrogenerated luminescence is in many cases very weak and emitted from a large area, i.e., the radiance of the source is low. Interesting information is also given by the temporal behavior of light pulses. Together these characteristics create difficult problems in constructing the apparatus for spectral measurements. Low radiance means that the light cannot be effectively focused into the spectrograph slit. Hence at very low light levels we cannot use the multiplex advantage of an OMA. Even less efficient is the conventional scanning monochromator equipped with a photomultiplier. We have constructed an instrument which is based on the use of an interference filter and photon counting.[6] For scanning purposes there is commercially available a circular wedge interference filter for the range from 400 to 700 nm (OCLI, CA). By continuous rotation of the filter, a low-resolution spectrum can be measured during one revolution, but only if dc excitation is used. With pulsed excitation as in the present case, we come to the main difficulty. Rotation of the filter wheel and the timing of the pulses should be exactly mutually synchronized. If in addition the temporal behavior is to be measured, this must be carried out in the boxcar mode. One period of the pulse train is divided into several "slices". During one rotation of the filter, photon pulses are collected only during one excitation slice, and the memory addresses where the counts are averaged are determined by the slice number and the position of the filter, i.e., wavelength. With the present instrumentation the period can be divided into maximally 32 slices. Due to the memory limitations, 16 spectra with different time delays can be recorded. The time resolution of the instrument is 500 s.

The instrument in principle is a two-channel photon counter in which one channel collects pulses while the contents of the counters of the other channel are being transferred to the computer memory. What makes the instrument and the control program rather complicated is the combination of the boxcar mode with the exact synchronization with the rotation of the filter wheel.

Optically the instrument can be compared with a simple monochromator with ca. 10 mm wide slit with no focusing optics. Because of the low luminance and large emitting area focusing is of little avail. The optical throughput compared with a typical spectrograph with 25 μm slit as used, e.g., in the connection with a typical OMA configuration is ca 300-500 times better. Taking into account the multiplex advantage of a 1024-diode image intensified array, the S/N ratio may still be 10 times better. A typical run takes 5 minutes, which means that using a state-of-the-art OMA system the recording time for the same S/N ratio would be a whole working day.

Electroluminescence on oxide-covered metal electrodes is a method which can be used for the determination of both inorganic[7] and organic[8] compounds. The highest sensitivity is obtained with thallium(I) and terbium(III) which can be determined on an aluminium electrode at less than 0.01 ppb level in aqueous solution. Also copper(II) comes close to that level. It is interesting to note that metal ions like Cu^{2+}, Hg^{2+} and Pb^{2+} which are not inherently fluorescent give an intense EL spectrum.

Presently analytically more interesting is the electroluminescence of organic compounds, because in that area many applications can easily be seen. Fig.4 shows the EL and fluorescence spectra of 4-amino-1-naphthalenesulfonic acid. It can be seen that the spectra are closely related and most probably they have the same emitting species. This is not always the case and we have examples where the EL and fluorescence spectra are considerably different. This may be due to a electrolytic process which generates a new emitting species, or often complexation with the metal oxide surface of the electrode. In any case, most fluorescent organic compounds we have tried give EL at aluminium electrode in the presence of some oxidizing agent like peroxydisulfate. The phenomenon is typical for aqueous solu-

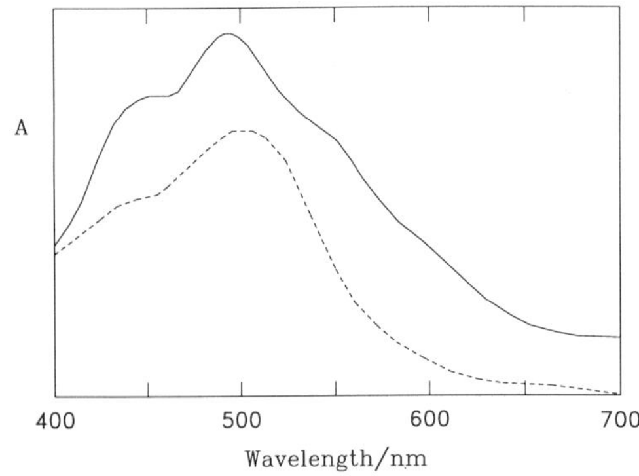

Fig.4. Electroluminescence spectrum (solid line) and fluorescence emission spectrum (dotted line) of 4-amino-1-naphthalenesulfonic acid in aqueous solution (10^{-5} M).

tions, so that, e.g., acetonitrile which otherwise is a good solvent for the classical annihilation electrochemiluminescence, effectively quenches the EL at oxide-covered electrodes. This is bad news for those who expect a simple HPLC detector on this principle, because acetonitrile is one of the most popular eluents. Fortunately micelles have no adverse effect to the EL, and thus micellar chromatography which is now becoming as a commonplace tool for chromatographists, is possible with the EL detection.[9]

Conclusions

Photoelectroanalytical methods and their development have not been in the mainstream of analytical chemistry in recent years in spite of the advances in physical photoelectrochemistry. The high information content of the photoelectrochemical methods and the very high sensitivity of certain analytical determinations based on these methods should give impetus for further developments in this field.

Acknowledgements

Financial support from the Academy of Finland is gratefully acknowledged.

References

1. A.O. Patil, A.J. Heeger and F. Wudl, Chem. Rev., **88** (1988) 183.
2. J. Roncali and F. Garnier, J. Phys. Chem., **92** (1988) 833.
3. J.J. Kankare and A. Paren, to be published.
4. R. Lange and K. Doblhofer, J. Electroanal. Chem., **237** (1987) 13.
5. J.J. Kankare, D.E. Ryan and B.J. Fürst, Can. J. Chem., **55** (1977) 1193.
6. S.V. Pihlajamäki and J.J. Kankare, Anal. Instrum., **15** (1986) 171.
7. K. Haapakka, J.J. Kankare and S. Kulmala, Anal. Chim. Acta, **171** (1985) 259.
8. K. Haapakka, J.J. Kankare and O. Puhakka, Anal. Chim. Acta, **207** (1988) 195.
9. K. Haapakka, J.J. Kankare and K. Lipiäinen, Anal. Chim. Acta, **215** (1988) 341.

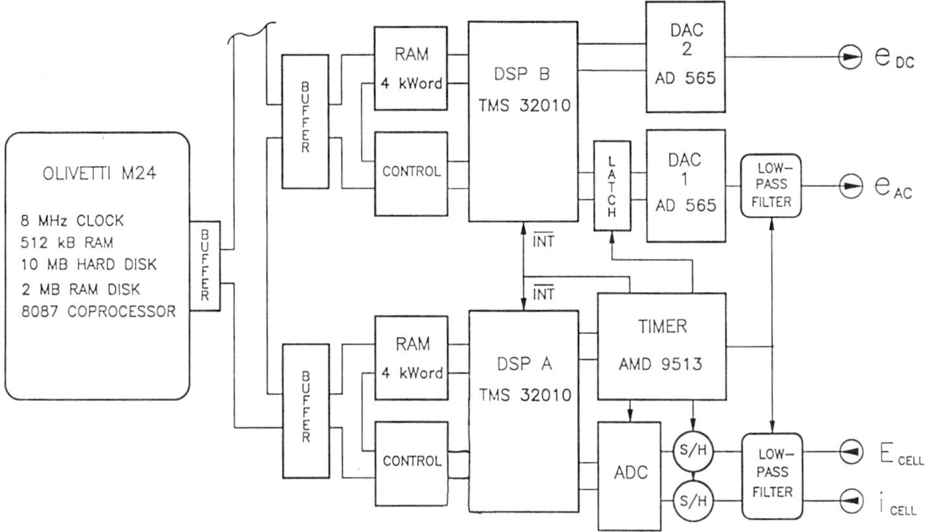

Fig.4. A schematic diagram of the instrument.

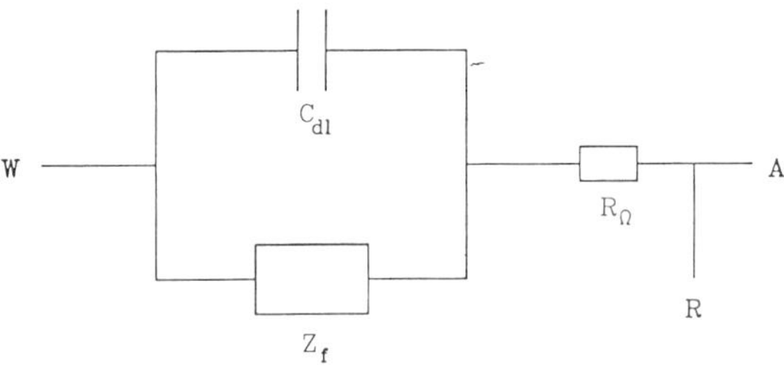

Fig.5. Dummy cell with three components: Solution resistance R_Ω, double layer capacitance C_{dl}, and the faradaic impedance Z_f, which in many cases is represented by a resistance, the charge transfer resistance R_{ct}.

multiplied by the transfer function) at up to 10 frequencies in real time. It is possible, of course, to use more than 10 frequencies in an experiment but, in order to restrict time consuming operations during an actual run, only 10 frequencies are shown on the screen. The acquired data are stored in a file on a 2 MB RAM disk together with information about the experimental parameters used. Separate software is used to analyse the data after the completion of the experiment.

Data Analysis

The case considered here will be that of an uncomplicated electrode reaction controlled by charge transfer kinetics and diffusion only. There are mainly two techniques used for analysis of the acquired admittance/impedance data. First, it is possible to construct different kinds of plots and then, by fitting simple functions like straight lines or parabolas and studying slopes and intersections, one can find good estimates of the values of the components in an equivalent circuit. Several plots based on variation in frequency, concentration, and dc potential, have been described by Sluyters-Rehbach and Sluyters.[7] A simple, but also noise-sensitive, method has been used by de Levie et al.[8] This method will be described in some detail here since it is a fast and reasonably accurate way to find estimates to be used in the second technique, described below.

The impedance of a typical equivalent circuit (like the one in Fig.5) is given by

$$Z_{\text{cell}} = R_\Omega + \frac{1}{i\omega C_{\text{dl}} + Y_{\text{f}}} \tag{3}$$

where R_Ω is the solution resistance, C_{dl} the double layer capacitance, and Y_{f} the faradaic admittance. i denotes $\sqrt{-1}$ and ω is the angular frequency. It can now be shown that by letting $\omega \to \infty$ the cell impedance tends to R_Ω. In a plot of imaginary impedance, Z_{cell}'', versus real impedance, Z_{cell}', see for example Fig.6, it is possible to fit a function to the experimental points and extrapolate to infinite frequency.* The point where this curve intersects the Z_{cell}'-axis gives the solution resistance. Once this value is found it can be subtracted from the cell impedance. The next step is to convert the impedance into admittance, yielding the electrode admittance,

$$Y_{\text{el}} = Y_{\text{f}} + i\omega C_{\text{dl}} \tag{4}$$

Here a plot of Y_{el}''/ω versus Y_{el}'/ω once again gives an opportunity to extrapolate towards infinite frequency, and the intersection with the Y_{el}''/ω-axis gives C_{dl}. Another subtraction yields the faradaic admittance/impedance. In case of a simple electrode reaction, controlled only by charge transfer kinetics and diffusional mass transfer, this impedance is given by

$$Z_{\text{f}} = R_{\text{ct}} + \sigma\omega^{-1/2}(1 - i) \tag{5}$$

and the cotangent of the phase angle, $\cot\phi = Z_{\text{f}}'/Z_{\text{f}}''$, is given by

$$\cot\phi = 1 + \frac{R_{\text{ct}}}{\sigma}\omega^{1/2} \tag{6}$$

R_{ct} is the charge transfer resistance while $\sigma\omega^{-1/2}(1 - i)$ is the Warburg, or mass transfer, impedance.[8] Eqs.5 and 6 can now be used to study the electrode reaction in detail. For example, a $\cot\phi$ versus dc potential plot will aid in finding the charge transfer coefficient, while a $\cot\phi$ versus $\omega^{1/2}$ plot serves to find the heterogeneous rate constant. Details on these examples can be found in the literature.[5]

The second alternative in data analysis is a technique called complex nonlinear least squares fitting, CNLS. Here the values of the components of the equivalent circuit are changed iteratively until the calculated admittance/impedance of the circuit approaches the experimentally acquired data. The convergence of the calculation, i.e., the finding of a minimum of the sum of squares, depends on the accuracy of the data, the choice of a model, the accuracy of the initial guesses on component values, weighting, and the frequency range. The CNLS program used in this work was the LOMFP program written by Macdonald and coworkers.[9] Initial guesses used in this work were those values found by the extrapolation-subtraction method described above.

Results and Discussion .

In order to test the instrument and the data analysis procedures we made a number of experiments on dummy cells, for example the one shown in Fig.6. Here we made use of the possibility of time-domain averaging, the ac excitation was repeated 8 times and the data were averaged. The dummy cell was enclosed in a Faraday cage and the noise level was quite low, reducing the need for frequency-domain averaging, i.e., averaging of admittance data, after Fourier transformation and calculation of the admittance.

In analysing the data the following results were obtained using the extrapolation-subtraction method:

$$R_\Omega = 44.5\ \Omega, \; R_{\text{ct}} = 467.8\ \Omega \text{ and } C_{\text{dl}} = 0.33\ \mu\text{F}$$

The CNLS method, on the other hand, gave the following results:

$$R_\Omega = 52.1\ \Omega, \; R_{\text{ct}} = 465.2\ \Omega \text{ and } C_{\text{dl}} = 0.33\ \mu\text{F}$$

---

* Note that we define, in accordance with ref. 7, the complex impedance as $Z = Z' - iZ''$.

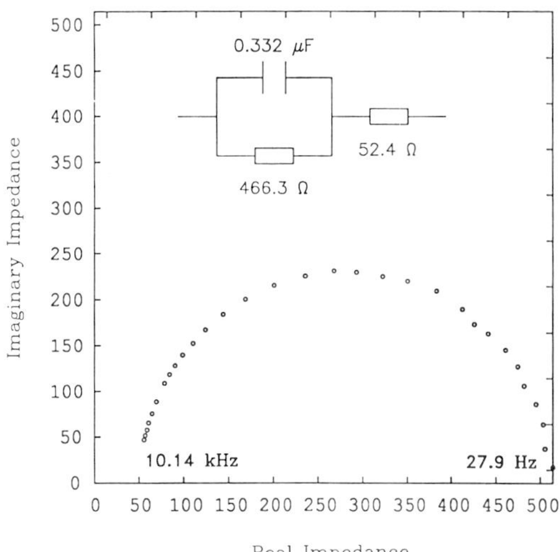

Fig.6. A complex plane plot of the impedance of a dummy cell at frequencies ranging from 27.9 Hz to 10.14 kHz. The components of the dummy cell have the following values: $R_\Omega = 52.4\ \Omega$, $R_{ct} = 466.3\ \Omega$ and $C_{dl} = 0.332\ \mu F$.

It is evident that the CNLS method does a better job than the extrapolation-subtraction method. In this case, however, one could have achieved a better result with the extrapolation-subtraction method by using higher frequencies. The problem of finding a good value of R_Ω would have been largely eliminated in this way. A more serious disadvantage is that this method tends to enhance noise. In a real experimental situation more noise is present and, in our opinion, CNLS is here without doubt the method of choice. The extrapolation-subtraction method is still of value as a method to find initial guesses needed when using a CNLS-program.

As an example of a practical situation where the instrument has been used, the study of an electrically conductive polymer, poly(3-octylthiophene), is representative. This polymer is electrochemically polymerized on a Pt-disc, which then is used as a working electrode in a conventional three-electrode electrochemical cell. In this example glassy-carbon was used as an auxiliary electrode, while the reference electrode was a Ag/AgCl (3M KCl). The supporting electrolyte was 0.1 M LiBF$_4$ in propylene carbonate.

It was possible to obtain the impedance of the cell containing the polymer coated electrode at about 20 frequencies simultaneously during a stepwise dc potential scan. The use of a dc potential staircase made it possible to study the impedance of the system at different oxidation states of the polymer film. This provided valuable information about the kinetics of the electron transfer at the polymer/metal interface, as well as about the diffusion of the charge compensating ions in the film. A typical impedance plot at a dc potential of 1.2 V vs. Ag/AgCl is shown in Fig.7. Also shown, in the same figure, is the impedance of the equivalent circuit (shown as an insert in Fig.7) we used to represent the cell. The values of the components in the circuit were found by CNLS fitting. Note that the impedance spectrum is composed of two experiments to cover the whole frequency range of interest (0.6 Hz − 17 kHz).

The linear branches of the curve shown in Fig.7 are most likely due to diffusion of charge compensating ions in a film of restricted thickness. The large semicircle is characterized by the domination of the C_2 and R_2 components while the small semicircle (not always present) is usually accounted for by the C_1 and R_1 components, but the physical meaning of this is still ambiguous. The interpretation of the data has been discussed in more detail elsewhere.[10]

When a multiple-frequency excitation signal is used, the impedance at all frequencies is obtained at the same time, which is a considerable advantage when studying systems with fairly low stability. If the impedance is measured separately at every frequency the noise-level is usually lower, but the system may change during the time passed from the recording of

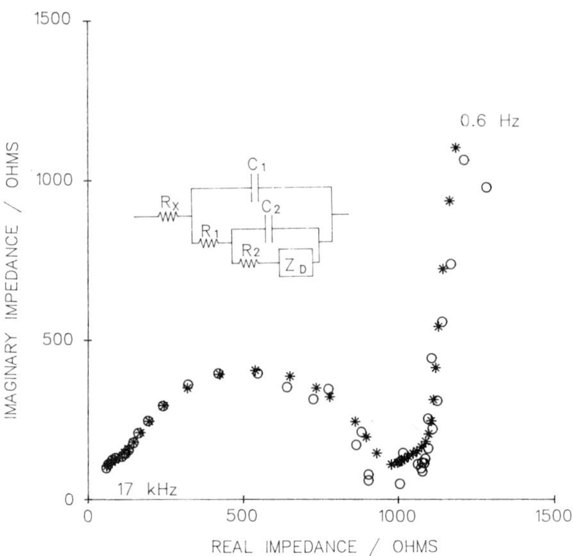

Fig.7. Impedance plot of a polymer coated Pt-electrode in 0.1 M LiBF$_4$/propylene carbonate. Estimated film thickness = 0.5 μm. E_{dc} = 1.2 V vs. Ag/AgCl. Circles = experimental data. Stars = calculated data using the CNLS fit to the circuit shown in the insert, where Z_D is a diffusion impedance.

the impedance at the first to the last frequency. In the experiment described above time-domain averaging was used in order to reduce the noise-level of the high-frequency data while the low-frequency data, 0.6–25 Hz, were not averaged at all due to requirements on the dc potential scan speed. Frequency-domain averaging was not used in the example shown but in other experiments it has been used with good results.

Conclusions

We see many advantages in the use of the FT-FAM technique, many of which have already been mentioned. The most important one is speed, especially the very important fact that in this technique one measures the impedance at each frequency simultaneously, thereby rendering less importance to eventual time-dependent phenomena in the cell. Also the increased data processing capability of the instrument, e.g., posibilities for real time display of experimentel curves etc. deserves mentioning. We would also like to stress the importance of portability and ease of use, consequences of the use of modern microprocessors together with flexible software. We predict increasing importance of this aspect as the technical development continues, making this technique truly competitive.

References

1. R.J. Schwall, A.M. Bond, R.J. Loyd, J.G. Larsen and D.E. Smith, Anal. Chem., **49** (1977) 1797.
2. D.E. Smith, in: "Fourier, Hadamard, and Hilbert Transforms in Chemistry", A.G. Marshall (Ed), Plenum Publishing Co, New York, 1982, Chap. 15.
3. M. Wasberg and S.O. Engblom, Abstracts L29 and P39, Abstracts of Scientific Computing and Automation (Europe), 13–15 May 1987, Amsterdam, The Netherlands.
4. S.C. Creason, J.W. Hayes and D.E. Smith, J. Electroanal. Chem., **47** (1973) 9.
5. D.E. Smith, in: "Electroanalytical Chemistry", A.J Bard (Ed.), Vol. **1**, Marcel Dekker, New York, 1966, Chap. 1.
6. D.E. Smith, Anal. Chem., **48** (1976) 221A.
7. M. Sluyters-Rehbach and J.H. Sluyters, in: "Electroanalytical Chemistry", A.J. Bard

(Ed.), Vol. 4, Marcel Dekker, New York, 1970, Chap. 1.

8. R. de Levie, J.W. Thomas and K.M. Abbey, J. Electroanal. Chem., **62** (1975) 111.
9. J.R. Macdonald and L.D. Potter Jr., Solid State Ionics, **23** (1987) 61.
10. J. Bobacka, M. Grzeszczuk and A. Ivaska, Abstract C8.13, Extended Abstracts of the First International Symposium on Electrochemical Impedance Spectroscopy, 22–26 May 1989, Carcans, France.

PROCESSOR-CONTROLLED FAST POTENTIOSTAT

J. Kankare and J. Lukkari

Department of Chemistry, University of Turku
SF-20500 Turku, Finland

Introduction

Modern electrochemical experiments often require recording of fast transients or generation of complex potential waveforms. The former problem is encountered in step experiments where instantaneous response of the system to an abrupt potential (or current) change is studied. The other group consists of acmethods and different kind of pulse techniques used both in basic research and analysis. All these usually necessitate the use of a computer to control the measurement.

The basic component of electrochemical instrumentation is the potentiostat. The use of a single computer or microprocessor to control voltammetric instrumentation is well established and many digitally controlled potentiostat configurations have been described,[1-6] but the limited capabilities of microprocessors may render their use difficult and a minicomputer system is quite expensive for control purposes. On the other hand, the use of two computers, a microprocessor or a microcomputer for realtime control of the experiment and a host computer for data evaluation, offers several advantages.[7,8] In addition, the large multi-user minicomputer can be replaced by a personal computer to give the operator full control over the system.[7]

As the commercial instruments are usually expensive and not versatile enough, we have constructed a microprocessor-based fast potentiostat and linked it to an already existing minicomputer in our laboratory.

System Description

Computer System and Software

Fig.1 shows the general configuration of the system hardware. The core of the system is a Rockwell RM 65 microprocessor board containing a 6502 CPU and a 6522 VIA (Versatile Interface Adapter), which provides two 8-bit parallel I/O-data ports, four control lines, and two programmable counter-timers.

The RM 65 bus is connected through an Asynchronous Communications Interface Adapter module (ACIA, Rockwell) to the AUX-port of a CIT-101 video terminal (C.Itoh Electronics, Inc.), which communicates independently with our PDP 11-type SM 5 minicomputer. The computer uses tailor-made FORTRAN libraries for data handling, presentation and permanent storage. They also take care of all interactive operations and calculate the experimental parameters, including complex potential waveforms, which are then transferred into processor RAM. All the time-critical measurement programs are written in an assembler language[9] and reside in PROM. They control the potentiostat (POT) and the function generator (FG). The programs can be used to perform either experiments with continuous waveform or potential step experiments. Data collection frequencies up to 18 kHz are attainable which enables measurements to be performed in a submillisecond time scale. For program development the microprocessor can be replaced by a Rockwell AIM 65 microcomputer.

Contemporary Electroanalytical Chemistry, Edited by A. Ivaska *et al.*
Plenum Press, New York, 1990

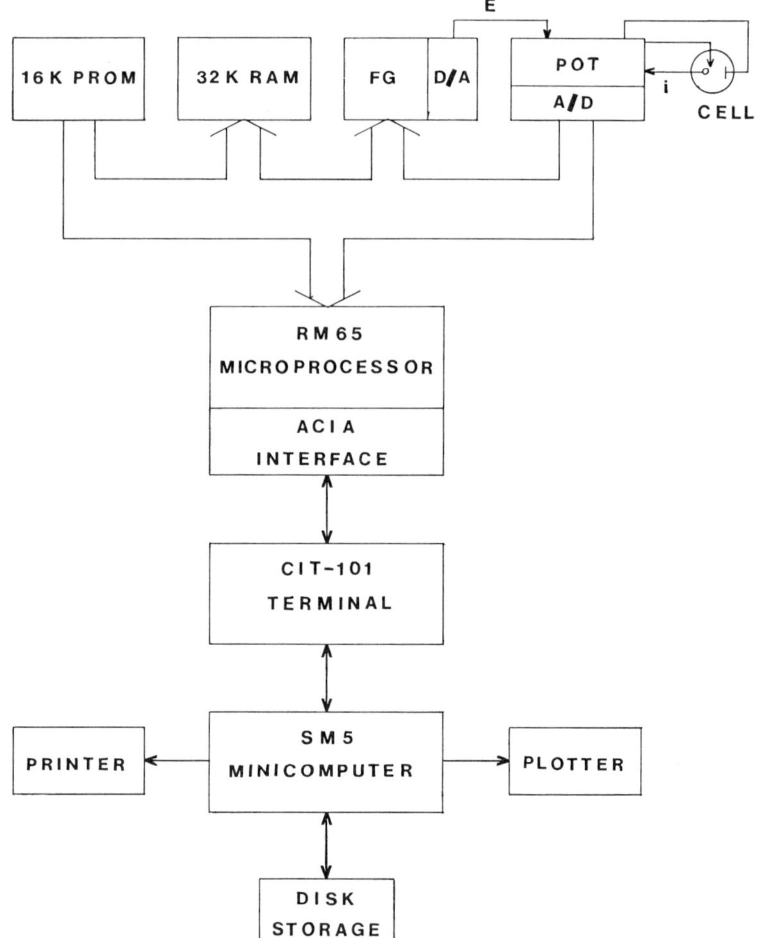

Fig.1. The general configuration of the system hardware.

Potentiostat and Data Acquisition

The potentiostats are generally based on two designs, i.e., the single (or differential) amplifier type and the adder type.[10] Our potentiostat (Fig.2) is basically of the single amplifier type but with a summing input and an iR compensation. Ramaley et al. have studied different configurations of the control amplifier, its feedback loop and the current follower and they found the configuration in Fig.2 to be superior in terms of stability, response and noise.[5] The small capacitors between the auxiliary and reference electrodes and in the feedback loop of the current follower have been shown to enhance the stability of the system.[11] The potentiostat also contains an analogue integrator with various time constants (not shown).

From the current follower the signal is transferred to an autoranging circuitry based on TCA 965 window discriminators (Siemens) (Fig.3). The preamplifiers cover the span of four decades and the reference voltages on the discriminators have been so chosen that the output of the discriminator is set to logical 1 when the absolute value of the input voltage is less than 10 V. The EXOR-circuit (4070) then selects the voltage whose absolute value lies between 1 V and 10 V. The output of the first latch (4042) is used to control the electrical switches (AD 7512) which pass this properly amplified signal to the sample-hold (Burr-Brown SHC 80). Since the gains of the preamplifiers are not changed their slewing does not contribute to the signal. Our autoranging system resembles that of Salt and Lunney.[12] but a variety of other autorangers have been described and an overview is provided by Ramaley et al.[13]

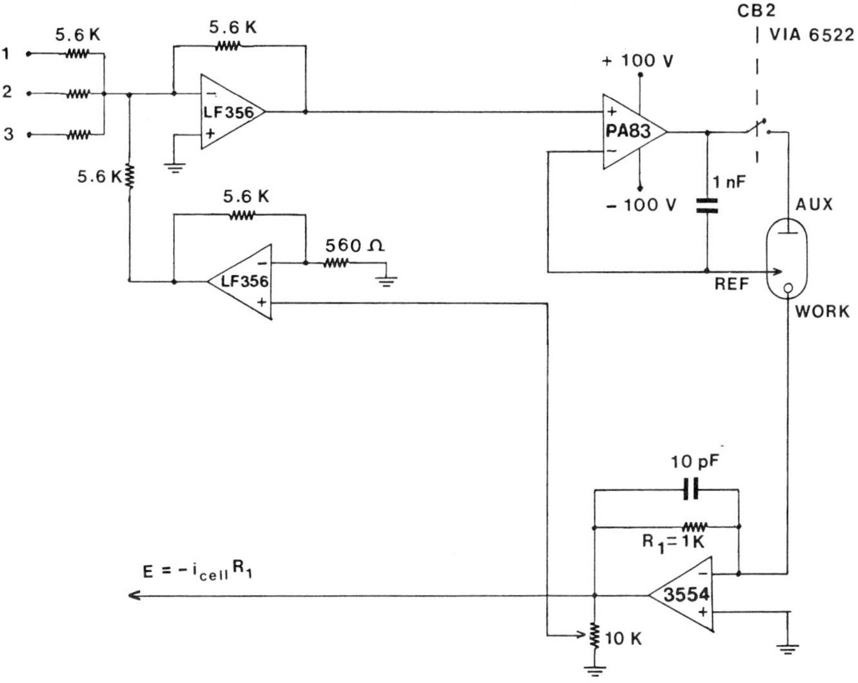

Fig.2. The potentiostat. Operational amplifiers: LF 356, Burr-Brown 3554, and Apex PA83.

Data acquisition timing is handled by one of VIA's timers, which contains the time interval of data points. VIA is programmed to a free running-mode and the time-out interrupt is enabled. The interrupt line is directly connected to the sample-hold, the A/D-converter (Burr-Brown ADC 80) and the latches, which enables the interrupt to sample data without any time lag. A data point is taken when the interrupt line goes low (1 in Fig.3) forcing the S/H to the hold mode. At the same time the output of the first latch is locked and a convert command-pulse is sent to the A/D. The same pulse also allows the output of the autoranger, which constitutes the exponent part of data, to be transferred through the second latch and be locked at its output. After data conversion is complete (approx. 22 μs) the status of the A/D-converter goes low (2 in Fig.3) signalling to the VIA and the microprocessor that data can be read. Reading of data causes the interrupt line to go high again (3 in Fig.3) forcing the S/H to the sample mode and enabling the first latch to follow the output of the autoranger. After the next time-out the cycle begins anew until a predetermined number of data points has been read.

The Function Generator

Our function generator is of conventional design consisting of a VIA, three latches and a D/A-converter (Burr-Brown DAC 800-I), whose output is fed to an operational amplifier (AD 509) (Fig.4). VIA is programmed to a pulse output-mode, where writing of the lower bits of a potential causes VIA to generate a clock pulse for the latches, which transmit the potential to the D/A-converter. The output range of the converter-amplifier system can be manually selected.

Discussion

Our primary aim has been to construct a convenient electrochemical measuring system for the characterization of conducting polymers. These polymers have a long conjugated backbone and are usually insulating in the reduced state.[14] Upon oxidation they become electrically conducting and positive charges appear in the backbone in form of polarons or

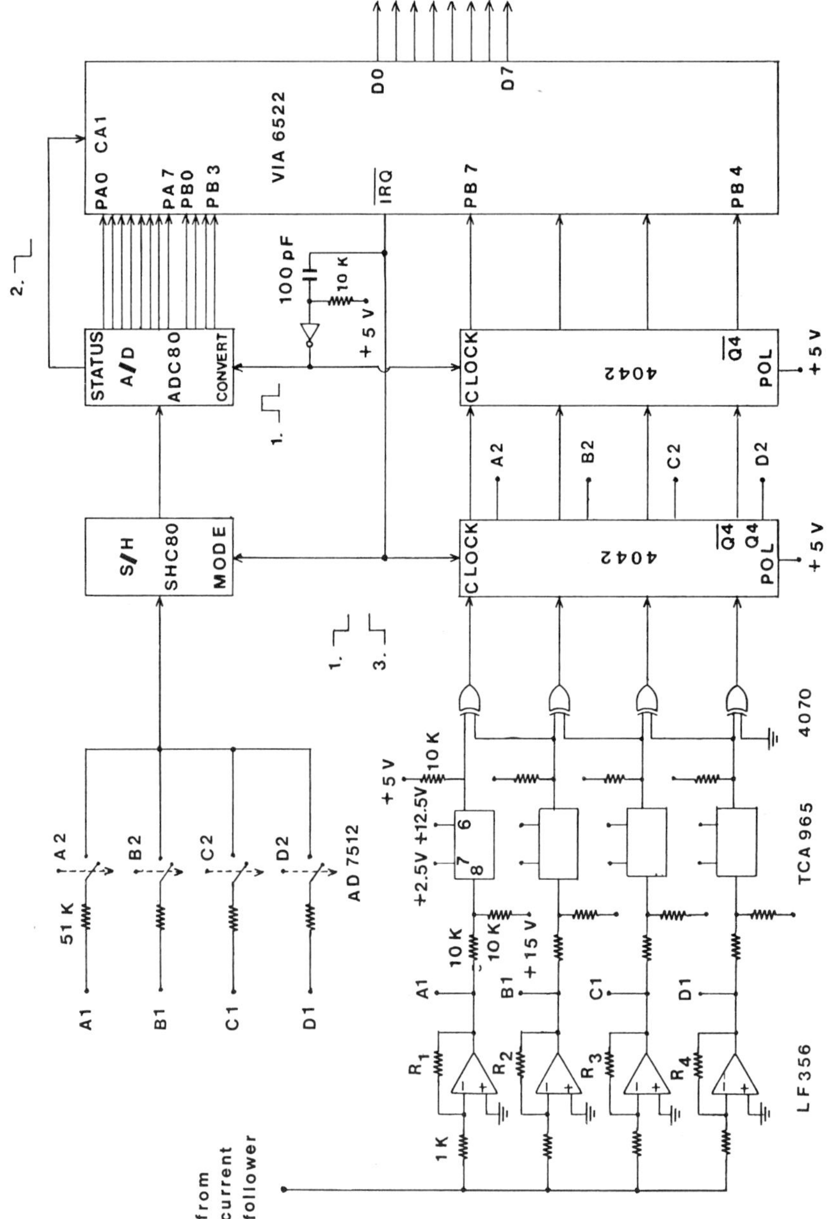

Fig.3. The autoranging (to the left) and data acquisition systems. R1 = 1 K, R2 = 10 K, R3 = 100 K, R4 = 1 M.

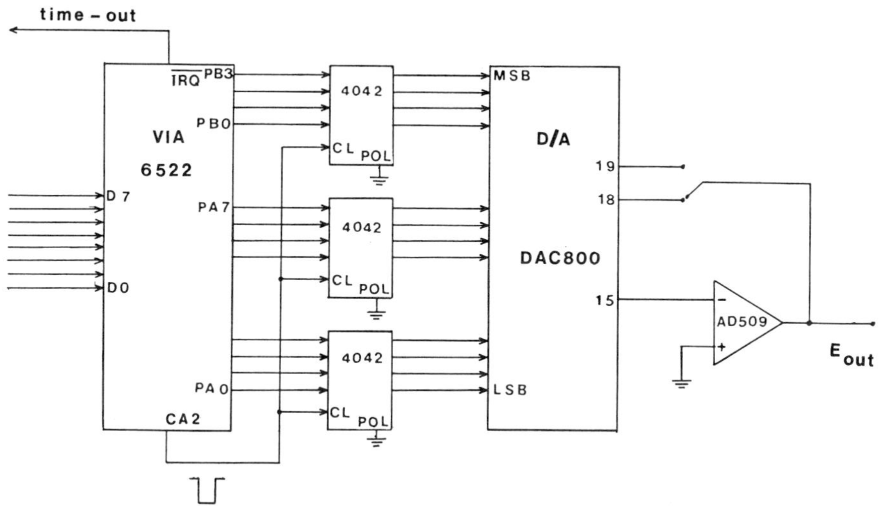

Fig.4. The function generator.

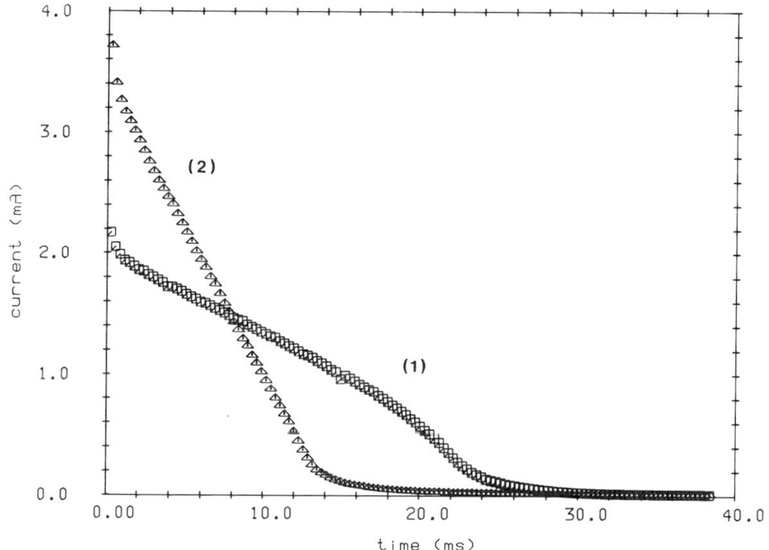

Fig.5. The response of a Pt/p3MeT, ClO_4 electrode to a potential step from 0.60 V to -0.60 V vs. Ag/Ag$^+$ in 0.1 M acetonitrile solutions of $(n\text{-Bu})_4NClO_4$ (1) and $(n\text{-Bu})_4NBF_4$ (2). Electrode area is 0.5 mm^2.

bipolarons. To compensate for these charges there must be a net influx of anions from the adjacent solution. Cations also diffuse but their role in the oxidation-reduction process is still largely unsolved.

We have used our potentiostat system in a study of these basic phenomena with poly-3-methylthiophene (p3MeT)[15] as a model polymer. The film was galvanostatically polymerized onto a Pt-electrode in a 0.1 M acetonitrile solution of 3MeT and $(n\text{-Bu})_4NClO_4$ using a current density of 2 mA/cm^2. The total charge was 100 mC/cm^2 which corresponds to an approximate film thickness of 100 nm.[16] Some chronoamperometrical results with different anions are shown in Fig.5. As can be seen the time scale of the reduction process is relatively small. The sharp decrease below 1 ms is a capacitive peak but the faradaic current does not follow the thin-layer modification of the Cottrell equation.[17] This is undoubtedly due to the influence of the electric field which enhances the migration of the ions in the film. Different anions also behave in a way that has not yet been rationalized.

Fig.5 shows that our potentiostat system is capable of handling fast transient measurements. It should also be noted that although the current scale extends over several decades the autoranging system produces a smooth curve.

References

1. L. Ramaley, Chem. Instrum., **6** (1975) 119.
2. O.R. Brown, Electrochim. Acta, **27** (1982) 33.
3. D.W. Paul, T.H. Ridgway and W.R. Heineman, Anal. Chim. Acta, **146** (1983) 125.
4. H.J. Wieck, G.D. Heider Jr. and A.M. Yacynych, Anal. Chim. Acta, **166** (1984) 315.
5. P. Jayaweera and L. Ramaley, Anal. Instrum., **15** (1986) 259.
6. M. Wasberg and P. Sárkány, Talanta, **34** (1987) 757.
7. P. Baecklund and R. Danielsson, Anal. Chim. Acta, **154** (1983) 61.
8. L. Ramaley, Anal. Instrum., **15** (1986) 101.
9. R. Zaks, Programming the 6502, Sybex Inc., 1980, U.S.A.
10. M.C.H. McKubre and D.D. MacDonald, Comprehensive Treatise of Electrochemistry, Vol. **8**, (1984) 1.
11. E.R. Brown, D.E. Smith, and G.L. Booman, Anal. Chem., **40** (1968) 1411.
12. A.D. Salt and D. Lunney, Anal. Chem., **52** (1980) 2237.
13. P. Jayaweera and L. Ramaley, Anal. Instrum., **15** (1986) 241; L. Ramaley and D.P. Surette, Chem. Instrum., **8** (1978) 181.
14. G.K. Chandler and D. Pletcher, Spec. Period. Rep. Electrochem., **10** (1985) 117.
15. P. Marque, J. Roncali and F. Garnier, J. Electroanal. Chem., **218** (1987) 107.
16. J. Roncali and F. Garnier, Nouv. J. Chim., **10** (1986) 237.
17. E.M. Genies and J.M. Pernaut, Synth. Met., **10** (1984/85) 117.

SMOOTHING OF AC POLAROGRAPHIC DATA BY FFT FILTERING

Johan Bobacka and Ari Ivaska

Laboratory of Analytical Chemistry
Åbo Akademi University
SF-20500 Turku-Åbo, Finland

Introduction

In ac polarography the measured signal often contains random fluctuations, usually referred to as noise. When the faradaic current is studied the noise becomes more and more disturbing when the concentration of the electroactive species decreases, and close to the detection limit the noise starts to mask the faradaic signal.

These random fluctuations may have several origins. Thermal noise is a type of white noise which arises from random thermal perturbation of the distribution of electrons in a resistor.[1] Thus the solution resistance, charge-transfer resistance as well as the Warburg impedance might introduce this kind of noise (and some excess noise) to the ac polarographic data. When using a dropping mercury electrode (DME) or a static mercury drop electrode (SMDE) the variations in drop size result in variations in the measured ac current. In addition to these "internal" sources some "external" sources like static electricity and induced currents from surrounding electrical equipments might introduce noise to the measured ac signal.

The thermal noise from the resistors can provide useful information about kinetic and diffusional parameters[1] and should not be overlooked, but usually the noise is regarded as a nuisance because it makes it difficult to calculate parameters like peak potential, peak height and width at the half height. Fortunately there are several methods to enhance the signal-to-noise ratio. In addition to ensemble averaging (which is often very time-consuming) smoothing can be obtained using analog and digital filters.[2] Digital filters offer some advantages over the analog ones[3] and due to the availability of fast Fourier transform algorithms (FFT algorithms)[4] the FFT filtering technique is an advantageous alternative and has been used in electrochemistry[2,5] as well as in other methods of chemical analysis.

Theory

One-Dimensional FFT

The FFT filtering technique is based on three mathematical operations:
1. Fast Fourier transformation (FFT) of the noisy ac polarographic data to yield the Fourier spectrum which is the frequency representation of the data.
2. Multiplication of the Fourier spectrum by a filter function, e.g., elimination of the Fourier components that represent the noise.
3. Inverse fast Fourier transform (FFT^{-1}) of the modified Fourier spectrum to yield the smoothed ac polarographic data.

The theory for the calculation of the discrete Fourier transform and algorithms for the FFT are well described in the literature[4] and will not be discussed further in this report. In this study the Cooley-Tukey FFT algorithms are used.

The most frequently used filter function is the rectangular filter. This means that the low-frequency components of the Fourier spectrum are unaffected while the high-frequency components (representing the noise) are set to zero. The cut-off point of the filter can be freely

Contemporary Electroanalytical Chemistry, Edited by A. Ivaska *et al.*
Plenum Press, New York, 1990

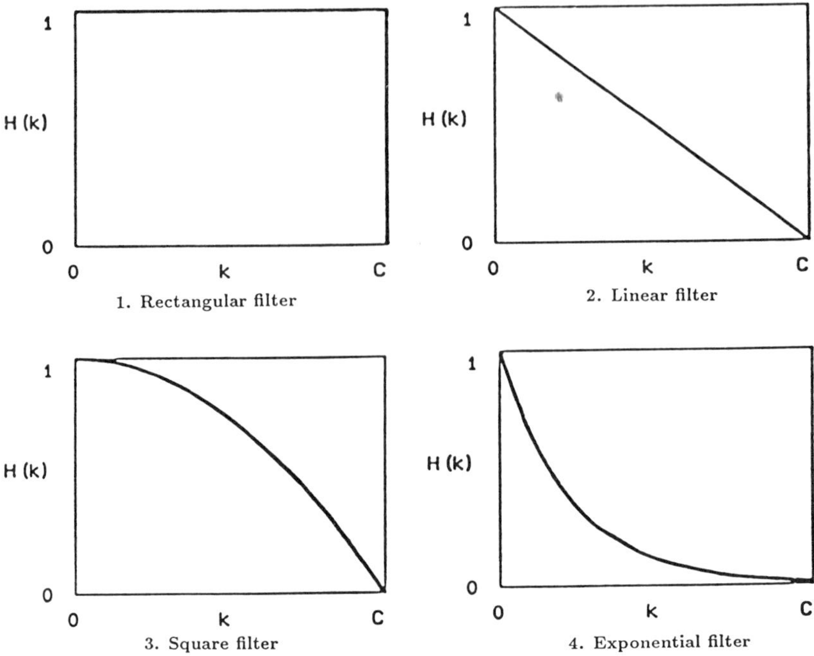

Fig.1. Graphical description of the four filter functions studied.

chosen to yield the best smoothing result. The abrupt truncation by the rectangular filter is however known to introduce sidelobes to the smoothed ac polarograms. Use of a "matched filter" will reduce the sidelobes and enhance the signal-to-noise ratio but the smoothed signal will be broader.[3]

In this study four different filters were investigated using simulated ac polarograms with noise added (by random number generation). The four filters used are presented in Table 1 and graphically described in Fig.1.

Table 1. Equations describing the four filters used in this work. $H(k)$ = filter function, C = cut-off frequency and k = index of the Fourier components.

Filter	$0 \leq k < C$	$k \geq C$
1. Rectangular filter	$H(k) = 1$	0
2. Linear filter	$H(k) = 1 - \frac{k}{C}$	0
3. Square filter	$H(k) = 1 - \frac{k^2}{C^2}$	0
4. Exponential filter	$H(k) = e^{\left(\frac{-k}{\sqrt{C}}\right)}$	0

In this work we study the following electrochemical reaction:

$$Ox + ne^- \rightleftharpoons Red \tag{R1}$$

If R1 is a reversible reaction the admittance (A_{rev}) can be expressed as:

$$A_{rev} = \frac{n^2 F^2 A C^* \sqrt{\omega D} \Delta E}{4RT \cosh^2(j/2)} \tag{1}$$

where $\quad j = \frac{nF}{RT}(E_{dc} - E^r_{1/2})$.

The symbols in the equations above have their usual meanings.[6] The normalized admittance (A_{norm}) can be calculated from equation (1) by defining the numerator divided by $4RT$ equal to one:

$$A_{norm} = \frac{1}{\cosh^2(j/2)} \tag{2}$$

Equation (2) was used to calculate the simulated ac polarograms which all had a peak height equal to 1 due to the normalization. The smoothed ac polarograms were compared to their noise-free versions with respect to peak height and peak area.

Two-Dimensional FFT

When using a multi-frequency excitation signal it is possible to obtain several ac polarograms in one single dc scan, and this makes it desireable to smooth the ac polarographic data using two-dimensional FFT filtering. The main advantage of this method is that the data can be smoothed in the frequency direction and in the dc potential direction at the same time.

Two-dimensional FFT filtering is an extension of the one-dimensional case and both are based on the same FFT algorithms.[7] The M noisy ac polarograms (N points each) can be represented as a M*N matrix (M rows and N columns). First the FFT is performed on each row to yield an intermediate matrix and then the columns of the intermediate matrix are transformed. The result is a two-dimensional Fourier spectrum (also an M*N matrix) which is multiplied by a two dimensional filter and inverse transformed to yield the smoothed ac polarographic data. In this study 8 ac polarograms of 256 points each were smoothed simultaneously using the two-dimensional FFT filtering technique.

Experimental

Reagents

The solutions of Cd^{2+} in 1 M NaCl were prepared using $Cd(NO_3)_2 \cdot 4H_2O$ (MERCK, pro analysi), NaCl (MERCK, pro analysi) and high-purity water which was obtalned by passing distilled water through a Milli-Q water purification system. A supporting electrolyte of 1 M NaCl was used in all the experiments.

Apparatus

The computer programs for FFT filtering were written in the MODULA-2 language using an Olivetti M24 PC which was connected to the FT-FAM device used. The Metrohm 663 VA stand with a saturated Ag/AgCl reference electrode and a glassy carbon auxiliary electrode was used in the dropping mercury electrode (DME) mode. All the dc potentials in this work are measured agaist the Ag/AgCl reference electrode.

Procedure

Dissolved oxygen was removed by bubbling N_2 through the solution for 10 minutes before the ac measurements and for 4 minutes after each addition of the standard. The dc potential scan was performed from -475 mV to -725 mV using an increment of 1 mV and the scan rate was 0.58 mV/s. When recording the ac polarograms separately an excitation signal of 97.92 Hz was used. The multi-frequency excitation signal was composed of a base

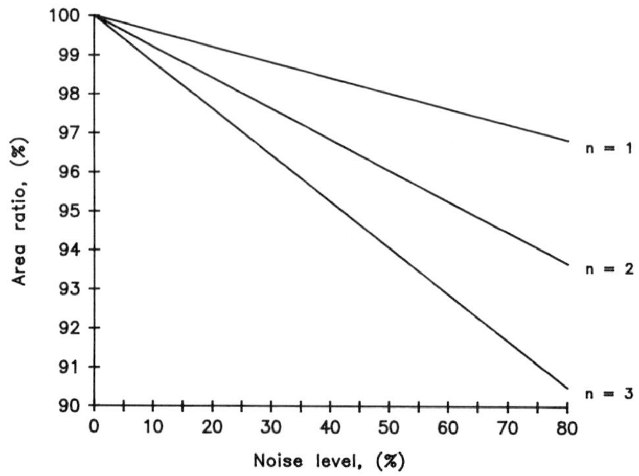

Fig.2. Area ratio versus noise level, where n = number of electrons in R1.

frequency of 19.6 Hz and the following frequencies: 58.8, 97.9, 137.1, 176.3, 215.4, 254.6 and 293.8 Hz. These frequencies are odd harmonics of the base frequency.

The ac polarographic data from the measurements were stored in the memory of the computer for further analysis and smoothing. When necessary, background correction was performed prior to the FFT filtering. When smoothing the ac polarographic data there were two possibilities:

1. The in-phase and the quadrature currents were regarded as complex signals composed of a real part (in-phase current) and an imaginary part (quadrature current) and both were smoothed simultaneously.
2. The in-phase and the quadrature currents were both regarded as real signals which were smoothed separately.

When using a multi-frequency excitation signal the in-phase and the quadrature currents were smoothed simultaneously.

Results and Discussion

Simulated ac Polarograms

Simulated ac polarograms with noise levels from 0 to 80 % were smoothed. Here "noise level" means the peak-to-peak value of the noise divided by the peak height (=1) of the ac polarogram.

The peak area of the smoothed ac polarogram as the function of noise level is shown in Fig.2. The area ratio is given in percentage and is defined as the area of the smoothed ac polarogram divided by the area of the noise-free peak. As can be see from Fig.2 the peak area decreases with increasing noise level. If the cut-off at a certain noise level was chosen too low the peak became broader and lower but the peak area still remained constant.

The peak height was found to decrease nonlinearly with increasing noise level and was strongly dependent on filter function and cut-off. Fig.3 describes how the four filter functions studied affect the peak height in smoothing simulated ac polarograms at different noise levels. The cut-off was chosen to minimize the sum square error between the smoothed ac polarograms and the noise-free ac polarograms. The peak height ratio is defined in the same way as the area ratio.

As can be seen in Fig.3 the exponential filter causes a decrease in the peak height (and increase in the peak width due to the constant peak area) even when the noise level is 0 % and should not be used when smoothing ac polarograms. The rectangular filter is found to give the best result. This is obviously due to the fact that the shape of the ac

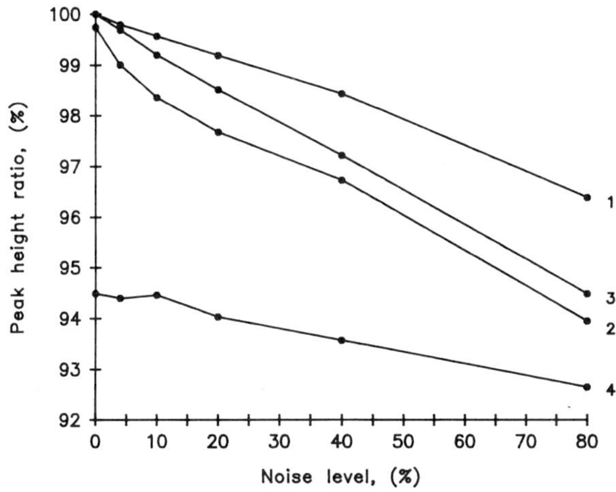

Fig.3. Peak height ratio at different noise levels after smoothing simulated ac polarograms (n=1) and using the four filters described in Table 1 and Fig.1.

polarogram is mainly described by the low-frequency Fourier components and is not affected by the rectangular filter if a proper cut-off frequency is chosen. The peak height is a very important parameter in ac polarography and therefore the rectangular filter is advantageous. The sidelobes introduced by this filter can be minimized by choosing a small dc potential range. This strategy was used in this study when smoothing experimental ac polarograms obtained in Cd^{2+} solutions.

Experimental ac Polarograms

Ac polarograms of 10^{-5} and 10^{-6} M Cd^{2+} were recorded using an excitation signal of 97.92 Hz.

When the FFT filtering technique is used it is possible also to obtain information of the noise by studying the Fourier spectrum. If the magnitude of the Fourier components that describe the noise is the same over the entire high-frequency range it may be concluded that the noise is pure white noise. In our studies, however the magnitudes of the high-frequency Fourier components were found to vary slightly with the frequency. This indicates that the noise in our studies was not pure white noise but had a definite origin. On the basis of this kind of information it might be possible to draw conclusions about the source of the noise. This, however, was not further investigated in our study.

According to C.A. Bush[8] it is possible to get an approximate value of the rms noise of a circular dichroism signal by calculating the standard deviation of the high-frequency components of the Fourier spectrum. The shapes of circular dichroism signals as well as ac polarograms are mainly described by the low-frequency Fourier components. This similarity was taken advantage of and the same method was used to calculate the noise level of the experimental ac polarograms studied in this work.

The noise level of the in-phase current varied between 3.0 and 3.6 % when the Cd^{2+} concentration was 10^{-5} M. At 10^{-6} M Cd^{2+} concentration, which is close to the detection limit for the DME, the noise level was between 31 and 41 %. The noise level of the total current (= the vector sum of the in-phase and the quadrature current) was found to be 4.4 – 9.6 % for 10^{-5} M Cd^{2+} and 42 – 165 % for 10^{-6} M Cd^{2+}. This indicates that the total current contains more noise than the in-phase current when using a DME.

The result of the FFT filtering strongly depends on the choice of the cut-off frequency. Fig.4 shows the peak height, the peak area and the width at the half height of the in-phase current at different cut-off frequencies. The ac polarogram is obtained in 10^{-5} M Cd^{2+} solution.

As can be seen in Fig.4, smoothing can severely affect the ac polarogram if the cut-off

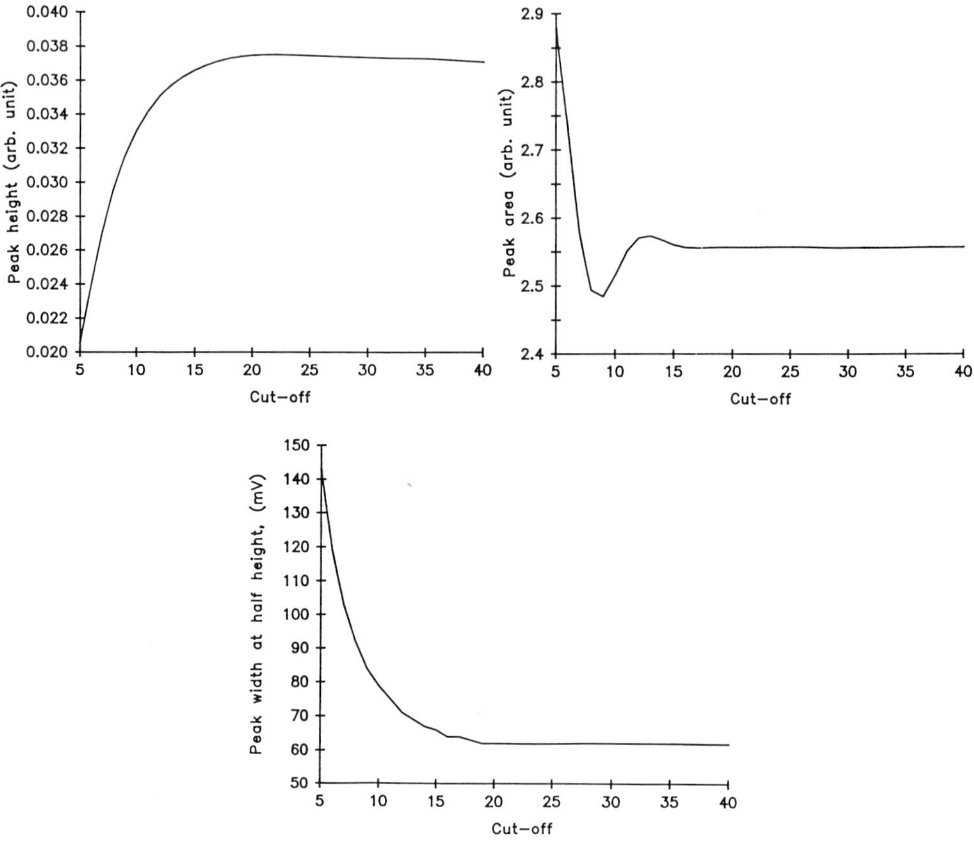

Fig.4. Peak height, peak area and width at half height at different cut-off when smoothing an ac polarogram of 10^{-5} M Cd^{2+} and using the rectangular filter.

is chosen too low. In this experimental case even the peak area is affected by the choice of cut-off, which could not be observed when smoothing the simulated ac polarograms. If the cut-off is chosen too high the noise will not be fully removed. The task is to find a cut-off that removes the noise as completely as possible but also in such a way that the peak is deformed as little as possible. In our case a cut-off of about 20 gives the best result for this particular ac polarogram. Usually the Fourier spectrum gives a good idea of a proper choice of cut-off.

Ac polarograms of Cd^{2+} were studied at the following concentrations: 1.0, 1.8, 2.6, 3.4, 4.2 and 5.0 μM. Fig.5 shows the ac polarograms (in-phase component) corresponding to the concentrations 1.0, 2.6 and 4.2 μM before and after smoothing (rectangular filter, cut-off = 18).

The peak areas of the studied ac polarograms were calculated by numerical integration using the "Simpsons formula". The change in the peak area due to the smoothing varied between 0.03 and 0.7 % for the 6 ac polarograms studied ($[Cd^{2+}] = 1 - 5 \mu$M). The peak area as well as the peak height of the smoothed ac polarograms versus $[Cd^{2+}]$ are shown in Fig.6.

Fig.6 shows that the peak height gives a more linear relationship than the peak area. The correlation coefficients of the linear regression lines of the peak height and the peak area are 0.9997 and 0.9977 respectively. This indicates that the peak height of the smoothed ac polarograms is a useful parameter in quantitative analysis. The reason that the straight lines in Fig.6 does not go through origin is obviously due to an incomplete removal of the background current.

The two-dimensional FFT filtering technique was tested on ac-polarograms of 10^{-3}

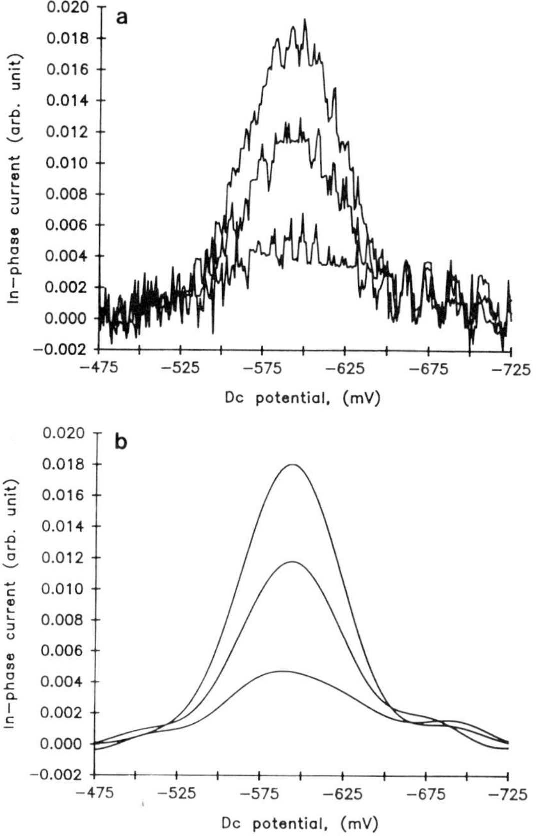

Fig.5. Ac polarograms obtained in Cd^{2+} solutions with $[Cd^{2+}] = 1.0$, 2.6 and 4.2 μM. Excitation signal = 97.92 Hz. (a) Original ac polarograms. (b) Smoothed ac polarograms using the rectangular filter with cut-off = 18.

M Cd^{2+}. Due to the limited data storage capacity of the MODULA-2 language only 8 ac polarograms of 256 points each could be smoothed simultaneously.

The choice of cut-off differ from the one-dimensional case because the Fourier spectrum consists of a M*N matrix. If M is approximately equal to N the Fourier components describing the signal are located to the four "corners" of the matrix while the components representing the noise are in the "middle". In this case the cut-off can be chosen as described in Fig.7.

In our case N >> M (256 >> 8) and the Fourier components of the signal, s are located to the two "ends" of the Fourier spectrum. In this situation good results can be obtained by choosing the cut-off equal in every row as is shown in Fig.8.

Fig.9 shows the original ac polarographic data obtained by a multi-frequency excitation signal and Fig.10 represents the same data after two-dimensional filtering. The cut-off was chosen equal (=8) in every row of the Fourier spectrum as is described in Fig.8.

The advantages of the two-dimensional FFT filtering technique could be better used if the number of ac polarograms (M) was increased. Then it could be possible to obtain better smoothing in the frequency direction than was obtained in this case.

Conclusions

As shown in this work ac polarograms can be smoothed by using digital filtering based on the FFT (Fast Fourier Transform). The noise could be removed and smooth polarograms

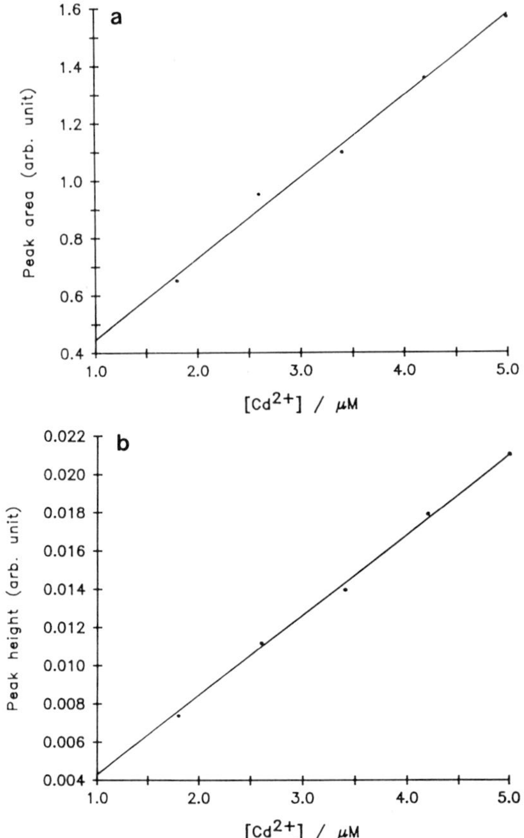

Fig.6. Peak area (a) and peak height (b) of smoothed ac polarograms versus $[Cd^{2+}]$. Excitation signal $= 97.92$ Hz. The filtering is done with the rectangular filter and cut-off $= 18$.

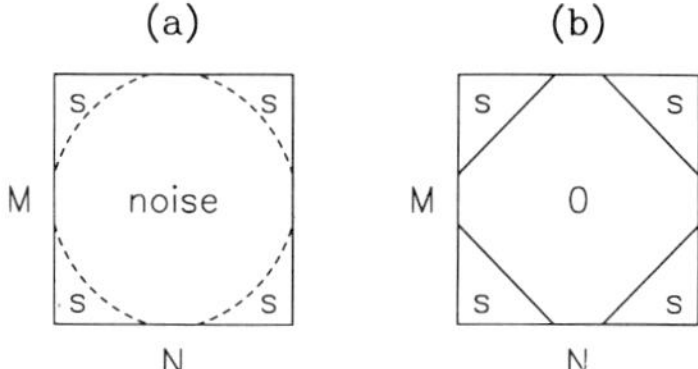

Fig.7. Matrix representation of a Fourier spectrum and the choice of cut-off when $N = M$. The Fourier spectrum of the ac-polarographic data (a) is multiplied by a filter function to yield the modified Fourier spectrum (b) where the components describing the signal, s are left unchanged while the components representing the noise are set to zero.

obtained. The advantage of smooth polarograms is that the characteristic parameters such as summit potential, peak width at half height and peak area can then automatically be recorded. The advantage of the two dimensional filtering is that smoothed polarograms at several frequencies can simultaneously be obtained. This makes the automatic recording of the characteristic parameters possible at all the frequencies studied. The change of these parameters as fuction of frequency gives information of the reaction mechanism and kinetics.

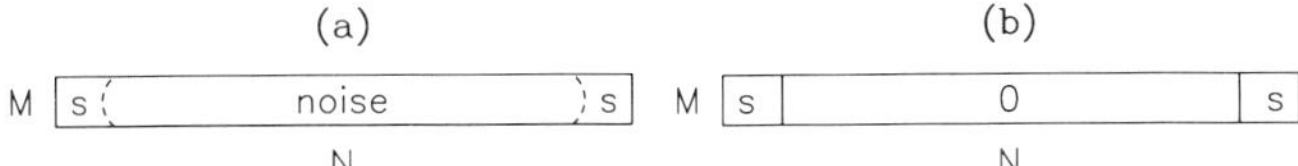

Fig.8. Matrix representation of a Fourier spectrum and the choice of cut-off when N >> M. (a) Fourier spectrum of the ac polarographic data. (b) Fourier spectrum after multiplication by the filter function (compare with Fig.7).

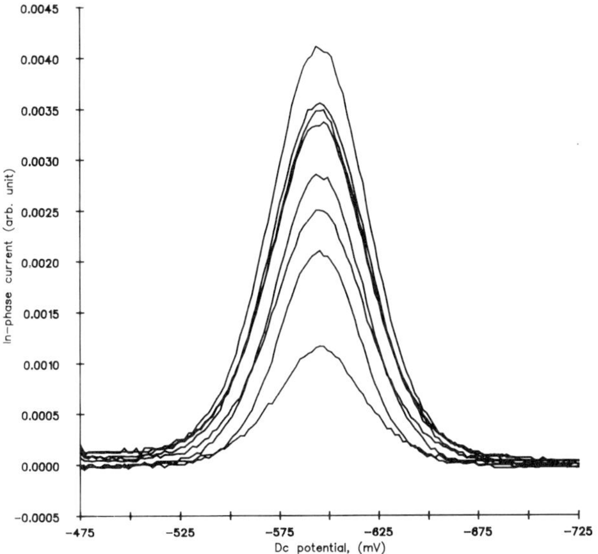

Fig.9. Ac polarograms of 10^{-3} M Cd^{2+}. Data obtained by a multi-frequency excitation signal composed of the following frequencies: 19.6, 58.8, 97.9, 137.1, 176.3, 215.4, 254.6 and 293.8 Hz.

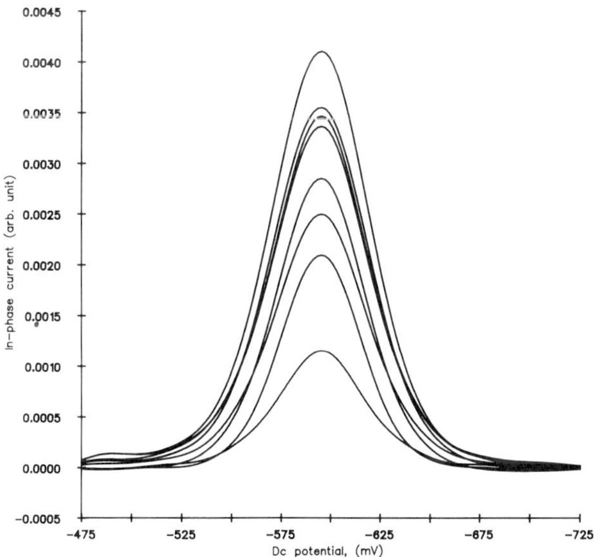

Fig.10. The same data as in Fig.9 after two-dimensional smoothing with cut-off = 8 in every row of the Fourier spectrum.

References

1. A. Bezegh and J. Janata, Anal. Chem., **59** (1987) 494A.
2. J.W. Hayes, D.E. Glover and D.E. Smith, Anal. Chem., **45** (1973) 277.
3. G. Horlick, Anal. Chem., **44** (1972) 943.
4. C.S. Burrus and T.W. Parks, in: "DFT/FFT and Convolution Algoritms. Theory and Implementation", John Wiley & Sons, New York, 1985.
5. D.E. Smith, Anal. Chem., **48** (1976) 517A.
6. A.J. Bard and L.R. Faulkner, in: "Electrochemical Methods, Fundamentals and Applications", John Wiley & Sons, New York, 1980.
7. R.M. Mersereau and D.E. Dudgeon, in: "Proceedings of the IEEE", 63, April (1975), 610.
8. C.A. Bush, Anal. Chem., **46** (1974) 890.

REVERSE PULSE VOLTAMMETRY AT MICROELECTRODES. NEW POSSIBILITIES IN ANALYTICAL CHEMISTRY

Zbigniew Stojek

Department of Chemistry, University of Warsaw
PL 02-093 Warsaw, Poland

Introduction

Reverse pulse voltammetry (rpv) is a very useful electroanalytical technique in the cases where current measurements of direct reduction or oxidation of substrated is complicated due to poorly defined waves. It is also very useful in investigations of chemical reactions coupled with electrode processes. The potential waveform is presented in Fig.1.

The use of rpv (as well as npv) at solid electrodes of regular size, combined later with interpretation of the obtained data, is not straightforward due to the fact that after each consecutive pulse the current concentration profiles at the electrode surface deviate progressively from initial conditions. To overcome this problem the electrode is slowly rotated[4] or the solution stirred before application of the next reverse pulse.[5]

When solid submicroelectrodes are used there is no need of stirring the solution between pulses. The initial conditions are easily regained due to steady state current which develops at the microelectrode surface after relatively short time, during the delay period τ.

Results

In the calculations it was assumed that $D_{ox} = D_{red}$ and the process is of Nernstian type. The total current passing through the electrode is described by[8]

$$\frac{I}{nF} = \int_0^t f\left(4D(t - u)/r^2\right) \frac{d\left(1/(1 + e^{-\xi(u)})\right)}{du} du \tag{1}$$

where

$$f(z) = 1 + 0.71835z^{-1/2} + 0.05626z^{-3/2} - 0.00646z^{-5/2}, \; for \; z > 0.88$$

and

$$f(z) = \left(\frac{\pi}{4z}\right)^{1/2} + \frac{\pi}{4} + 0.094z^{1/2}, \; for \; z < 1.44$$

To calculate the current flowing under conditions of rpv, the method used by Weland et al.[6-7] (in the case of square wave voltammetry) was applied. The derived equations for the normalized currents measured at the end of the delay time, I_{FC}, and at the end of the pulse time, I_{RP}, can be written as:

$$\frac{I_{FC}}{4nFC \times Dr} = C_0 f[(k+1)p - p_{tp}]$$

$$- \sum_{j=0}^{k-1} \left((C_j - C'_j)f[(k-j)p] + (C'_j - C_{j+1}) + f[(k-j)p - p_{tp}]\right)$$

$$\frac{I_{RP}}{4nFC \times Dr} = C_0 f[(k+1)p]$$

Contemporary Electroanalytical Chemistry, Edited by A. Ivaska *et al.*
Plenum Press, New York, 1990

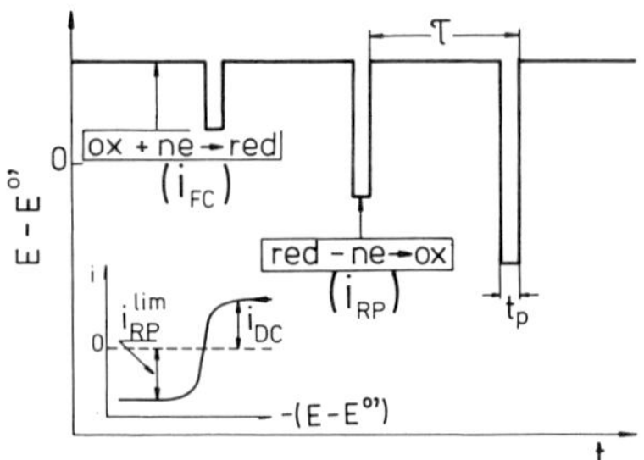

Fig.1. The potential waveform applied in rpv.

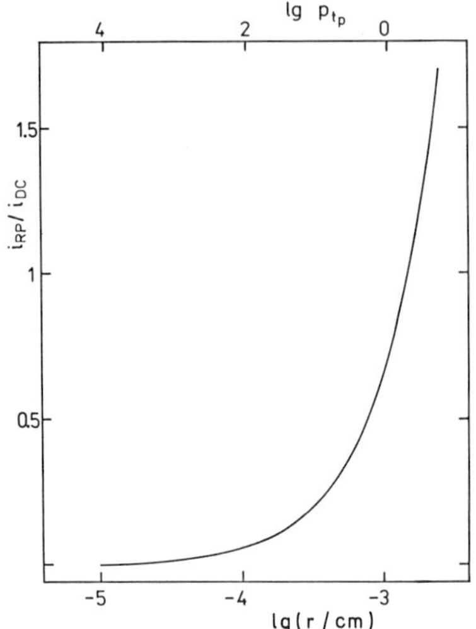

Fig.2. The dependence of $\lim I_{RP}/I_{DC}$ on $\log P_{tp}$. Alternative axis: $\log r$; then $t_p = 30$ ms, $\tau = 4.03$ s, $r = 0.00125$ cm.

$$-\sum_{j=0}^{k-1}\left((C_j - C_j')f(k-j)p + p_{tp} + (C_j' - C_{j+1})\right.$$

$$\left. + f(k-j)p - (C_k - C_k')f(p_{tp})\right) \tag{2}$$

where

$$p = 4D\tau/r^2$$

$$p_{tp} = 4Dt_p/r^2$$

$$C_k = 1/\left(1 + e^{(-nF(E_r - E^0)/RT)}\right)$$

$$C_k' = 1/\left(1 + e^{(-nF(E_r + kE_s - E^{0\prime})/RT)}\right)$$

$$\sigma = 4Dt/r^2$$

Table 1 presents the theoretical data obtained for a microelectrode and for an electrode with a relative large area. Note, that I_{FC} is practically constant and close to 1 in the first case, and substantially changes and deviates from its initial value in the second case. Also, I_{DC} current is constant in the microelectrode case, while it diminishes with potential when electrodes of large area are used.

The dependences of $\lim I_{RP}$, $E_{1/2}$, ∂I_{FC} and I_{RP}/I_{DC} on p_{tp} $(=4Dt_p/r^2)$ are presented in Fig.2 and 3. The calculations also showed that for microelectrodes, at the proper range of r, I_{RP}/I_{DC} is virtually independent of $\tau - t_p$ and number of potential steps. Where τ is the delay time. Experimental curves confirmed the theoretical predictions.

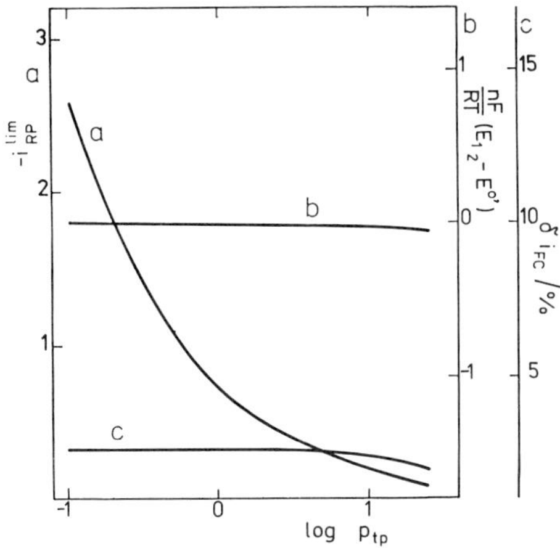

Fig.3. ∂I_{FC}, $n(E_{1/2} - E^{0\prime})$ and $\lim I_{RP}$ plotted vs. $\log p_{tp}$. $p > 100$.

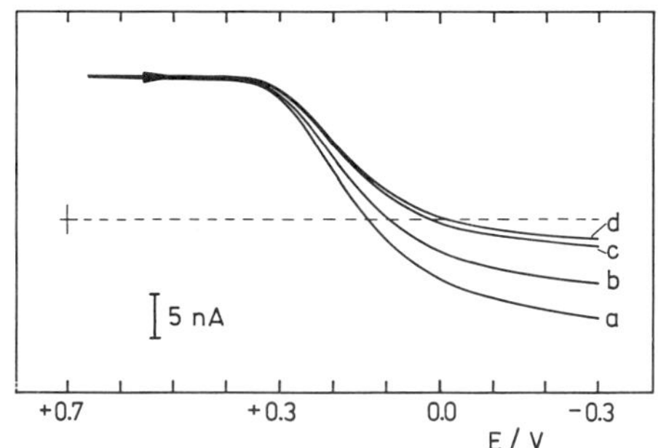

Fig.4. Experimental RPV curves of 2 mM ferrocene in acetonitrile containing 0.1 M TBAP; t_p = 60 (a), 120 (b), 600 (c), 1000 ms (d), sampling time = 20 ms, r = 12.5 μm.

References

1. J. Osteryoung and E. Kirova-Eisner, Anal. Chem., **52** (1980) 62.
2. J. Osteryoung, D. Talmor, J. Hermolin and E. Kirova-Eisner, J. Phys. Chem., **85** (1981) 285.
3. S. Kashti-Kaplan, J. Hermolin and E. Kirova-Eisner, J. Electrochem. Soc., **128** (1981) 802.
4. Z. Karpiński and R.A. Osteryoung, J. Electroanal. Chem., **164** (1984) 281.
5. Z. Karpiński, Anal. Chem., **58** (1986) 209.
6. D.P. Whelan, J.J. O'Dea, J. Osteryoung and K. Aoki, J. Electroanal. Chem., **202** (1986) 23.
7. K.E. Atkinson, in: "Introduction to Numerical Analysis", Wiley, New York, 1978.
8. K. Aoki and J. Osteryoung, J. Electroanal. Chem., **160** (1984) 335.

MULTIPLE SENSOR ARRAYS: ADVANTAGES AND IMPLICATIONS

Dermot Diamond

School of Chemical Sciences
National Institute for Higher Education
Glasnevin, Dublin 9, Ireland

Instrument Intelligence

An intelligent analytical instrument is one that can "understand" its environment to some degree.[1] The basic components are illustrated in Fig.1. An array of sensors enables the instrument to monitor its environment. Raw data from the sensors is converted into a suitable form by an interface before being passed to the microprocessor. This typically involves signal amplification, offsetting and digitisation.

The microprocessor interprets the incoming data using memory resident routines, displays it in a suitable form, and takes appropriate action if required. Getting signals into a computer is, in these days, relatively straightforward. Interface boards are now available at prices ranging from around \$500 to many \$1000's depending on the specification. Most are designed for use with the IBM (or IBM compatible) range of microcomputers, either as internal boards slotted directly into one of the internal expansion board slots, or as external boards connected to the computer via an RS-232 (serial) or IEEE/GPIB (parallel) data transfer link. Facilities available on these boards can include multi-channel data aquisition, on-board memory, variable data sampling rate, programmable gain and offset, variable voltage output, and low-level signal conditioning. To complement these impressive hardware specifications, powerful software data interpretation/manipulation packages have been developed which can enhance signal purity, carry out complex mathematical operations, and display the data in a multitude of different forms. Pattern recognition routines are available which can, in some applications, decipher the raw experimental data, unusually by best fit using carefully defined algorithms or by comparison to a "look-up" table or data library.

Facilities may also be available to enable the instrument to expand a stored knowledge base through interaction with an experienced expert in the field. When an unrecognised data pattern is confronted, a question and answer routine is initiated with the expert who may then add his or her interpretation of the situation to the instrument knowledge base. In this way, the instrument gradually gains experience as its knowledge expands. The software required to perform this task is known as an "Expert System".[2,3] Besides passively monitoring its environment, the instrument may take an active role in controlling various conditions through actuators (valves, pumps, motors, heaters).

Expert Systems and Chemical Sensors

Compared to instrumentation aimed at physical or electronic measurements, expert systems have made relatively little impact on chemical instrumentation. For example, a recent search of the European Community (EC) database EUROKOM (which was set up to facilitate links between research groups throughout EC countries) using the keywords "Sensor Arrays/Artificial Intelligence" revealed forty eight entries, all of which were concerned with

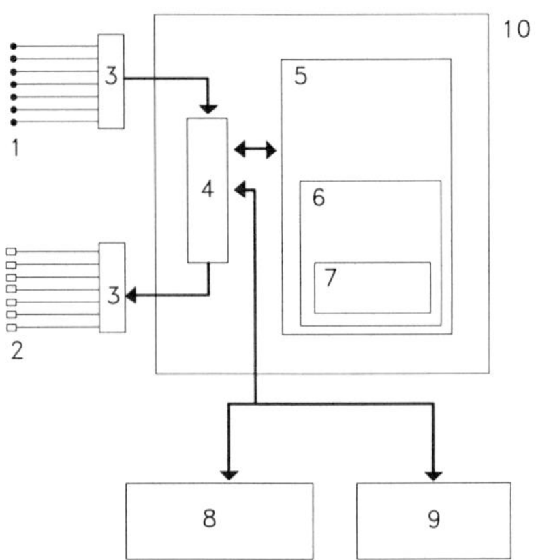

Fig.1. Schematic layout of an intelligent instrument. 1. Sensors, 2. Actuators, 3. Input/output interface, 4. Microprocessor, 5. Memory, 6. Expert system shell, 7. Knowledge base, 8. External expert, 9. Periferals (VDU, printer, mouse etc.), 10. Microcomputer.

physical sensors, mainly optical sensors. Similarly, a computer search through Chemical Abstracts using the keyword "sensor arrays" gave thirty three responses, twenty one of which were related to optosensors and two to temperature measurement. Of the ten involving the use of chemical sensor arrays, nine were concerned with gas sensing, and only one with the use of ISEs for blood analysis.[4] A search using the term "intelligent instrumentation" produced three results, one each involving GC/MS spectral analysis,[4] polarography[5] and NMR optimisation.[6]

These figures together suggest that there is virtually no collaboration between chemical sensor specialists and information technologists. Given that the computing power, memory capacity, data aquisition system and expert system shells are available, why is it that there has been so little progress towards the development of intelligent chemical instruments? There are several reasons, some inter-related, which have contributed to this state of affairs.

(a) The performance of chemical sensors relative to physical sensors: In order to allow correlation between stored information and experimental data using complex algorithms, the sensors used to build an intelligent instrument must be extremely stable during extended use, and respond in a highly reproducible manner to changes in their environment. In addition, they should be of small dimensions, robust and have a fast response. In terms of design, ISFETs (Ion-Selective Field Effect Transistors) are probably the best suited chemical sensors for this type of work as they are small, multi-channel devices based on integrated circuit technology.[7]

Unfortunately, these devices tend to have a short effective life-time and suffer from drift. This is due, in part, to gradual penetration of the sensor by the sample (particularly at the PVC/encapsulate seams) and a gradual leaching of the active ingredient (typically a low-molecular weight ionophore dissolved in a PVC plasticiser) into the sample.[8]

Design modifications have extended the lifetime and stability of ISFETs, for example the suspended mesh gate approach initiated by Blackburn and Janata[9] and the ionophore doping technique recently reported by Bezagh et al.[10] In the latter case, seams were avoided by completely coating a dual-gate FET with PVC and subsequently doping the gate regions with ionophore solutions sensitive to Na^+ and Cl^-. Although a useful life-time for these devices of about two months was reported, the sensitivity of each sensor decreased by 4 mV (Na^+) and 5 mV (Cl^-) for a ten-fold concentration change over a period of three weeks. Drift was reported at 0.05 mV/hour in each case. While this performance compares favourably with other chemical sensors currently available, significant improvements will be needed if they are to be used along with an expert system (for monovalent ions, a 0.1 mV error in voltage measurement results in a 0.4 % error in the estimated ion activity). Although variations in sensitivity and baseline drift can be compensated for automatically using software-controlled calibration routines and appropriate circuitry, there is no real substitute for stable, sensitive sensors with extended lifetimes.

(b) The smaller apparent market potential of intelligent chemical instrumentation: Optical pattern recognition systems have many enormously lucrative applications, for example;
 i intelligent missiles capable of recognising flight paths and targets;
 ii security systems which can recognise faces, photographs, handwriting, or fingerprints;
 iii digitisation of text (including handwriting), photographs of video film for desktop publishing;
 iv position recognition by machines in manufacturing environments.
Investment in opto-sensors thus reflects the potential financial returns which may be obtained.

Other areas where sensor array technology is well developed include exploration seismology, sonor and radar applications, image reconstruction from radio telescope data and tomographic imaging (using X-ray, ultrasound and microwave sources).[11] Indications are that the sensor market is set to undergo a period of great expansion. For example, it has been predicted that by 1991, the U.S.A. biosensors market will be worth around $365M, compared to $14.4M in 1986. The main applications will be in healthcare, but other markets will open up in the process industries, environmental monitoring and agriculture.[12] Investments in chemical sensor development should show a similar expansion.

(c) Traditional scientific links: Physicists, electronic engineers and computer scientists all broadly speak the same language. Chemists have been somewhat distanced from the microelectronics/microcomputer revolution. Indeed a sizeable proportion have viewed the inevitable introduction of microcomputers into chemistry departments with suspicion, disinterest, and in extreme cases, open hostility. However, given the general usefulness of microcomputers (wordprocessing, electronic communication, spreadsheets, databases etc.), and the preponderance of microcomputer-based instrumentation, attitudes are gradually becoming more positive.

Advantages of Multi-Sensor Approach

(a) Analytes are identified on the basis of their effect on the electrode array. Reliability is therefore improved. Individual sensors need not be overly selective.

(b) Signals arising from interferents can be distinguished from those caused by analyte ions. For instance, lipophilic cations which penetrate the membrane phase via interaction with the membrane solvent will tend to affect all the electrodes.

(c) Fluctuations in signals due to variable physical conditions can be compensated by monitoring physical parameters such as temperature, pressure, flow-rate etc.

(d) A number of analytes can be monitored simultaneously.

The ultimate aim of this type of approach is to produce a greater degree of confidence in the performance of chemical sensors. This should in turn lead to more applications, which should stimulate further investments.

Performance Priority in Future Sensors

An intelligent instrument requires information from many sources in order to gain as complete a picture of its environment as possible. This is a completely different situation from the normal single-analyte measurements carried out using, for example, an ion-selective electrode (ISE) and high-impedance voltmeter. The main priority with these traditional measurements is to be sure that the electrode signal is overwhelmingly dominated by the primary ion, so that signal variations can be related with some certainty to changes in analyte activity. Hence the search for highly selective ionophores has dominated ISE research over the past twenty years. However, no electrode is completely specific, and the range and extent of interference from other ions largely determines the practical usefulness of an ISE.

In contrast, the expert system/multi-sensors approach requires only that the response pattern of the sensor array be unambigously assigned to individual ions. This "fingerprinting" of ions makes deduction of a sample composition theoretically possible, and also removes the burden of producing individually highly selective electrodes. Indeed, some response to a range of ions could be a bonus, as long as each ion produces its own characteristic response pattern in the electrode array. The feasibility of using moderately selective chemical sensors to quantify more than one analyte simultaneously has been demonstrated in work published recently by Beebe et al.[4] An array of electrodes was used to measure Na^+ and K^+ activities at blood serum levels through interpretation of the array response pattern with a simplex-type algorithm.

A similar approach has been adopted to analyse gaseous mixtures using an array of moderately selective piezo-electric sensors.[13] Commercial interest in this approach is growing, as demonstrated by the recent launch of an Analogue Interface Expert System called 'ANNIE' by the UK company Intelligent Applications Ltd.

Improving Data Quality

What factors must be addressed in order to improve the quality of data obtained from chemical sensors?

Reproducibility

Reproducible behaviour among batches of sensors can only be expected if the sensors are manufactured in an identical manner. At present, most chemical sensors are hand-made, either by university research groups, or by small specialist firms. While the quality of the sensors produced is often high, there is no doubt that improvements could be made if advanced manufacturing methods and technology were applied. This underpins the urgent need for much closer industrial and academic links. One possibility is for flexible pilot manufacturing plants to be set up, which could undertake precommercial small-scale manufacturing of new sensors. Careful attention must also be paid to the purity of materials used in sensor production, as the presence of even trace impurities can often have a marked effect on sensor behaviour. As such, the introduction of ISE-grade materials by Fluka BV Ltd, is to be welcomed.

Lifetime

Sensors with lifetimes extending to at least months are required. While progress with ISFETs has been slow, it is encouraging to see collaboration between major groups in an effort to overcome this problem.[10] With less emphasis on selectivity, a wider range of gate sensor materials may be considered for ISFETs, concentrating instead on compability with FET/encapsulant materials. For example, ion-selective glasses, although generally less selective, may graft more firmly than the PVC/ionophore membranes currently favoured. Three-dimensional fast ion conductors (e.g. NASICON) have also been shown to produce working ISFETs.[14] Advances in materials science may produce entirely new substances with potential as solid state sensors (e.g. new super conductors or conducting polymers). A strategy which has already been tried, with some success, is to attach the ionophore in the PVC/liquid membrane sensors covalently to the polymer backbone, thus preventing a gradual loss of the active component into the sample solution.[15]

Stability

Drift in ISEs and ISFETs arises from three main sources viz,[16]

a) Leaching of membrane components from the sensor to the sample: The best way to prevent loss of active components by gradual dissolution is to bind them covalently into a giant structure such as a glass polymer. Hence a solid-state sensor would seem desirable.

b) Temperature related effects: Temperature related drift may be reduced by introducing a temperature sensor into the array, and compensating the individual sensor output

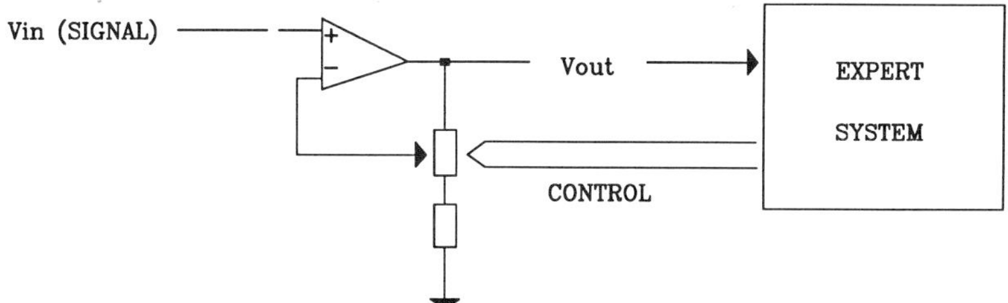

Fig.2. Automatic compensation for drift. This circuit is useful for correcting a signal which has drifted off-scale. When the digitised signal crosses a pre-set threshold, an offset voltage is applied, using a digital-to-analogue converter (DAC), to the inverting input of a differential amplifier, where it is subtracted from the signal. Drift may also be measured periodically with the sensor exposed to a standard solution and corrected via the DAC.

for fluctuations in the measured temperature. Once again, this will work best if the sensor responds in a reproducible manner to temperature changes.

c) Variations in reference electrode junction potential: The generation of a stable reference electrode junction potential is extremely important, as fluctuations will affect the output of the sensor array. Recently progress has been reported with macro reference electrodes.[17] A free flowing junction was used, with extremely slow flow rates ($0.5 - 2.0$ $\mu l/h$). Measurements taken with the electrode were very reproducible in comparison to commercial reference electrodes with ceramic plug and ground glass junctions. The reference electrode was successfully applied for measurements of K^+ and Na^+ in undiluted blood serum over a period of several months without any visible signs of clogging of the liquid junction bore. Miniaturisation of this type of electrode should be possible using modern microlithographic techniques. It is interesting to note that a flow rate of 0.1 $\mu l/h$, a 10 ml reservoir would have the capacity to maintain an electrolyte bridge for over one year.

d) Redesign the measuring circuitry (see, for example, the operational Transducer of Optrode developed by Sibbald).[18] Combinations of analogue and digital circuitry can provide facilities for correction of baseline drift and compensation for decreasing sensor sensitivity. Drift is corrected by integrating the sensor signal over a set time interval when no variations are anticipated (e.g. when in contact with a calibration solution) and comparing to previous values. Correction can be applied by using software routines to modify the value of the digitised signal or by offsetting the analogue sensor signal using DAC (digital-to-analogue converter) as shown in Fig.2. Decreasing sensitivity can be compensated automatically by increasing the gain of a digitally controlled amplifier (Fig.3).

Automatic Identification of Sensor Malfunction

Signals emanating from damage or faulty sensors will obviously cause enormous problems for intelligent instruments, so a facility for automatic sensor malfunction will be necessary in critical applications. Methods of doing this include:

i activating a polling routine to compare the signal obtained from a number of identical sensors. A sensor which consistently fails to agree with the others can be switched out of the array.

ii checking sensor resistance or leakage current using integrated circuitry and software routines. A very low resistance with PVC membrane ISEs indicates membrane rupture, whereas an extremely high resistance suggests loss of active components or the formation of an impermeable coating on the membrane.[19,20]

iii exposing the sensor(s) preiodically to standard solutions and checking the response. This

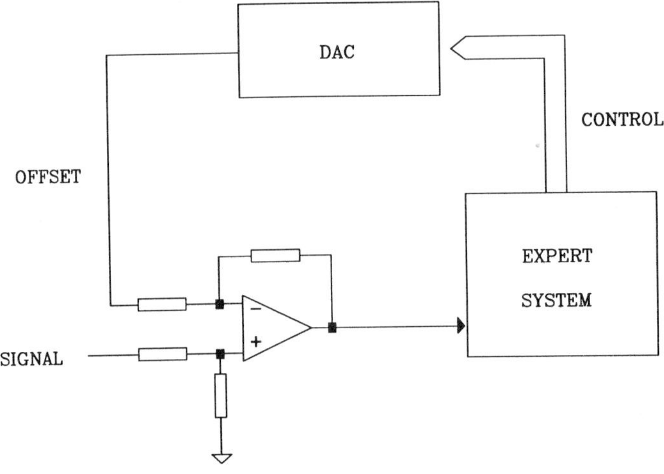

Fig.3. Automatic compensation for decreasing sensitivity. The sensor is
periodically exposed to standard solutions and the response compared
to stored values. The expert system calulates the amount of compensation required and adjusts a digital potentiometer accordingly. The
potentiometer controls the gain in a simple non-inverting operational
amplifier circuit.

facility will also be required to compensate for drift and sensitivity variations as discussed
above.

Conclusions

The coming decade will undoubtedly see an enormous expansion in the area of intelligent chemical instrumentation. Instrumental specifications will include automated features
such as fault detection, calibration and temperature compensation, with the more advanced
versions having the capacity to interpret sample composition. Increasing confidence in the
results obtained from these instruments should generate more applications, which will in turn
stimulate more progress. A movement towards solid state sensors will be required in order to
obtain the vital extended sensor lifetimes and long-term stability.

Given the complex media often involved in chemical analysis, it is likely that the production of efficient chemical sensors will not be easy. However, the new flexibility in sensor
specifications introduced by developments in computing and electronics may make the task
a little easier.

References

1. N. Ford, in: "How Machines Think: A General Introduction to Artificial Intelligence",
   Whiley, New York, 1987.
2. "Building Expert Systems", F. Hayes-Roth, A. Waterman and D. Lenat (Eds),
   Addison-Wesley, Berks., 1983.
3. A.M. Harper and S.A. Liebman, J. Res. Natl. Bur. Stand., (U.S.) **90** (1985) 453.
4. K. Beebe, D. Uerz, J. Sandifer and B. Kowalski, Anal. Chem., **60** (1986) 66.
5. P.R. Fielden, R.N. Carr and C.F. Oduoza, in: "Electrochemistry, Sensors and Analysis",
   M.R. Smyth and J.G. Voss (Eds), Anal. Chem. Symp. Ser., **25** Elsevier, 1986, p. 55.
6. H. Jazayeri-Rad and M.A. Browne, J. Phys. E. Sci. Instrum., **20** (1987) 643.
7. G.F. Blackburn, in: "Biosensors, Fundamentals and Applications", A.P.F. Turner, I.

Karube and G.S. Wilson (Eds), Oxford University Press, Oxford, ch. 26, p. 481, 1987.

8. A. Sibbald, A.K. Covington and R.F. Carter, Med. Biol. Eng. and Computing, **23** (1985) 329.

9. G.F. Blackburn and J. Janata, J. Electrochem. Soc., **129** (1982) 2580.

10. K. Bezagh, A. Bezagh, J. Janata, U. Oesch, A. Xu and W. Simon, Anal. Chem., **59** (1987) 329.

11. "Array Signal Processing", S. Haykin (Ed.), Prentice-Hall, New Jersey, 1985.

12. Chem. Brit., **24** (1988) 114; quoting The Biosensor Market in the U.S., Frost and Sullivan, Sullivan House, 4 Grosvenor Gardens, London SW1W 0DH.

13. G. Haugen and G. Hieftje, Anal. Chem., **60** (1988) 23A.

14. M. Kleitz, J.F. Millon-Broadz and P. Fabry, Solid State Ionics, **22** (1987) 295.

15. L. Ebdon, A.T. Ellis and G.C. Corfield, Analyst, **104** (1979) 730.

16. B.J. Birch and T.E. Edmonds, in: "Chemical Sensors", T.E. Edmonds (Ed.), Blackie, London, ch. 3, p. 75, 1988.

17. R.E. Dohner, D. Wegmann, W.E. Morf and W. Simon, Anal. Chem., **58** (1986) 2585.

18. A. Sibblad, Sensors and Actuators, **7** (1985) 23.

19. G.J. Moody and J.D.R. Thomas in ref. 16, ch. 9, p. 221.

20. F. Regan and D. Diamond, paper presented at the Research and Development Topics meeting, Analytical Division, Royal Society of Chemistry, at NIHE Dublin, March 1989.

SIMULTANEOUS ESR-ELECTROCHEMICAL INVESTIGATIONS AT SOLID ELECTRODES

Lothar Dunsch

Institute of Polymer Technology
Academy of Sciences of the G.D.R.
Dresden, G.D.R.

Introduction

Within the last two decades Electron Spin Resonance-(ESR) spectroscopy has become a standard experimental technique in electrochemical research.[1-5] The main interest was in the field of electrochemical generation of radicals to characterize their structure by ESR spectroscopy or to prove their presence in electrode reactions. The studies have been extended to the kinetics of radical reactions and the set up of reaction mechanism, to the solvation phenomena in radical electron densities and to radical conformation and ion complex structure. The latest development is the study of the electrode materials and their surface layers in electrochemical systems by simultaneous ESR spectroscopic and electrochemical measurements, e.g., of polymer modified electrodes.

Further research in this field has enlarged the number of techniques and problems used and solved by the combination of ESR spectroscopy and electrochemical techniques. In general both potentiostatic as well as galvanostatic techniques are applied in such a combination. Among the techniques are:
- cyclic voltammetry
- chronoamperometry
- chronopotentiometry
- coulometry
  potentiostatic electrolysis
- galvanostatic electrolysis

On the other hand several new ESR-techniques and radical stabilization methods were included in *in situ* electrochemical studies, e.g.:
- rapid scan technique
- signal accumulation
- spin-trap technique
- spin-label techniques (spin label and spin probes)

In this way a lot of possibilities in the study of electrochemical systems exists by the combination of electrochemical and ESR spectroscopic techniques.

Therefore the performance of the equipment for such studies is of high importance. Especially the cell construction has to meet a number of demands derived from the special characteristics of both ESR spectroscopy and electrochemical techniques where some of the demands are contradictory. These problems are the main topic in the experimental section of this paper.

Experimental

The most common types of ESR spectrometer are working in the X-band to meet most of the experimental requirements in different applications of ESR spectroscopy. The X-band

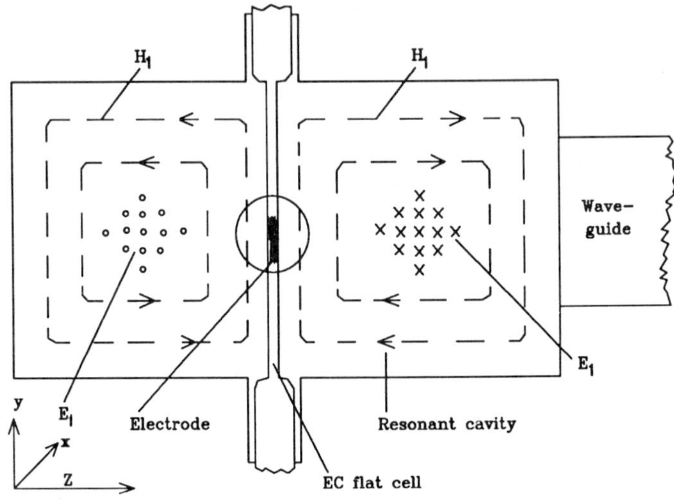

Fig.1. Scheme of a H 102 retangular resonant cavity.

spectrometer used in the present study was the ERS 221 (Zentrum für wissenschaftlichen Gerätebau, Berlin) equipped with both the H 102- and the TE 110-cavity as well as with a rapid scan unit and a magnetometer (MJ 110 R, Radiopan Poznan).

The characteristics of the ESR cavity (Fig.1) implies the basic restrictions in the construction of an electrochemical cell for ESR-measurements. Like any other probes the electrochemical cell is to be mounted in the center of the resonant cavity where the magnetic field has its maximum. Lossy samples like electrolyte solutions have to be restricted into the z-direction to avoid high dielectric absorption which lowers the quality-factor of the cavity and the sensitivity of the measurement or makes the measurements impossible at high absorption. Therefore a flat cell with a 0.3 to 0.5 mm thickness of the solution layer must be used. The cell has to be made of quartz because of the "sucking in" effect of that material which improves the sensitivity by a factor of 2. This geometry gives the limitations of the electrochemical conditions: low electrolyte volume, high cell resistance and small electrodes. For the last fact even further restrictions exist.

The resonant cavity is equipped with modulation coils of limited diameter which results in a smal area of the homogeneous magnetic field H_m (full circle in Fig.1, center of the cavity) with a diameter of at least 10 mm. Therefore the working electrode has to be centered in this area within the flat cell. The maximum thickness of the electrode has to be between 0.1 to 0.2 mm to allow an electrolyte layer of sufficient thickness at both sides of the electrode surface. The inner-diameter of the sample holder is limited to 10 mm and therefore the inner width of the electrode to about 8 mm. Thus the maximum surface of a smooth working electrode might be 2×80 mm². The electrode surface can be enlarged by the use of a metallic mesh. Therefore the general equipment of the ESR-electrochemical cell comprises a platinum mesh (0.1 mm wire diameter). In some cases micro meshes (Buckbee-Mears Comp., St. Paul) of nickel were used as working electrode material. A further increase in the surface area can be reached with a carbon fibre bunch electrode (12 cm² surface area). In general the carbon ESR signal makes the use of this material in the *in situ* ESR-electrochemical work impossible. The problems are discussed in the next section. A special carbon fibre (RK Textiles), however, was found to be usefull in ESR-electrochemical studies.

The electrical resistance within the flat cell causes a potential drop at the working electrode. The potential drop is in the order of 0.5 V and even higher. Therefore the use of a reference electrode to control the electrode potential is limited to electrodes with a length of 4 to 6 mm to avoid a significant potential drop which would cause further electrode reactions. Furthermore the reference electrode must be mounted in the closest contact (without electrical contact troughout all experiments)to the working electrode. A small silver wire fixed at the working electrode by an isolating silicon rubber strip and covered by a silver chloride layer was

60

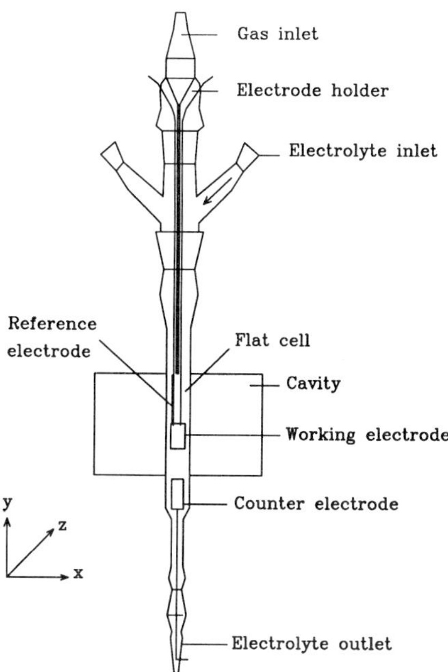

Fig.2. The new cell construction for simultaneous ESR and electro-chemical measurements.

used (Fig.2) as the reference electrode the potential of which must be controlled separately in every test solution.

The counter electrode was in each case a platinum mesh and mounted downstream within the flat cell at the edge of the cavity. This is a compromise to assure a widespread use of the cell construction. In this way the reaction products at the counter electrode do not interfere the ESR-signal and the reaction at the working electrode (care has to be taken with gas bubbles). On the other hand the contact to the working electrode is also close enough to avoid large cell resistances.

The gas inlet tube has its end at the opposite edge of the cavity. Thus the stability of the resonance condition during the ESR work is not distorted and the inert gas atmosphere can be maintained during long term experiments. A continuous solution inlet is thus also possible and flow through experiments can be done using the electrolyte outlet.

The cell construction is mounted in two adjustable sample holders at the top and the bottom of the flat cell area. It is of great importance for a good adjustment to use both these sample holders to correct the flat cell axis at both sides of the cavity. As the adjustment is the most time consuming preparation step of each measurement the construction is optimized to allow to exchange the working electrode without a new adjustment. The special electrode holder with both the working and the reference electrode can be taken off in the adjusted state and can be replaced by the next holder with a new electrode or the renewed electrode surface (especially with modified or amalgamated silver electrodes). The use of a slotted cavity allows the application of an irradiation equipment. As the working electrode is at the axis of the irradiation source and mounted in a quartz cell within a very thin absorbing electrolyte layer the effect of, e.g., UV-irradiation on the electrode behavior can be also studied with this electrochemical cell construction under *in situ* conditions of ESR spectroscopy. In this work, the UV-irradiation device HBO 200 (Carl Zeiss, Jena) was used.

Additional data concerning experimental conditions are: the use of double distilled water and analytical grade supporting electrolyte substances (Na_2SO_4, H_2SO_4, KNO_3) as well as purified nitrobenzene, p-phenylenediamine and aniline. The isometronidazoles were tested by thin layer chromatography to prove the absence of by-products.

The styrene maleic anhydride copolymer (Leuna-Werke) was dried and recrystallized

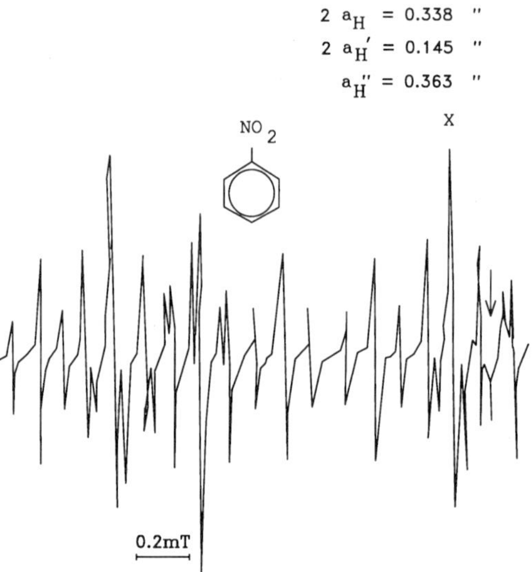

Fig.3. Low field part of the ESR-spectrum of nitrobenzene radical anion. The center of the spectrum is indicated by an arrow, the line used for time dependent measurements by a cross.

before it was used in the synthese of the spin-labeled polymer according to the reference[6] under the use of 4-hydroxytetramethylpiperidineoxyl (TEMPOL) prepared by a standard procedure.[7] The measurements were done at room temperature in deaerated solutions.

Results

The nitrobenzene reduction is a useful test system for new cell constructions and electrode configurations. It can be used to improve the cell adjustment and to a number of test measurements. The low field part of the ESR-spectrum of the nitrobenzene anion radical is shown in Fig.3.

The radical anion was formed under potentiostatic conditions at an amalgamated silver electrode at -0.4 V. The splitting constants calculated by a computer simulation are in good agreement with literature data[8] as it is expected for aqueous solution. If the magnetic field is fixed at the line indicated by a cross, the radical formation and decay can be studied under potentiostatic controll by applying a slight electrolyte flow through to have constant nitrobenzene concentrations (Fig.4).

The two curves at different time scales reflect the possibilities of this technique. First a constant radical formation can be maintained in the electrochemical cell which is a basic condition for further studies in the radical reaction of this anion radical. The second point is that there exists a short radical concentration increase after the end of the potential step. This increase might be attributed to the reaction of the dianion formed at a certain extent at the electrode with initial nitrobenzene now available for such a reaction under the formation of further radical anions (synproportionation). After 2.5 seconds the radical decay will dominate because no further dianions are available.

The advantages of the potentiostatic control are demonstrated with isometronidazole and its halide derivatives. The ESR-spectrum of the radical of this nitrocompound, which is currently used as a radiosensitizer,[9] is given in Fig.5.

This nitrogen centered radical can be studied under galvanostatic and potentiostatic conditions. Like with nitrobenzene under potentiostatic conditions a constant radical concentration is established but no increase will be found at the end of the potential step. Contrary to this method the galvanonstatic radical formation creates different concentration - time - profiles dependent on the current applied. No constant radical concentration is found. At

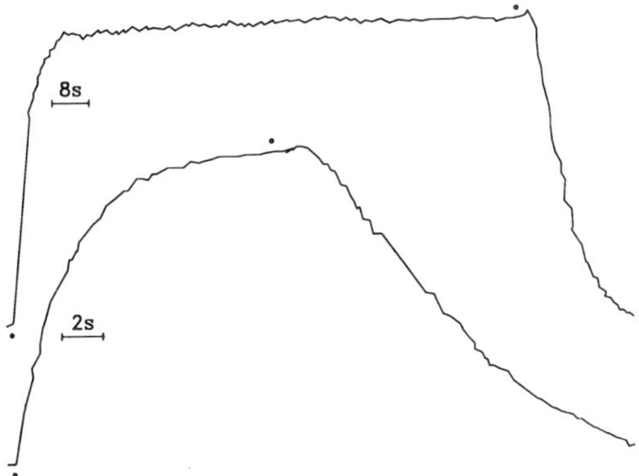

Fig.4. Nitrobenzene anion radical formation at −0.4 V under potentio-static control at the silver amalgam electrode. The points at the curves indicate the start and the end resp. of the potential step.

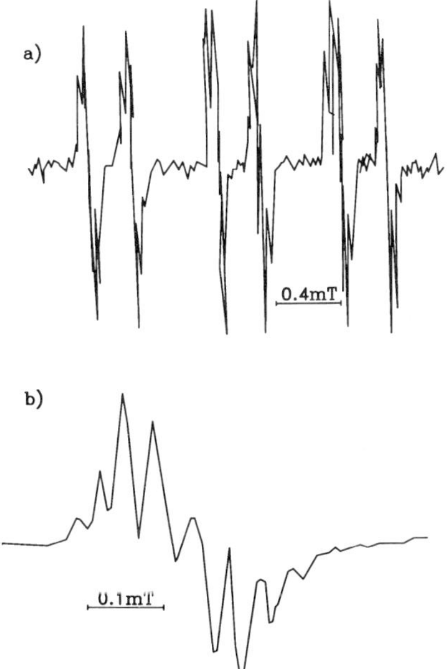

Fig.5. ESR-spectrum of the isometronidazol anion radical. a) whole spectrum b) resolution of the low field line.

medium current density quasi-stationary conditions can be reached which might be in some cases satisfactory for studies of radical structures.

To continue the study of the isometronidazole series the halide derivatives at the 5-position were electrolysed under potentiostatic conditions. The iodine derivate give a six line spectrum after an induction period looking like a non-resolved isometronidazole radical (Fig.5) with a similiar nitrogen splitting constant. The behaviour of the bromine derivative is more complex with a pronounced time dependence. Within the first 20 minutes of the po-tentiostatic electrolysis a badly resolved spectrum with changing relative intensities is found. This indicates the superposition of spectra belonging to radicals of different life time. The spectrum found after 20 minutes is a simple three line spectrum which is also found with the chlorine derivative already at the very beginning of the electrolysis. These results indi-cate that the splitting off of the halide substituent dominates in the radical structure. With

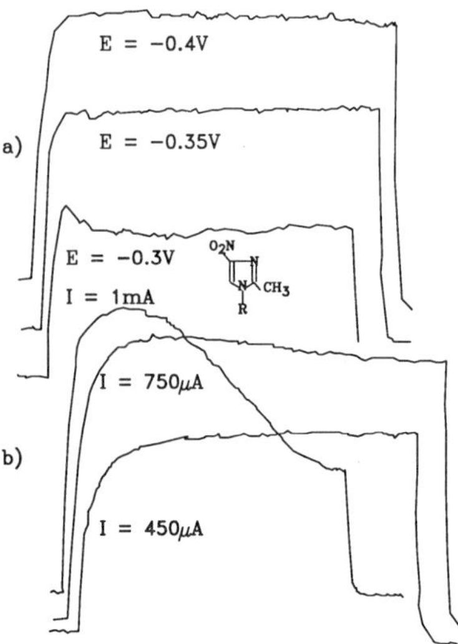

Fig.6. Time dependence of the isometronidazole radical ion concentration. a) potentiostatic conditions b) galvanostatic conditions.

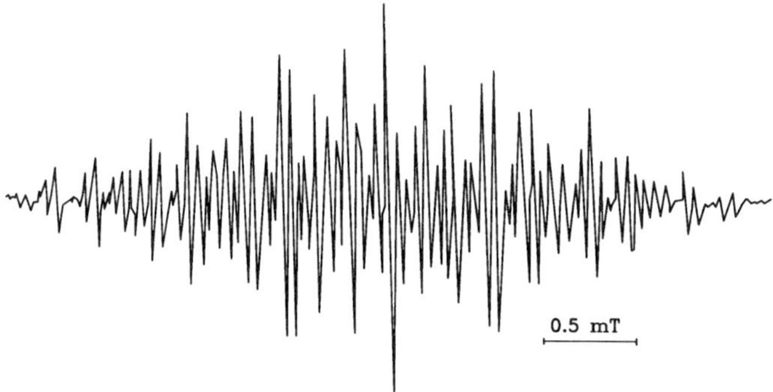

Fig.7. ESR-spectrum of the p-phenylenediamine cation radical formed at carbon fibre electrode at +0.35 V in 0.1 M Na_2SO_4.

iodine the substituent is easily abstracted and the radical of the basic compound is found. The bromine abstraction is somewhat slower resulting in two types of radicals existing in the initial stage of the electrolysis. The chlorine derivative is stable even in the anion radical form and results in the nitrogen centered radical.

The use of carbon fibre electrodes was possible by using a fibre type which has nearly no carbon resonance. In general an intensive single line is found in the ESR spectrum of carbon fibre[10] the shape and intensity of which are dependent on the pyrolysis conditions. Testing a couple of carbon fibres of various origin we found a sample with nearly no ESR signal. The advantage of this carbon electrode is that it can be used both in the cathodic and anodic potential regions. For the cathodic reduction nitrobenzene was used once more again giving a very intensive ESR spectrum of the radical anion. The anodic oxidition was tested with p-phenylenediamine at a potential of +0.35 V. The spectrum of the cation radical is shown in Fig.7.

Both these results prove that the carbon fibre electrode can be used for sensitive ESR-electrochemical measurements. The problem with the carbon electrode is the stability with respect to the ESR spectroscopy. It has been found during this study that the change between anodic and cathodic treatment of the carbon electrode creates the carbon signal at such a level that can be detected under conditions of the highly resolved spectrum. It is the change from

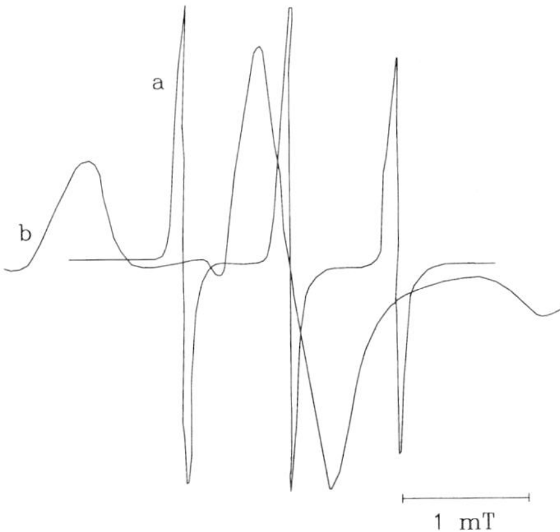

Fig.8. ESR-spectrum of 4-hydroxy-2,2,6,6-tetramethylpiperidine -N-oxyl in solution. a) free rotation, correlation time -10^{-11} s b) restricted mobility, correlation time -10^{-7} s.

anodic to cathodic conditions that is responsible for the appearence of the carbon resonance. The surface layer of the carbon electrode obviously alters its structure in flow conditions or in the absence of an applied electrode potential where the spin intensity decreases. The line width of the carbon signal changes between 0.4 (low spin concentration) to 0.18 mT (increased spin concentration).

Taking this behaviour into account it is preferable to use two types of carbon fibre electrodes for ESR spectroscopy - one for cathodic and the other for anodic processes. Care must be taken in the field of signal accumulation where a background signal abstraction is preferable with carbon electrodes.

To demonstrate the usefullness of this new ESR-electrochemical cell construction in the field of polymer modified electrodes [11] two examples are given.

The first is the application of spin labels[12] in maleic acid copolymers for the modification of electrodes. The following structure was used:

The influence of the relaxation time of this stable radical is given in Fig.8.

The line shape of the radical is a measure for the mobility of the spin probe and at least of the polymer chain where the label is fixed. Comparing the ESR spectra of spin labeled styrene maleic acid copolymer in solution (Fig.9a) and on the electrode surface layer of nickel micro mesh (Fig.9b) it can be concluded that the surface layer on the polymer modified electrode has two regions of different mobility. The more restricted mobility of one part of the polymer might be found in contact with the solid electrode and the mobile polymer chains in contact with the solution. This problem might be further studied by ESR tomography.

The second example is the polyaniline modified electrode known for some time in detail.

65

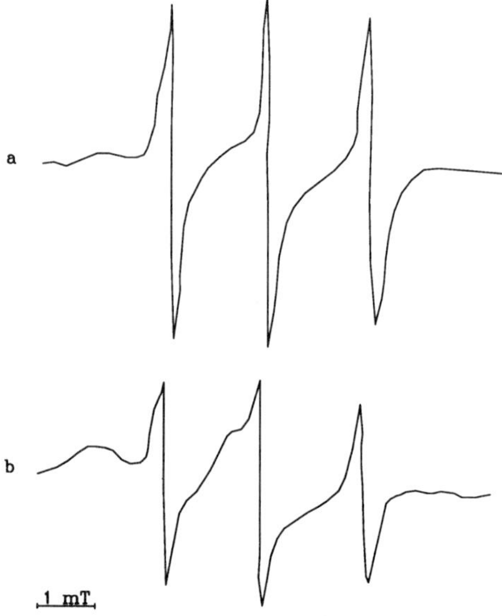

Fig.9. ESR spectrum of spin-labeled styrene-maleic acidcopolymer. a) in aqueous solution b) on a nickel micro mesh electrode.

The stable configuration of polyaniline has the following structure:[13-14]

The potentiostatic electrolysis at a platinum electrode at +1.0 V results in a polymer layer on the electrode surface. After an induction period the electrode surface layer gives an ESR signal with one line of 0.15 mT linewidth (Fig.10). The linewidth is independent on the radical concentration and the type of the cation in the supporting electrolyte. If the polyaniline film is in contact with aniline in the abscence of an applied potential, the line intensity decrease with time indicating a further reaction of the polymer with the aniline. Under UV-irradiation the free spin can be renewed without reaching its initial intensity (Fig.10b).

If the polyaniline structure can react with benzoquinone which is the final oxidation product of aniline in acidic aerated solution, a broad ESR signal of low intensity is found indicating the chemical change of the polymer structure (Fig.11). Even the UV-irradiation can not alter the radical concentration of the new polymer structure.

Conclusion

The different fields of application have demonstrated the advantages of the new ESR-electrochemical cell for simultaneous studies in radical formation and reaction at solid electrodes as well as in studies of polymer modified electrodes.

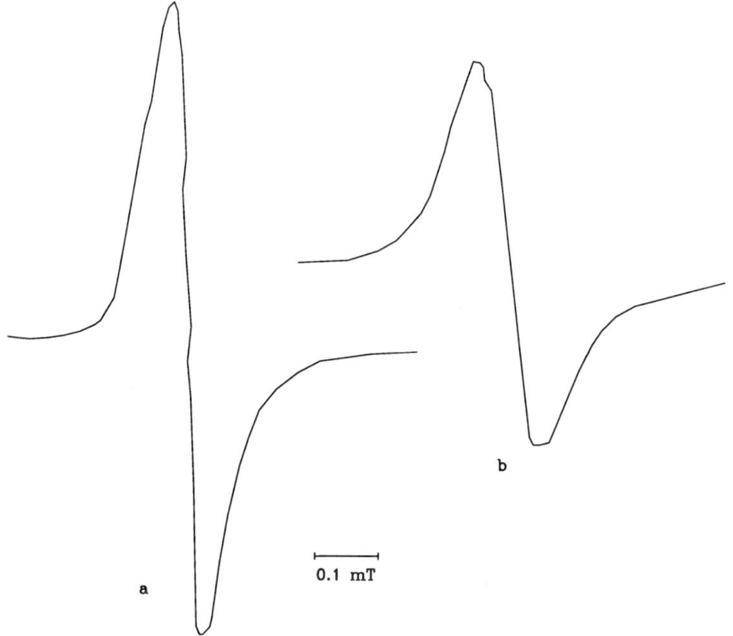

Fig.10. ESR-spectrum of a polyaniline modified electrode. a) anodic oxidation at +1.0 V, b) renewed spins under UV-irradiation.

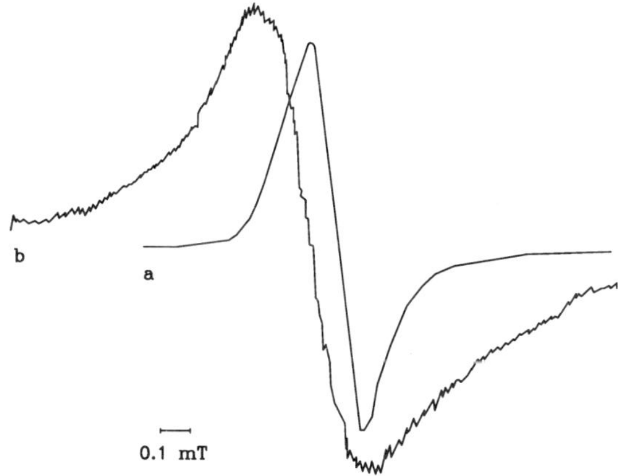

Fig.11. ESR-spectrum of a polyaniline modified electrode before (a) and after (b) the reaction with benzoquinone (The amplification ratio of a to b is 20).

References

1. H.-G. Gilde, in: "Methods in Free Radical Chemistry", Vol. **3**, E.S. Huyser (Ed.), Marcel Dekker, New York.
2. B. Kastening, in: "Electroanalytical Techniques and Applications", H.W. Nuernberg (Ed.), New York, Interscience 1974.
3. I.R. Goldberg and A.J. Bard, in: "Magnetic Resonance in Chemistry and Biology", J.N. Herak and K.J. Adamic (Eds), Marcel Dekker, New York, 1975.
4. T.M. McKinnay, in: "Electroanalytical Chemistry", Vol. **10**, A.J. Bard (Ed.), Marcel Dekker, New York, 1977.

5. H. Wittchen, Thesis FU Berlin-W., 1980.

6. M. Andres, Z. Veksli, R. Vukovic and D. Fles, J. Polym. Sci. Chem., **22** (1984) 2559.

7. B.J. Gaffney, in: "Spin labeling. Theory and Application", L.J. Berliner (Ed.), Vol **1**, pp. 184-238.

8. K. Scheffler and H.B. Stegmann, in: "Electronenspinresonanz", Berlin, Heidelberg, New York, Springer, 1970, p. 401.

9. P. Wardman, Current Topics in Rad. Res. Quart., 11 (1977), pp. 347-398.

10. L.S. Singer, in: "Ultrahigh-Modulus Polymers", A. Ciferri and I.M. Ward (Eds), London, Appl. Science, 1979, pp. 251-277.

11. L. Dunsch, in: "Polymermodifizierte Elektroden", L. Dunsch (Ed.), Dresden, ITP 1987, pp. 141-157.

12. "Spin labeling. Theorie and Application", L.J. Berliner (Ed.), Vol. **1+2**, New York, Academic Press 1976 and 1979.

13. L. Dunsch, Thesis, Freiberg 1973; L. Dunsch, J. Electroanalytical Chemistry, **61** (1975) p. 61, L. Dunsch, J. prakt. Chemie, **317** (1975) p. 409.

14. L. Dunsch, in preparation.

INDUSTRIAL APPLICATIONS

INDUSTRIAL APPLICATIONS AND PERSPECTIVES OF ELECTRO-ANALYTICAL METHODS

E. Pungor, Zs. Niegreisz and L. Pólos

Institute for General and Analytical Chemistry
Technical University of Budapest, 1111 – Hungary

Introduction

All fields of modern industry that depend on the composition of products has to use the methods of analytical chemistry. In earlier periods of this century only the final products were analysed. In this respect the analysis of pharmaceutical products was important as well as the material characterisation in metallurgy. Later on manufacturers realised the financial importance of process control in addition to the analysis of the products. Based on process analysis the technology can be controlled in order to ensure the required product properties. Metal industry was among the first to implement process analysis.

Starting from the middle of this century, the demands on process analytical methods have become increasingly better defined. It became clear that the efficiency of the analytical process control depends on the ratio of the time from sampling to action, to the so called process time. The smaller this ratio is, i.e., the higher the number of analyses and actions made in a technological period, the higher will the controllability of the process be.

The controllability factor, R, defined as:

$$R = \sqrt{\frac{s_p^2 - s_c^2}{s_p^2}} \tag{1}$$

where s_p is the standard deviation of the product quality without control and s_c is the corresponding value after control, expresses the extent of control.

If $R = 1$, the controllability is extremely good and there is a complete control, i.e., $s_c = 0$. If $R = 0$, there is no control. As concluded by Leemans,[1] a realtionship exists between the controllability factor and the times of different actions expressed as fractions of the process time or technological period in a non-buffered system as follows:

$$R = ke^{-\frac{t_{sf}}{t_p}} e^{-\frac{t_a}{t_p}} e^{-\frac{t_c}{t_p}} e^{-\frac{t_s}{t_p}} \tag{2}$$

where:

t_p = the process time of technology
t_{sf} = the time between subsequent samplings
t_a = the analysis time
t_c = the control action time
t_s = the sampling time
k = factor expressing also the accuracy of the analysis

In a buffered case the mathematical treatment is more complicated.[2]

It is obvious that analytical process control has to be devised and organized depending on the process time or technological period time.

It is clear, that research and development plays an essential part in modern industrial production first of all in the design of new products and technologies. In the choice of the electroanalytical techniques to be used in industry three different levels in accordance with the industrial demands have to be considered. Monitors used to control continuous technologies, quality control systems and analytical systems used in research and development all have to be treated separately. On all three levels the demands on the analytical technique to be used is of basic importance.

It is clearly shown by equation (2) that the accuracy requirements on the analytical monitors are determined by the product of exponential factors. It is not necessary to use analytical techniques of higher accuracy than that determined by the equation, as they would unnecssarily increase production costs. For the other two categories it is very important to achieve the highest possible accuracy.

It should also be mentioned that environmental protection and control are to be considered as parts of industrial analysis. For this reason in this paper attention is given to analytical problems of environmental control, too, and electroanalytical techniques used in this field are also outlined.

Industrial monitors used for control purposes can be implemented in different ways depending on 1) the corrosion interaction between the process stream analysed and the sensor, 2) on the degree of contamination of the medium by the sensor and 3) by the additives necessary to the operation of the sensor.

In an in-line system the sensor is built directly into the reactor. The advantage of such a system is the high speed of information provision which causes, that $R \rightarrow 1$. In-line methods have already found wide application (determination of pH, use of conductometry and fibre optic systems etc.).

The speed of information provision is lower if an on-line method is used, namely, if the sensor is situated in a bypass. The method has the advantages of easier calibration and the possibility of using the optimum medium for a particular sensor. A wide range of methods can be used for on-line analysis, the classical example being the use of process chromatographs, but other techniques, among them electroanalytical methods have also been applied.

A different technique is the at-line method which is a link between on-line and off-line methods. Sampling and analysis are similar for both the at-line and off-line systems but an at-line analytical system is installed close to the production line, and by time-sharing it suites well for analyses of several important samples according to a program prepared in view of the technological safety.

Demands on Monitors

Before dealing with applications of electroanalytical methods in in-line and on-line process control, we wish to emphasize some prerequisites to be fulfilled by monitors:
- long-term stability and reproducibility
- feasibility of calibration without dismantling the monitor
- fast response in comparison with the fluctuations of the process monitored
- readily identifiable sensor signals that can be assigned to a single component.

Industrial Measurement Techniques

In view of the above features, oscillometric (conductometric) and potentiometric sensors seem to be best suited for use in industrial and environmental monitors among the electroanalytical techniques. Oscillometric and conductometric sensors fulfil the first and third requirement, the technical problems in connectin with calibration are not too difficult. There are, however, problems with the fourth requirement, because impedance and the complex dielectric constant both depend on all constituents present, and because these properties do not only depend on one single factor but on several factors. The electrical equivalent circuit for the oscillometric capacities is given in Fig.1. The expression of the h.f. conductance, G, is as follows:

$$G = \frac{\omega^2 R C_s^2}{1 + \omega^2 R^2 (c + C_s)^2} \tag{3}$$

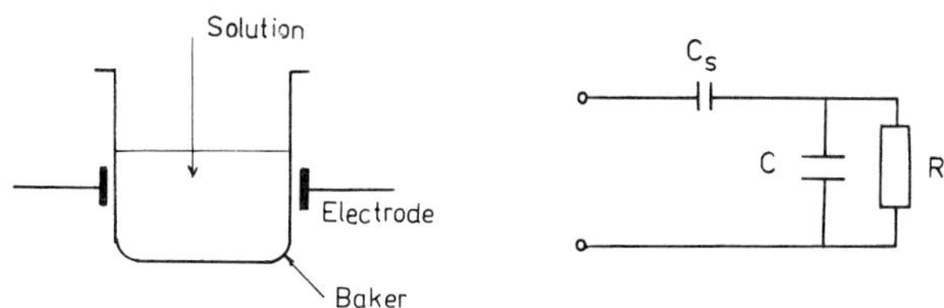

Fig.1. Electrical eqivalent circuit for the oscillometric capacitive cell.

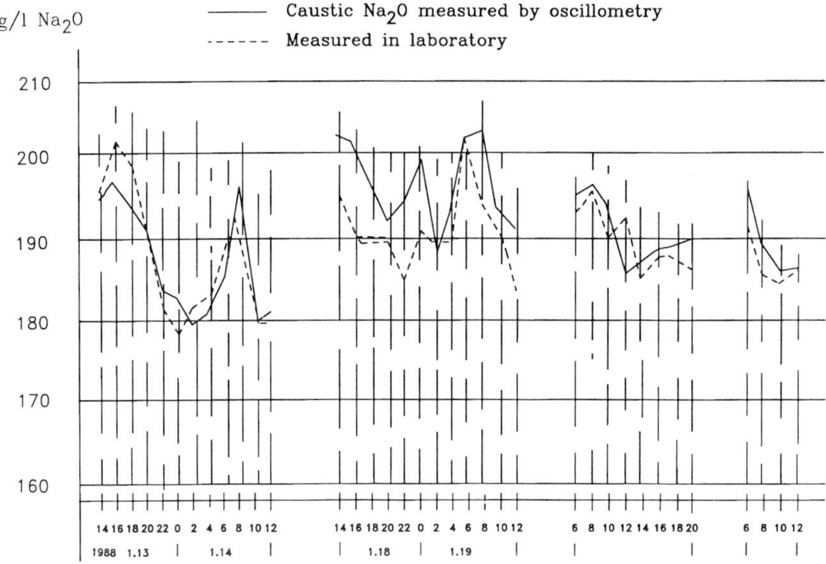

Fig.2. Industrial use of oscillometric remote control of caustic Na$_2$O in Bayer process.

where ω is the angular frequency, R the resistance of the solution in the capacitive cell, c the measuring capacity of the cell, C_s the stray capacity. Conductometry and oscillometry can be used in monitors only in cases where the signal is dependent mainly on changes essentially in one parameter only and the matrix effects can be neglected. We have constructed together with F. Valló[3] a system for controlling the concentration of NaOH for the Ajka Alumina Works (Hungary) using the oscillometric principle. In the capacitive type cell the measuring electrodes are not in galvanic contact with the NaOH solution thus preventing corrosion whicn therefore have no effect on the life-time of the electrode system. As shown by data in Fig.2, there is a good agreement between the results of in-line and laboratory measurements.

Oscillometry can also be used in environmental monitors, in following the changes in the total salt content in solutions. Capacitive type cells offer the additional advantage of having two sensitive concentration ranges: one at low and one at high concentrations.

Electrodeless measuring techniques have a number of varieties. One interesting variety is based on the principle of transformers, induction being dependent on the composition of liquid in the inside of the transformer. An equipment has been produced for industrial purposes by Beckman Instruments Co,[6] based on the results published by Griffiths.[4,5] A similar principle was used by Maurer[7] and Foxboro Co[8] constructed an apparatus to monitore the pH in flotation systems.

Although the dielectric constant has no effect on the signal if dc conductivity is measured, the signal is the result of a collective property instead of depending on the concentration of a single component. Despite this fact, several conductometric industrial monitors have been constructed. If it can be ensured that the concentration of only one component changes

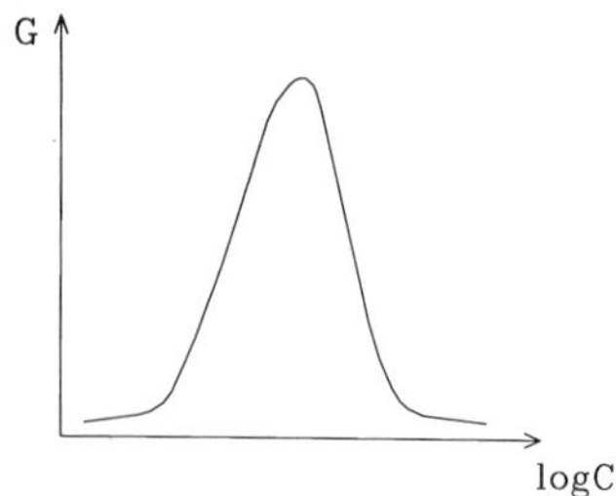

Fig.3. The real part of admittance (conductance) as a function of log concentration.

during a process in the reactor, the signal becomes selective. Measuring systems of this type have been used for a long time. In a cell constructed by Barendrecht and Janssen[9] four electrodes were used to avoid electrode polarization, the potential drop at a constant current being measured across two electrodes being placed in the electric current stream. Philips has constructed a similar cell.[10] Weidemann et al.[11] constructed a cell with concentric electrodes, with a shielding electrode between the two working electrodes for analysis of slags.

For monitoring applications only such potentiometric sensors can be used which, during contacting with the medium measured, do not lose their active component or are not fouled by an insulating layer which slows down or even ceases the electrode response function.

Among electroanalytical methods voltammetry is perhaps the most sensitive to different adverse effects and therefore their use in monitors is connected with many problems.

Of the few electroanalytical monitors the ones used in the chloralkali industry are worth mentioning.[12] Sulphate was determined in brines. Off-line conductometry was used to determine sulphate in the concentration range $25 - 500$ mM with Ba^{2+} as titrant, or Pb^{2+} as titrant when potentiometric measurement was used. These methods can, however, not compete with infrared spectrophotometry in this application. Water was determined in chlorine gas by coulometry with 100 % current efficiency. In this case the analyzer should be installed very close to the production plant.

Another field of application of electroanalytical techniques is the monitoring of copper ion concentration in the electrolyte during copper raffination[13] by using ion-selective electrode in the measurement at concentrations around 5×10^{-4} M (Fig.4 and 5). Interference by Fe(III) was eliminated by addition of ascorbic acid. The reproducibility was reported to be very good and the potential drift was 4 mV during 17 hour. Turbulent flow was used to ensure cleaning of the electrode surface and for reducing potential drift. For comparison atomic adsorption spectrometry was used. The potentiometric method proved to be superior.

Industrial pH analyzers have been discussed on a scientific level by Jola.[14] He described systems with glass and antimony electrode, and compared amperometric and spectrophotometric methods in chlorine determination.[15]

An interesting field is the use of oxide electrodes with different composition (e.g. ZrO_2, TiO_2, YO_2) for determing oxygen in molten metals. An important type of solid electrolyte oxygen sensors is the stabilized ZrO_2, first used by K. Kinkhola and C. Wagner[16] about 30 years ago. It is widely used in determination of oxygen in molten metals such as steel, aluminium, copper or in the control of metallurgical atmospheres and regulation of the air/fuel ratio in cars, boilers and furnaces. As the operation temperatures of the sensor is high, between 600 and 1000 °C, it is best used in cases where the temperature of the system to be monitored is high.

Formally the sensor can be considered as an oxygen concentration cell:

$$p_{O_2'} \; Pt \; ZrO_2 \; , \; CaO \; Pt \; p_{O_2''}$$

Zirconia stabilized by calcia is an oxygen ion conducting solid electrolyte. The emf of the cell:

$$emf = \frac{RT}{4F} \ln \frac{p_{O_2'}}{p_{O_2''}}$$

where $p_{O_2'}$ and $p_{O_2''}$ are partial pressures of oxygen on the two sides, the partial pressure on

74

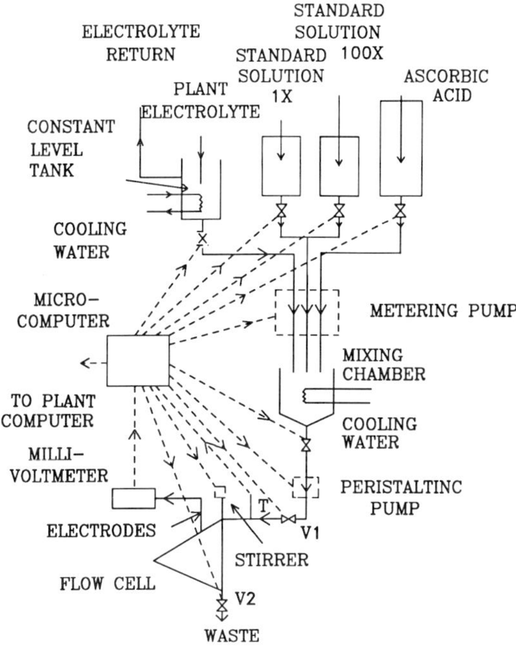

Fig.4. Schematic flow diagram showing microcomputer control for on-line monitoring of Cu(II) in plant electrolyte.

one side can be determined if the partial pressure of oxygen on the other side is kept constant.

On applying a voltage to the two electrodes, oxygen is pumped from the cathode side to the anode side by "electrochemical pumping":

$$O_2 + 4\,e^- \rightarrow 2\,O^{2-}$$

and

$$2\,O^{2-} \rightarrow O_2 + 4\,e^-$$

are the cathodic and anodic reactions, respectively. Porous platinum layers are deposited on both sides of the zirconia solid electrolyte. The cell (Fig.6) is surrounded by a furnace, and it is heated to a constant temperature according to a program. In using the electrochemical pumping technique, the oxygen partial pressure can be detrmined from the pumping time, and from the current or at a constant current from the voltage. The third type, the oxide semiconductor type oxygen sensor is based on the resistance change of the oxide due to changes in the partial pressure of oxygen.

A good survey of this field was presented by T. Takeuchi at the 2nd international meeting on chemical sensors in Bordeau, France, 1986.[17] In Fig.7 some forms of the three types are presented along with the working principle, based on the paper mentioned. A SO_2 sensor based on a similar principle has been developed by H. Torvela.[18] The sensor produces a Nernstian response.

Electrochemical monitors have also been constructed for use in environmental control. Methods were developed for oxidizing alkanes, first of all methane in non-aqueous solution[19,20] (Fig.8).

Electrolytic oxidation takes place at a platinized working electrode supported on a porous teflon membrane. Of the solvents studied butirolactone and propylene carbonate proved to be best. The effect of the teflon content of the working electrode on the course of reaction was also studied, the optimum content was found to be 35 %. The effects of water content, temperature and gas flow rate were also investigated.

The determination of fluoride in air is very important in areas near aluminium industry. We have constructed an instrument using potentiometry with a fluoride selective electrode for continuous measurement of hydrogen fluoride in stack gases by absorbing them in a suitable

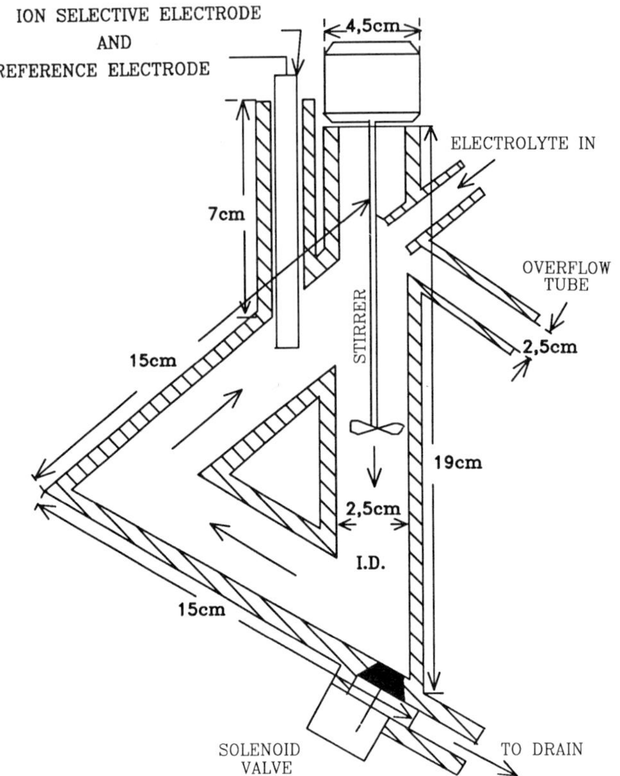

Fig.5. Flow through cell for Cu(II) determination.

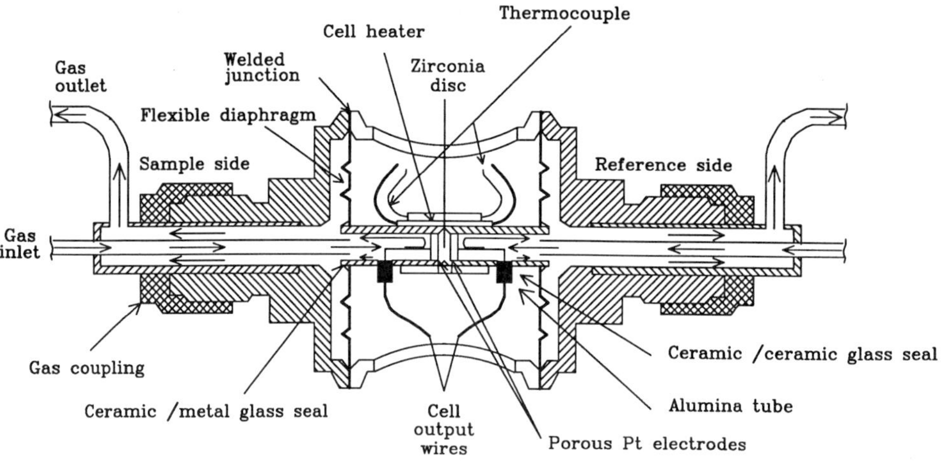

Fig.6. Schematic of the miniature oxygen sensor.

solution until the concentration of fluoride in the solution reaches a preset value.[21,22] Similar problems have been solved by Deng et al.[23]

Cserfalvi et al.[24] described a pH monitor built into the process control system of a surface water clarification plant in which automatic calibration was also used. The scheme of the monitor is shown in Fig.9.

The conductivity of waste water gives information on quality of the waste water. For this purpose an oscillometric monitor with good long-term stability and operational safety

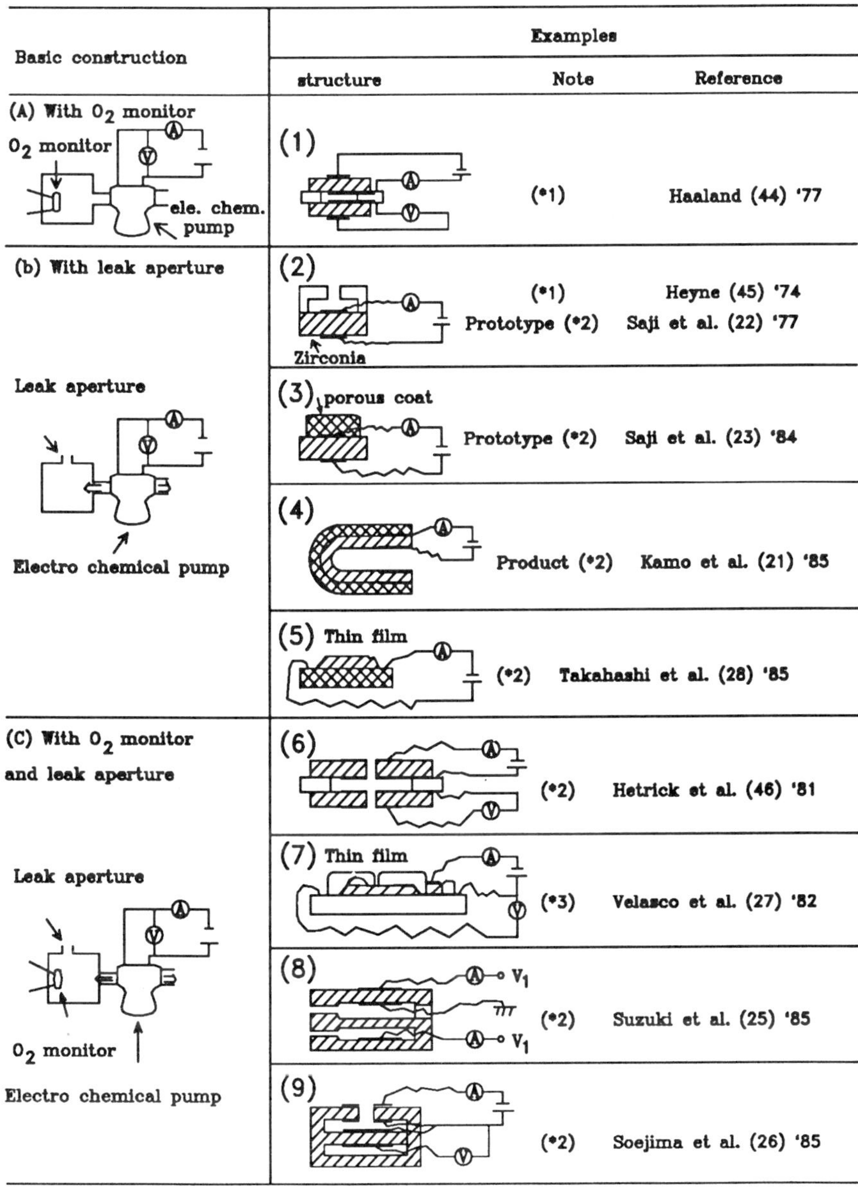

Fig.7. Various electrochemical pumping type oxygen sensors.

was constructed and used in controlling waste water in alumina production plants.[25]

The problems associated with waste water treatment increase exponentially with the size of the plant. This is the reason why manufacturers like BASF give great importance to waste control.[26] Depending on the needs, on-line and off-line methods are equally used. Special attention is given to the determination of poisonous substances (Fig.10).

In addition to on-line and in-line analysis, very often off-line systems are used in industry. Intermediates and products are analysed with laboratory analyzers. An example of this is the equipment constructed for the content uniformity analysis and dissolution rate determination of pharmaceutical tablets.[27] The tablet is dissolved under well-defined conditions and the solution analysed using a flow technique[28] with voltammetric detector (Fig.11).

The innumerable laboratory methods used for industrial analysis are outside the scope of this paper. However, we wish to mention a laboratory on-line control technique developed

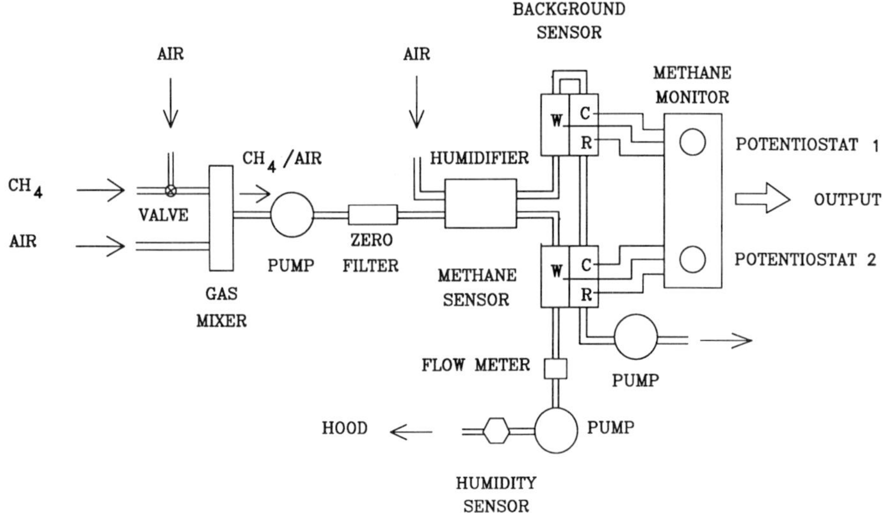

Fig.8. Schematic drawing of the experimental set-up.

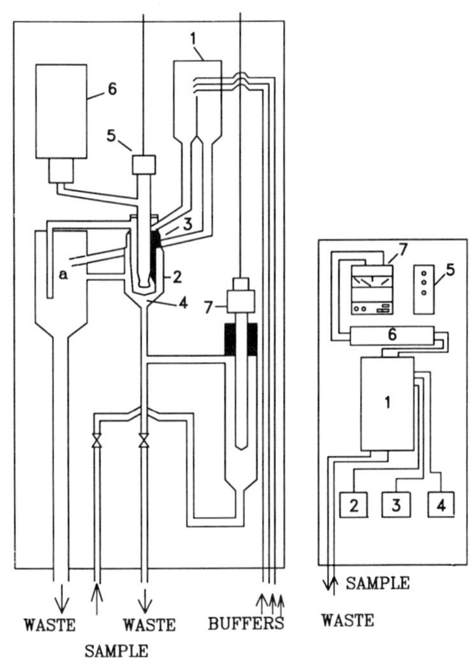

Fig.9. a) Schematic diagram of the measurement unit: 1. solution dispensers, 2. measuring cell, 3. inner cell, 4. membrane, 5. combination glass electrode, 6. reference filling solution, 7. temperature sensor.
b) 1. Block diagram of the pH monitor: 1. measurement unit, 2. buffer I, 3. buffer II, 4. rinsing solution, 5. local control, 6. preamplifier, 7. readout/transmitter.

by S.B. Nagy[29] based on dielectric constant measuremnt which may be used for industrial control. The technique is suitable for automatic control of distillation and evaporation processes (Fig.12).

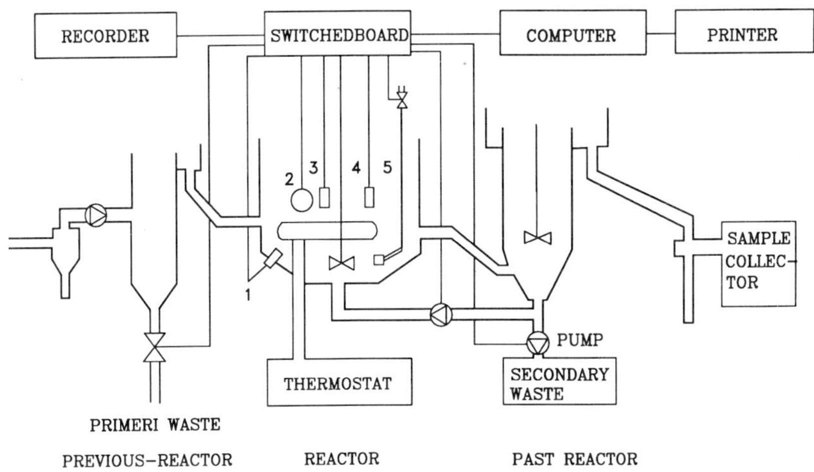

Fig.10. Control system in BASF.

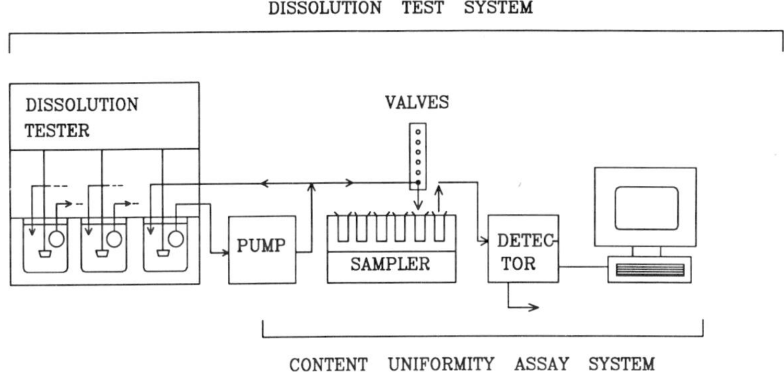

Fig.11. Dissolution test system.

Application in Industrial Research

A number of electroanalytical methods are used in industrial research. In this paper we wish to direct attention to an interesting application.

The method developed by M.M. Faktor et al.[30,31] is widely used in semiconductor industry. Based on the same principle an equipment was constructed by POLARON for the British Post. The scheme of the system is shown in Fig.13. The equipment is suitable for the continuous and automatic determination and is used to monitore the impurity profile in the surface layer of semiconductors with a resolution of about 10 nm in the range $1 - 50$ μm. Using this equipment operating on electrochemical principle the number of charge carriers can be determined in the surface layer. In addition to this n- and p-type semiconductors can be distinguished (Fig. 14).

The priciple is as follows: the surface of the semiconductor which is in contact with an etching solution containing Tiron or KOH is illuminated and the voltage – capacitance characteristic is measured. The capacitance can charge in a range extending over $5 - 6$ orders of magnitude corresponding to an impurity (charge carrier) concetration of $10^{13} - 10^{19}$ cm^{-3} (Fig.15). Without illumination only the charge distribution of p-conductors can be studied,

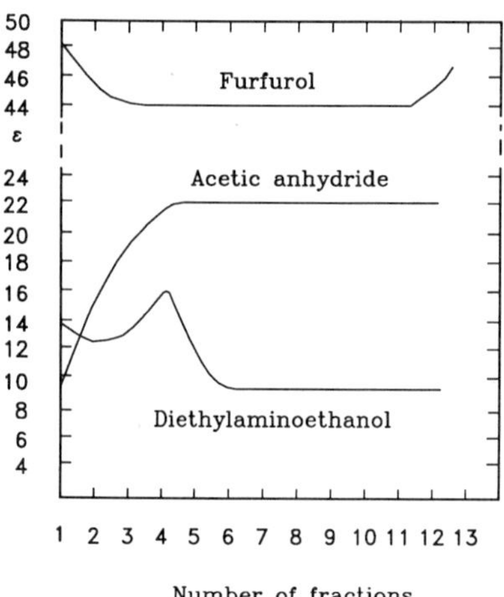

Fig.12. Distillation curves.

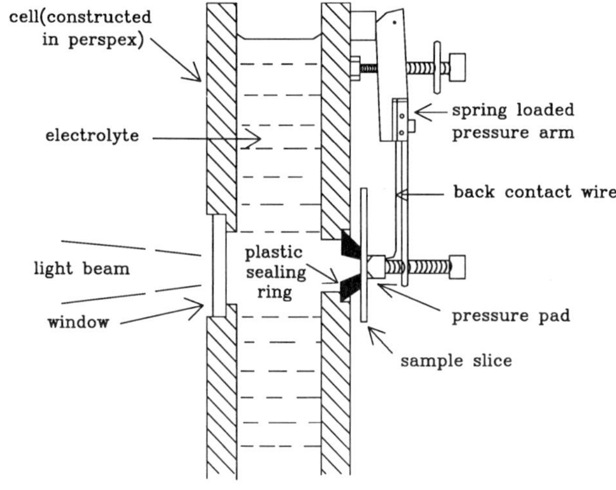

Fig.13. The electrochemical cell of the Post Office Profile Plotter. The sample slice is pressed onto the sealing ring (copreme or polyethylene) having a diameter of 3.5 mm. Ohmic contact is made by a current pulse flowing through the two black contact wires. Light having a photon energy higher than the energy band gap of the semiconductor illuminates the electrolyte-semiconductor interface by passing through the window and the transparent electrolyte.

while illumination is necessary in the case of n-conductors. In special cases Ferenczi et al.[32] succeede in reducing the detection limit to 10^{11} cm^{-3} for GaAs doped with phosphorus and nitrogen.

Electroanalytical methods find application in modern fermentation industrym but rather on the research level than in plant technology control. In most cases the pH, pO$_2$ and in some instances other components are determined using enzyme electrodes. However, we do not wish to deal with this field in detail.

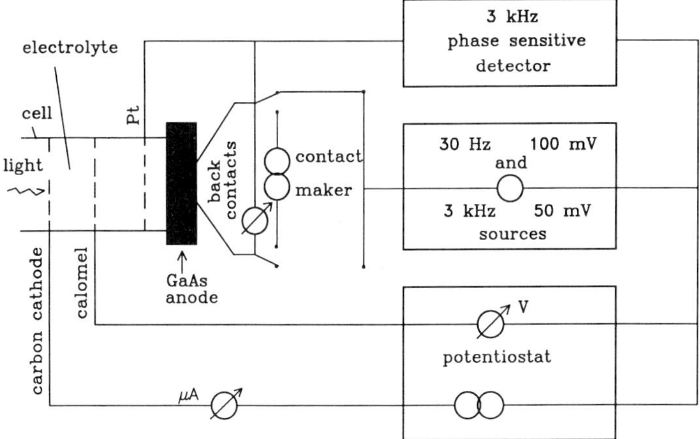

Fig.14. Schematic representation of the connections of the electrolyte cell.

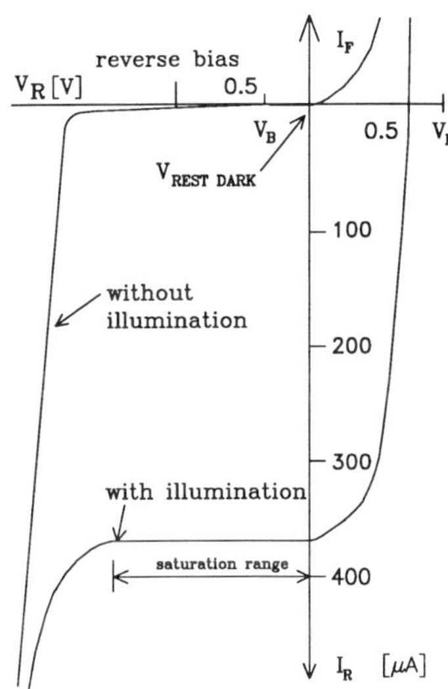

Fig.15. The typical current-voltage characteristics of the Schottky diodes formed as the interface between 10 % KOH and n-type GaAs.

Perspective of Electroanalysis

Scientific papers clearly show the great development on the theory of electrochemistry and electroanalysis over the past decade, and the widening of the fields of application. These trends are emphsized by Bard[33] who outlined the great development in sensor technology, but also discussed electrochemical cybernetica for which a fast development is foreseen and scanning tunnelling electrochemistry which opens up new vistas for surface analysis (Fig.16).

Of the achievements of scientific research only those find industrial applications for which there is a demand and which are likely to provide reliable results under industrial conditions. In our opinion good new sensors and sensor arrays, like ISFETs and CHEMFETs, biosensors have great future prospects. Accordingly, remarkable efforts are being expended in sensor research and development. The wide range of compounds that can be measured with biosensors has been summarized by Mullen and Evans.[34] Compounds that can be measured by electrochemical methods are listed in Table 1. There is a large volume of litterature on the combination of ion-selective electrodes and biosensors.

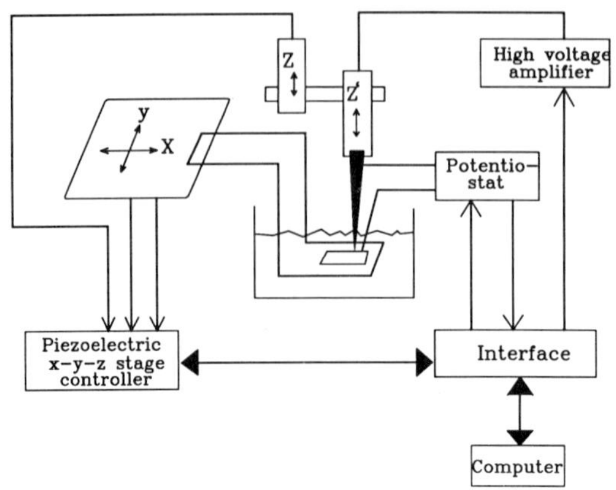

Fig.16. Design of a scanning electrochemical microscope.

Table 1. Biosensors

Measuring tecnique	Biological component
Potentiometry	Enzyme, antibody, liposome
Amperometry	Enzyme, antibody, receptor molecule
Piezoelectricity	Antibody, enzyme
Fibre optic technique	Enzyme, antibody

Obviously there are electrodes with poor characteristics with respect to selectivity, life time and response time. Although arrays of such electrodes may be applied if no other sensors are available, but this has the danger of providing false results. We do not agree with the statement made by Beebe et al.[35] that chemometrics is capable of improving the quality of results provided by poor sensors. Apart from this, we wish to emphasize the importance of chemometrics in research, development and industry.

The data given by Reeves[36] help answering the question as to why the industry requires an ever increasing number of monitors for process control. He shows that the analytical information produced by monitors and used to control the process reduces process costs. I think this is an important stimulus for both fundamental research and monitor development.

References

1. F.A. Leemans, Anal. Chem., **43** (1971) 36A.
2. A. Rijnsdorp, Anal. Chim. Acta, **190** (1986) 33.
3. E. Pungor, B. Tòth and F. Vallò, Hung. Patent Nr. 167038.
4. V.S. Griffiths, Anal. Chim. Acta, **18** (1958) 174.
5. V.S. Griffiths, Talanta, **2** (1959) 230.
6. Beckman, Electrodes Conductivity Systems, Catalogue.
7. O. Maurer, CZ-Chemie Technik, **2** (1973) 117.
8. W. Musow and A. Bolland, Instrum. Min. Metall. Ind., **11** (1984) 205.
9. E. Barendrecht and N.G.L.M. Janssen, Anal. Chem., **33** (1961) 199.
10. F.G. van Luyk and S.M. de Veer, Philips Industrie-elektronik, Sonderdruck, 1169.
11. J. Weidemann, H. Boss and D. Radke, Deutsches Patent, DT 2328959 A1.
12. H. van den Dolder, Anal. Chim. Acta, **190** (1986) 25.
13. A.M. Bond, H.A. Hudson, P.A van den Bosch, F.L. Walter and H.R.A. Exelby, Anal. Chem., **55** (1983) 2071.
14. M. Jola, Anal. Chim. Acta, **190** (1986) 67.
15. M.J. Madou and K. Kinoshita, Electrochim. Acta, **29** (1984) 411.

16. K. Kiukkola and C. Wagner, J. Electrochem. Soc., **104** (1957) 308.
17. T. Takenchi, Proceedings 2nd International Meeting on Chemical Sensors, Bordeaux, France, July 7–10, 1986, p. 69.
18. H. Torvela, Kemia – Kemi, **14** (1987) 533.
19. T. Otagowa, S. Zaromb and J.R. Stetter, J. Electrochem. Soc., **132** (1985) 2951.
20. T. Otagowa, S. Zaromb and J.R. Stetter, Sensors and Actuators, **8** (1985) 65.
21. A. Hrabéczy-Páll, K. Tóth, E. Pungor and F. Valló, Anal. Chim. Acta, **77** (1975) 278.
22. A. Hrabéczy-Páll, F. Valló, K. Tóth and E. Pungor, Hung. Sci. Instr., **41** (1977) 55.
23. M. Deng, H. Zang and Sh. Shu, Fenxi Huaxue, **12** (1984) 946.
24. T. Cserfalvi and I. Mosó, Anal. Chim. Acta, **190** (1986) 271.
25. E. Pungor, in: "Conductometry and Oscillometry", Van Noitrand, 1967.
26. W.S. Haltrich, Gas, Wasser, Abwasser, **65** (1985) 300.
27. Zs. Fehér, Gy Horvai, G. Nagy, Zs. Niegreisz, K Tóth and E. Pungor, Anal. Chim. Acta, **145** (1983) 41.
28. E. Pungor, Zs. Fehér and G. Nagy, Anal. Chim. Acta, **51** (1970) 417.
29. S.B. Nagy, Dielektromos Méréstechnika (Müszaki Kiadó).
30. T. Ambridge and M.M. Faktor, Inst. Phys. Conf. Ser. No. 24, 1975, pp. 320–330.
31. M.M. Faktor and J.L. Stevenson, J. Electrochem. Soc., **125** (1978) 621.
32. G. Ferenczi, P. Krispin and M. Somogyi, J. Appl. Phys., **54** (1983) 3902.
33. A. Bard, Anal. Chem., **59** (1987) 347A.
34. W.H. Mullen and G.P. Evans, International Labmate, **13** (1988) 15.
35. K. Beebe, D. Merz, J. Sandifer and B. Kowalski, Anal. Chem., **60** (1988) 66.
36. P. Reeves, Anal. Chim. Acta, **190** (1986) 45.

EXAMPLES OF ELECTROMETRIC METHODS APPLIED TO PROCESS AND PRODUCT CONTROL IN THE EXPLOSIVES INDUSTRY

Jan Asplund

Nobel Chemicals AB
S-691 85 Karlskoga, Sweden

General

In the explosives industry, there is a constant demand for simple, fast and reliable methods of analysis for a very wide range of applications. The methods should be such that standard equipments can be used, and sensitive parts like electrodes should withstand extended use with no deterioration or need for replacement. Preparation of samples has to be simple, and clearly defined. The determination must be rapid, and it must be possible to calculate the results in an unambiguous way. The assessment should be such that new personnel can carry out it with no change in the results.

The electrometric method to match most closely those requirements is voltammetry. Among the advantages of voltammetry are that it can be used for determination of inorganic as well as organic electroactive compounds, i.e., compounds that either can be reduced or oxidized, and that these compounds can be determined over a wide range of concentration.

Knowing the main reaction, i.e., the electrode reaction, and possible side reactions, the most suitable experimental conditions can be chosen to resolve a wide variety of problems.

This article describes how electrometric methods are used in the explosives industry to:
- analyse process acids for product contents and nitrous acid of concentration
- determine the composition of propellant, secondary explosives and initiating explosives
- determine the concentrations of stabilizer and stabilizer derivatives in propellant, and to determine nitrogen dioxide in aged propellant.

As a general background, a short description of nitrations (process acids), explosives and stabilizers is given.

The electrometric methods are discussed with respect to electrode reactions, working electrodes, solvents, electrolytes, pH, masking agents and the conditions of the determination of multi-component systems with completely or partially overlapping voltammograms.

Nitration

Usually a distinction is made between three types of nitration:
- C-nitration, where a nitro group is attached to a carbon atom producing a nitro compound:

$$-\overset{\displaystyle |}{\underset{\displaystyle |}{C}} - NO_2$$

- O-nitration, where a nitro group is attached to an oxygen atom producing a nitrate ester:

$$-\overset{\displaystyle |}{\underset{\displaystyle |}{C}} - ONO_2$$

Contemporary Electroanalytical Chemistry, Edited by A. Ivaska *et al.*
Plenum Press, New York, 1990

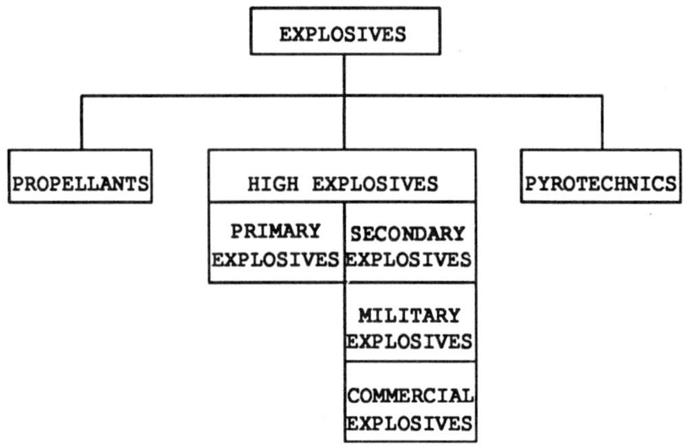

Fig.1. Classification of explosives.

– N-nitration, where a nitro group is attached to a nitrogen atom producing a nitramine:

$$- \overset{|}{\underset{|}{C}} - \overset{}{\underset{|}{N}}NO_2$$

Nitration plays an important part in the preparation of explosives. The most common military and commercial explosives, like 2,4,6-trinitrotoluene or trotyl (TNT), cyclotrimethylene trinitramine or hexogen (RDX), glycerol trinitrate or nitroglycerine (NG), pentaerythritol tetranitrate (PETN), cellulose nitrate or nitrocellulosa (NC) and many others are produced by nitration. Nitrating reactions are also employed to produce many commercially important non-explosive organic chemicals and intermediates.

The following nitrating agents are mainly used industrially for the direct introduction of nitro groups:
– concentrated nitric acid
– mixtures of concentrated nitric acid and concentrated sulfuric acid or oleum in varying proportions

The process acid, i.e., nitrating acid and residual acid, will dissolve the product to a greater or lesser extent. During nitration, small quantities of nitrous acid (HNO_2) are usually formed.

Explosives

An explosive is a substance or mixture of substances that can undergo a fast chemical reaction without external oxygen supply, thereby liberating large amounts of energy; normally liberating hot gases and fumes. A large number of substances may be defined as explosives, but in practice compounds containing oxygen, nitrogen and oxidizable elements (fuel) like carbon and hydrogen are used. Exceptions, like azides, are also used. In Fig.1, a classification of the explosives is given. In the following, military secondary explosives, propellants (powder) and initiating explosives are treated.

Propellants (Powder)

The propellants (powder) used for military weapons is smokeless powder. Depending on composition, a differentiation is made between single base propellant, double base propellant and triple base propellant.

Single base propellant consists mainly of cellulose nitrate and a stabilizer, and some single base propellants also contain small quantities of glycerol trinitrate. In addition, single base propellants may be surface treated with centralite I (diethyl diphenyl urea,) dibutylphthalate, camphor, dinitrotoluene and other phlegmatisers.

Double base propellants consists of cellulose nitrate, glycerol trinitrate, plasticizers (phthalates) and a stabilizer.

Triple base propellant consists of cellulose nitrate, glycerol trinitrate or diglycol dinitrate, nitroguanidine and a stabilizer.

All propellant types may also contain small amounts of different additives, like combustion catalysts.

Secondary Explosives

The most important military secondary explosives are either pure substances, e.g., 2,4,6-trinitrotoluene, cyclo-trimethylene trinitramine, cyclo-tetramethylene tetranitramine, pentaerythritol tetranitrate or mixtures of these compounds. They may also contain phlegmatisers (wax, oil).

Initiating Explosives

The initiating explosive initiates the detonation of the main charge or the secondary explosive. The most common initiating explosives are lead azide, mercury fuminate and lead nitroresorcinates.

Stabilizers

Diphenylamine and certain derivatives of urea, N,N'-diethyl-N,N'-diphenyl urea (centralite I) and N'-methyl-N,N'-diphenyl urea (acardite II) are used as stabilizers in propellant.

These compounds will substantially increase the shelf life of propellant through binding decomposition products like free acid and nitrogen oxides, which are formed during the slow decomposition of the cellulose nitrate. The stabilizer is transformed into relatively stable nitro and nitroso derivatives.

Electroactive Components, Reactions and Working Electrodes

The electrode reactions in voltammetry are usually composed of several part reactions, and the rate determing step may be an electron transfer, a diffusion controlled reaction or a chemical reaction. The electrode reaction will often be more complicated than the simple reaction formulas usually given, which in most instances will be sufficient for describing the total reaction.

In Fig.2, the potential ranges are given where a number of compounds topical to the explosives sphere are reduced/oxidized. The potential ranges are also given where carbon and mercury electrodes can be employed as working electrodes in Fig.2. The reactions that are thought to occur at the electrode and on which the method of analysis is based, are briefly discussed for the different components.

Nitrate Esters

The nitrate ester group is irreversibly reduced in water/alcohol solutions in the potential range -0.2 V to -1.3 V at a dropping mercury electrode, via a two electron reaction:

$$RONO_2 + 2e^- + H_2O \rightarrow ROH + NO_2^- + OH^- \tag{1}$$

When the concentration of alcohol is increased, the reduction is displaced towards more negative potentials. The reduction of polynitrate esters will occur at more positive potentials when the number of nitrate ester groups in the molecule is increased. The change will be approximately 0.1 V per each new nitrate ester group. Under certain conditions (water/methanol solutions), the reduction of the separate nitrate ester groups can be detected.

Nitro Compounds

The nitro group, having a strong tendency to absorb electrons, is reversibly reduced at a dropping mercury electrode. The reduction usually takes place in two steps, to amine via

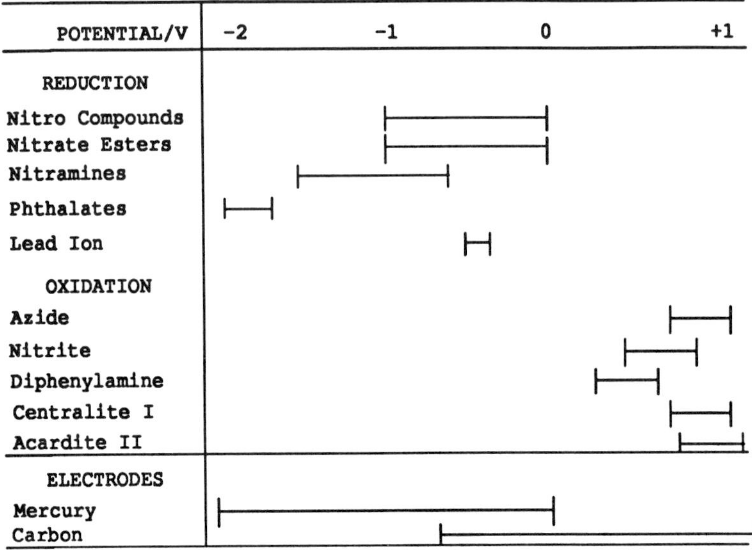

Fig.2. Potential ranges for the reduction/oxidation of number of compounds and functional groups, found in the explosives area.

hydroxyl amine, with 4 and 2 electrons respectively, according to the following reactions:

$$RNO_2 + 4e^- + 4H^+ \rightarrow RNHOH + H_2O \qquad (2)$$

and

$$RNHOH + 2e^- + 2H^+ \rightarrow RNH_2 + H_2O \qquad (3)$$

The reaction mechanism for the reduction of aromatic nitro compounds is more complicated, and depends not only on the number of nitro groups but also on their relative positions in the molecule, other functional groups and of course also on the pH of the solution and the solvent. Under certain conditions, the reductions of the separate nitro groups in aromatic polynitro compounds can be distinguished, and it is also possible to demonstrate the gradual reduction via hydroxyl amine. The reaction mechanism will be still more complicated when the molecule contains more than one benzene ring, as in nitro derivatives of diphenylamine.

In this context it is only of interest to note that polarography is highly sensitive in the determination of nitro compounds when 6 or, possible 4 electrons per nitro group participate in the electrode reaction.

Nitramines

Simple alifatic nitramines are reduced in two steps in neutral and alkaline solutions, according to the reactions:

$$=N-NO_2 + 2e^- + 2H^+ \rightarrow\ =N-NO + H_2O \qquad (4)$$

and

$$2 =N-NO + 4e^- + 3H_2O \rightarrow 2 =NH + N_2O + 4OH^- \qquad (5)$$

where two separate waves or peaks can be distinguished.

In acid solution, nitramines are reduced by 6 electrons, giving amines. Only one wave or peak can be distinguished. The reaction mechanism for the reduction of heterocyclic nitramines, like cyclo-trimethylene trinitramine (hexogen) and cyclo-tetramethylene tetramine (octogen) is complicated and occurs in several steps. In alkaline solutions, e.g. cyclo-trimethylene trinitramine is reduced in the first step via a two electron reaction.

88

Phthalates

Phthalates are reduced in two steps at comparatively negative potentials, approximately -1.6 V and more negative. For analytical purposes, the first reduction step is used:

$$\text{(phthalate diester)} + 4\,H_2O + 4\,e \longrightarrow \text{(phthalide)} + 2\,ROH + 4\,OH^- \tag{6}$$

The half wave potential moves towards more negative values and the diffusion current decreases slightly when the molecular weight increases. Also in the determination of phthalates, polarography has a high sensitivity.

Metal Ions

A metal ion, or rather the hydrate complex of a metal ion in an aqueous solution, is reduced at a dropping mercury electrode according to the reaction:

$$M^{z+} + ze^- + Hg \rightarrow M(Hg) \tag{7}$$

where $M(Hg)$ represents the amalgam formed at the drop surface.

In the presence of a complexing agent, forming a stable complex with the metal ion, the reduction of the metal complex will be:

$$ML_n + ze^- + Hg \rightarrow M(Hg) + nL \tag{8}$$

The reduction of the metal complex occurs at a more negative potential than that of the free metal ion.

Azides

The oxidation of the azide ion at different metal and carbon electrodes follows the reaction:

$$2N_3^- \rightarrow 3N_2 + 2e^- \tag{9}$$

which is a pH dependent reaction because hydrogen azide is a weak acid.

Nitrite

The nitrite ion, NO_2^-, or nitrous acid, HNO_2, can be oxidized at different metal and carbon electrodes. In a slightly acid or neutral solution (pH 3.5 to 8.5), the oxidation will occur according to the following reaction:

$$NO_2^- + H_2O \rightarrow 2H^+ + NO_3^- + 2e^- \tag{10}$$

The carbon paste electrode is proved to be superior to both the glassy carbon and the platinum electrodes as a working electrode. Using a platinum electrode, the oxidation is inhibited due to the formation of a platinum oxide layer.

Diphenylamine, Centralite and Acardite

The stabilisers diphenylamine, centralite and acardite can be oxidized by strong oxidation agents such as potassium dichromate in concentrated sulfuric acid, forming coloured products. According to I.M. Kolthoff and L.A. Sarver,[1] a bluish violet compound called diphenyl benzidine is formed when oxidizing diphenyl amine. The oxidation can be illustrated by the following reactions:

$$2\,\text{(diphenylamine)} \longrightarrow \text{(diphenyl benzidine)} + 2e + 2H^+ \tag{11}$$

$$\text{(oxidized form)} + 2e + 2H^+ \tag{12}$$

No information exists as to what rections take place when centralite and acardite are oxidized.

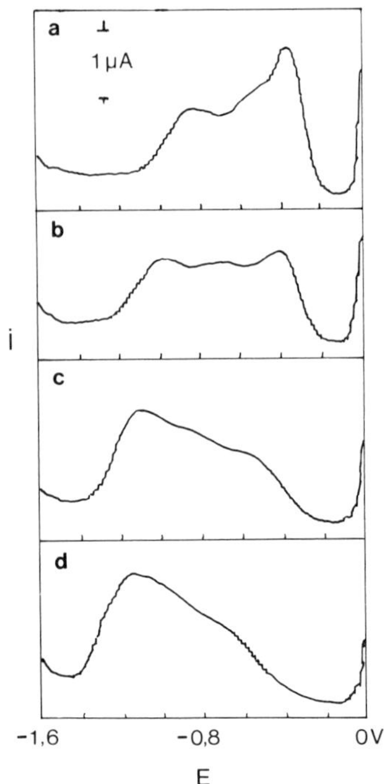

Fig.3. Differential pulse polarograms of 0.1 mM glycerol trinitrate in water/methanol solutions, using 0.1 M ammonium acetate electrolyte. Methanol content: a = 20 %, b = 40 %, c = 60 % and d = 80 %.

Solvents

Voltammetric determinations of inorganic components are carried out mainly in aqueous solutions, while determination of organic compounds employs organic solvents, like alcohols, dioxane, acetonitrile, dimethyl sulfoxide, or mixtures of water and an organic solvent.

The composition of the solvent has an essential influence on the reversibility of the electrode reaction. From an analytical point of view, solvents giving reversible electrode reactions are to be preferred. The form of a differential pulse polarogram is dependent on the composition of the solution, which in turn will have an influence on the possibilty of getting a successful multi-component determination. As an example, the differential pulse polarograms for glycerol trinitrate (Fig.3) and 2,4,6-trinitrotoluene (Fig.4) in water/methanol solutions, are shown. 0.1 M ammonium acetate is used as the electrolyte.

Fig.3 shows that the differential pulse polarograms for glycerol trinitrate are very dependent on the methanol content of the solution. Increasing methanol content broadens the polarogram and shifts it towards more negative potentials. The peak height is decreased at the same time. At a methanol content of appr. 40 %, three distinct but overlapping peaks are obtained; one peak per nitrate ester group. If the methanol content is further increased, the three peaks again will coincide, and the peak height will increase. A conceivable explanation for the fact that glycerol trinitrate will give three peaks is that the molecule is asymmetric and that the nitrate ester group in position 2 differs from the groups in position 1 and 3.

It is evident from the differential pulse polarograms for 2,4,6-trinitrotoluene in Fig. 4 that one peak is obtained for each one of the three nitro groups. The polarograms are not influenced by the methanol content of the solution to the same extent as those for glycerol trinitrate, but the peak heights change slightly and the peaks move towards more negative values when the methanol content is increased.

A comparison between the polarograms in Fig.3 and 4 shows that the reduction of a nitro group also will produce a peak considerably narrower and more well-defined than the reduction of a nitrate ester group will do. This is caused partially by the fact that the nitro

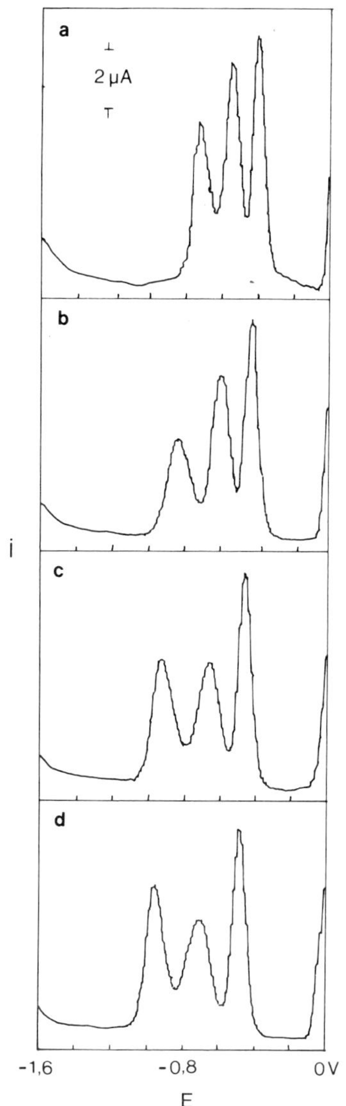

Fig.4. Differential pulse polarograms of 0.13 mM 2,4,6-trinitrotoluene in water/methanol solutions, using 0.1 M ammonium acetate electrolyte. Methanol content: a = 20 %, b = 40 %, c = 60 % and d = 80 %.

group is reduced using six electrons and the nitrate ester group using two electrons, and partly because reduction of nitrate ester group is irreversible.

It can be seen in the Fig.4 that the peak heights of the individual nitro groups are different, due to the fact that the reversibility of the reduction reactons varies. The dc polarography waves are of the same height but have slightly different slopes, proving that the reduciton of the nitro groups proceeds in slightly different ways.

Electrolyte and pH

The pH of the solution is an important parameter in voltammetric determination of organic compounds, partly because of the fact that hydrogen ions often take part in the electrode reaction and partly because the electroactive compound may take part in a hydrogen

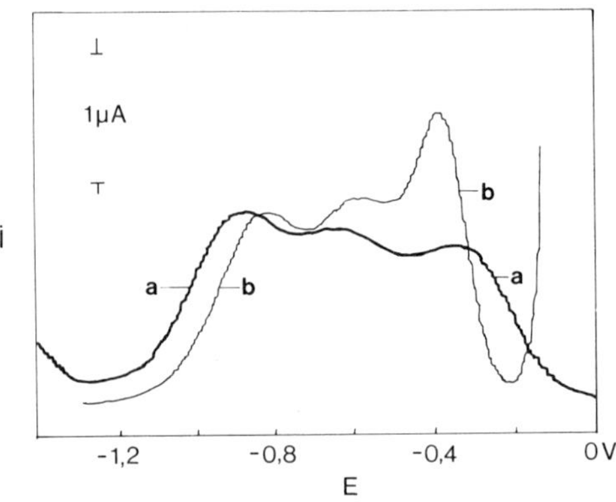

Fig.5. Differential pulse polarograms of 0.1 mM glycerol trinitrate in 60 % methanol, using a) 0.1 M ammonium acetate, b) 0.1 M tetramethyl ammonium bromide electroyte.

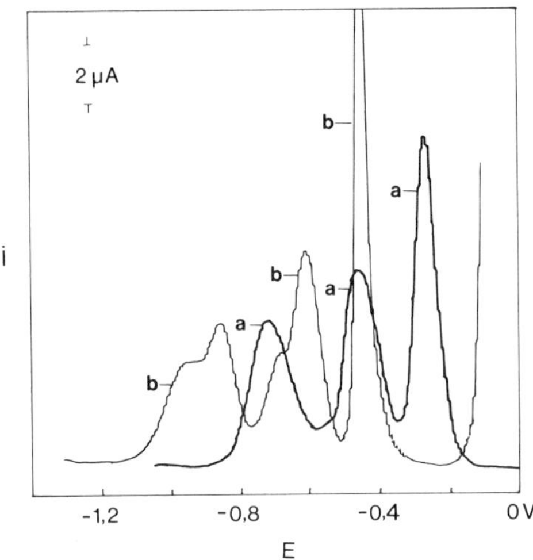

Fig.6. Differential pulse polarograms of 0.1 mM 2,4,6-trinitrotoluene in 60 % methanol, using a) 0.1 M ammonium acetate, b) 0.1 M tetramethyl ammonium bromide electrolyte.

ion dependent equilibrium reaction.

As the pH of the solution may influence the course of the electrochemical reaction as well as the reaction potential, it will, in most cases, be necessary to add a suitable buffer to the solution.

To demonstrate the influence of buffered and non-buffered solutions differential pulse polarograms of glycerol trinitrate and 2,4,6-trinitrotoluene, are shown in Fig.5 and 6 respectively. The electrolytes are 60 % methanol solution with 0.1 M ammonium acetate (buffered solution) and 0.1 M tetramethyl ammonium bromide (non-buffered solution) electrolytes.

Fig.5 and 6 show that the polarograms differ, depending on which electrolyte is used, due to the fact that the solution will have a different pH value in the immediate proximity of the electrode than in the main portion of the solution.

It might be of interest to note that, when using tetramethylene ammonium bromide electrolyte, two double peaks are obtained (one peak being completely overlapped by a maximum) for 2,4,6-trinitrotoluene, indicating that the reduction of the nitro groups proceeds step by step via hydroxyamine to amine. This also means that the polarograms will be more difficult to interpret in a quantitative determination.

Multi-Component Determinations

Using differential pulse polarography, it is possible to analyse mixtures of several electroactive compounds with completely or partially overlapping polarograms, provided that the analytical conditions can be chosen in such a way that sufficiently large differents in the differential pulse polarograms can be achieved for the different compounds.

It should be noted that the term multi-component system refers not only to mixtures of different compounds but also to molecules with several electroactive groups like polynitroaromates, polynitrate esters, polynitramines, nitrobenzaldehyde and nitrobenzoic acid.

The total current at a given potential for a solution containing several components that can be reduced is equal to the sum of the currents for the separate components, i.e., the currents are additive, providing that no chemical reactions occur between the compounds, and so the statement below can be given for the total current:

$$i = \sum_i k_i \times c_i \tag{13}$$

In Fig.7, the schematic differential pulse polarogram for a two-componenet system A + B is shown, together with the differential pulse polarograms for the separate components A and B. The differential pulse polarographic current for the two-component system A + B at the potentials E_1 and E_2 (Fig.7A) can be expressed as follows:

$$i_1 = k_{A1}c_A + k_{B1}c_B \tag{14}$$

$$i_2 = k_{A2}c_A + k_{B2}c_B \tag{15}$$

where $k_{A1}...k_{B2}$ are the constants for A and B at the potentials E_1 and E_2. The concentrations c_A and c_B can be calculated from the equations (14) and (15), using known constant values.

When analysing a two-component system, the maximum accuracy will be achieved when measuring the differential pulse polarographic current at potentials where the difference between the values of the constant k will be as large as possible.

In the limiting case $k_{A1} = 0$ and $k_{B2} = 0$ (or $k_{A2} = 0$ and $k_{B1} = 0$), the two-component system is simplified, giving two separate and independent differential pulse polarographic determinations, i.e., the differential pulse polarograms of A and B are completely separated (Fig. 7C) or partially separated (Fig.7B).

In the limiting case $k_{A1}/k_{A2} = k_{B1}/k_{B2}$ the system is indeterminate and neither c_A nor c_B can be determined using differential pulse polarography.

Multi-component analysis with dc polarography is based on completely separated polarographic waves. The separation of two polarographic waves is dependent on the reversibility of the electrode reaction, and also on the concentration of the electroactive compounds, and the number of electrons taking part in the electrode reaction. In Fig.8, schematic polarographic waves are shown when determining two compounds in the same solution, assuming that the components are present in the same concentrations and that the electrode reaction is either a reversible one-electron reaction or a reversible two-electron reaction.

Example 2. Determination of dibutylphthalate (DBP), nitroguanidine (NiGu) glycerol trinitrate (NG) in propellant.[4]

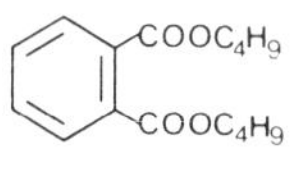

DBP 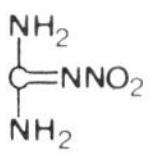   NiGu 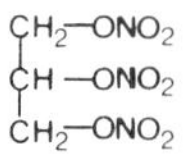   NG

The differences between the half wave potentials is assumed to be 100 mV and 200 mV. It can be seen that the difference in half wave potentials should be > 200 mV to produce separation of two waves, in the case that the electrode reaction is a one-electrone reaction. When the electrode reactions are two-electrode reactions, a smaller difference in half wave potentials will be sufficient to separate the polarographic waves.

Irreversible electrode reactions need a greater difference in half wave potentials in the determination of multi-component systems. Dc polarography might in some instances be preferred to differential pulse polarography in multi-component analysis (v. example 1).

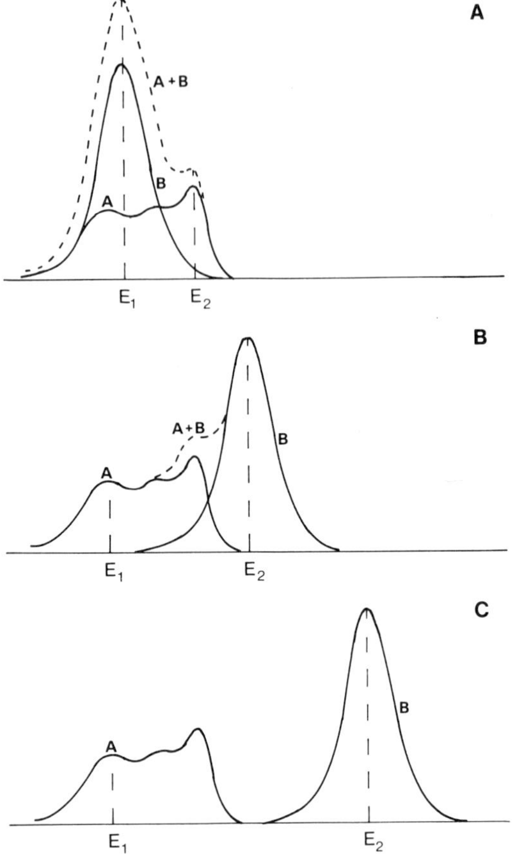

Fig.7. Schematic differential pulse polarograms for the two-component system A + B.

Applications

Example 1. Determination of the product content (2,4-dinitrotoluene and 2,4,6-trinitrotoluene) and nitrous acid in process acids from the 2,4,6-trinitrotoluene process.[2,3]

2,4,6-trinitrotoluene is prepared by nitrating toluene or o-nitrotoluene, using mixtures of nitric acid and oleum. The nitration is carried out in a continuous process in a series of nitrating reactors, through which toluene or the product phase and the nitrating acid flow in counterflow. In spite of the nitrating process being continuous, the reaction occurs in three distinct steps, corresponding to the introduction of the first, the second and the third nitro group in the molecule. Each reaction step demands specific conditions, e.g., nitrating acid composition and nitrating temperature. In Table 1, approximate compositions of the acid and product phases are given, for the first and the last reactor, and one of the intermediate reactors. The nitrating acid composition and the nitrating acid/product quotient differ from one reactor to another. In the last reactor a surplus of SO_3 will be found. Nitrous acid is formed during the nitration. The temperature in the nitration reactors is kept sufficiently high for the product phase to exist as in liquid form. This means that the acid and product phase consist of a more or less homogenous liquid. The composition of the nitrating acid is calculated on the product free portion of the sample.

To make a polarographic determination of the product content and the composition of the product phase with respect to 2,4-dinitrotoluene and 2,4,6-trinitrotoluene, a suitable sample is dissolved in γ-butyrolactone. A portion of the dissolved sample is diluted to a suitable concentration using a 60 % methanol solution containing 0.1 M ammonium acetate electrolyte. For the polarographic determination, in this case, the dc technique is preferred in

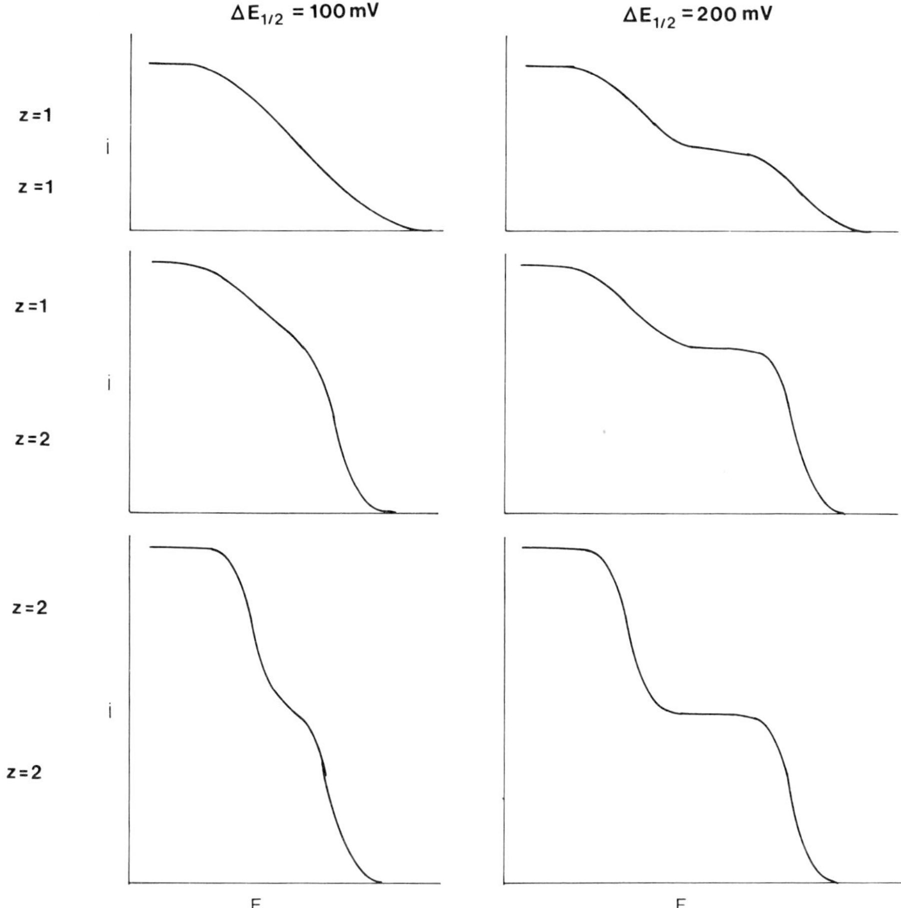

Fig.8. The influence of the difference in half wave potentials (ΔE) and the number of electrons taking part in the electrode reaction (z) on the separation of polarographic waves.

stead of differential pulse polarography, in spite of the latter having greater selectivity. The differential pulse polarogram for a mixture of 2,4-dinitrotoluene and 2,4,6-trinitrotoluene contains 5 peaks of different heights, which makes the evaluation of the polarogram more complicated. The dc polarogram, on the other hand, will give three waves of equal height for 2,4,6-trinitrotoluene (B) and two waves for of equal height 2,4-dinitrotoluene (A), as shown in the schematic polarograms in Fig.9. The equal heights of the 2,4,6-trinitrotoluene waves are explained by the facts that the same number of electrons are exchanged in the reduction of the three nitro groups, and that the diffusion cofficient is identical, the molecule being the same one.

The same facts apply to 2,4-dinitrotoluene. The wave heights are, however, different for the two compounds, in spite of their occuring in the same concentrations, which can be explained by the fact that the diffusion coefficient is different for the two compounds.

Considering Fig.9, it can be noted that 2,4,6-trinitrotoluene can be determined entirely independently of 2,4-dinitrotoluene at the potential E_1 and that the latter can be determined in a two-component system at the potential E_2.

Fig.10 shows a dc polarogram when determining the product content in a process acid, containing appr. 13 % 2,4-dinitrotoluene and appr. 87 % 2,4,6-trinitrotoluene.

The nitrous acid concentration is determined by diluting the sample to a suitable con-

Table 1. Approximate compositions of nitrating acids and product phases from the first, last and one of the intermediate nitration reactors in the trinitrotoluene process.

Acid/Product	Reactor		
	1		n
Process acid, %	100	80	85
H_2O, %	25	1	
SO_3			10
H_2SO_4, %	72	78	75
HNO_3, %	3	16	15
HNO_2, %		5	1
Product, %	0.2	20	15
Mononitrotoluene, %	77		
Dinitrotoluene, %	18	20	1
Trinitrotoluene, %	5	80	99
Temperature, °C	50	80	100

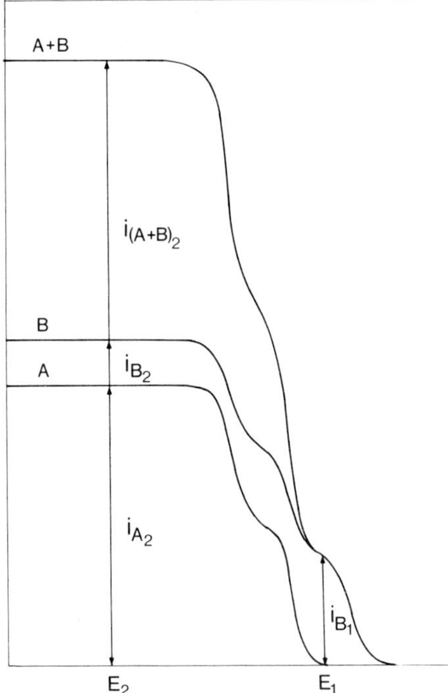

Fig.9. Schematic dc polarograms of 2,4-dinitrotoluene (A) and 2,4,6-trinitrotoluene (B).

centration with 0.1 M ammonium acetate solution and obtainig a voltammogram, using 10 mV/s scan speed and a carbon paste electrode (Fig.11).

In Table 2 some results are given from the analysis of process acid from the 2,4,6-trinitrotoluene process with respect to product content and nitrous acid.

The composition analysis of propellant normally is preceeded by extraction, where low molecular weight organic compounds are separated from the cellulose nitrate and inorganic componenets. The extraction agent usually employed is diethyl ether or methylene chloride.

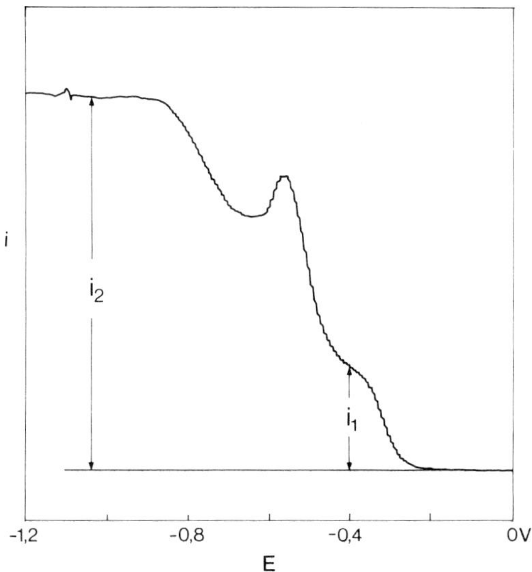

Fig.10. Dc polarograms of the analysis of a sample, containing appr. 13 % 2,4-dinitrotoluene and appr. 87 % 2,4,6-trinitrotoluene.

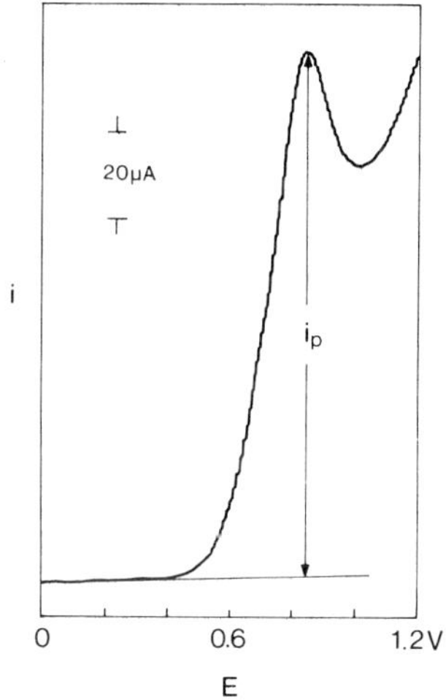

Fig.11. Voltammograms from the determination of 0.1 mM nitrous acid (HNO_2) using 0.1 M ammonium acetate electrolyte.

An analysis of propellant containing nitroguanidine will be more difficult, nitroguanidine being insoluble in these solvents. Hot water will, however, dissolve nitroguanidine from the propellant.

As the solubility of nitroguanidine is higher in water than in methanol, an aqueous solution with low methanol content (20 %) is used for the polarographic determination. Dibutylphthalate, nitroguanidine, (nitramine: $=NNO_2$) and glycerol trinitrate are reduced within different potential ranges (cf. Fig.2).

Consequently, these compounds can be determined fully independently as three one-component systems, even when present at different concentrations. Fig.12 shows a differ-

Table 2. Results from analysis of samples from a 2,4,6-trinitrotoluene process with five nitration reactors with respect to product and nitrous acid contents, and dinitrotoluene and trinitrotoluene contents in the product phase.

Reactor	HNO$_2$, %	Product, %	Product phase	
			Dinitrotoluene, %	Trinitrotoluene, %
1				
2	0.57	1.5	50.0	50.0
3	2.28	11.0	48.8	51.2
4	1.08	14.1	10.2	89.8
5	0.10	18.8		100
1				
2	0.79	1.3	47.8	52.2
3	2.08	7.3	43.3	56.7
4	1.10	6.9	17.2	82.8
5	0.20	16.5		100

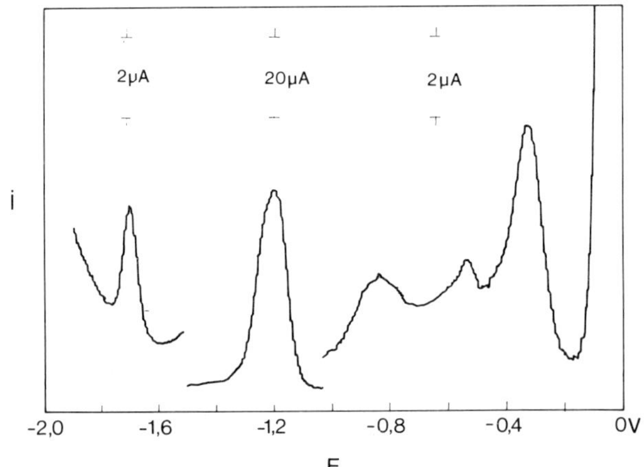

Fig.12. Differential pulse polarogram of 0.2 mM glycerol trinitrate, 0.67 mM nitroguanidine and 0.04 mM dibutylphthalate in 20 % methanol solution, using 0.1 M tetramethyl ammonium bromide electrolyte.

ential pulse polarogram for a propellant containing approx. 21 % glycerol trinitrate, 35 % nitroguanidine and 6 % dibutylphthalate. The cellulose nitrate content of the propellant is approx. 35 %. The figure demonstrates that the three compounds will yield totally separated differential pulse polarographic peaks, despite of the fact that different sensitivities have been used when recording the differential pulse polarographic peaks. Table 3 shows some results of analysis for a nitrogunidine propellant.

Example 3. Determination of glycerol trinitrate (NG), 2,4-dinitrotoluene (DNT) and 2,4,6-trinitrotoluene (TNT) in propellant.

Table 3. Results from the determination of glycerol trinitrate, nitroguanidine and dibutylphtalate in propellant.

Glycerol trinitrate %	Nitroguanidine %	Dibutylphtalate %
21.29	31.84	5.58
21.41	31.90	6.03
21.28	30.22	5.87
21.26	29.85	5.90
21.61	30.54	6.13

Propellant in its simplest form, single base propellant, consists of colloidal cellulose nitrate stabilised with, e.g., diphenylamine. To improve the ballistic properties, the propellant may be surface treated with nitrotoluenes and phthalates, among other substances. Moreover, may be added to the basic powder paste small amounts of glycerol trinitrate.

Nitrotoluenes and nitrate esters are reduced in the same potential range. By carefully selecting the methanol concentration of the solution, a mixture of two nitrotoluenes and one nitrate ester can be determined. The polarograms are influenced by the methanol content of the solution, and also by the fact that the nitrate ester group is reduced by 2 electrons and the nitro group by 6 electrons, and that the nitrate ester group reduction is strongly irreversible.

Glycerol trinitrate (N) will give a polarogram with three strongly overlapping peaks, 2,4,6-trinitrotoluene (T) a polarogram with three separate peaks and 2,4-dinitrotoluene (D) a polarogram with two separate peaks (cf. Fig.13). A differential pulse polarographic determination of these three compounds will involve 8 electroactive components or groups. A quantitative determination is quite feasible, measuring the currents at three carefully monitored potentials. The polarograms in Fig.13 indicate that glycerol trinitrate can be determined completely independent of 2,4-dinitrotoluene and 2,4,6-trinitrotoluene when the current is measured at the potential E_3. 2,4,6-trinitrotoluene can be determined from a two-component system measuring the current at the potential E_1, and 2,4-dinitrotoluene can be determined from a three-component system measuring the current at the potential E_2.

Fig.14 shows a differential pulse polarogram for a propellant sample containing appr. 5 % glycerol trinitrate, 1.5 % 2,4-dinitrotoluene and 1.5 % 2,4,6-trinitrotoluene and in Table 4 some results from analyses of propellant samples are given.

Table 4. Results from the determination of glycerol trinitrate, 2,4-dinitrotoluene and 2,4,6-trinitrotoluene in propellant.

Sample	Glycerol trinitrate %	2,4,6-trinitrotoluene %	2,4 dinitrotoluene %
1	4.85	1.16	1.49
	4.59	1.20	1.50
	4.90	1.19	1.53
2	4.48	1.35	1.17
	4.56	1.37	1.13
	4.54	1.37	1.18

Military explosive compositions in their most simple form consist of either a mixture of octogen and trotyl, octol, or hexogen and trotyl, hexotol. By dissolving the sample in a suitable solvent (γ-butyrolactone), the octogen/hexogen and trotyl are only partially reduced within the same potential range (cf. for Fig.2), and the determination can be carried out for two independent one-compound systems.

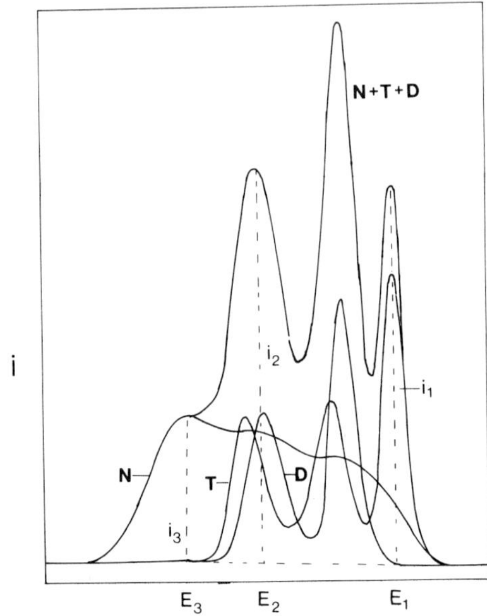

Fig.13. Schematic differential pulse polarograms of glycerol trinitrate (N), 2,4,6-trinitrotoluene (T) and 2,4-dinitrotoluene (D) in 60 % methanol solution, using 0.1 M ammonium acetate electrolyte.

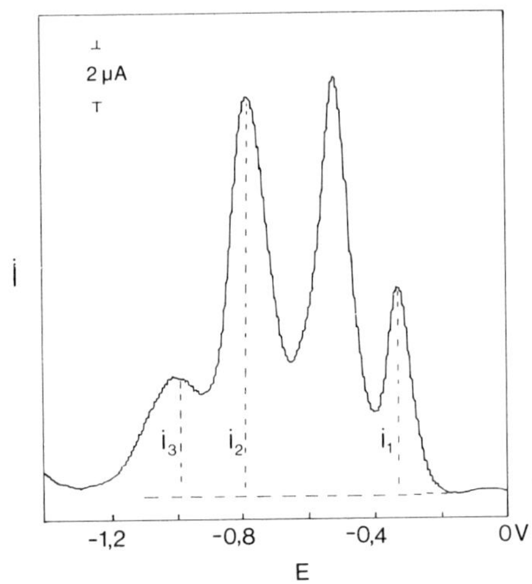

Fig.14.   Differential pulse polarograms of glycerol trinitrate, 2,4-dinitrotoluene and 2,4,6-trinitrotoluene in 60 % methanol solution, using 0.1 M ammonium acetate electrolyte.

Fig. 15 shows a differential pulse polarogram for an octol sample containing appr. 70 % octogen and 30 % trotyl. The polarogram shows four distinct peaks. The peak at appr. -0.4 V emanates solely from trotyl, and the peak at appr. -1.3 V from octogen. The peaks at appr. -0.5 V and -0.9 V originate from both trotyl and octogen. The current is measured at the potentials -0.4 V and -1.3 V. In Table 5 some results from analyses of octogen and trotyl in octol are given.

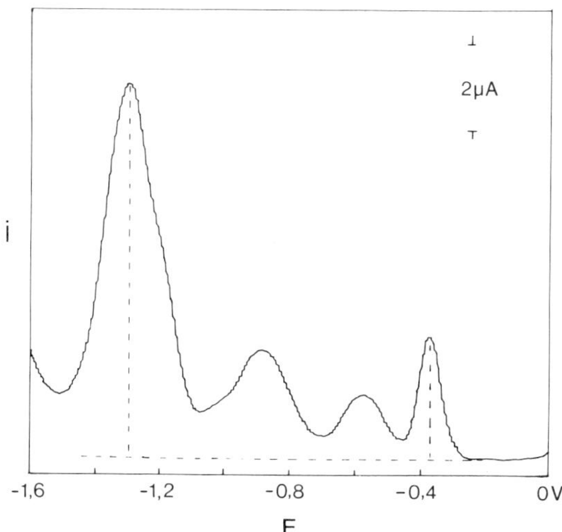

Fig.15.   Differential   pulse   polarogram   of   octogen   and
2,4,6-trinitrotoluene in 65 % methanol solution, using 0.1 M ammonium acetate electrolyte.

Example 4. Determination of cyclo-tetramethylene tetranitramine (octogen, HMX) and 2,4,6-trinitrotoluene (trotyl, TNT) in explosives. [4]

HMX   TNT

Table 5. Results from determination of octogen
and 2,4,6-trinitrotoluene in octol.

Octogen, %	2,4,6-trinitrotoluene, %
70.5	30.1
69.9	30.1
70.7	29.6
70.4	30.1
71.4	28.9
69.9	28.4

Example 5. Analysis of lead-2,4-dinitroresorcinate (PbDNR).[5]

PbDNR

Lead ions and nitro groups are normally reduced within the same potential range (cf.
Fig.2). The half wave potential for a metal ion is displaced towards more negative values
when the metal ions is transferred to a metal complex. By transferring lead to a PbNTA

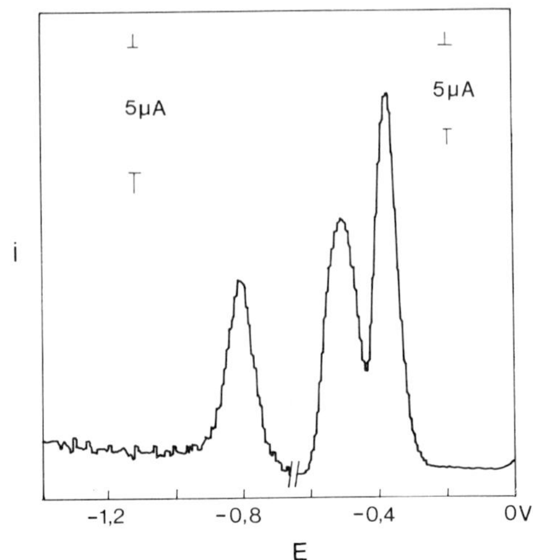

Fig.16.   Differential pulse polarogram of the analysis of
lead-2,4-dinitroresorcinate, 2.4 mg / 100 ml, in the presence of 0.013 M
NTA and 0.065 M ammonium acetate at pH 6.52.

or PbEDTA complex, separate differential pulse polarograms are obtained for lead and the nitro groups of the resorcinate. Another benefit of complexing agents when analysing lead-2,4-resorcinate, which is poorly soluble in a pure aqueous solution, is that the solubility can be increased by complexation. A nitro group being reduced with 6 electrons and divalent lead with two electrons, either as free lead ions or in a complex, the polarographic wave for a nitro group will be approximately 3 times higher than that of lead.

Fig.16 shows a differential pulse polarogram of lead-2,4-dinitroresorcinate in the presence of NTA. It can be seen that the lead peak is fully separated from the nitro peaks and that a higher sensitivity has been used in recording of the lead peak. In Table 6 results from the analysis of lead-2,4-dinitroresorcinate are given. This method has also been used in analysis of other lead nitroresorcinates.

Table 6. Results from the analysis of
lead-2,4-dinitrorescorcinate in the
presence of NTA.

Lead, %	2,4-dinitrorescorcinate, %
51.02	48.52
53.61	48.05
49.52	48.23
50.97	47.03
52.79	47.98

Example 6. Analysis of lead azide.[6]

Lead azide, $Pb(N_3)_2$, can be analysed by voltammetry, oxidizing azide ions at a carbon paste electrode and reducing lead ions at a dropping mercury electrode at pH 4.6 in an acetate buffer solution. Lead azide is poorly soluble in a pure aqueous solution, but can be dissolved when acetate ions are present. In Fig.17 a differential pulse voltammogram for the determination of azide at a carbon paste electrode is shown. Some results from the analysis of lead azide are given in Table 7.

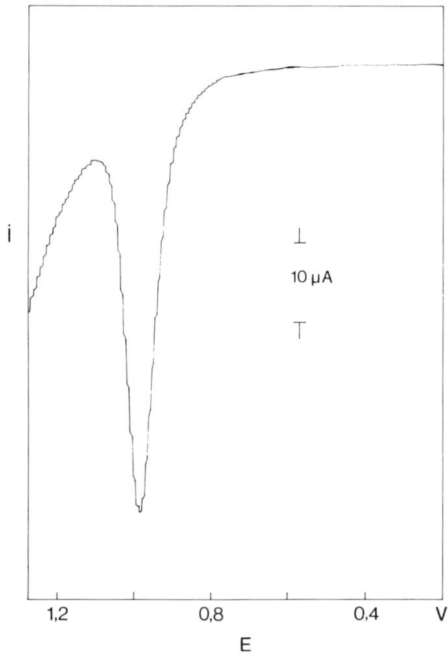

Fig.17. Differential pulse voltammogram from the determination of 0.065 mM N_3^-.

Table 7. Results from the analysis of lead azide.

Lead, %	Azide, %
70.78	27.83
71.56	28.84
71.17	30.37
69.54	28.24
72.57	29.11

Example 7. Determination of stabilisers, stabiliser derivatives and NO_x in propellant.[7-11]

Nitrate esters in propellant, e.g. cellulose nitrate, are slowly decomposed, the weak nitrate ester bond ($RO-NO_2$) being broken up producing nitrogen dioxide and aldehydes:

$$RO - NO_2 \rightarrow RO^{\cdot} + NO_2 \tag{16}$$

The nitrogen dioxide formed can react further with organic material, forming nitrogen oxide and other gases, e.g., N_2O, CO, CO_2 and H_2O. The total decomposition reaction can be written:

$$RO - NO_2 \rightarrow NO_x, \quad x = 1 \text{ or } 2 \tag{17}$$

In the presence of air, the NO formed can be oxidized to NO_2, which will catalyse and therefore accelerate the decomposition process.

In order to prevent those reactions, stabilizers are added to the propellant. The function of the stabilizer is to react sufficiently fast with the nitrogen dioxide formed, thereby preventing the gas formed to take part in the catalytic decomposing reaction with more nitrate ester groups.

The first change which occures during the ageing of a diphenylamine stabilized propellant is the forming of N-nitroso diphenylamine, 4-nitro diphenylamine and 2-nitro diphenylamine. During continued ageing, higher nitro derivatives are also formed, like 2,4'-dinitro diphenylamine, 4,4'-dinitro diphenylamine and 2,4,4'-trinitro diphenylamine.

To give an idea of the stability of an aged propellant, the relative amounts of the stabilizer and its derivatives in the propellant can be measured. The occurence of nitro- and nitroso diphenylamine derivatives is not a real proof that the propellant is unstable, indicating however that less than the original amount of stabilizer is available.

The nitroso and nitro derivatives of diphenylamine do reduce at a dropping mercury electrode, but diphenylamine is not reducable.[7] Diphenylamine, like other stabilizers, e.g., acardite, centralite and lower nitroso and nitro derivatives of diphenylamine, can be oxidized at carbon electrodes.[8]

In Table 8, the peak potentials for a number of stabilizers are present. Diphenylamine is the stabilizer that is oxidized most readily and then come centralite and acardite. Diphenylamine is oxidized at a potential allowing a voltammetric determination to be carried out even in a methanol/water solution. Nitroso and nitro diphenylamine derivatives are oxidized at more positive potentials.

Table 8. Peak potentials in voltammetric determination of a number of stabilizers.

Stabilizer	Peak potential/V
Diphenylamine	+0.45
N-nitrosodiphenylamine	+1.25
2-nitrodiphenylamine	+0.80
Centralite I	+0.80
Acardite II	+0.90

Coloured products are formed during the oxidation, centralite giving bright red and acardite dark red oxidation products.

Differential pulse polarography will not give any greater sensitivity in the quantitative determination of mixtures of different nitroso and nitro derivatives of diphenylamine, but may be used for the identification of separate derivatives.

Differential pulse polarograms for a number of nitroso and nitro diphenylamine derivatives are shown in Fig.18. Each nitroso and nitro group will give one peak, and derivatives with several groups will give one peak per separate group. Even 2,4,4',6-tetranitro diphenylamine will give four well defined separate peaks.

A nitro group in the 2 position is more easily reduced than a group in the 4 position. As to the reduction of nitroso groups, it emerges that a C-nitroso group is much easier reduced than a N-nitroso group.

Using electrochemical detection in high-performance liquid chromatography, nitroso and nitro diphenylamine can be selectively detected in the presence of other reducible compounds and high concentrations of diphenylamine.[9,10] Electrochemical detection presents roughly the same response for similar compounds or compounds with the same electroactive groups, while the spectrophotometric response is dependent on the molecular coefficient of absorption for the compounds at the chosen wavelength.

The electrochemical detector chosen is a thin-layer cell with two glassy carbon working electrodes (cf. Fig.19). At the first electrode W_1, the potential of which being -1.0 V vs. Ag/AgCl, nitrodiphenylamine derivatives are reduced according to:

$$Ar-NH-Ar-NO_2 + 4H^+ + 4e^- \rightarrow Ar-NH-Ar-NHOH \tag{18}$$

In the analysis, the signal from the working electrode W_2, having a potential of $+0.6$ V, is used. Here, hydroxylamine is oxidized via a 2-electron reaction:

$$Ar-NH-Ar-NHOH \rightarrow Ar-NH-Ar-NO + 2e^- + 2H^+ \tag{19}$$

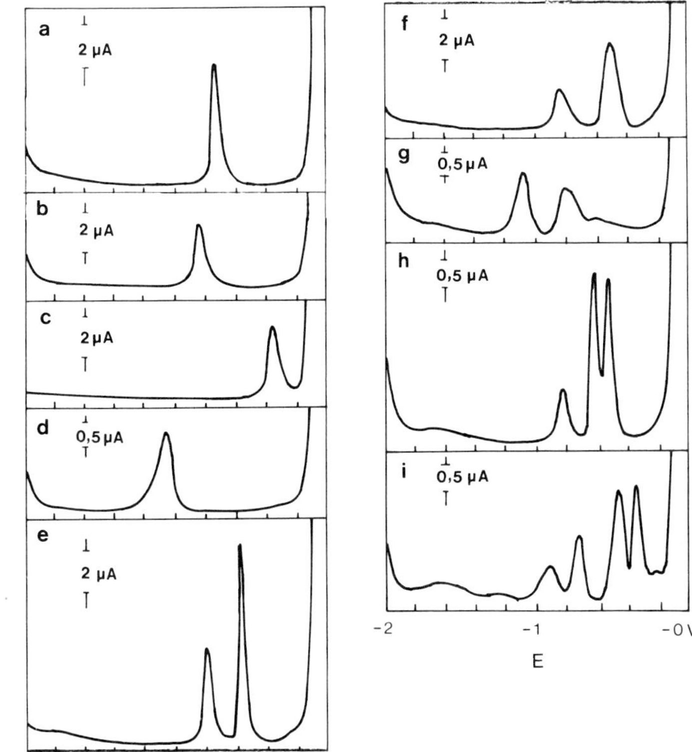

Fig.18. Differential pulse polarograms from the determination of selected nitroso and nitro diphenylamine derivatives.

a: 2-nitrodiphenylamine
b: 4-nitrodiphenylamine
c: 4-nitrosodiphenylamine
d: N-nitrosodiphenylamine
e: 2,4-dinitrodiphenylamine
f: 2,4-dinitro-4-hydroxi-diphenylamine
g: N-nitroso-4-nitro-diphenylamine
h: 2,2',4-trinitrodiphenylamine
i: 2,4,4',6-tetranitrodiphenylamine

At the potential +0.6 V the diphenylamine will not interfere to any great extent, as its oxidation takes place at more positive potentials. The introduction of nitro and nitroso groups in the diphenylamine molecule will yield derivatives more difficult to oxidize than diphenylamine.

Fig.20 shows a chromatogram from the determination of 4-nitroso-2-nitrodiphenylamine, 4-nitrodiphenylamine and 2-nitrodiphenylamine in the presence of a high diphenylamine concentration. The method can also be used for a selective determination of nitroso and nitro diphenylamine derivatives when high concentrations of nitrate esters are present.

The nitrogen dioxide formed in propellant can be analysed by collecting the gas in a cartridge containing Florisil coated with diphenylamine.[11] The nitrogen dioxide reacts with diphenylamine, forming N-nitroso and nitro diphenylamine derivatives. The diphenylamine derivatives are eluted from the cartridge, analysed using polarography, and the amount is related to the amount nitrogen dioxide formed. N-nitroso diphenylamine is used as a reference, this compound being responsible for more than 90 % of the liberated nitrogen dioxide. The dc polarographic analysis measures the diffusion current at the potential −1.10 V.

Results from the analysis of some single base propellant, having undergone forced ageing, are shown in Fig.21. The loss in weight for the propellant has also been determined. It

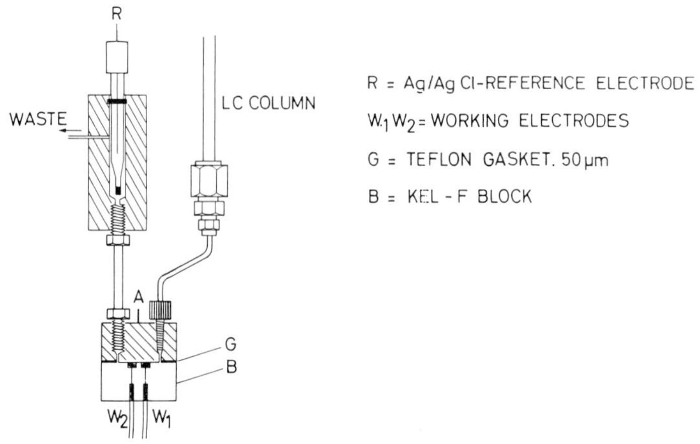

Fig.19. Electrochemical cell with two electrodes.

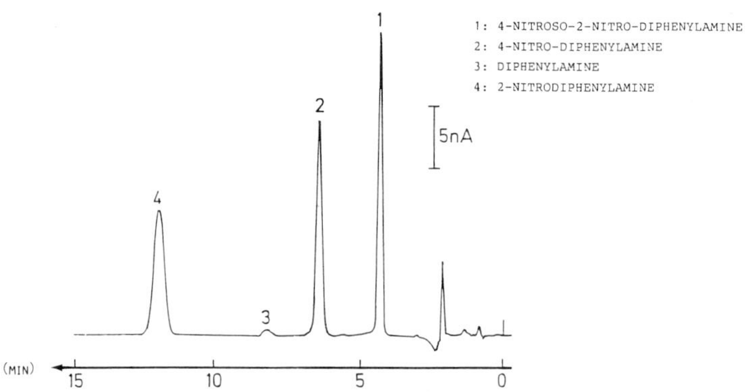

Fig.20. Chromatogram of electochemically detected nitro.

can be seen that a loss in weight of, e.g., 2 % corresponds to the formation of appr. 2 mg NO_2/ 1 g of propellant. Moreover, a direct relation between the loss in weight and the nitrogen dioxide formed can be found. The results also indicate that a single base propellant stabilized with acardite is more stable than a propellant that has been stabilized with diphenylamine or centralite. It should, however, be noted that the propellants examined contain different amounts of stabilizers.

After modifying the cartridge, this method also can be used for field determinations and for the analysis of nitrogen dioxide, e.g., in ammunition plants and in storages.

Conclusions

Voltammetry is proved to be a accurate and selective method for the analysis of electroactive components in the explosives area. The method can be used for quantitative as well as for qualitative determinations.

The introduction of electrochemical detectors in high-performance liquid chromatography can be seen as the great break-thorugh for the determination of trace amounts of electroactive components in mixtures. The developement potential in this area is judged to be very great and interesting, especially when taking into account the advantages the three-dimensional evaluation of a chromatogram can offer.

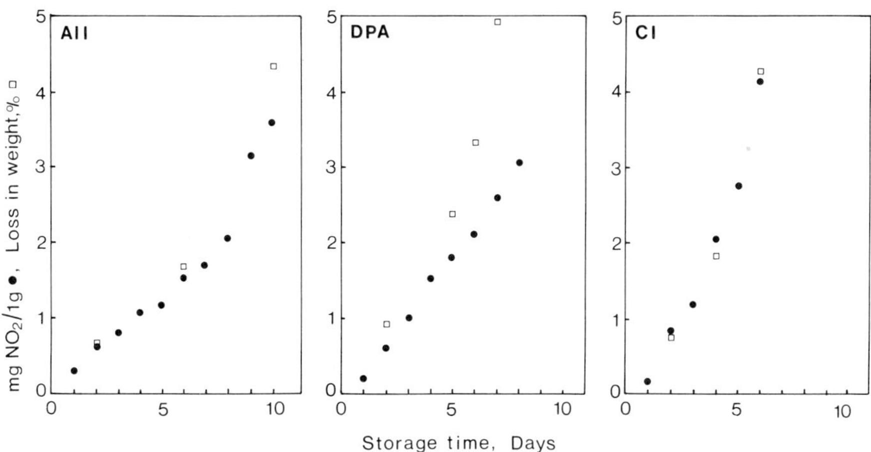

Fig.21. Results from the determination of loss in weight and NO_2 in propellant having undergone forced ageing.

A II: single base propellant stabilized with 0.9 % acardite II

DPA: single base propellant stabilized with 1.1 % diphenylamine

C I: single base propellant stabilized with 0.6 % centralite I

References

1. I.M. Kolthoff and L.A. Sarver, J. Am. Chem. Soc., **52** (1930) 4179.
2. J. Asplund, in: "Electrochemistry, Sensors and Analysis", M.R. Smyth and J.G. Vos (Eds), Anal. Chem. Symposia Series **25** Elsevier, Science Publishers B.V., Amsterdam, 1986, p. 91.
3. J. Asplund, Anal. Chim. Acta, **206** (1988) 137.
4. J. Asplund, Propellants, Explosives and Pyrotechnics, **11** (1986) 69.
5. J. Asplund, J. Ener. Mat., **4** (1986) 339.
6. J. Asplund, in the Proceedings of the ninth International Pyrotecnics Seminar, Colorado Springs, 6-10 August 1984, p. 15.
7. J. Asplund, in the Proceedings of Internationale Jahrestagung 1983, Karlsruhe, 29 June – 2 July 1983, p. 313.
8. A. Bergens, K. Lundström and J. Asplund, Talanta, **32** (1985) 893.
9. A. Bergens and J. Asplund, in the Proceedings of the 17th International Annual Conference of ITC 1986, Karlsruhe, 25-27 June 1986, pp. 12-1
10. A. Bergens, J. Chrom., **410** (1987) 437.
11. J. Asplund, in the Proceedings of the Symposium on Compatibility of Plastics and other Materials with Explosives, Propellants, Pyrotechnics and Processing of Explosives, Propellants and Ingredients, Long Beach, 27-29 October 1986, p. 305.

APPLIED POLAROGRAPHY AND VOLTAMMETRY IN DAY-TO-DAY ENVIRONMENTAL ANALYSIS, POSSIBILITIES AND LIMITATIONS

Pierre M. Bersier[1] and Jacques Bersier[2]

1. Central Analytical Laboratory, Ciba-Geigy Ltd.
   CH-4000 Basle, Switzerland
2. Central Research Laboratories, Ciba-Geigy Ltd.
   CH-4000 Basle, Switzerland

Introduction

"The natural resources of the earth including the air, water, land, flora and fauna and especially representative samples of natural ecosystems must be safeguarded for the benefit of present and future generations through careful planning or management, as appropriate".[1]

The contamination of the environment by natural and especially man-made pollutants is widespread.

The volume of hazardous waste generated annually in the US is 266 million tons; the chemical industry generates 68 percent of this. In England 5 million pounds of "potentially hazardous" waste is disposed of.[2]

Unlike EPA, the British environmental control agency does not use the term "hazardous waste" preferring the term "potentially hazardous." The linguistic nuance is said to reflect the way the British dispose of their waste. Instead of treating hazardous waste one way and municipal waste another, the UK places both hazardous and non hazardous waste in the same landfill, calling the process cogeneration. Cogeneration in the right combination can be beneficial.

Awareness of the need to monitor environmental quality in both urban areas and at the workplace has meant that analyses are being carried out with increasing frequency. The problem of environmental protection is not new. These problems were discussed already by Hippocrates (460–377 B.C.) in his book on "Air, Water and Environment".[3] The legislation dealing with the effects of human activities on the environment is not a product of the last decade. It was already existent in the far preindustrial times. Some highlights of environmental legislation are summerised in Fig.1.[4]

Of the 7 million chemical compounds known today, about 60000 to 70000 are actually in current use. The quantity of compounds and their metabolites which reach the environment yearly (predominantly via the atmosphere as exhaust gases and wastewater) is estimated to be about 65000; of these 65000 (under the best circumstances) only a fraction of the especially harmful 4000 to 6000 are actually detected. Heavy metals, pesticides, mineral oils and organic pollutants (solvents) all belong to this group.

Sulfur dioxide and NO_x, which also act as a major precursor of acid rain as well as ecotoxic heavy metals and metalloids (Cd, Pb, Hg, Cu, Ni, Zn, As, Se, etc.) have gained considerable significance with regard to the ecotoxic burden on vegetation, soil, natural waters, and as air pollutants.[5] The SO_2 emission in the F.R.G. in 1984 had reached the limit of 3.5 million tons a year.[6]

The annual deposition of mercury is 20 t a year over the whole F.R.G.. It may be estimated that during the last decade or so this amount of mercury has been added yearly

Contemporary Electroanalytical Chemistry, Edited by A. Ivaska *et al.*
Plenum Press, New York, 1990

1240	edict of Friedrich II on the conservation of air, water and land
1273	law against the use of coal in London
1306	execution of a blacksmith because of the use of coal
1340	ban of coal for forging in Zwickau
1464	closure of copper and lead works in Cologne because of air pollution
1550	installation of smoke and flyash cabins in the metal-works in Joachimsthal
1627	Corpus Juris Civilis Justinianei : Aere corrumpere non licet
1863	England: Akali, etc., Works Regulation Act and Orders

Fig.1. High-lights of environmental legislation.[4]

to the soil, vegetation and inland water sources. The annual deposition of the highly toxic methylmercurychloride (LC 50 = 5 mg/kg) in the F.R.G. is estimated to 0.5 tons a year.[7]

In Table 1A, the average values and maximum contents of As, Cd, Cr, Pb, Cu, Hg and Zn in unfiltered Rhine water, as measured in Basle and in Lobith at the German/Dutch border are summerised.[8]

Table 1A. Typical concentrations of inorganic and organic pollutants measured in the Rhine river (unfiltered water).

Element	in Lobith (1983) average* μg/l	measured maximum* μg/l	in Basle average** μg/l	measured in Lobith (1983) average* μg/l
As	3	15	-	
Cd	0.5	1.4	0.2	
Cr_{tot}	7	27	1	
Cu	8	23	2	
Pb	9	24	0.5	
Hg	0.1	0.3	0.02	
Zn	54	120	31	
Organochloro pesticides				0.01
Organophosphorous compounds (total)				0.5

* Meijers et al.,[8] ** unpubl. data

Maximum concentrations of organic trace impurities (pollutants) in the Rhine river at Lobith between January and June 1984 are summerized in Table 1B.[8]

Already "clean" water poses an extremely complex qualitative and quantitative problem to the analytical chemist. Grob and Grob[9] detected not less than 136 substances in Zürich drinking water.

High-quality, effective analytical chemistry thus assumes a crucial role, 1) in describing the current status (quantitative and qualitative recording of the environmental situation), and 2) in clarifying ecotoxicological effects, e.g., biomethylation of metals, such as Hg, Sn, Pb, etc., and mobilisation of the metals from sediments, solids, etc. by anthropogenic complexing agents, such as nitrilotriacetic acid (NTA) etc.

Further tasks of environmental analysis are:

i) the quantitative and qualitative control of pollutant emission in the environment

ii) the elucidation of the path of the emitted pollutant from the source of the pollutant to the interaction with humans

Table 1B. Highest concentrations of organic pollutants measured in the Rhine at Lobith (German / Dutch border). Period of January – June 1984 (μg/l).

Substance	Maximum Value μg/l
Trimethylcyclohexane	3.9
$C_8H_{18}/$, alcohol	3.0
polysaccharides	5.7
α,α-Dimethylbenzoylmethanol	5.6
Dioctylphtalate	3.8
Nitrobenzene	7.0
Quinoline	11.0
Dimethylaniline	4.0
o-Anisidine	12.6
Trimethyltiophosphate	7.7
Triethylphosphate	3.5
N,N-Dibutylformamide	17

iii) the monitoring/study of the conversion and transformation of pollutants as well as their natural reactions

iv) the discovery and elucidation of processes for the prevention of the emissions, or if necessary, of processes for their neutralization.[10]

In this short article practical examples from *i)* and *iv)* will be discussed in some more details in Section II.

Fundamentally, trace analysis is possible with a range of different procedures. In the organic pollutant determination, chromatographic methods such as GC, GC-MS, HPLC and, more recently immunoassay[11a] and to a greater degree, the coupling of HPLC with electrochemical detection (HPLC-EC) (see below) take precedence (cf.[11])

Today, the field of trace metal analytical chemistry is dominated by a few single-, oligo- and multielement techniques having high detection power:
- atomic absorption spectrometry (AAS) with and without flame as single element techniques, and
- polarography and voltammetry, particularly modern pulse and square-wave methods, as oligo-substance techniques, are of greatest importance and range of applicability.[11]

Atomic emission spectroscopy with inductive coupled exitation (ICP-AES), although quite costly, is important for multielement determination with high sample rate. Neutron activation analysis (NAA) is a powerful detection method but costly in terms of both financial and work expenditures. X-ray fluorescence (XRF) methods are perfect multielement methods with high sampling rate. ICP-MS is also applied.

I. Polarographic, Voltammetric and Hybrid Techniques (HPLC-EC) Applied in Modern Environmental Analysis

Direct Methods

Electrochemical methods proposed in the past years in environmental analysis are shown in Table 2.

Electroanalytical techniques for identification and determination of pollutants in matrices of environmental significance, which have central importance are:
- *i)* pulse and square wave polarography and voltammetry
  (developed by Barker in the 50's[12])
- *ii)* inverse or stripping methods using different voltammetric modes
  (differential pulse, square wave and linear sweep)
- *iii)* on-line electrochemical detection after HPLC separation (HPLC-EC) which combines high separation power with the inherent high sensitivity of electrochemical techniques;

high separation power with the inherent high sensitivity of electrochemical techniques;
and in combination of electrochemical detection with flow injection analysis (FIA).
Probably as many laboratories now use HPLC-EC (few having formal training in electrochemistry) as use straight electrochemistry.[13]

iv) polarographic adsorption and tensammetric techniques. (cf.[91,92])

Modern polarographic and voltammetric methods possess a scope of applicability beyond almost all other modern instrumental methods in quantitatively determining traces of
inorganic, organometallic and organic compounds in the trace and ultra trace region.

Table 2. Electrochemical methods proposed in
environmental analysis. The methods most
frequently used are underlined.

Dc polarography
Ac polarography
Normal pulse polarography/voltammetry

Differential pulse polarography/voltammetry
Reverse pulse polarography
Linear sweep voltammetry

Square wave polarography/voltammetry
Electrocapillary curve methods
Polarographic adsorption analysis

Ac-tensammetry
Square wave tensammetry
Pulse tensammetry
Kalousek commutator technique
Derivative chronopotentiometry

Anodic/ cathodic stripping voltammetry
Adsorptive stripping voltammetry
Potentiometric stripping techniques
Amperometric titration

Combination HPLC (FIA) and electrometric detection

Thus modern electrochemical methods are applicable to all "toxic" elements, metalloids,
(anions and cations), and also to a very large number of organic compounds encountered not
only in environmental monitoring but in many other fields. Meites and Zuman have listed the
polarographic and voltammetric behaviour of a great number of substances.[14] A multitude of
the EPA's "priority" toxic pollutants[15] as well as inorganic, organometallic and organic pollutants specified in the Common Market Countries "Priority List"[16] along with compounds in
current use listed as water hazardous substances by West German Environmental Protection
Agency,[17] are polarographically/voltammetrically active. The reader is also referred to the
publications of the "BUA" (Beratergremium für umweltrelevante Altstoffe, GDCh, F.R.G.),
(cf.[146])

It is noteworthy that over 60 to 70 % of the substances named in the "Fourth Annual
Report on Carcinogens" are polarographically active.[18] Classes of compounds of environ-
mental interest that can be determined by polarographic/voltammetric, tensammetric and
HPLC-EC methodology are listed in Table 3.

Table 3. Classes of molecules (examples of organic pollutants) amenable to polarographic/voltammetric and HPLC-EC determination.

Class	Example	Applications	References
halogenated aromatics	X-R	TCB in waste/natural waters bromoxynil, ioxynil	19 20
nitro	$R-NO_2$	nitro-containing agrochemicals in air, water nitrosubstituted polynuclear aromatic hydrocarbons (NO_2-PAH) in diesel soot/exhaust munitions in waters drugs in foodstuff, feed	21 – 23 24, 25 26 – 30 31 – 33
nitroso	R-NO RN-NO	nitrosamines in waters	10, 34, 35
hydroxylamines	R-NHOH		
azo	-N=N-	dyes in waters/effluents	36
amines	$R-NH_2$	benzidine in air, soil, waters	37, 38 – 40
carbamates phenylurea S-containing compounds	R-SH	agrochemicals in waters herbicides CS_2 in waters mercaptans in water butanthiol, cyclohexanthiols in air	41 – 43,75 44 45 46
heterocyclic compounds		cynuric chloride in air triazine agrochemicals/ paraquat, diquat in waters	47 48 – 51 10, 52, 53
hydocarbons	$CH_2=CHCN$ $RC_6H_4CH=CH_2$ $CH_2=CHCl$	monomers /polymers in waters in air	54, 55
hydroxyl	phenols	phenols in waters pentachlorophenol in air antioxidants in foodstuff	56 – 59 60 61
carbonyl	$>C=O$ -COOH	polymers formaldehyde in air chloracetaldehyde in air lactic acid in waters	 62 – 66 67 68
quinone		9,10 anthraquinone in waters dithianon, food dyes in effluents	69 70 36
carboxyl carboxylate		phthalate esters in wastewaters	71
peroxo, peroxides		in polymer product/effluents	54
halogenated hydocarbons	R-X	hexachlorocyclohexane insecticides/chlorofos 1,2-dibromoethane DDT, heptachlor, aldrin in soils, tobacco, aspargus food	72, 73 74
P-containing compounds	$>P=O$, $>P=S$	organophosporous compounds in waters / soils foodstuffs, crops	 75,76,77,78 79 – 82
As-containing compounds	arsenials	feedstock	83
sequestrants		nitrilotriacetic acid/ EDTA in waters	84 – 86 87 – 89

(continued)

Class	Example	Applications	References
surface active compounds		alkylbenzene sulphonates detergents/surfactants /oils in waters	90 91, 92
organometallic compounds		organolead in air/waters organo tin in waters/food organo mercury in waters/food organo tellur in air	93/94 – 98 99 – 101/102, 103 104 – 107 108
miscellaneous compounds	polysaccharides	in waters	92

Indirect Methods

Many compounds which are neither reduced nor oxidized in the available potential range or for which the signals acquired are not suitable for analytical purposes can be converted into electroactive substances via chemical or electrochemical functionalisation.[109,110]

Nitration, for instance, has been used for the indirect assay of such pollutants in waters.[90,111–116] Benzene can be determined in water after displacing it by passage of nitrogen into the nitration mixture; phenol is determined in an analogous manner after extraction into ether (cf. references in ref. 111).

The determination limit of carbaryl (1-naphthyl-N-methylcarbamate) which is 0.2 mg/l by the direct method can be lowered to 0.05 mg/l after nitration of the extracted samples.[82] Glyphosate (N-phosphonomethylglycine) traces in water[96,115] and in crops and soils are determined after conversion to a nitroso derivative.[96]

Carbaryl

Glyphosate

Parathion (I) and paraoxon (II) can not be determined simultaneously by direct polarography (cf.[78]) Pd(II) catalyses the hydrolysis of parathion (I) but not of paraoxon (II). Thus (I) can be determined by measuring the p-nitrophenol (III) formed after addition of Pd(II), whereas (II) is measured by its reduction peak.[77]

In real samples it is necessary to determine the p-nitrophenol present, by measuring its reduction peak before adding Pd(II).

Some determinations based on catalytic and tensammetric waves as well as on amperometric titrations can be classified as indirect methods (cf.[116])

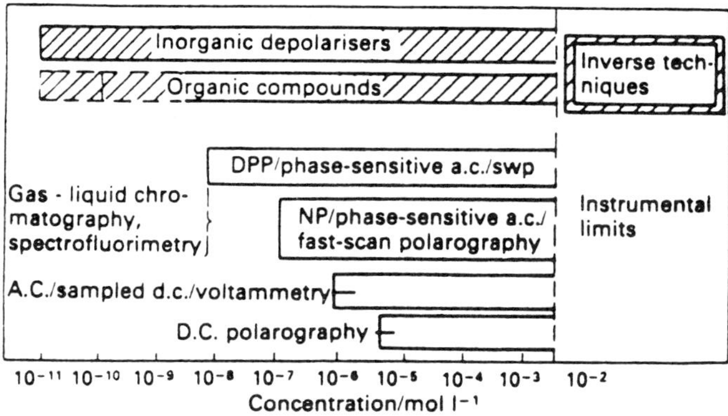

Fig.2. Ranges of practical usefulness of polarographic and voltammetric techniques.[13]

Features of Modern Polarographic and Voltammetric Methods

Some of the main features of modern polarography and voltammetry, such as sensitivity, selectivity, specificity and rapidity, which are of importance for the practical analytical chemist are

- large useful concentration range for reducible and oxidizable inorganic, organometallic and organic species (10^{-3} to 10^{-11} M)
- larger linear range of signal versus depolarizer/analyte concentration than most other instrumental methods; for instance, the concentration limit of Cd in drinking water is 5 mg/l, whereas in soils it is 3 to < 200 mg/kg; for As = 50 mg/l and 20 to < 8000 mg/kg.[117]

A linear range of 0.01 to 50000 mg/l for simultaneous determination of Cu, Pb, Cd, Zn and Ni/Co, respectively has been reported.[118]

- speciation of inorganic species on the basis of complex lability (cf.[119]) Distinction between different oxidations states of the same element: Cr(III)/Cr(VI), As(III)/As(V), Sb(III)/Sb(V), etc.
- multi-element, multi-species determination in a single run. Ex: determination of Pb(II), and the highly toxic methyllead together or Cu, Pb, Cd, Zn, Ni, Co, together with auto-pH change and "metal specific reagent" addition. (i.e. dimethyl glyoxime (DMG) for Ni/Co).
- application to small sample volumes due to the high sensitivity of modern polarographic/voltammetric techniques. The determination of Cu, Pb, Cd, Zn in less than 1 mg of suspended particulate matter obtained from a 50 ml pond water sample after separation by centrifugation and wet digestion with $HClO_4/4HNO_3/HF(2:10:10)$ is feasible.[120]
- general matrix effect immunity to samples with high (an)ionic content; e.g. heavy metals directly determined in seawater without the removal of the NaCl.
- non-destructive technique allows the concentration calculation either by calibration curve for more-or-less fixed sample matrices or quantification of results by standard addition methods in situations were there is the possibility of some matrix variation or where the linearity of the calibration has to be qualified for each depolariser in the sample matrix.
- possibility for full on-line automation of the technique for process monitoring and control, including a certain degree of sample pre-treatment, such as dilution, pipetting or standard reagent addition.
- cost of equipment is not prohibitively high. Polarography ranks among cheap methods. Systems range from simple easy-to-use analogue instruments with potentiometric recorder output to fully automatic digital systems, capable of anything up to multi-element determinations and total on-line operations, all software controlled, and user-defined.

In the extreme region of 10^{-9} M and less with concentrations in the ppt range the sensitivity of direct polarographic and voltammetric methods (Fig.2) and other modern instrumental methods are not, in general, sufficient (cf.[13])

Thus, concentration methods play an important role in environmental chemistry, also in electrochemical trace techniques.

Carbon disulfide, for example, can be determined by dpp in water, in concentrations as low as 1 μg CS_2/l, using a purging step to preconcentrate the CS_2 in ethanol/diethylamine with the subsequent formation of diethyldithiocarbamate. The polarographic procedure is substantially more sensitive than other reported methods.[45]

Current "wet chemical" preconcentration methods include: extraction, absorption in wash solutions or concentration on filters, chromatographic concentrations on plates and columns, etc.

An important aspect of voltammetric methods is the ease with which it is possible to preconcentrate, *in situ*, certain dissolved species from the solution onto or in the working electrode at a defined potential. These so-called inverse- or stripping techniques* have become of great importance in environmental analysis for the determination of very low levels of inorganic, organometallic and organic pollutants as low as 10^{-10} M and less. Gustavsson [121] has described a differential pulse anodic stripping voltammetric procedure employing a commercial Teflon-embedded gold disc electrode and a self manufactured epoxy resin cast gold disc electrode for the determination of mercury concentration in the region of 10^{-11} M in sea water. The mercury concentrations found in nonanthropogenic sea waters are extremely low (about 10^{-11} M (2 ng/l)) but could even then be harmful to marine organisms.[121] Recent developments in ultra-trace stripping analysis utilize an adsorptive approach.[122,123]

Adsorptive stripping voltammetry at mercury or solid electrodes provides very sensitive determinations with detection limits in the 10^{-8} to 10^{-11} M range for surface active compounds or their (metal)conjugates that cannot normally be accumulated electrochemically.[122,123]

A practical example of this is the determination of pesticides (DNOK, dinobutone, prometryne, ametryne)[23] and of TCB[19] in waters (Table 4).

Table 4. Limits of detection of different organic pollutants. Comparison of direct dpp and AdSV (ppb).

Compound	dpp	AdSV		References
DNOK	1.5	0.1	4.9×10^{-10} M	22,23
Dinobutone	16.5	0.6	1.9×10^{-9} M	22,23
Ametryne	7.1	0.18	7.9×10^{-10} M	22,23
Prometryne	-	0.95	3.9×10^{-9} M	22,23
TCB	-	4	1.6×10^{-8} M	19

AdSV = Adsorptive Stripping Voltammetry

In view of the determination limits of 10^{-10} M and less presented here and in the literature the questions immediately facing the practical industrial analytical chemist in day-to-day work relate to

1) the practical workability and 2) the relevance of such ultra trace methods in comparison to the given government requirements, for instance. They, however certainly are of relevance for the study of the ecotoxical situation.

Regarding the workability: the common constituents in ambient air are aerosols composed mainly of solid and liquid particulate matter which can cause serious contamination problems at trace and ultra trace levels. The values in Table 5 demonstrate the significance of "clean room" conditions for precise and accurate determinations at these levels.[124] Thus, for prevention of foreign contamination through Pb, Cu, Zn the "clean room" technology is a must.[124,126]

---

* The term "inverse" reflects the essence of the method more exactly than does "stripping", since the analytical signal recorded is determined not by the solution composition, but by that of the electrode.[125]

Table 5. Influence of laboratory environment on the determination of some trace elements in seawater samples.[124]

elements	amount of element, $\mu g/l$*	
	clean lab acidified samples pH 2.7	conventional lab acidified samples pH 2.7
Cd	0.05 ± 0.01	0.06 ± 0.02
Cu	0.81 ± 0.04	7.22 ± 1.95
Pb	0.63 ± 0.03	1.72 ± 0.82
Zn	1.71 ± 0.05	4.07 ± 2.59

* Water samples were kept in the clean room but were exposed to the different laboratories during acidification and determination steps.

Table 6. Recommended / lawful maximum values (in $\mu g/l$) for metal contaminants in drinking water in comparison to the detection limits of inverse voltammetry.

Element	Voltammetric Detection Limit	Switzerland* 1980	W.Germany**	U.S.A.* 1975
As(III)	0.1	50	40	50
Ba(II)	0.04	-	-	1000
Pb(II)	0.001	50	40	50
Cd(II)	0.0002	5	5	10
Cr_{tot}	0.02	20	50	50
Cu(II)	0.002	1500	-	-
Hg(II)	0.002	3	1	2
Se(IV)	0.1	-	-	10
Ag(I)	1.0	200	-	50
Zn(II)	0.02	1500	-	-

* Merian, Metalle in der Umwelt,[117] ** Trinkwasserverordnung, 22.05.1986

The detection limits for "toxic" metals lie in general an order of magnitude or more, lower than even the limiting value for drinking water (Table 6). On the other hand, practice shows that the determination limit[13] in actual samples lies easily a factor of 5 to 10 and more, higher than the detection limits (Table 7). This observation is not unique and is also valid for other instrumental methods.

Table 7. Comparison of detection limits (c_d) and determination limits (c_D) using AdSV.

Compound	literature values		own measurements	
	detection $\mu g/l$	determination $\mu g/l$*	detection $\mu g/l$	determination $\mu g/l$**
Ametryne	0.18	2.3	2.3	36

* 15 ml of supporting electrolyte solution was mixed with 5 ml of relatively polluted river water (Vltava river, near Charles Bridge, Prague [23])
** Rhine river water (Cf. Fig.3A and 3B)

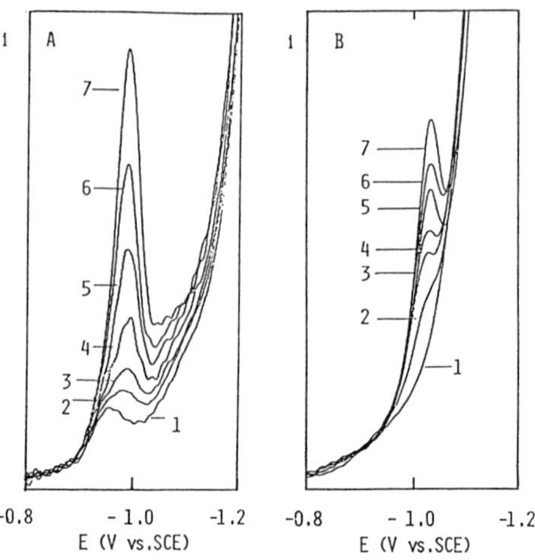

Fig.3. AdSV curves of ametryne in (A) bidest
water/ Britton-Robinson-buffer, pH 3.75, (B) in Rhine water/Britton-
Robinson-buffer, pH 3.75. t_{acc} = 30 sec, E_{acc} = −0.7 V (vs. SCE).[155]

The voltammograms of ametryne in model solutions (Fig.3A), and measured in Rhine water, Fig.3B, as well as the relationship of peak hight (i_p) vs. analyte concentration is shown in Figs 4A and 4B.

Drawbacks (disadvantages) of polarographic and voltammetric methods compared with other methods are:

1) high quality requirement of the digestion processes when total metal contents are being determined. For natural waters UV digestion with sulphuric acid/hydrogen peroxide or microwave digestion are often sufficient. UV takes a longer time, but needs less sulphuric acid (typically two hours with only 100 ml sulphuric acid per 20 ml sample); microwave digestion uses more acid, but digestion times are said to be ca. 5 to 10 min.

2) low selectivity and specificity for organic species, in comparison with chromatographic methods. High selectivity can be achieved by the combination of separation techniques, primarily HPLC with high-sensitivity potentiostatic detection (HPLC-EC).

3) supply of commercially available high-performance, fully automated methods is still limited.

Sampling, sample storage, sample treatment, treatment of vessels, etc. which are regarded as equal unit process contributors in terms of importance in all techniques, are not discussed here in any depth.

Thus modern polarography is a powerful technique that offers sub-part per million sensitivity with low cost instrumentation. It enables a very wide range of elements and compounds to be analyzed; to distinguish between the different oxidation states of a species, the technique is unexelled, it also provides a powerful tool for routine determinations and for speciation studies. Interferences and contaminations are usually obvious, but generally of a different nature to those encountered in AAS for example.

Yet, are polarography and voltammetry used in practical day-to-day environmental analysis?

In the book by Keith: "Identification and Analysis of Organic Pollutants in Air"[127] out of the 473 pages only a short chapter of 13 pages, reporting a preliminary study of the potential of electrochemical methods in the analysis of the mixtures of polycyclic aromatic compounds (PAC), is devoted to polarography and voltammetry.

Crompton[128] in his new book (1987): "Comprehensive Organometallic Analysis" dedicates one chapter (Chapter 9) to the analysis of organometallic compounds in the environment with over 150 pages and 710 references. Only few polarographic papers are quoted.

Horwitz[129] remarked: "Polarography is not a popular technique in the general laboratory .... although it is becoming more prevalent. It has its strong advocates but the majority

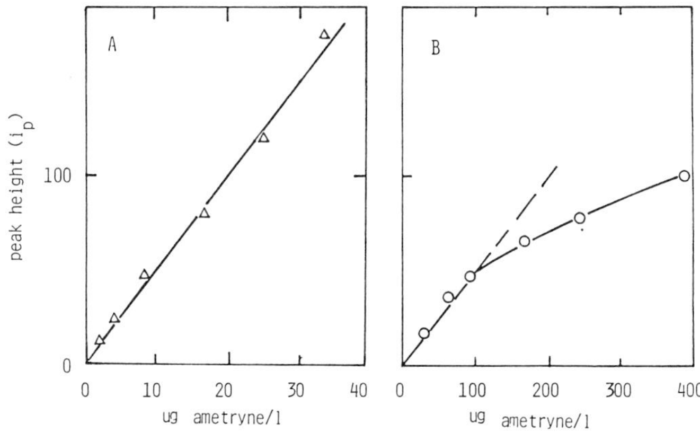

Fig.4. AdSV peak height vs. concentration of ametryne, measured (A) in bidest. water, (B) in Rhine water.[155]

of analysts have not taken the trouble to become informed about the simplicity and advantages of this technique". In a recent meeting on environmental analysis in F.D.R., no mention was made of electrochemistry.[129a] In the survey of "The past, present, and future of trace analytical chemistry examined at a recent NBS symposium"[130] polarography or voltammetry are not mentioned.

In other words people are not aware of modern polarography and voltammetry. High sensitivity, simplicity, broad application, low cost are qualities of particular value in environmental analysis, which envolves mass applications of instrumental techniques and large numbers of monitoring stations and a widespread network.

One of the limiting factors with regard to the application of polarography and associated techniques is probably due to chemist's educational experiences, which results in the association of voltammetry with complex equations and bulky equipment and has also been connected by some authors[13] with the routine use of mercury and mercury electrodes. Mercury is unjustly marked as one of the most insidious hazards within the laboratory, although there does not seem to be a corresponding general appreciation of its toxicity at least not in textbooks or in the literature of some equipment manufacturers.

At room temperature (25°C) the equilibrium concentration of mercury vapour is about 20 mg/m^3 or 200 times the threshold limit value of 0.1 mg/m^3 recommended as a maximum atmospheric concentration for normal work schedule by the American Conf. of Governmental Hygienists. For information on the toxicity of mercury and safe handling of mercury the reader is referred to the manual of the 646 VA processor by Metrohm, Chapter 6.4, and references therein.

Mercury, however, is only as toxic as the user permits. The actual risks for those exposed can be minimised: when working in a well ventilated area, preferably the polarographic setups should be placed in a fume hood; by use of catch trays beneath the polarographic cells, and by a minimum of care and common sense excercised in handling mercury. The mercury content of urine samples of people working in the polarographic laboratory in Basle compares with values of individuals not exposed to mercury. Nevertheless, periodic medical check-ups are recommended for those potentially exposed.

II. Practical Application of Polarographic/Voltammetric, Tensammetric Techniques and Hybrid Methods (HPLC-EC)

Few modern books mainly from Eastern countries,[10,131−135] exist, but several recent review articles discuss the application of electrochemical methods in environmental analysis. [3,41,136−141]

Practical application of polarography in an environmental situation was first introduced in 1933. Prochazka[10,142] developed a method to monitor the purity of water after purification by coagulation, based on the suppression of the oxygen maximum by colloid. An automatic instrument, the "Coagulograph" operated on this principle and was used to monitor the Prague water supply before World War II.

Determination of Inorganic and Organometallic Pollutants

Metals and metalloids are characterized by special ecochemical features. They are not biodegradable, but undergo a biochemical cycle during which transformations into more or less toxic species occur. They are accumulated by organisms and cause increased toxic effects in man and mammals after long-term exposure. The most important source of uptake is food and the significant pathways of toxic metals into food chains is through the atmosphere from which the metals are introduced to ecosystems by dry and wet deposition. Wet deposition through rain and snow is of particular importance because it provides favorable conditions for uptake by vegetation and water.[143]

The polarographic and voltammetric trace and ultra trace analysis of

Ag, As, Au, Br, Bi, Cd, Cl, Co, Cr, Cu, Eu, Fe, Ga, Hg, I, In, Mn, Mo, Ni, Pb, Pd, Pt, Sb, Se, Sn, Tl, U, V, W, Yb, Zn etc., and of anions (NO_2^-, NO_3^-, CN^-, halogens, SO_x^{z-} etc.)

in the significant environmental areas: air, precipitation, aquatic media, soils, food and human, has received considerable attention, especially from Nürnberg, Brainina, Bond, Florence, Kalvoda, van den Berg, Valenta, Wang et al. The reader is referred to a recent review paper on inverse techniques applied in water analysis by Brainina et al.[144] with 186 references. All elements of environmental significance are amenable to polarographic or voltammetric determination. Metals which are of particular relevance to fresh waters are Cd, Cr, Cu, Pb, Hg, Ni, Zn.[145]

The reader is referred to the original literature. Numerous review articles provide comprehensive discussions. In this context see also references,[146] discussing the most important legislation guidelines relating to the quality of drinking water, natural waters, wastewater and sewage sludge.

The polarography and voltammetry of organometallic compounds up until the early 80' has been reviewed by Crompton.[128]

From an inquiry made by the author in 1986[13] on the application of polarography and voltammetry in official and recommended methods it can be deduced that the main effort at present is made in Europe. Examples of applications are listed in Table 8.

Table 8. Polarographic – voltammetric determinations of inorganic species in official and recommended methods (cf.[13])

Ion	Fields of application	Standard, country
Heavy metals	Water	DIN, F.R.G.
	Foods*	LMBG
Sn	Wastewater, food*	DEV
Pb, Cu, Zn, Cd, Tl, Ni, Co	Water (surface, ground, dinking, snow, rain)*	DIN
Pb, Cu, Cd	Surface water*	79/869 EWG, EWG
Pb, Zn, Cd, Se	Water[&]	Spain
Toxic metals	Natural waters*	U.S.S.R.
O_2	Water, blood	Italy
Heavy metals	Different waters, blood, urine, air[&]	Hungary
Cu Pb, Cd, Zn, Ni	Food[&]	

* Methods in preparation, [&] Existing methods

In Switzerland, polarography has been adopted as an official analytical method for monitoring water quality. These analyses are applied to wastewater, river and portable water, and also for general quality monitoring of foodstuffs.[147,148] Stripping voltammetry and a number of other methods have been accepted by the Australian Environmental Protection Authority for regulatory testing by industry, also using EPA data. In the U.S.A. progress in standard methods for examination of waters and wastewater is said to have been stopped.[13]

In view of the large differencies in toxicity even within the same substance classes a rapid identification of inorganic and organometallic pollutants is of critical importance: tetraoctyl tin exhibits an LC50 of 50000 mg/kg; in comparison the LC50 of triethyl tin is 4 mg/kg; Sn (II) is not toxic; Hg, Cd, Tl, Ni ions etc., are highly toxic.

Each different physico-chemical form of an element has a different toxicity, so analysis of a water sample for the total metal concentration alone, e.g. Sn, does not provide sufficient information to predict toxicity. If the main species of a river water containing 40 μg/l dissolved copper is ionic Cu(II) few organism will survive, while the same amount Cu adsorbed on colloidal particles will have no or only little effect on aquatic life.[119] The toxicity of many substances is thus modified by water quality. For example, heavy metals are usually more toxic to fish in soft water than in hard water. The presence of suspended organic matter, can affect the observed toxicity when pollutants are adsorbed or form complexes. This has been noted for pesticides and for heavy metals.

The comparison of the value of Cu traces found in a wastewater by direct polarographic assay and by atomic adsorption after acidic digestion (Table 9) revealed that the Cu is present mostly in the ionic form or as a very weak complex (complex constant equal or less than $10^{5.7}$) and not (as expected) as a strongly complexed and therefore harmless species.[155]

Table 9.   Determination of Cu in wastewater samples by polarography and atomic absorption. Polarographic determination: anodic stripping voltammetry = determination of ionic and weakly complexed Cu species. Atomic absorption = determination of Cu_{tot}.

Sample	Anodic stripping voltammetry. mg/l	Atomic adsorption mg/l
38.12	0.07	0.1
38.13	3.2	4.0
38.15	1.2	2.4
38.08	6.5	7.8
38.11	15	14
38.12	19	20
38.13	23	24
38.14	33	33

Lipid-soluble metal complexes such as copper xanthates (from mineral flotation plants), copper 8-hydroxyquinolinate (agricultural fungicide) or alkyl-mercury compounds are particularly toxic forms of heavy metals because they diffuse rapidly through a biomembrane and carry both metal and ligand into the cell.[119]

The degree of toxicity of a pollutant thus depends on the quantity and also, to a considerable extent, on the form in which it is present. The measurement of the total concentration of a trace element provides little or no information about its bioavailability or its interactions with sediments and suspended particles. Questions central to inorganic environmental analysis were described as follows by Taylor (1979):[149] "One of the major errors in attempts to relate data obtained from laboratory experiments to the situation in the field is the failure to take into account the chemical form in which the material is present. The questions which must be asked in all cases are: What proportion of the material present in the test solution is harmful? Will this biologically active material exist in the natural environment? If so, for how long?"

The analytical chemist is thus increasingly being called to determine not only amounts of elements but also their chemical form. According to Florence:[150] "It is inevitable that in the near future, water-quality legislation for heavy metals will include statements relating to their speciation. It may be possible to tolerate higher concentrations of some metals as long as the labile fraction is below a certain limit. In the same way, future may require the bioavailable fraction of trace elements and vitamins be measured as a part of the routine analysis of foods".

The physico-chemical characterisation of the species in the case of, for example, the determination of the form in which it is present, represents a substantially more demanding field of trace analysis than total trace contents. According to Florence:[119] "speciation analysis of an element in a water sample may be defined as the determination of the concentrations of the different physico-chemical forms of the element which together make up its total concentration in the sample".

Electrochemical methods have some distinct advantages for speciation measurements.[97,119,150-152] Electrospeciation has been applied to (or is potentially applicable to) about 30 elements.[119] Electrochemical methods can be highly sensitive, which is essential when analyzing natural waters, where the total concentration is often in the mg/l or ng/l range.

Operative conditions can be adjusted so that only metals with rate of dissociation of their complexes, within a desired range, are included in the electroactive fraction. Conditions that can be adjusted to achieve selectivity are deposition potential, electrode rotation rate, solution stirring, pulse frequency, potential scan rate, temperature, pH, etc. As electrochemical techniques require much less sample handling than other speciation methods, such as solvent extraction, dialysis or ultrafiltration,[119] the potential sources of contamination are highly reduced. An in depth discussion of the pro and cons of electrochemical speciation is far beyond this article. Theoretical aspects and applications have been covered in great detail by Nürnberg, Florence et al., cf.[119] [150,151] and references therein.

In addition to achieving speciation on the basis of complex lability, polarography can be used to distinguish between different valency states of the same element in solution,e.g. Fe(III)/Fe(II), Cr(VI)-Cr(III)-Cr(II), Tl(II)/Tl(I), Sb(V)/Sb(III), Se(VI)/Se(IV), As(V)/As(III), V(V)/V(IV), Eu(III)/Eu(II), U(VI)/U(IV), Sn(IV)/Sn(II) and Ce(IV)/Ce(III).

Change in the oxidation state of an element can have a profound effect on bioavailability and toxicity.[150] Fe(II) is biologically the main usable form of iron, Fe(III) in excess can cause stomach lesions.[154] When chromium appears as a pollutant in a sample of river water or in a Cr complex dye, its valency is of relevance in ascertaining the relative toxicity of the sample. Trivalent chromium is an essential element in mammalian systems, hexavalent chromium is considered to be a moderate to severe industrial hazard.[13]

The author[155] has developed a method for the determination of traces of Cr(VI) in Cr(III)-complex dyes, down to 0.1 ppm Cr(VI) based on differential pulse polarographic determination of the Cr(VI) after preconcentration of the Cr(VI) traces from the aqueous Cr-complex dyestuff solution by anion-exchange resin (AG1-X4);[155] the adsorption and elution are made in opposite direction, thus minimising elution volume, contact time (oxidation of resin), and band spreading (cf.).[156] The setup used is shown in the Fig.5.

An electrochemical process for recycling Cr(VI) traces in mother liquors and wastewater based on accumulating Cr(VI) traces on an anion-exchanger and reducing the eluted Cr(VI) to Cr(III) electrochemically was evaluated in the authors laboratory.[155]

Electrochemical recycling of Sn(IV/II), Ce(IV/III) and Cu(II) was also studied.[155]

The disposal of metals, e.g., through deposition after chemical or physical treatment, is senseless, since this only results in a local change in the environmental problem. Thus, the only long-term solution is the recycling (recovery) of metals, e.g., through electrochemistry, chromatographic processing, etc., and subsequent feedback into the process.

A large petroleum refinery produces up to 1000 tons a day of sulphur, corresponding to nearly 1100 tons a day of highly toxic H_2S. The total emitted H_2S to the environment must be less than 100 kg a day necessitating highly effective desulphurisation processes of the emitted gases. The authors have extensively used polarography for the assay of vanadium(V) and quinone additives in liquors of the Stretford process used for the desulphurisation of gases,[155] see also.[157] The left side of the reaction sequence in Fig. 6 stands for the oxidation

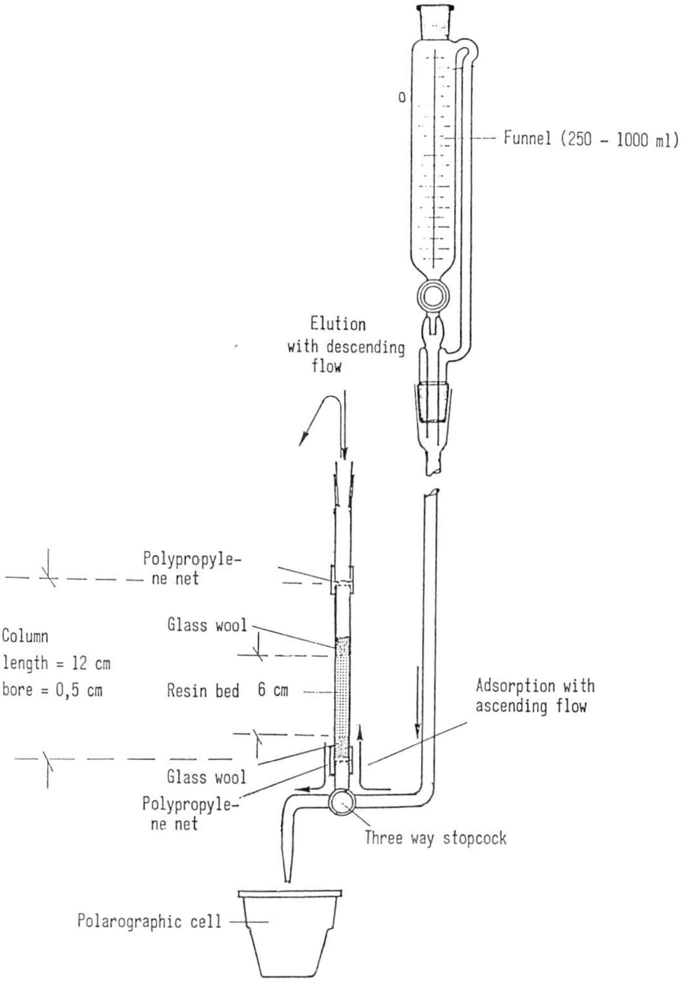

Fig.5. Determination of traces of Cr(VI) in Cr(III) complex dyes. Diagram of the preconcentration set-up. Adsorption and elution in opposite flow.[155]

of the sulphide to sulphur by vanadium(V) in the adsorption and reaction step, the right side shows the oxidation step of vanadium(IV) in the presence of a quinoid additive.

The OSHA permissible exposure limit of H_2S is 20 ppb, the ACGIH threshold limit value is 10 ppm. No exposure and limit values for sodium vanadate(V) ($NaVO_3$) and anthraquinone disulphonic acid has been established.[157] (ACGIH = American Conference of Governmental Industrial Hygienists). Polarography proved to be very useful for monitoring the vanadium and additive concentration and for testing the efficacy of the different anthraquinone additives proposed as oxygen carrier in the reoxidation step of V(IV) to V(V), and for the study of the stability of V(V) and V(IV) ions as the function of the pH of Stretford liquor and as the function of time. Fig.7 illustrates the behaviour of a quinone additive and vanadium (V/IV) in the Stretford liquor: 1) before addition of HS^-, 2) after addition of HS^-, 3) after purging with oxygen for 15 min (curve 3) an increased vanadium wave is observed.

Ni concentrations measured in air, grass and water by adsorptive stripping voltammetry (AdSV), by AAS and X-ray fluorescence (XRF) are compared in Table 10.[158]

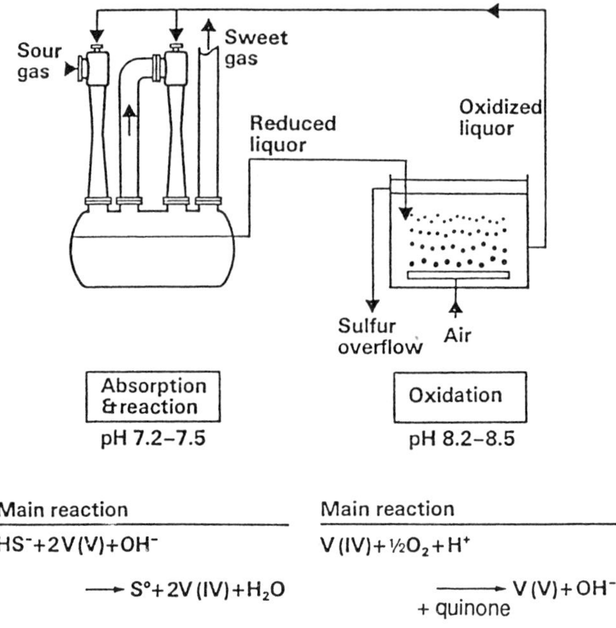

Fig.6. Desulphurisation process.

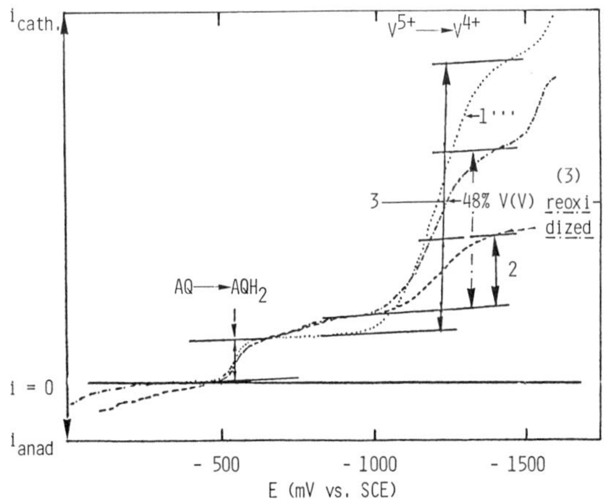

Fig.7. Sampled dc polarograms of Stretford-liquor containing an quinoid additive[155] (1) before adding HS^- (2) after adding HS^- (0.66 g/l) (3) after purging with oxygen for 15 min.

Organic Field

Perusal of the literature of the last 10 years shows that polarography and voltammetry have been used in the organic field, particularly in the pharmaceutical and biological field;[13] for this period, however, more than 250 papers were published, discussing the application of these techniques and of HPLC-EC to the determination of metallorganic (few) and organic pollutants (nearly 400 compounds) in areas of environmental relevance (aquatic bodies, soils, air, foodstuffs and biological matrices).

Classes of molecules of potential organic pollutants amenable to polarographic, voltammetric and HPLC-EC determination are listed in Table 3.

There are only a few existing methods, or methods in preparation, in official and rec-

Table 10. Determination of Ni(II) in air, grass and water by adsorptive stripping voltammetry (AdSV), by AAS and X-ray fluorescence (XRF). Comparison of methods.[158]

Sample	Ni-concentrations		
	AdSV μg/g	μg/g	method
Air-dust I	1708	1463 ± 577	AAS
Air-dust II	62	-	
Flyash I	585	615	AAS
Flyash II	51	50	XRF
Grass	2	2	AAS
Rain	4.6 μg/l	-	-
	4.7 μg/l*	-	-
Air	0.2–1.2 μg/l	-	-

* measured in an independent lab.

ommended directives. Polarography has been proposed as an official method for German standard methods (DIN) for the determination of nitrilotriacetic acid (NTA) in water, wastewater and sludge.[159] NTA appears to be (in detergents for laundry purposes) the most suitable substitute for sodium tripolyphosphate to prevent the eutrophication of natural waters,[84] although the justification for its use has been questioned.[160] Suspicion in the U.S.A. that NTA causes mutagenic and teratogenic effects (cf. also)[161] in experimental mice and rats resulted in its prohibition as a laundry detergent. The introduction of complexing agents, such as NTA, may affect the distribution of metals among various parts of aquatic ecosystems.[84,154] The complexing agent remobilises heavy metals from the suspended matter and from recent sediments and enhances the dissolved level of metals from which the uptake of (toxic) metals by organisms mostly occurs.[84] The mobilisation of metal will increase their transport down the river into the estuaries and coastal waters. Thus, enhanced NTA levels pose problems. It has been estimated that NTA levels in the Rhine in Holland may vary from 30 to 600 μg/l.[84,154,161a] Fig.8 shows the voltammogram of Rhine water (Basle) spiked with Bi(III), 380 ppb NTA and 780 ppb EDTA using the indirect Bi(III) differential pulse voltammetric procedure developed by Voulgaropoulos et al.[84] Limits of detection in natural waters of approximately 0.2 μg/l for NTA and 0.1 μg/l for EDTA, using a preconcentration time of 2 min at the h.m.d.e., were reported.

There has also been extensive discussion on the influence of EDTA on the equilibrium of natural systems and biological and medicinal consequences of chelation of metals in the body (cf.[89] and references therein).

The contamination of the aqueous environment from degradation and metabolism of organisms and man-made organic pollutants, such as detergents, mineral oils, and organic compounds is widespread. According to Grob and Grob[9] who studied the organic pollution of waters in the area of Zürich (which has no important chemical industry) the most widespread material among industrial pollutants found in all kinds of water is automobile fuel. The most important mechanism of contamination seems to be that of evaporation into the atmosphere, from which it is transferred to the water by direct surface contact or, with maximum efficiency by falling rain.[9] Solvents such as trichloro- and tetrachloroethylenes, as well as chlorinated aromatic hydrocarbons, are second in importance. Tetrachloroethylene is often found to occur at the highest concentration in natural water. Diesel oil and related materials containing mainly saturated hydrocarbons, do not play any considerable role in organic pollution in the Zürich area. The heavier water pollutants were plasticicers, such as phthalate esters and diphenylether. Among substances of natural origin are C_{15} and C_{17} alkenes and alkanes which are always present during summer; terpenes of any functionality (hydrocarbons, aldehydes, ketones, alcohols) are present in surface waters, though not in

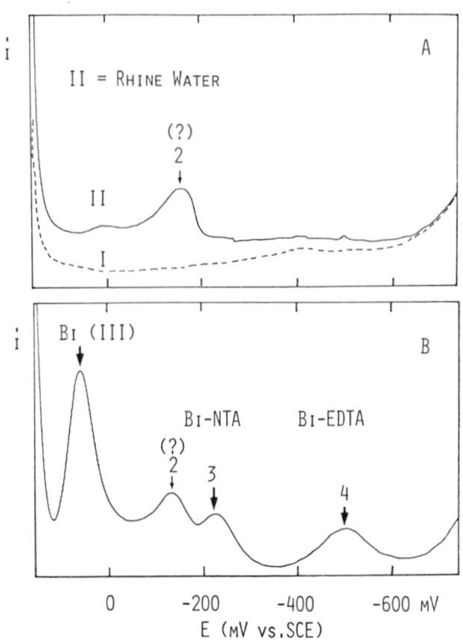

Fig.8. (A) Voltammogram of bidest.water + 2 % ascorbic acid. pH 2 (adjusted with $HClO_4$), curve (I). Curve (II): Rhine water + 2 % ascorbic acid, pH 2 (adjusted with $HClO_4$). (2) unknown impurity.[155] (B) Voltammogram of Bi^{3+} (1), BiNTA (3) and Bi-EDTA (4) measured in Rhine water + 2 % ascorbic acid, pH 2 (adjusted with $HClO_4$). $Bi^{3+} = 7 \times 10^{-6}$M, + 380 ppb NTA + 780 ppb EDTA. Working electrode = hanging mercury drop. Scan = 5 mV/s.

important concentrations; sulphur compounds, indicate anaerobic breakdown of proteins.

Among above mentioned compounds and those listed in Table 1 B, numerous substances are amenable to polarographic, voltammetric or HPLC-EC determination (cf. also Tab.3).

Tanaka and Takeshita[71] have developed a dpp procedure for the determination of the total phthalate esters as phthalic acid over the range of 0.3 to 30 μg/l in crude and treated wastewaters. Detection limits for phthalic acid = 5×10^{-7} M. The polarography of halogenated aliphatic hydrocarcarbons has recently been reexamined by Meyer et al.[162,163] Tokoro et al.[74] studied the electrochemical reduction of 1,2 dibromoethane (EDB) used as fumigant and added as a scavanger to leaded petrol with dc, normal pulse and reverse pulse polarography, voltammetry and coulometry. EDB can be determined directly from the reduction current $E_{1/2} = -1.42$ (vs. SCE) in 0.1 M tetraethylammonium perchlorate or indirectly from the limiting current for the anodic oxidation of mercury in the presence of the reduction product, bromide. The detection limit is ca 1 μM.

Kalvoda[91] has reviewed different electrochemical methods: suppression of polarographic maxima, eletrocapillary measurements, Kalousek commutator technique, differential pulse polarography and tensammetric methods for the determination of oily substances in various types of waters. These compounds are neither reducible nor oxidizable but are surface active. The study of oil fractions has shown that the Kalousek technique responds only to higher levels of dissolved petroleum fractions (0.02 to 100 mg/l) found only in pollution sources such as harbors and refinery effluents.[164,165] Recent results indicate that the water pollution with petroleum can be determined by differential pulse tensammetry.[166] Good calibration graphs were obtained with Diesel oil at a concentration from 30–50 μg to 0.5 mg/l and with Saratov petroleum from 0.1–0.2 to 3.3 mg/l.[166]

Polarographic adsorption techniques and tensammetric methods (for further details the reader should consult two recent review papers[91,92]) are non-specific and mostly do not permit analysis of mixtures of substances, and thus, the results correspond only to the overall content of the surface active pollutants present in the water sample. Often, however, the determination of the surface activity of water is more important than the determination of

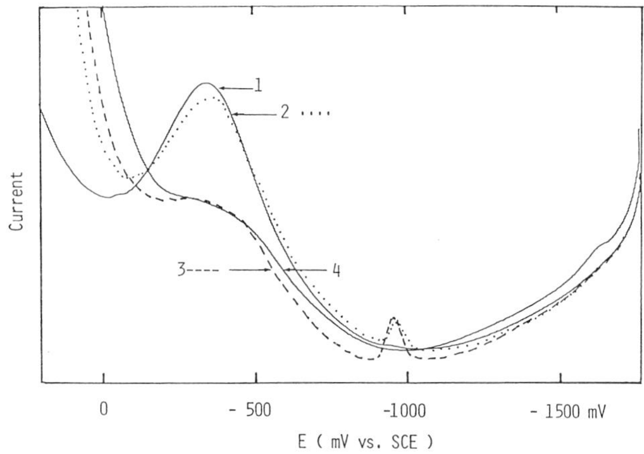

Fig.9. Ac tensammograms of (1) doubly distilled water, (2) tap water, (3) process water, (4) Rhine water (Basle).[155] Applied alternating voltage = 15 mV (r.m.s.) at 75 Hz, 15 ml test water + 2 ml 1 N NaClO$_4$.

concentrations of the individual surface active compounds, because of the toxic effects of surface active compounds on living organisms. Surfactants, due to their hydrophobic nature and lypophilic action, are dangerous to river, lake and marine organisms because they dissolve in the cellular membrane and may destroy it.[10]

Methods based on the maximum suppression are simple, rapid, and nondestructive and sensitive to water soluble surfactants, such as proteins, polysaccharides and glycoproteins. Attempts are being made to increase selectivity of such procedures, e.g., by using a variety of maxima accuring at various potentials.[116]

The surfactant activity is expressed in terms of activity of an arbitrarily chosen surfactant, e.g. Triton X-100, by means of calibration graphs.

Fig.9 shows ac tensammetric curves of double distilled, tap, process and Rhine river (Basle) waters as measured in the authors laboratory as compared to distilled water (curve 1, Fig.9). A significant depression of the current is observed for both the process and Rhine water.

A comprehensive account up to 1970 of the electroanalysis of agrochemicals is given in the book by Nangniot.[76] Birch and Hart reviewed papers until 1979.[41] The recent review by Smyth and Smyth[136] "Electrochemical Analysis of Organic Pollutants" provides numerous modern references in this field.

The polarographic and voltammetric determination of chlorinated compounds,[172] S-triazines, 2,4-AD-herbicides, nitro-containing pesticides, paraquat, diquat, carbamate pesticides, phenylurea herbicides, organophosphorous compounds, in air,[60,167−169] waters,[21−23,41−44,48−52,72,73,76−78,113,168,170,175,176] soils,[171−173] foodstuffs,[20,53,70,79,82,116,174−176] and biological matrices,[177−181] has been demonstrated.

S-triazines, for instance, are reported to have a long persistence and to have contaminated wells and streams.[48,182,183] Traces of procyazine and atrazine have also been found in human urine.[184]

The electrochemistry of triazines is well documented.[116] Polak and Volke[51] described a dpp procedure for the determination of symmetrical and asymmetrical triazines (desmetryne, methoprotryne, prometryne, aziprotryne, terbutryne, atrazine, cyanazine, simazine, terbutylazine, metribuzine) in water. The herbicides are extracted into chloroform from the sample at pH 8 to 9. The detection limit in 0.02 M H$_2$SO$_4$ or acetic acid buffered solution at pH 2.5 is given as 10^{-7} M.

In Fig.10B are shown the curves of Rhine water spiked with 100 to 600 ppb atrazine using direct dpp, as compared to model solutions (Fig.10A). Passing the water samples (500 ml) through a Sep Pak C$_{18}$, however, allows much lower determination limits, of the order of 8 to 20 ppb, even in Rhine water, as shown by the Figs 10C and 10D.

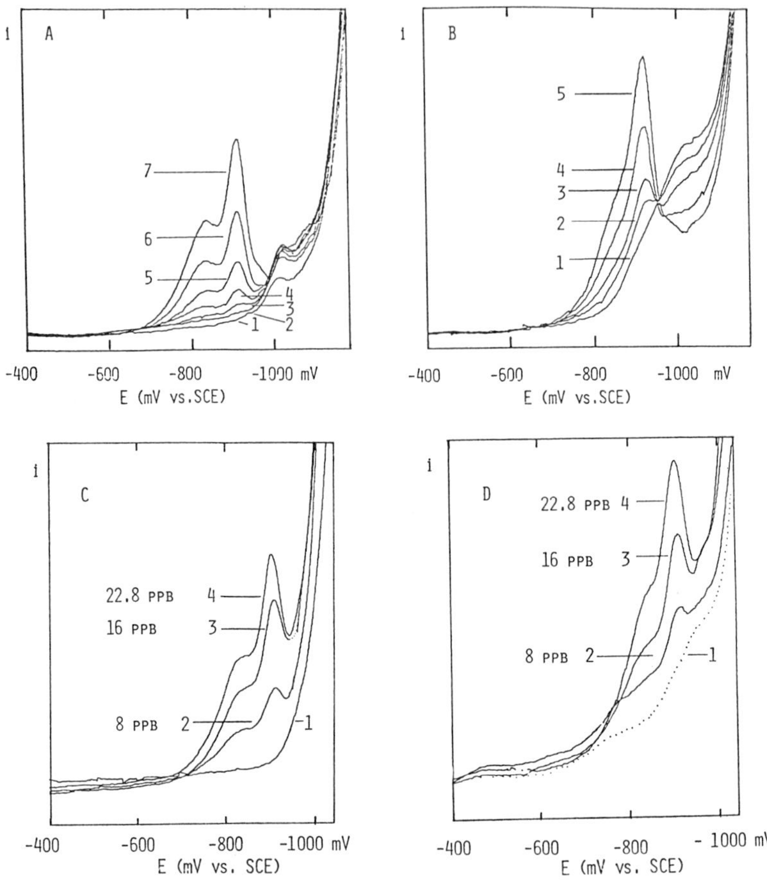

Fig.10. Differential pulse polarographic curves of atrazine measured in doubly distilled water, (B) in Rhine water.

(A) curve (1) sample before addition of atrazine, (2-7) after standard additions of 20, 50, 100, 200, 400 and 600 ppb atrazine to the polarographic solution.

(B) curve (1) Rhine water before addition of atrazine, (2-5) after addition of 100, 200, 400 and 600 ppb atrazine to the polarographic solution. Polarographic solution = 12 ml water sample + 0.12 ml 1 M HCl/KCl.[155]

Differential pulse polarograms of (C) doubly distilled water, (D) of Rhine water after pretreatment of the water samples with Sep Pak C18. (1) sample, (2) (3) (4) sample + standard additions of 8, 16 and 22.8 ppb atrazine to the polarographic solution. Polarographic solution = 12 ml doubly dist.water + 0.12 ml 1 M KCl/HCl.[155]

Kalvoda[185] and Benidakova and Kalvoda[23] have recently reported detection limits of 1 ppb for prometryne and 0.18 ppb for ametryne, repectively, using the adsorptive stripping technique; this approach is sensitive but must be combined with a separation step for real applications.[48] The detection limits found by these authors for ametryne were about 10 times higher (cf. Table 7). A swept-potential electrochemical detector, operating in the square wave voltammetric mode is used to detect mixtures of simazine, atrazine, cyanazine, propazine and anilazine after separation on a reverse-phase resin column. The cell used was a jet cell with a mercury drop working electrode, using acetonitril/water 60:40 (V/V) (as the mobile phase) with a final apparent pH of 1.02. The limits of detection are below 1 ng (injected).[48]

Organochlorines have been largely superseded by organophosphorous insecticides, such as azinphos methyl, fenitrothion, malathion and parathion, or, more recently still, by the carbamate insecticides, aldicarb, carbaryl and carbofuran,[145] as both groups have shorter environmental persistence and lower propensity for bioaccumulation. They do, however, have anticolinesterase activity and are potentially dangerous to users, although this is minimal if proper precautions are taken.[145] Polarographic activity is only conferred upon phosphoric esters containing a reducible functional group, most often nitro, but also double bonds; perhaps most applications of polarography for the determination of nitro-containing agrochemicals have come from nitrophenylesters, e.g. parathion, fenitrothion, etc., (cf. Rowe and Smyth).[75] Scope and limitations of the polarographic assay in this group are well documented.[76,186]

Photolytic derivatisation in LC-EC (HPLC-hν-EC) has been proposed for the trace assay of organophosphorous agricultural chemicals.[81] The detection limits in the ppm range are said to be sufficiently low that the method can be applied to crop extracts at levels routinely encountered.

Clark et al.[79] have compared UV and reductive amperometric detection of ethyl- and methylparathion in surface water and green vegetables using HPLC. Concentrations, less than 10 ng/ml were readily measured using a column concentration procedure.

Phosphoric acid esters with $>$P=O and $>$P=S but no polarographically reducible or oxidizable group produce well-formed desorption peaks using linear sweep voltammetry (lsv),[76,170] Fig.11B. Lsv voltammograms of the Rhine water prior and after adding methidathion and iodofenphos are shown in Fig.11A.

Cathodic stripping voltammetry was proposed for the assay of dithiodialkylphosphoric acid pesticides on tea leaves after hydrolysis to dithiodialkylphosphates. O,O-diethyldithiophosphate yields at -0.9 V (vs. Ag/AgCl) in NaOH/NaHCO$_3$ a well-formed stripping peak. Linearity between the stripping current and concentration was observed in the range of 10^{-6} to 10^{-9} M.[80]

Decolourisation of Dye Effluents

The extensive literature material on the electrochemical treatment of wastewater to remove inorganic and organic pollutants, e.g. organic dyes and surfactants, is evidence of the efficacy of such procedures (cf.[187-190] and the original literature).

One of the strongest points in favour of electrochemical techniques, is that in the case of metal ions and metal complex loaded wastewaters, the metal ions or complexed ions are reductively removed from solution and extracted in the most desirable form, e.g. as pure metals, as films or as powders.

Electrochemical reduction or oxidation can destroy organic compounds and can frequently render a toxic compound less toxic or harmless or biologically degradeable; on the other hand highly toxic compounds, such as chlorinated aromatics, phenols or NCl$_3$ can be formed.

Electrolytic purification of wastewaters from production of direct or reactive dyes has been the subject of many investigations,([188-191] and references therein).

Dying solutions and wastewaters are usually purified by using conventional methods such as: decolourisation by chemical means, coagulation by lime, aluminium or iron salts, adsorption on activated carbon, and by biological oxidation techniques.

The technique of cathodic reduction/anodic oxidation of wastewaters containing dyes is a relatively new technique which has drawn the attention of investigators in Japan, China, U.S.A. and (especially) in the U.S.S.R. Electroanalysis is thus assigned an important role in elucidation of these processes.

Figs 12A, 12B show the differential pulse polarograms of a dye wastewater (A) before electrolysis at a percolated graphite electrode and (B) after electrolysis. The reduction wave of the dye is greatly reduced, which is also demonstrated by decolourisation of the wastewater and the spectrograms (Fig.12C). Less than 20 ppm ionic or weekly complexed copper was determined.[155]

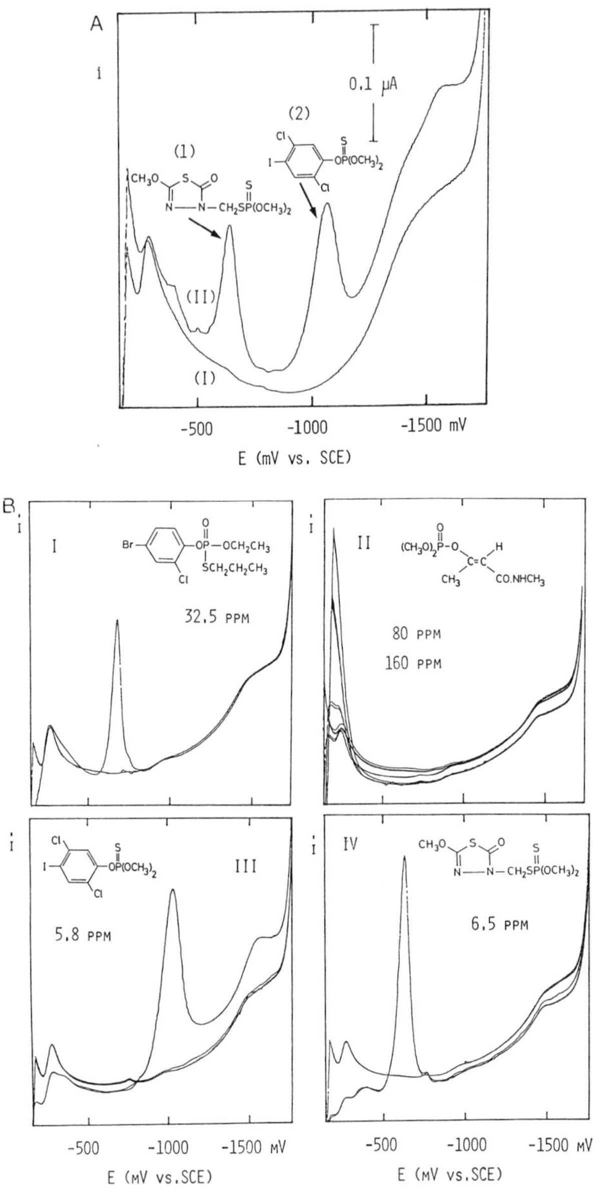

Fig.11 (A) Linear sweep voltammograms of (I) Rhine water/0.2 M NaOH/50 % MeOH, (II) after addition of (1) 2.7 ppm methidathion and (2) 3.4 ppm iodofenphos. Scan = 200 mV s^{-1} (measured in the author's laboratory).[155] (B) Linear sweep voltammograms of (I) profenphos, (II) monocrotophos, (III) iodofenphos and (IV) methidathion, measured in 0.2 M NaOH/50 % MeOH (v/v); Scan = 200 mV/s (measured in the author's laboratory).[155]

III. Tasks of Polarography, Voltammetry and Hybrid Methods (HPLC-EC, FIA.-EC) in Day-to-Day Environmental Analysis; Conclusions

The days where a polarographic method was always the last resort and where it had to prove its efficacy several times over are coming to a close. On the other hand, overoptimistic views predicting a bright future for the polarographic and voltammetric techniques are out of

130

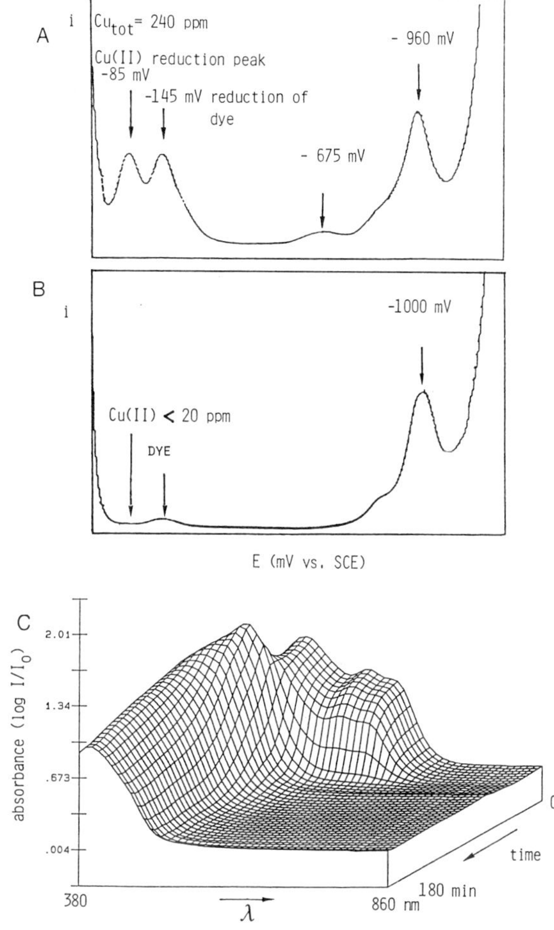

Fig.12. (A) Differential pulse polarogram of a dye waste water before electrolysis at a percolated graphite electrode. (B) after electrolysis.[155] (C) Absorbance - time - profile of the dye wastewater measured on-line during the electrochemical reduction at a percolated graphite electrode using fibre optic and guided-wave spectrometer [155]

place. Over 20 years in the industry promoting the application and adaption of polarographic and voltammetric techniques to given problems and the developement of quantitative polarographic and voltammetric procedures for inorganic, organometallic and organic compounds have shown that numerous realistic analytical problems in the industrial and technical, and thus in environmental monitoring can be solved using these techniques.

In the authors opinion, the main tasks of direct polarography and voltammetry in the environmental fields are:

1) in routine analysis, the determination of inorganic species along with selected organometallic and organic pollutants and this in areas where polarographic and voltammetric techniques have been proven to be the best.

2) speciation tasks, including the determination of the "complexing capacity" which is a measure of the concentration of heavy metals that can be discharged to a waterway before free metal ions appear, due to natural and synthetic complexing agents present in the

water, (cf.[119,192] and the numerous other papers by Nürnberg).

3) <u>control measurements</u> of other higher-performance instrumental methods (cross-correlations). In Table 11 the comparison of emission spectrographic, flame-less atomic absorption and inverse pulse voltammetric data of Cu, Pb, Cd, and Zn in a NBS reference for trace elements in water are shown.

Table 11. Comparison of emission spectrographic (ES), flameless atomic absorption (FAA) and anodic stripping voltammetric data (ASV) of Cu, Pb, and Zn in NBS 1643.[155]

Ion	NBS ng/g	ES ng/g	FAA ng/g	ASV ng/g
Cu	18 ± 2	21	14	18.4
Pb	27 ± 1	28	20	26.7
Cd	10 ± 1	5	6.1	9.1
Zn	72 ± 4	-	-	(63)

With the exception of Zn the inverse (pulse) voltammetric values are close to the NBS values, while the atomic absorption values tend to be too low. The low AAS values obtained initiated the voltammetric control in the authors laboratory.

An interlaboratory test for heavy metals in Elbe water carried out with 33 research groups employing different analytical methods (AAS, photon activation, polarography, inverse voltammetry, ICP, XRF, NAA), showed a coefficient of variation between 7 and 50 % for most elements; higher values were obtained for Cr and As (130 %), Cd (80 %), Hg (77 %), Ni (72 %), Br (65 %), Pb (60 %). These rather high ranges of variation indicate various sources of systematic errors in the different working groups. It is obvious that a well reproducible method does not yet imply accurate results.[193]

In environmental analysis, the accuracy of the results assumes major importance, for legal and ecotoxicological reasons.

Hybrid techniques such as HPLC-EC or electrochemical detection in combination with flow injection analysis (FIA) will foreseeably play a much more important role in the environmental analysis.

It must always be kept in mind that the increased sophistication in instrumental design should not be used as a substitute for detailed chemical knowledge required to generate meaningful data from the equipment. In this respect polarography is in the same league as all other instrumental techniques.[194] Along with the choice of method, the correct degree of analytical-economic commitment, practical demand and appropriate expenditure must not be disregarded. An elevated analytical efficiency should not be coupled with such an elevated apparatus and cost expenditure that a reasonable relationship no longer exists between the costs and the benefits.

Nürnberg[195] remarked quite correctly that many institutions without any or with only insufficient analytical expertise start buying expensive instrumentation and performing analyses of chemical pollutants. In this manner the already significant amount of false data could be increased dramatically by a wave of irrelevant results within the next few years. As accurate results will constitute a minority, they might frequently be drowned in the sea of false data.

"Environmental analytical chemistry is a matter for analytical experts and much too important to be performed by dilletantes" (Nürnberg).[195]

Acknowledgements

The authors gratefully acknowledge the skillful technical assistance of H.G. AWenzel and J-P. Worch. We thank Dr. J. Burmicz and Dr. P. Rach for useful discussion.

References

1. Declaration on the Human Environment - The United Nations Conference on the Human Environment, Stockholm, June 1972, through.[10]
2. Chem. Marktg. Rept., **233** (9) (1988) 7+15.
3. R. Kalvoda and R. Parsons, Electrochem. Res. Dev. Proceedings of the UNESCO Forum, 1984, Publ. (1985) 103, C.A. **105** 69015n.
4. H. Bretscher, G. Eigenmann and E. Plattner, Chimia, **32** (1978) 173.
5. R. Kammel, Vortr. Rheinisch-Westfäl. Akad. Wiss. Nat. Ing.- Wirtschaftswiss., **316** (1983) 7-30;C.A. **98** 184907d.
6. J. Divisek and L. Fürst, Fresenius' Z. Anal. Chem., **317** (1984) 324.
7. R. Ahmed, K. May and M. Stoeppler, Sci. Tot. Environ., **60** (1987) 249.
8. A.P. Meijers, R. De Groot and R. Den Hartog, Gas, Wasser, Abwasser, **65** (1985) 279.
9. K. Grob and G. Grob, J. Chromatogr., **90** (1974) 303.
10. R. Kalvoda, in: "Electroanalytical Methods in Chemical and Environmental Analysis", Plenum Press, New York and London, SNTL, Publishers of Technical Literature, Prague (1987).
11. M. Stoeppler and W. H. Nürnberg, in: "Metalle in der Umwelt, Verteilung, Analytik und biologische Relevanz", E. Merian (Ed.), Verlag Chemie, Weinheim, 1984, Analytik von Metallen und ihrer Verbindungen, p. 45.
11a. M. Vanderlaan, B.E. Watkins and L. Stanker, Environ. Sci. Technol., **22** (1988) 247.
12. G.C. Barker, G.W.C. Milner and H.I. Shagolsky, Proceed. Congr. Mod. Anal. Chem. in Industry, St. Andrews, 1957, p. 199.
13. P.M. Bersier, Anal. Proceed., **24** (1987) 44.
14. L. Meites and P. Zuman, in: "Organic Electrochemistry", Vol. I - VI, CRC Press, Boca Raton, U.S.A., 1977.
15. L.H. Keith and W.A. Telliard, Environ. Sci. Technol., **13** (1979) 416.
16. Amtsblatt der Europ. Gemeinschaften, C. 176.25, 14. Juli 1982, Richtlinien des Rats 76: 464, EWG,(=EEC Black List).
17. Katalog wassergefährdender Stoffe, Stand Mai 1985, Umweltbundesamt, BRD, Mai, 1985, LTWS-Nr 12.
18. Fourth Ann. Rep. Carcinogens, U.S. Dept. Health. and Human Services, Washington, D.C. 1985, trough,[10] p. 16.
19. N.K. Lam and M. Kopanica, Anal. Chim. Acta, **161** (1984) 315.
20. W. Buchberger, H. Malissa and K. Winsauer, Microchim. Acta, **I** (1984) 53.
21. J. Polak, Chem. Listy, **77** (1983) 306.
22. H. Benadikova, M. Popl and V. Jakubickova, Collect. Czech. Chem. Commun., 48 (1983) 2636.
23. H. Benadikova and R. Kalvoda, Anal. Lett., **17** (1984) 1519.
24. S.M. Rappaport, Z.L. Jin and X.B. Xu, J. Chromatogr., **240** (1982) 145.
25. Z. Jin and S.M. Rappaport, Anal. Chem., **55** (1983) 1778.
26. G.C. Whitnack, in: "Identification and Analysis of Organic Pollutants in Water", L.H. Keith (Ed.), Ann. Arbor Sci., Ann. Arbor, 1976, Chapter 18, p. 265.
27. G.C. Whitnack, Anal. Chem., **47** (1975) 618.
28. K. Bratin, P.T. Kissinger and C.S. Bruntlett, J. Liquid Chromatogr., 4 (1981) 1777.
29. M.P. Maskarinec, D.L. Manning, R.W. Harvey, W.H. Griest and B.A. Tomkins, J. Chromatogr., **302** (1984) 51.
30. V. Mejstrik, Z. Sagner, L. Drzkova, J. Jandera and M. Matrka, Cesk. Hyg., **31** (1986) 513; C.A. **106** 97450q.
31. O.W. Parks and R.C. Doerr, J. Assoc. Off. Anal. Chem., **69** (1986) 70.
32. T. Schmidt, W. Tomberg and H. Buening-Pfaue, Z. Lebensm. Unters. -Forsch., **180** (1985) 53.
33. T. Schmidt and H. Buening-Pfaue, Deutsch. Lebensm-Rundschau, **81** (1985) 239.
34. C. Bighi, C. Locatelli and F. Pulidori, Metodi Anal. Aque, 1 (1985) 10.
35. M.B. Thomas, H. Msimanga and P.E. Sturrock, Anal. Chim. Acta, **174** (1985) 287.
36. P.M. Bersier, unpublish. results.
37. J.R. Rice and P.T. Kissinger, Environ. Sci. Technol., **16** (1982) 263.

38. V. Concialini, G. Chiavari and P. Vitali, J. Chromatogr., **258** (1983) 244.

39. D.N. Armentrout and S.S. Cutié, J. Chromatogr. Sci., **18** (1980) 370.

40. R.M. Riggin and C.C. Howard, EPA-600/4-82-022, NTIS.

41. B.J. Birch and J.P. Hart, in: "Polarography of Molecules of Biological Significance", W.F. Smyth (Ed.), Academic Press, London, 1979, Chapter 7, p. 205.

42. J.L. Anderson, K.K. Whiten, J.D. Brewster, T-Y. Ou and W.K. Nonidez, Anal. Chem., **57** (1985) 1366.

43. W.J. Mayer and M.S. Greenberg, J. Chromatogr., **208** (1981) 295.

44. M.W.F. Nielen, G. Koomen, R.W. Frei and U.A.Th. Brinkman, J. Liquid Chromatogr., **8** (1985) 315.

45. H. Hu-Ch., Anal. Chim. Acta, **107** (1979) 387.

46. V. Stara and M. Kopanica, Collect. Czech. Chem. Commun., **50** (1985) 42.

47. V. Stara M. Kopanica and J. Jenik, Anal. Chim. Acta, **147** (1983) 371.

48. D.S. Owens and P.E. Sturrock, Anal. Chim. Acta, **188** (1986) 269.

49. N. Totir, S. Marchidan, C. Volanschi, N. Cimpoeru, R. Andrei and I. Funduc, Rev. Roum. Chim., **22** (1977) 137; C.A. **87** 28416k.

50. C.E. McKone, T.H. Byast and R.J. Hance, Analyst, **97** (1972) 653.

51. J. Polak and J. Volke, Cesk. Farm., **32** (1983) 282; Electroanal. Abstracts, **22** (1984) 212.

52. J. Polak and J. Volke, Chem. Listy, **77** (1983) 1190.

52a. W.F. Smyth, Anal. Proc., (1982) 82.

53. F. Erb, J. Dequidt, A. Philippo and P. Thomas, Ann. Falsif. Expert. Chim. Toxicol., **74** (Nr 794) (1981) 71.

54. V.N. Dmitrieva, O.V. Meshkova and V.D. Bezuglyi, Zhur. Anal. Khim., **30** (1975) 1406.

55. S.R. Betso and J.D. McLean, Anal. Chem., **48** (1976) 766.

56. J.V. Geil, J. Schaefer and H. Kraenzler, Gewaesserschutz, Wasser, Abwasser (Aachen), **67** (1983) 229.

57. D.E. Weisshaar, D.E. Tallman and J.L. Anderson, Anal. Chem., **53** (1981) 1089.

58. R.E. Shoup and G.S. Mayer, Anal. Chem., **54** (1982) 1164.

59. W.F. Smyth, in: "Polarography of Molecules of Biological Significance", Academic Press, London, 1979.

60. D.A. Bagon and C.L. Warwick, Chromatographia, **16** (1982) 290.

61. W.P. King, T.J. Kuriakose and P.T. Kissinger, J. Assoc. Off. Anal. Chem., **63** (1980) 137.

62. A.D. Pickard and E.R. Clark, Talanta, **31** (1984) 763.

63. P.M. Williams, I.R. Whiteside and T.P. Jones, Int. Environ. Saf., (1981) (4) 15.

64. J.C. Septon and J.C. Ku, Am. Ind. Hyg. Assoc. J., **43** (1982) 845.

65. I. Eskinja, Z. Grabaric, B.S. Grabaric, M. Tkalcec and V. Merzel, Mikrochim. Acta, **III** (1984) 215.

66. Z.V. Zaitseva, E.K. Prokhorova and R.M.F. Salikhdzhanova, Zh. Anal. Khim., **33** (1978) 1823.

67. R.G. Williams, Anal. Chem., **54** (1982) 2121.

68. M. Kopanica, V. Stara and J. Jenik, Vodni Hospod., B, (1983) 49; C.A. **98** 204061d.

69. J.O. Broenstad, K.H. Schroeder and H.O. Friestad, Anal. Chim. Acta, **119** (1980) 243.

70. W. Buchberger and K. Winsauer, Microchim. Acta, **(II)** (1980) 257.

71. K. Tanaka and M. Takeshita, Anal. Chim. Acta, **166** (1984) 153.

72. L.Ya. Kheifets and N.A. Romanov, Khim., Tekhnol. Vody, 1, (1979) 42; C.A. **93** 24501g.

73. N.A. Sobina, L.Ya. Kheifets and N.A. Romanov, Zh. Analit. Khim., **33** (1978) 137.

74. R. Tokoro, R. Bilewicz and J. Osteryoung, Anal. Chem., 58, (1986) 1964.

75. R.R. Rowe and M.R. Smyth, in: "Polarography of Molecules of Biological Significance", F.W. Smyth (Ed.), Academic Press, 1979, Chapter 8, p. 229.

76. P. Nangniot, La Polarographie en Agronomie et en Biologie, J. Duculot (Ed.), Gembloux, 1971.

77. J.H. Mendez, R.C. Martinez and J.S. Martin, Anal. Chem., **58** (1986) 1969.

78. M.R. Smyth and J.G. Osteryoung, Anal. Chim. Acta, **96** (1978) 335.

79. G.J. Clark, R.R. Goodin and J.W. Smiley, Anal. Chem., **57** (1985) 2223.

80. T. Peng and R. Lu, Huanjing Kexue Xuebao, **1** (4) (1981) 285; C.A. **96** 99267j.

81. X-D. Ding and I.S. Krull, J. Agr. Food Chem., **32** (1984) 622.
82. J. Davidek, in Anal. Chem. Symp. Ser., **2** W.F. Smyth (Ed.), Elsevier, 1980, p. 399.
83. M. Andrews and W.E. Geiger Jr., Anal. Chim. Acta, **132** (1981) 35.
84. A. Voulgaropoulos, P. Valenta and H.W. Nürnberg, Fresenius' Z. Anal. Chem., **317** (1984) 367.
85. P.W.W. Kirk, R. Perry and J.N. Lester, Intern. J. Environ. Anal. Chem., **12** (1982) 293.
86. F. Guerrieri and G. Bucci, Anal. Chim. Acta, **167** (1985) 393.
87. Z. Stojek, M. Ciszkowska and J. Osteryoung, J. Electroanal. Chem., **195** (1985) 405.
88. Z. Stojek and M. Ciszkowska, Pol. J. Chem., **60** (1986) 567.
88a. Z. Stojek and J. Osteryoung, Anal. Chem., **53** (1981) 847.
89. M. Ciszkowska and Z. Stojek, Talanta, **33** (1986) 817.
90. J.P. Hart, W.F. Smyth and B.J. Birch, Analyst, **104** (1979) 853.
91. R. Kalvoda, Pure & Appl. Chem., **59** (1987) 715.
92. P.M. Bersier and J. Bersier, Analyst, **113** (1988) 3.
93. T. Batonnier, Ann. Fals. Exp. Chim., **75** (1982) 15.
94. D.J. Hodges and F.G. Noden, Pwc. Int. Conf. Manag. and Control of Heavy Metals in the Environment, London (1979), 408.
95. J. Riley and J.V. Towner, Marine Poll. Bull., **15** (1984) 153.
96. H.O. Friestad and J.O. Broenstad, J. Assoc. Off. Anal. Chem., **68** (1985) 76.
97. K.J. Mann and T.M. Florence, Sci. Total Environ., **60** (1987) 67.
98. P.J. Hayes and M.R. Smyth, Anal. Proc., **23** (1986) 34.
99. P. Kenis and A. Zirino, Anal. Chim. Acta, **149** (1983) 157.
100. K. Hasebe, Y. Yamamoto and T. Kambara, Fresenius' Z. Anal. Chem., **310** (1982) 234.
101. T.L. Shkorbatova, D.A. Kochkin, L.D. Sirak and T.V. Khavalits, Zh. Anal. Khim., **2** 6, (1971) 1521.
102. M.D. Booth and B. Fleet, Anal. Chem., **42** (1970) 825.
103. H. Woggon, H. Saeberlich and W.J. Uhde, Fresenius' Z. Anal. Chem., **260** (1972) 268.
104. I. Ioneci, I. Tanase and C. Luca, Anal. Lett., **18** (A8) (1985) 929.
104a. A.M. Bond, J.R. Bradbury, G.N. Howell, H.A. Hudson, P.J. Hanna and S. Strother, J. Electroanal. Chem., **154** (1983) 217.
105. J. Ireland-Ribert, A. Bermond and C. Ducauze, Anal. Chim. Acta, **143** (1982) 249.
106. O. Evans and G.D. McKee, Analyst, **113** (1988) 243.
107. W.A. MacCrehan and R.A. Durst, Anal. Chem., **50** (1978) 2108.
108. E.P. Makeeva and G.I. Krivda, Gig. Tr. Prof. Zabol., (1982) (5) 54; Electroanal. Abstracts, **21** (1983) 4800.
109. P. Zuman, in: "Organic Polarographic Analysis", Pergamon Press, Oxford, London, New York, Paris 1964.
110. J. Volke and M. Slamnik, in: "Pestic. Anal.", K.G. Das (Ed.), Marcel Dekker, 1981, p. 175.
111. M. Brezina and P. Zuman, in: "Polarography in Medicine, Biochemistry and Pharmacy", Interscience Publ., 1958.
112. M. Adamovsky, Vodny Hospodarstvi, **16** (1966) 102; C.A. **65** 6908f.
113. A. Lechien, P. Valenta, H.W. Nürnberg and G.J. Patriarche, Fresenius' Z. Anal. Chem., **306** (1981) 150.
114. R. Engst, W. Schnaak and H. Woggon, Fresenius' Z. Anal. Chem., **207** (1965) 30.
115. J.O. Broenstad and H.O. Friestad, Analyst, **101** (1976) 820.
116. P.M. Bersier and J. Bersier, CRC Crit. Rev. Anal. Chem., **16** (1985) 81.
117. E. Merian, in: "Metalle in der Umwelt, Verteilung, Analytik und biologische Relevanz", Verlag Chemie, Weinheim, 1986.
118. P. Valenta, person. commun.
119. T.M. Florence, Analyst, **111** (1986) 489.
120. T. Miwa, L.T. Jin and A. Mizuike, Mikrochim. Acta, (**III**) (1984) 259.
121. I. Gustavsson, J. Electroanal. Chem., **214** (1986) 31.
122. J. Wang, in: "Stripping Analysis, Principles, Instrumentation and Applications", Verlag Chemie, 1985.

123. W.F. Smyth, in Anal. Chem. Symp. Ser., **25** M.R. Smyth and J.G. Vos (Eds), Elsevier, 1986, p. 29 and references therein.

124. B. Adeloju and A.M. Bond, Anal. Chem., **57** (1985) 1728.

125. Kh.Z. Brainina, Talanta, **34** (1987) 41.

126. H. Siegenthaler, Swiss Chem., **8** (1986) 54.

127. N.K. Taylor, K.D. Bartle, Ch. Gibson, D.G. Mills and D. Servante, in: "Identification and Analysis of Organic Pollutants in Air", L.H. Keith (Ed.), Butterworth Publ, 1984, Chapter 15, p. 243.

128. T.R. Crompton, in: "Comprehensive Organometallic Analysis", Plenum Press, New York, 1987.

129. W. Horwitz, J. Assoc. Off. Anal. Chem., **62** (1979) 1251.

129a. Perkin Elmer, Workshop Umweltanlytik, Wasser-Boden-Luft, 18-21.01.1988, Hinterzarten, F.R.G.

130. M. Warner, Anal. Chem., **59** (1987) 1311 A.

131. V. Krizan, in: "Analyza Ovzdusia, Analysis of the Atmosphere", Alfa-SNTL, Bratislava and Prague, 1981.

132. M.D. Manita, R.M.F. Salikhdzhanova and S.F. Yarovovskaya, in: "Modern Methods of Determining Atmospheric Pollution of Populated Places", Meditsina, Moscow, (1980) C.A. **94** 161995j.

133. J.O'M. Bockris, in: "Environmental Chemistry", Plenum Press, New York and London, 1977.

134. "Electrochemistry and the Environment" (in Czech), J. Balej (Ed.), Academic Prague 1982, C.A. **97** 81737y.

135. G. Henze and R. Neeb, in: "Elektrochemische Analytik", Springer, Berlin, Heidelberg, New York, Tokyo (1986), pp. 272-289.

136. W.F. Smyth and M.R. Smyth, Pure and Appl. Chem., **59** (1987) 245.

137. W.F. Smyth and M.R. Smyth, Rep. Eur., **7623** (1982) 323.

138. M.R. Smyth, P.J. Hayes and D. Dadgar, Anal. Chem. Symp. Ser., **25** M.R. Smyth and J.G. Vos (Eds), Elsevier, 1986, p. 37.

139. W.F. Smyth, D. Dadgar and M.R. Smyth, Trends Anal. Chem., **1** (1982) 215.

140. M. Kopanica, Sci. Tot. Environ., **37** (1984) 83.

140a. R. Kalvoda, Sci. Tot. Environ., **37** (1984) 3.

141. W.F. Smyth and J. Vaneesorn, Int. Labmate, **11** (1986) 41.

142. R. Prochazka, Chimie & Industrie Special No. 281, (1933) C.A. **27** 5608, (1933).

143. R. Ahmed, K. May and M. Stoeppler, Sci. Tot. Environ, **60** (1987) 249.

144. Kh.Z. Brainina, L.I. Roitman, R.M. Khanina and N.A. Gruskova, Khim. i Tekhnol. Vody, **7** (1985) 27.

145. J.M. Hellawell, Environ. Pollution, **50** (1988) 61.

146. Some references to the most important legislation guide lines relating to the quality of drinking water, water, wastewater and sewage sludge:

- **Switzerland:**
  · Schweizer Lebensmittelbuch, Bern, 1985.
  · Verordnung über Abwassereinleitungen. Der Schweiz. Bundesrat, Bern, 1975.
  · Klärschlammverordnung. Der Schweiz. Bundesrat, Bern, 1981.
  · Richtlinien für die Untersuchung von Abwasser und Oberflächenwasser, Eidg. Dep. des Innern, Bern, 1983.
  · Verordnung über umweltgefährdende Stoffe (Stoffverordnung, Sto V) 09.06.1986, Änderung vom 21.09.1987, Bundeskanzlei, Bern.
- **EC:**
  · Councile Directive of 15 July 1980, Relating to the Quality of Water Intended for Human Consumption, Offic. Journal of the European Communities, No. L229/ 11-29 (80/778/EEC).
- **WHO:**
  · Guidelines for Drinking Water Quality, Vol. **1**, Recommendations, WHO, Geneva, 1984.

- **U.S.A.:**
  - Standard Methods for the Examination of Water and Wastewater, 16th (Ed.), American Public Health Assoc., Washington 1985 (publ. jointly by: Am. Publ. Health Assoc., Am. Water Works Assoc., Water Pollution Control Federation).
  - Methods for Chemical Analysis of Water and Wastes. EPA, Cincinnati 1979 (EPA-600 4-79-020).
- **F.R.G.:**
  - Trinkwasserverordnung. Bundesgesetzblatt Nr. 16, 15.02.1975.
  - Wasserhaushaltsgesetz. Bundesgesetzblatt Nr. 128, 16.10.1976.
  - Trinkwasser Verordnung, Bundesgesetzblatt, Jahrgang 1986, Teil I, 22.05.1986.
  - Gesetzuber Abgaben für das Einleiten von Abwasser in Gewässer (Abwasserabgabengesetz). Bundesgesetzblatt Nr. 118, 15.09.1976.
  - Klärschlammverordnung. Bundesgesetzblatt Nr. 21, 26.06.1982.
  - Deutsche Einheitsverfahren zur Wasser-, Abwasser- und Schlammuntersuchung, 16. Lieferung, Verlag Chemie, Weinheim 1986. (Fachgruppe Wasserchemie der GDCh, Normausschuss Wasserwesen in DIN).
  - cf.: · BUA/GDCh, Altstoffbeurteilung, Ein Beitrag zur Verbesserung der Umwelt, Nov. 1987.
    - BUA: Umweltrelevante Alte Stoffe: Auswahlkriterien und Stoffliste, Verlag Chemie, Weinheim, 1986.
    - BUA: Umweltrelevante Alte Stoffe: Auswahlkriterien und zweite Stoffliste, Stand 07.10.1987.
    - BUA: VCI-Altstoffliste, Chem. Ind., Sonderheft 4, 88.

147. Schweiz. Lebensmittelbuch, Bern, 1985.
148. J. Burmicz, R. Grether and P. Rach, in: "Metrohm Monographs", Metrom Ltd., Herisau, Switzerland, 1983.
149. D. Taylor, Residue Reviews, **72** (1979) 33.
150. T.M. Florence, Talanta, **29** (1982) 345.
151. T.M. Florence and G.E. Batley, Crit. Rev. Anal. Chem., **9** (1980) 219.
152. T.M. Florence, Anal. Proc., **20** (1983) 552.
153. M.L.S. Gongalves, L. Sigg and W. Stumm, Environ. Sci. Technol., **19** (1985) p. 141, and references therein.
154. D.H.S. Richardson, Environmental Pollution (Series B), **10** (1985) 261.
155. P.M. Bersier, unpubl. results.
156. J.F. Pankow and G.E. Janauer, Anal. Chim. Acta, **69** (1974) 97.
157. T.W. Trofe, D.A. Dalrymple and F.A. Scheffel, Gas Research Institute, Contract No. 5083-253-0936, DCN 87- 218-043-16.
158. H. Braun and M. Metzger, Fresenius' Z. Anal. Chem., **318** (1984) 321.
159. P. Rach, person. commun.
160. H.A. Mottola, Toxicol. Environ. Chem. Res., **2** (1974) 99.
161. R. Perry, Environ. Sci. Technol., **15** (1981) 5.
161a. W. Salomons (1981), Proc. Int. Conf. Heavy Metals in the Environment, Amsterdam, CEP, Consultants, Edinburgh, p. 694, through.[84]
162. A. Meyer and G. Henze, Fresenius' Z. Anal. Chem., **327** (1987) 123.
163. A. Meyer and G. Henze, Fresenius' Z. Anal. Chem., **329** (1988) 764.
164. Z. Kozarac, B. Cosovic and M. Branica, J. Electroanal. Chem., **68** (1976) 75.
165. V. Zutic, B. Cosovic and Z. Kozarac, J. Electroanal. Chem., **78** (1977) 113.
166. R. Kalvoda and L. Novotny, Collect. Czech. Chem. Commun., **51** (1986) 1587.
167. C.J. Purnell and C.J. Warwick, Proceed. Anal. Div. Royal Soc., (1981) 151.
168. E. Koen, Khig. Zdraveopaz, **22** (1979) (3) 317, C.A. **92** 46487s.
169. R.M. Novik and N.I. Plyngyu, Teor. Prakt. Polyarogr. Metodov Anal., 19, Yu.S. Lyalikov and R.M. Novik (Eds), "Shtiintsa", Kishinev, U.S.S.R., 1973, C.A. **83** 1609f.
170. P. Nangniot, Anal. Chim. Acta, **31** (1964) 166.
171. M.L. Hitchman and S. Ramanathan, Anal. Chim. Acta, **157** (1984) 349.
172. H.C. Fang, Chung-kuo Nung Yeh Hua Hsueh Hui Chih., **24** (1986) 248, C.A. **106** 97955h.
173. V.N. Kavetskii and G.G. Andrienko, Zh. Anal. Khim., **41** (1986) 168.

174. G.K. Budnikov, G.S. Supin., N.A. Ulakhovich and N.K. Shakurova, Zh. Anal. Khim., **30** (1975) 2275.
175. G.S. Supin and G.K. Budnikov, Zh. Anal. Khim., **28** (1973) 1459.
176. M.A. Alavi and H.A. Ruessel, Fresenius' Z. Anal. Chem., **309** (1981) 8.
177. V. Stara and M. Kopanica, Anal. Chim. Acta, **159** (1984) 105.
178. M.R. Smyth and J.G. Osteryoung, Anal. Chem., **49** (1977) 2310.
179. G. Franke, W. Pietrulla and K. Preussner, Fresenius' Z. Anal. Chem., **298** (1979) 38.
180. M. Zietek, Mikrochim. Acta, **II** (1975) 463.
181. M. Zietek, Mikrochim. Acta, **II** (1976) 549.
181a. M. Zietek, Mikrochim. Acta, **II** (1975) 75.
182. R. Frank, G.J. Sirons and B.D. Ripley, Pest. Monit. J., **13** (1979) 120.
183. W.D. Hoermann, J.C. Tournayre and H. Egli, Pest. Monit. J., **13** (1979) 128.
184. M.D. Erickson, C.W. Frank and D.P. Morgan, J. Agric. Food Chem., **27** (1979) 740.
185. R. Kalvoda, Proc. Conf. Trace Analysis of Organic Compounds in Water, p. 106, Perdubice (1984).
186. P. Nangniot, Trends. Anal. Chem., **4** (1985) 155.
187. B. Fleet, Pergamon Ser. Environ. Sci., (1980) (3) 589.
188. I.G. Krasnoborod'ko, Vodosnabzh. Sanit. Tekh., (1986) (10) 4, C.A. **106** 89585x.
189. T.A. Kharlamova and N.I. Mitashova, Khim. Prom. (Moscow) 1986, (4) 206, C.A. **104** 229826n.
190. M.S.E. Abdo and R.S. Al-Ameeri, J. Environ. Sci. Health, A 22, (1987) 27.
191. Ding Zhou, Cai Wei Min and Wang Qun Hui, Water Sci. Technol., **19** (1987) 391.
192. H.W. Nürnberg, Chem.- Ing.- Technik, **51** (1979) 717.
193. A. Knoechel and W. Petersen, Fresenius' Z. Anal. Chem., **314** (1983) 105.
194. J.S. Burmicz, U. Hutter and V. Pfund, in: "Electrochemistry, Sensors and Analysis", M.R. Smyth and J.G. Vos (Eds), Anal. Chem. Symposia Series, **25** Elsevier, 1986, p. 83.
195. W.H. Nürnberg, Fresenius' Z. Anal. Chem., **317** (1984) 197.

THE USE OF ELECTROANALYTICAL TECHNIQUES IN BIOTECHNOLOGY

Bauke te Nijenhuis
International Bio-syntheties
Patentloon 3
2288 EE Rijswijk, The Netherlands

Biotechnology, by definition, is the integration of the basic disciplines biochemistry, genetics, microbiology and process technology in biological systems on behalf of industrial processes and the environment. This definition does not include only the activities of the traditional fermentation industry, but gives also entrances to new fields of biotechnology as commercialized plant tissue culture, its use for production of important plant related chemical compounds, the use of animal, more specifically mammalian cells for special purposes and the rapidly growing field of monoclonal antibodies.

The roots of biotechnology are in the classical fermentation industry, making use of micro-organisms like yeasts, bacteria and moulds and which can be seen as one of the oldest industrial activities in the world. Just think of the leavening of bread and the production of a wide variety of fermented foods and beverages and not to forget production of antibiotics in general, like penicillin and enzymes for industrial, pharmaceutical and food applications.

The definition of biotechnology means also that analytical chemistry in biotechnology covers the whole area from fundamental research uptill and including bioprocess analysis and control. In research and development in the laboratory phase a normal picture of all common techniques, including electrochemical applications can be seen. Either the electrochemical technique is used as it stands or electrochemistry is included in special analytical equipment such as chromatographs, titration equipment, etc. There is nothing special in that way of operation. The differences start with translation of the R & D results to pilot plant and production because of the specific nature of biotech- nological processes. Therefore this paper will concentrate on application in pilot plant and production scale, especially when fermentation equipment starts playing its role. There is also another part of the analytical matrix that should be concidered. It is known as the five eras of process analytical chemistry in the context of the description by Bruce Kowalski and coworkers.[1]

- off-line
- at-line
- on-line
- in-line
- non invasive

Kowalski[1] states that: "The goal of process analytical chemistry is to supply quantitative and qualitative information about a process. Such information can be used not only to monitor and control a process, but also to optimize its efficient use of energy, time and raw materials". In the area of process chemistry bioprocessing is the most rapidly growing area, even faster than the special materials for the electronics industry, so it is worthwhile to study this area very carefully. "The main bottleneck for on- and in-line process analysis and control is the continuing lack of sensors, especially biochemical sensors, for a good process control".[1] The treatment of this statement will form the major part of this paper.

The first two eras of process analytical chemistry require both manual sampling and analysing the sample afterwards either in a centralized laboratory or in an on-site laboratory. It can be understood that all kind of analytical methods can be used in this case and therefore is not of any interest in this paper. In biotechnology the last era is almost beyond question

Contemporary Electroanalytical Chemistry, Edited by A. Ivaska *et al.*
Plenum Press, New York, 1990

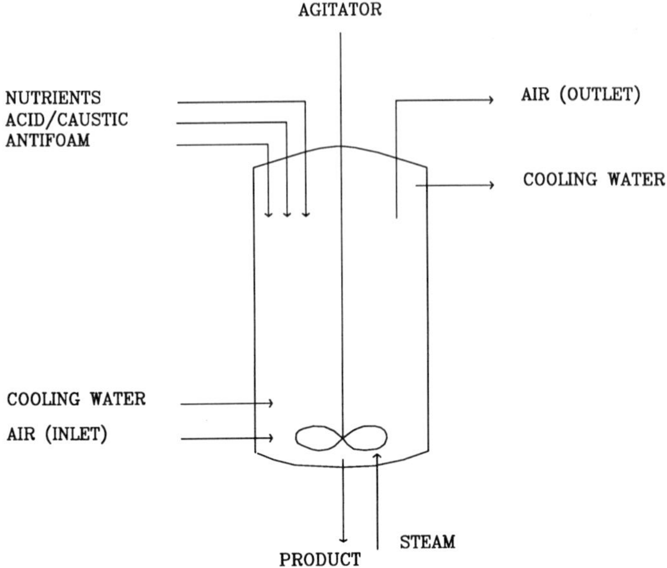

Fig.1. Fermenter.

so that we can concentrate on the on- and in-line analytical methods. The purpose of this paper can now be restricted to deal with the on- and in-line process analytical chemistry in bioreactors, especially in use for pilot experiments and production, taking into account what the contribution of electrochemistry could be.

Before going into more detail the specific properties of a fermentation process should be considered. A bioreactor or fermenter is described in Fig.1.

Production takes place in the fermenter which normaly is a stirred reactor with feedsystems and various instrumentation equipment. Nutritional compounds either in solution or as a slurry are added to the fermenter, an inoculum of the desired micro-organisms is added, the whole mash will be stirred, air for oxygen supply is passed through the mash and the fermentation process goes on. After the fermentation the products are isolated in a recovery process. The products are the micro-organism itself (e.g.yeast) or its metabolites like enzymes or antibiotics. Typical properties of a fermentation process are:
- complex matrix
- processtime 1-7 days or more
- monoculture
- aqueous systems
- variations during process (batchprocess!)

Due to the fact that the goal of a fermentation process is to produce an uncontaminated monoculture, high demands are necessary to prevent contamination. This does not only holds for addition of the various nutrition compouns, but also for the sampling systems and on-line analytical systems which have to be autoclavable.

Typical controls are given in Table 1.

Part of the information gathered is used for control purposes and automatic control (pH, temperature), part is used to monitor the process and leading to manual interferences into the process. Controllability of the fermentation process depends on how well the process is known and how well it can be modelled.

Measurements in the gas phase include oxygen and carbondioxide measurements to get insight in the metabolic processes in the fermenter and on gaseous compounds formed during the process. In the past special techniques have been developed for individual measurements of the various gases which in itself are still very usable, but nowadays it is generally accepted that on-line mass spectrometry is the best analytical technique. Of course, apart from sterility aspects, special precautions have to be made to prevent liquids and aerosols entering the inlet

140

Table 1. Typical fermentation controls.

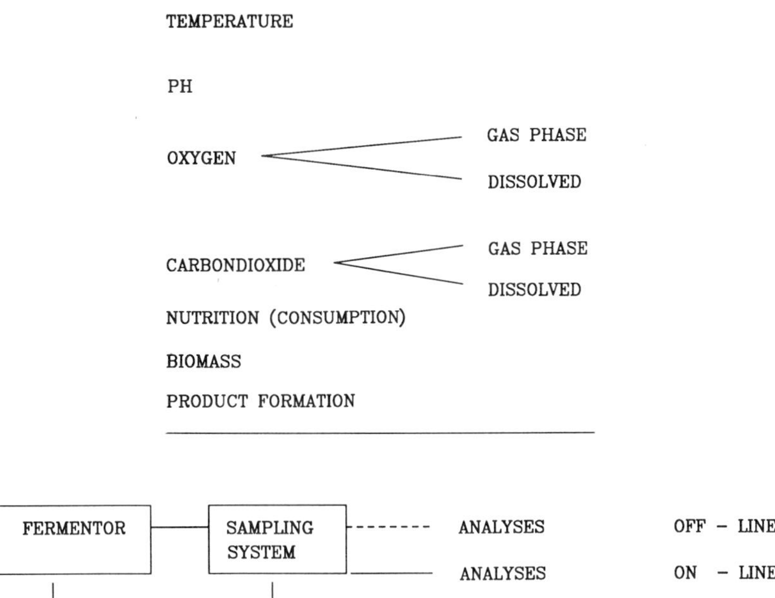

Fig.2. Analytical systems in fermentation control.

probe but in principle this technique is ready for use. Introduction of mass spectrometry into the industrial production plants has already started.

Measurements in the aqueous phase are described in Fig.2.

In on-line arrangements the following techniques are mainly used:
- mass spectrometry
- flow injection analysis
- HPLC
- Sensors

In the in-line arrangements application of sensors is exclusively the technique to be used.

Sensors are important analytical devices in biotechnology. Electroanalysis has concentrated itself in the past decades on the development of sensors for very diverging applications, in which biosensors form a major part. Biosensors can be defined as devices which convert biological events into electronic output signals. A more practical definition states that "biosensors are measuring in biological systems" and includes also pH and ion-selective electrodes as biosensor. Such a device brings together a biological element or receptor and a transducer resulting in an electronic signal. A lot of biosensors are known and Table 2 gives a complete overview of the various components being used in the formation of a biosensor.[2]

The most frequently used transducers, however, are:
- potentiometric
- amperometric
- optical
- calorimetric

The available biosensors in the wider practical sense are given in Table 3.

Based on the examples biosensors can be defined as a combination of a biological compound and a transducer brought together on a chip, resulting in miniaturizing the sensor.

The best developed biosensors in this sense are the amperometric biosensors of which

141

Table 2. Components that may be used to construct
a biosensor.

Biological elements	Transducers
organisms	potentiometric
tissues	amperometric
cells	conductometric
organelles	impedimetric
membranes	optical
enzymes	calorimetric
enzyme components	accoustic
receptors	mechanical
antibodies	"molecular" electronic
nucleic acids	
organic molecules	

Table 3. Some examples of available biosensors.

ion sensitive field effect transducers-(ISFETs)
chemically sensitive field effect transducers-(CHEMFETs)
amperometrical ferrocene based glucose sensor
potentiometric based area sensors (ISFET type)
Clark/ oxygen electrode

about 80 examples have been published now. More than 10 are commercially available now
and the most common ones are suitable to analyse
- glucose
- lactose
- lactate
- alcohol
- starch
- amylase
- lactate dehydroqenase (LDH)

Most of the applications of biosensors are in medical and clinical diagnosis with a spin-
off to food production, pollution control and biotechnology.

The enthusiasm in academic circles in biosensors is very high due to the fact that it
is an exciting science, which resulted in explosive growth in the R & D phase. However,
as has been demomstrated above there are relatively few products available for immediate
daily use and most of them are only suitable for the medical/clinical sector, and not in the
biotechnology area.

What is the reason that the application in the bioprocess industry is as remote as it
is today? Looking at the working principle and the unlimited possibilities to construct a
biosensor they could fit very well in the needs of the bioprocess analysis, especially because of
their specifity and sensitivity. Unfortunately they have to fit in one very severe demandment
of the bioprocess industry and that is autoclavibility or sterilisation. In biotechnology the
whole reactor has to be completely sterile inside, including all the adaptions, feeding lines, air
inlet and sampling tubes besause of the requirement that only the inoculated micro-organism
should grow and produce in a contamination free environment. That means that all on- and
in-line systems should be sterilized together with the sterilization of the fermenter itself or at
the moment when the device is used for analysis. Only in very limited cases the developers
have been succesfull in meeting this demand.

Besides the constraints of autoclavibility other disadvantages of the biosensors can be
defined, such as:
- low stability
- long respons time

- not suitable for continuous operation
- re-usability uncertain
- fouling by clogging of particulate matters
- recalibration hardly possible
- sensitivity for pH, ion strength, temperature.

The present biosensors can only be used, because of the presented constraints
- off-line and discontinuous
- on-line after a sampling treatment
- on-line behind a membrane.

These problems are fundamental and ask for a new generation of biosensors.

Conclusions

- biosensors are relatively cheap, easy to construct and can give important information on the bioprocess.
- the prospects of the present biosensors are very limited because of their fundamental properties, which result in inadequateness in biotechnoloqy.
- developments of cheap disposables, more types and miniaturization might be of importance for the medical and clinical sector, but can wait for biotechnology untill the mentioned constraints have been solved.
- a new generation of biosensors based on thermostable enzymes and/or biological compounds is needed for application in biotechnology.

Acknowledgements

The authors wishes to thank Mr. J. Raaymakers, Mr M. van Tilborg and Mr. C. van Dijk for their valuable discussions and information in the preparation of this paper.

References

1. J.B. Collis, D.L. Illman and B.R. Kowalski, Anal. Chem., **59** (1987) 624A.
2. ”Biosensors, Fundamentals and Applications”, A.P.F. Turner, I. Karube and G.S. Wilson (Eds), Oxford University Press, 1987.

POTENTIOMETRIC DETERMINATION OF COPPER IN VARIOUS PLATING BATHS

Adam Hulanicki,[1] Tomasz Sokalski[1] and Andrzej Lewenstam[2]

1. *Department of Chemistry of University Warsaw*
   PL- 02 093 Warsaw, Poland
2. *Laboratory of Analytical Chemistry, Åbo Akademi University*
   SF-20500 Turku - Åbo, Finland

Copper impurities at the concentration level 10^{-4} M in plating baths can cause deterioration of the physical properties of coatings.[1,2] Several analytical methods of determination of copper in presence of very large excess of other heavy metals are available but they are usually time consuming and susceptible to errors due to necessity of preliminary copper separation.

Application of ion-selective electrodes to determine copper by standard addition method in nickel plating bath was suggested by Frant.[3] Later Hulanicki et al.[4] using a copper ion-selective electrode proposed a method based on multiple standard addition in presence of a copper complexing agent to prevent a harmful influence of chloride ions.   In this work a similar method is used to determine copper also in zinc and cobalt baths.

Experimental

Instrumentation

The potential and pH measurements were performed with the pH-meters pHM-64 (Radiometer) and OP-206 (Radelkis), respectively. The following electrodes were used: copper selective 94-29A (Orion), glass pH-electrode 6.0216.000 (Metrohm) and a saturated calomel electrode K 401 (Radiometer). Copper standard solutions were added using the autoburette ABU-13 (Radiometer). Spectrophotometric measurements were performed with the spectrophotometer Spekol (Carl Zeiss, Jena). All measurements were performed at $25 \pm 1°C$. All reagents were of analytical grade. Double distilled water from quartz still was used throughout. The standard copper(II) solution was 1×10^{-2} M.

Analytical Procedure

To 25 ml sample of a plating bath containing 0.1 – 0.2 mg Cu, 1 ml of 1 M sodium thiosulphate solution was added, the solution was stirred for approximately 15 s. Then the stirrer was switched off and the potential was recorded using copper selective electrode until the drift was smaller than 0.1 mV/min (approx. 10 min). Then successively 5 times 0.2 ml of standard copper solution was added and after each addition the potential was measured in the same way as before.

The unknown copper concentration was calculated using the modified Gran function with iteration procedure.

Because the equation describing the electrode potential in the presence of thiosulphate ions is more complicated than the simple Nernst equation the usual Gran method must be modified. The Gran function has the form:[4]

$$G = 10^{E/S}(V + V_0)\left[\frac{n_L}{V + V_0} - A\left(\frac{C_x V_0 + CV}{V + V_0}\right)\right]^2 = 10^{E_0/S}(C_x V_0 + CV) \qquad (1)$$

Contemporary Electroanalytical Chemistry, Edited by A. Ivaska *et al.*
Plenum Press, New York, 1990

where E – electrode potential, E_0 – standard electrode potential, S – electrode slope, V – volume of standard copper solution added, V_0 – initial volume of the sample, n_L –number of moles of sodium thiosulphate, $A = 4D_I/D_{II}$, where D_I/D_{II} – ratio of diffusion coefficients of copper(I) and copper(II) thiosulphate complexes, C – concentration of copper standard solution, C_x – unknown copper concentration in the sample.

This function is complicated by the presence of the term containing the unknown concentration, C_x, in the square bracket on the left side of equation (1). This however can easily be solved by the following iterative procedure.[4]

By using the value $A = 3.5$ and substituting in the bracket $C_x = 0$, the first approximation of $\sqrt{G_i}$ is calculated and used as the coordinate in the Gran's standard addition plot versus corresponding V_i values. From those points using the linear regression method the first approximation of $C_x = C'_x$ is found as $C'_x = -V'_x CV_0^{-1}$.

In the second step the value C'_x is used as C_x and the procedure is repeated. The successive Gran's plots are calculated as long as the difference between the C_x values is smaller than the assumed accuracy. Usually the first approximation is sufficiently good, but it is recommended to check at least one more step.

Result and Discussion

The determination of copper by ion-selective electrode in presence of other metal ions, N, should be possible in the case when the selectivity coefficient:

$$K_{Cu,N}^{Pot} \leq \frac{C_{Cu}\Delta}{C_N 100} \qquad (2)$$

assuming the concentration response of the electrode in constant ionic strength solution, where C_{Cu} is the minimal concentration of copper, C_N is the maximal concentration of the main metallic component of the plating bats, and Δ is the maximal allowable systematic relative error caused by the interferent . When Δ is assumed as 1 %, C_N – 1.25 M and C_{Cu} – 1×10^{-5} M then $K_{Cu,N}^{Pot}$ should be smaller than 8×10^{-8}.

Most literature data,[5–8,12] however, give values which are either larger or at least of similar order than the calculated limiting value. Our recent studies[9] have indicated that those data are not correct and that true values of selectivity coefficients are close to those predicted by the thermodynamic model, i.e., to the ratio of solubility products of the respective sulphides (Table 1). This makes the determination of copper possible even in the presence of very large excess of other metal ions.

Table 1. Comparison of theoretical, experimental and literature values of coefficients for the copper selective electrode[9]

N^{2+}	$pK_{Cu,N,theor.}^{Pot}$	$pK_{Cu,N,exp.}^{Pot}$	$pK_{Cu,N,lit.}^{Pot}$
Ni^{2+}	15.1	16.8	$3.3^5, 3.7^6, 4.0^7, 3.0^8, 8.5^{12}$
Co^{2+}	14.9	16.7	8.5^{12}
Zn^{2+}	13.7	15.0	$3.4^5, 3.0^6, 8.5^{12}$

Another disturbing factor in the course of copper determination in plating baths is the presence of chloride ions in high concentrations. In its presence complicated redox and complexation reactions[10] influence the electrode response and result in irreproducible and unstable potentials of the indicator electrode. The more comprehensive study of the electrode behaviour in the presence of thiosulphate[10,11] has indicated that the parasitic effect of chloride can be completely eliminated. Therefore such conditions were selected for determination of copper in plating baths. Electrode potential in thiosulphate solution is given by the equation:[10]

$$E = E^0 + S\log\frac{C_{Cu^{2+}}}{(C_L - AC_{Cu^{2+}})^2} \qquad (3)$$

Table 3. Application of Gran's method with successive iteration. Synthetic cobalt plating bath.

V, ml	E, mV	$G' \times 10^3$	C'_x	$G'' \times 10^3$	C''_x
0.0	-93.9	0.960		0.942	
0.2	-81.0	1.581	1.0756	1.551	1.0766
0.4	-71.9	2.233	$\times 10^{-4}$	2.190	$\times 10^{-4}$
0.6	-64.5	2.945		2.889	
0.8	-59.0	3.598		3.529	

where $C_{Cu^{2+}}$ is the total concentration of copper, C_L the total concentration of thiosulphate.

This form of potential response equation implies a more complicated data processing by the modified Gran method, as indicated in the "Analytical Procedure".

Before analytical measurements the parameters in equation (3), namely electrode slope S and parameter A, had to be determined. This was done by making the calibration curve and fitting the relationship (3) to the experimental points by means of Simplex curve fitting algorithm. The values obtained in this way were 57.2 mV and 3.5 for S and A respectively. In Table 2 the results of the potentiometric determination of copper for different plating baths as well as for 1 M KNO_3 solution are presented. Copper concentrations in the samples was checked by a spectrophotometric method with diethyldithiocarbamate or with cuproine (2,2'-diquinolyl)[13] except for KNO_3 solution where known amount of copper was added. Composition of the plating baths was as follows:

synthetic Ni bath I [1]

$NiSO_4 \cdot 7H_2O$	1.20 M
$NiCl_2 \cdot 6H_2O$	0.18 M
H_3BO_4	0.51 M

synthetic Ni bath II [1]

$NiSO_4 \cdot 7H_2O$	0.71 M
$NiCl_2 \cdot 6H_2O$	0.74 M
H_3BO_3	0.65 M

industrial Ni bath

$NiSO_4 \cdot 7H_2O$	1.07 M
$NiCl_2 \cdot 6H_2O$	0.17 M
$MgSO_4$	0.12 M
K_2SO_4	0.06 M
H_3BO_3	0.65 M
brightener WXF	1 %

synthetic Zn bath [2]

$ZnSO_4 \cdot 7H_2O$	1.25 M
NH_4Cl	0.57 M
$NaC_2H_3O_2 \cdot 3H_2O$	0.11 M
Glucose	0.67 M

synthetic Co bath [2]

$CoCl_2 \cdot 6H_2O$	1.26 M
H_3BO_3	0.49 M

Table 2. Comparison of copper determination in plating baths by potentiometric standard addition technique and by spectrophotometry. Average values for n=4.

Sample	Copper content $\times 10^5$, M		Relative difference, %	Relative standard deviation, %
	Spectro-photometry	Potentiometry		
KNO_3	7.63	7.58	0.7	1.3
Ni synt. I	8.09	8.29	2.5	8.0
Ni synt. II	4.97	4.88	1.8	5.0
Ni ind.	2.38	2.58	8.6	5.2
Co synt.	10.8	10.7	0.9	0.9
Zn synt.	8.24	8.53	3.5	2.8

Table 3 and Fig. 1 show successive Gran's calculations for synthetic cobalt plating baths. The difference between first and second approximation of C_x is as small as 0.1 %.

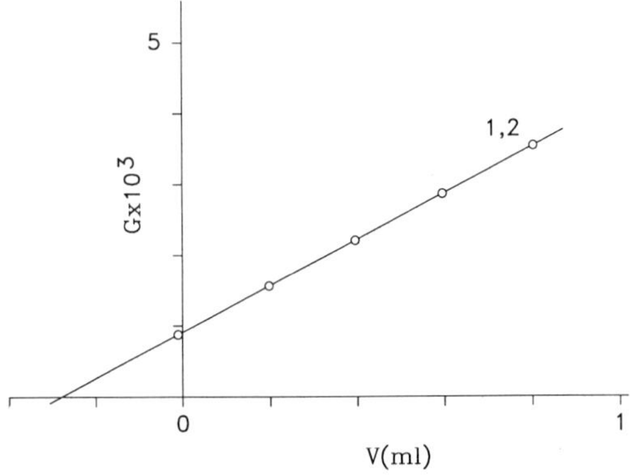

Fig.1. Gran's plot for synthetic cobalt bath: S = 57.2 mV, A = 3.5; (1) – $C_x = 0$, (2) – $C_x = C'_x$.

In all cases potentiometric results are in fair agreement with spectrophotometric ones. The relative difference is never larger than 10 % Also the values of the relative standard deviation are reasonably small, less than 10 %. The analytical samples contained various amounts of chloride ions and also some of them contained organic substances, this however has not influenced significantly the accuracy and precision.

Conclusions

The application of the copper selective electrode for detemination of copper in nickel, cobalt and zinc plating baths gives good results. In good agreement with very small values of selectivity coefficients of copper electrode towards Zn^{2+}, Co^{2+} and Ni^{2+} the interferences do not occur even in the presence of very large excess of ions mentioned. Addition of thiosulphate efficiently removes interferences of chloride ions, and creates favourable conditions for copper determination. Results of potentiometric determination of copper are close to those obtained spectrophotometrically.

Acknowledgements

This study was financially supported by the project CPBR - 01.17

References

1. Encyclopedia of Chemical Technology, R.E. Kirk and D.F Othmer (Eds), Vol. 5, The Interscience Encyclopedia, Inc., New York, 1950.
2. Modern Electroplating, F. Lowenheim (Ed.), John Wiley & Sons, Inc., New York, 1963.
3. M.S. Frant, Plating, 58 (1971) 686.
4. A. Hulanicki, T. Sokalski and A. Lewenstam, Chem. Anal., (Warsaw) 32 (1987) 601.
5. G.J.M. Heijne, W.E. van der Linden and G. den Boef, Anal. Chim. Acta, 98 (1978) 221.
6. J.M. van der Meer, G. den Boef and W.E. van der Linden, Anal. Chim. Acta, 76 (1975) 261.
7. M. Mascini and A. Liberti, Anal. Chim. Acta, 53 (1971) 202.
8. H. Hirata and K. Date, Talanta, 17 (1970) 883.
9. A. Hulanicki, T. Sokalski and A. Lewenstam, Mikrochim. Acta, III (1988) 119.
10. A. Lewenstam, T. Sokalski and A. Hulanicki, Talanta, 32 (1985) 531.
11. A. Hulanicki, A. Lewenstam and T. Sokalski, 4th Symposium on Ion-Selective Electrodes, Matrafüred, 1984, E. Pungor (Ed.), Pungor Akademiai Kiado, 1985, p. 459.
12. J. Pick, K. Tóth and E. Pungor, Anal. Chim. Acta, 65 (1973) 240.
13. Z. Marczenko, Spektrofotometryczne oznaczanie pierwiastków, PWN, Warszawa, 1979.

CONTROLLED-GROWTH MERCURY DROP ELECTRODE AND PERSPECTIVES IN PROCESS MONITORING APPLICATION

Zygmunt Kowalski and Jan Migdalski

Institute of Materials Science
Academy of Mining and Metallurgy
Cracow, Poland

Introduction

The concept of the dropping mercury electrode (DME) and the technical solution of the device for its realization as proposed by Heyrovsky (1922) have been applied in many laboratories. This kind of working electrode, the so-called polarographic detector, implies measuring of the current on a mercury drop whose surface is continuously expanding, while the maximum size of the drop is gravity controlled. A particular fraction of the current, the so-called charging current, which is connected with the changing of the drop area or with the changing of the potential of the electrode is independent of the concentration of the depolarizer. The charging current can be concidered as the background signal.

This basic polarographic detector has been improved over many years and the main objective has been to improve the measurement sensivity as well as to faciliate the use of the electrode in everyday laboratory practice. Reference[1] gives a very good revision of the efforts of many researches in this field.

When the new techniques such as ac, sw and pulse polarography were been developed, the polarographic detector in its conventional form appears to come back to everyday practice. These techniques allow measurements in which the capacity current is significantly depressed.

Efforts to improve the polarographic detector have been taken up again. Nowadays, however, some other features are required such as: credibility of data, reliability and accuracy, good signal-to-noise ratio and the possibility to remote control of the detector. The invention of the static mercury drop electrode (SMDE)[2,3] has turned away the attention of the constructors and users from the conventional DME. This is easy to understand when the principles of these new devices and their methodological consequences for electroanalysis are considered. In the case of the static mercury electrode device, the mercury drop is dispensed in time of hundreds of miliseconds (drop growing time) and then the outflow of mercury is stopped (constant size period) until mechanical dislodging. This kind of drop can be reproduced cyclically. The current is measured first when the drop has attained the required size. In this situation the capacity current arising from the expansion of the drop area is eliminated. As we know from an experimental study,[4] the dc polarography with a short drop time but with the application of SMDE yields performance similar to that associated with the pulse technique. In the absence of dA/dt term, the background current is much lower than that usually associated with DME.

In spite of these advantages the SMDE has a limited application in industrial process monitoring due to the troubles with, e.g., the drop recalibration. Another important limitation of the SMDE in comparison with the DME is that the data on the continuously changing fraction of the drop are not available.[4,10] It must be recalled that the shape of the current-time profile allows to distinguish some kinetic currents and to demonstrate also the presence of the adsorption phenomena. For this reason the application of the continuously growing drop electrode is justified in some experimental studies. Such measurements, in which both

Contemporary Electroanalytical Chemistry, Edited by A. Ivaska *et al.*
Plenum Press, New York, 1990

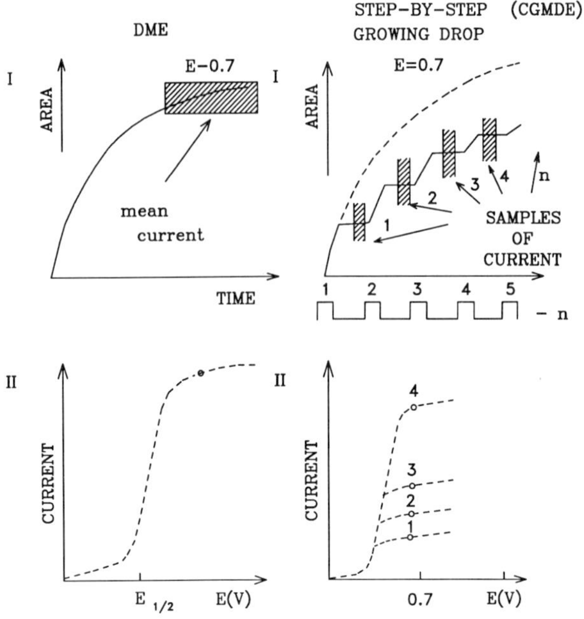

Fig.1. (I) Current-time-relation for one drop of DME (a) and CGMDE (b); (II) one mean current value (a), n-current sample values (b), derived from the top figures, sampled in off-pulse times.

the drop growing time and the constant size period of the drop are utilized, are possible when the Controlled Growth Mercury Drop Electrode (CGMDE) is employed. This aspect will be discussed in the present paper. Other features of the electrode, important from the point of view of process monitoring application, will also be considered.

Controlled-Growth Mercury Drop Electrode

The approach offered by the Controlled-Growth Mercury Drop Electrode (CGMDE) combines the advantages of both the conventional DME and the SMDE and has an additional advantage. This concept has already been published.[5,6] In this electrode the mercury outflow (drop growth) is controlled by a fast response valve, actuated by a pulse sequence generated by the pulse sequencer or a computer, which causes the drop to increase in the step by step mode. The drop size can be described operationally by specifying the number of pulses at the given pulse width. Hence the velocity of the growth of the drop can be controlled by the time intervals between the pulses.

The dynamics of the step-by-step growth of the drop depends on the time intervals between the pulses which operate the valve.

The above mode implies the possibility of current sampling, in the OFF pulse time of the pulse sequence which generates the drop. If this sampling is done with a delay long enough with respect to the cell time constant of the electrode, the sampled current will be free of the charging component.

As Fig.1 shows, the individual drop is generated by a package of pulses and (hence) the response is transmitted in the form of a sequence of current samples. Accordingly the values of the sequence of these current samples represent the dependence of the current upon time on a single drop and reproduce very well the features of the instantaneous currents (i-t curves) of a continuously growing single drop.

The shape of these quantized curves makes it posible to distinguish some kinetic currents, which may indicate the presence of adsorption phenomena.

The idea presented above offers a few interesting applications especially valuable for process monitoring where additional data can be expected.

Looking at Fig.1 we see that one drop lasting for instance for one second, when quantized as mentioned above, yields not just one mean current value as in the case of a continuously growing drop, but "n" samples, coming from different fractions of the drop. We can now reconstruct "n" polarograms, representing different stages of the step-by-step growing drop instead of one as is the case when the mean current is measured.

Each sequence of current samples generated by a package of pulses and processed in the computer gives a single number which represents the shape of the i-t curve.

A constant value of such a shape parameter is an evidence of the standard conditions under which the concentration measurement is done. The most frequent case which questions the credibity of the measurements is the sudden appearance of s.a.s. (surface active substance) or a sudden change in its concentration.

Because only at the later stage of the drop life time the current deviates most visibly from the diffusion – controlled value when s.a.s. is present in the solution, the shape parameter can be a supplementary warning about this undesirable effect. When having a set of polarograms it is possible to choose the most reliable of them.

If the proposed procedure should be applied in practice it would be necessary to check whether the exponents of parabola (shape parameter) are constant for different concentrations of the polarizer. An experiment carried out with cadmium confirmed this assumption. Table 1 presents exponents of barabola, b, ($y = a\,x^b$) for i-t curves of cadmium determined from the last 3, 4, 5, 6 or 7 current samples for two different concentrations. The concentrations differ from each other almost by 10 times. The next example shows how far the shape parameter can be helpful to detect the appearence of s.a.s in the solution under control.

Table 1. Exponents of parabola (shape parameters) presented for final fraction of the drop life time.

Exponent of parabola	1) 0.5 mM Cd(II), 0.5 KNO_3	2) 0.06 mM Cd(II), 0.5 KNO_3
b (3)	0.22	0.24
b (4)	0.24	0.24
b (5)	0.26	0.26
b (6)	0.31	0.31
b (7)	0.36	0.34

Fig.2 shows the i-t curve generated in 20 steps for 0.2 mM cadmium at the potential -0.7 V.

The change in the shape of this curve under the influence of s.a.s. has been determined in this case through an analysis of the initial period of the drop life (first seven steps) and of the final fraction generated in the last thirteen steps. Table 2 presents experimental data taken from i-t curves for cadmium. Current samples were taken with 150 ms delay after the drop growth was stopped. The ON pulse times were 10 ms. In these experiments the starting conditions were determined by pure 1 M KNO_3 as the supporting electrolyte.

In order to disturb these conditions, different amounts of Triton-X were added to the starting solution. The numbers listed in Table 2 (in the first and third columns) reflect the shapes of the i-t curve recorded:

i) within the first 7 steps, and

ii) within the last 13 steps

The current samples in columns 2 and 4 were read after the 7th and the 20th steps.

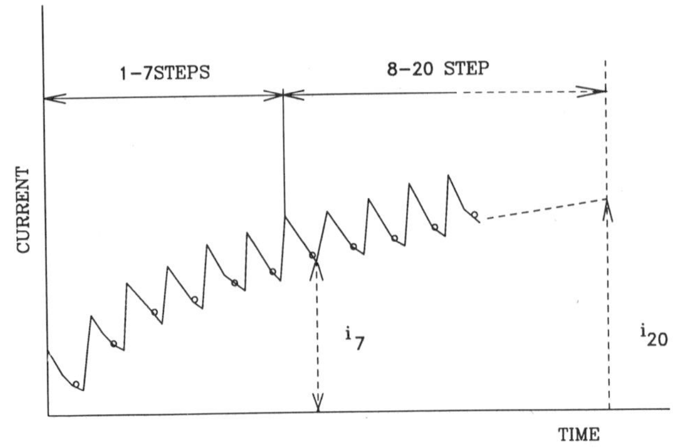

Fig.2. Current-time profile, drop size; 20 pulses pulse width 10 ms; Cd^{2+} 0.2 mM, KNO_3 0.5 M, potential −0.6 V.

Table 2. Effect of Triton-X on the shape of the i-t curve, and the diffusion controlled current in cadmium solution (0.2 mM).

Exponent 1	i_7 2	Exponent 3	i_{20} 4	Triton-X 5
0.20	0.63	0.30	0.88	0
0.19	0.62	$0.26_{(13)}$	$0.84_{(4.5)}$	$2 \times 10^{-4}\%$
$0.19_{(5)}$	$0.60_{(4.7)}$	$0.24_{(20)}$	$0.78_{(11)}$	$3 \times 10^{-4}\%$
$0.18_{(10)}$	$0.50_{(20)}$	$0.15_{(50)}$	$0.22_{(75)}$	$6 \times 10^{-4}\%$

These results confirm the already existing observations in this matter that in the later period of the drop life the current deviates more significantly from the diffusion controlled value than in the earlier period of the drop life. The decrease in the exponent value is significant (numbers in brackets represent the percentage changes with respect to pure solution).

The results listed in the columns 2 and 4 illustrate how far the limiting current has decreased. The correlation between columns 1, 2 and 3, 4 is obvious.

Finally, when amperometric measurements are carried out the decision must be taken automatically which set of the currents samples represents the diffusion current only. If the current samples are going to be used for a polarographic record the same principle is employed. Current samples for which the exponent value suggests any electrochemical complication are rejected. In the past much attention was given to studies of the i-t curves on a single drop as well as how the surface active substances influence the shapes of the curves. Worth noticing in this respect, according to the authors opinion, are a number of studies which sum up the more important developments.[7,8]

Our studies in the field where the CGMDE has been applied have confirmed all the earlier observations and justified continuation of the research work. The main advange is that the i-t curves can be registered free from the charging component and is promising in the analytical applications of the method.

A unique feature of the CGMDE is the possibility of automatic recalibration of the size of the electrode itself. From the process monitoring point of view, where the reproducibility of the experimental results is due to the advantage of automation – this aspect becomes especially significant. Here recalibration of the drop size can be achieved by generating the drop with a pulse sequence until it reaches the gravity controlled size. To make this problem more clear we shall discuss now the principle of recalibration as shown in Fig.3.

When testing the proposed method of calibration its full applicability could be con-

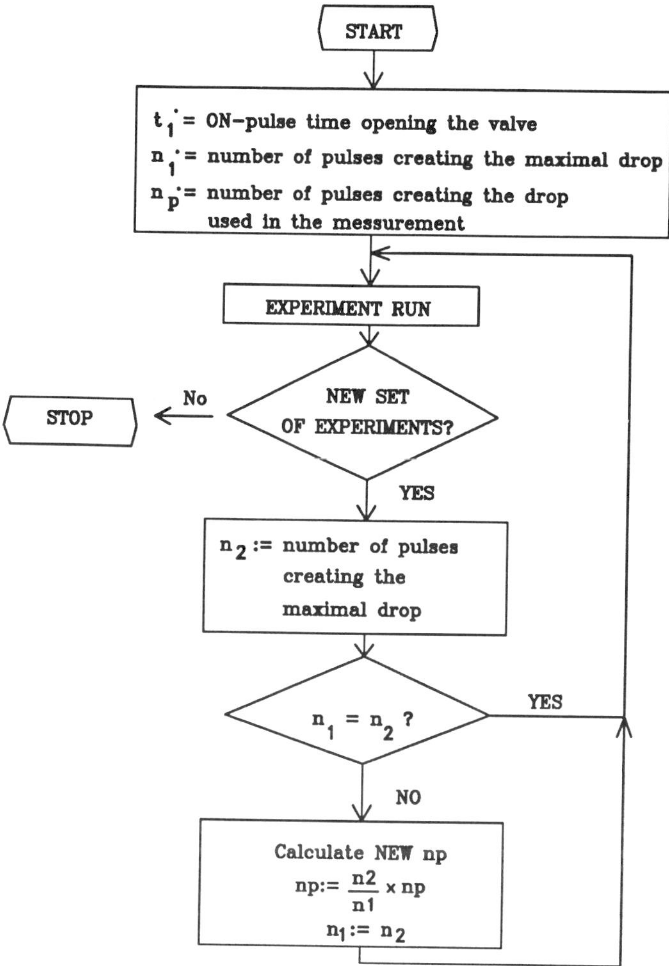

Fig.3. Flow diagram of the program for calibration of CGMDE.

firmed, as can be seen in Table 3. The procedures of calibration consist of the measurement of the number of pulses at which the drop has the maximal size n_1 - (gravity controlled size) and the choice of the drop of a desired size, e.g., n_p- starting size. To show the recalibration process the mercury level was lowered up down to 150 mm. Now the maximal drop was generated by 71 pulses (and not as before, by 54 pulses). The number of pulses needed to generate the drop of a starting size, will be now equal to $n_p \times \frac{n_2}{n_1}$, i.e., 34 pulses. The current measured on the previous starting drop generated by 26 pulses at the mercury level 195 mm was equal to 95 μA; when measured on a new drop generated by 34 pulses practically the same current was observed.

There is still another feature, a very essential one, regarding the possibility of increasing the accuracy of measurement, also resulting from the way the drop is generated. When the current is sampled after each step of the drop growth, and after all current samples are summed up a high value of the current is obtained when compared with the single sampling procedure at the end of the drop life. In this procedure we observe a certain similarity to the instrumental approach described earlier.[9] In this approach 50 % of the drop life time was integrated in digital operations and a very good S/N ratio was obtained when compared to the short sampling methods. In our method, however, the sampled currents after summation are free from the charging component, This is particularly important and gives the base line which is almost parallel to the x-axis even for a 10^{-6} M depolarizer.

Table 3. Quantitative ilustration of reliability and accuracy of the recalibration of the mercury drop size; 0.2 mM cadmium, with recording of the voltammetric curve.

	Measurement after calibration		
Mercury level mm	No of pulses for maximal drop n_1	No of pulses for applied drop n_p	μA for current voltage curve
195	54	26	95.1

	Measurement after recalibration		
New mercury level mm	No of pulses for maximal drop n_2	applied drop $n_p \times \frac{n_2}{n_1}$	μA for current voltage curve
150	71	34	95.6

When using the multi-sampling technique the repeatability of the signal was much better than in the case when only one current sample was taken. The same is valid also when the base current was measured. To illustrate the principles presented above, two current values were marked in the figure which show the trace of the polarographic wave in Fig.4, being the basis for the determination of the value of the limiting current similarly as in case of registering the full profile. The lower values of the current have been taken using a single sampling technique at the end of the drop life. The upper values of the current represent the number obtained using multiple sampling technique. The numbers in the brackets represent the range (in %) as a measure of reproducibility taken from 7 drops. This technigue of multi-sampling is very advantageous and can be recommended when CGDME is applied. The size of the signal is here growing evidently and at the same time the repeatability of the measurement is very high. In the opposite case when single sampling technigue is used, the size of the signal is much smaller and simultaneously the repeatability of signal and base line current are only $\sim$ 25 %.

A similar measurement, one reduced to two points, can be performed instead of registering the complete polarographic wave giving an evident gain of time.

Conclusions

The particular advantages of CGMDE found in computer aided experiments important for a process monitoring system are:
- parameters of the drop can be adjusted to the particular electrochemical conditions,
- i-t curve can be recorded totally free from the capacity component,
- the shape parameter of the i-t curve can be used as an additional parameter for analytical purposes,
- recalibration of the drop size may be done automatically,
- by using multi-sampling of the current, one can increase the size of the signal, and minimize the base line drift.

The above features are essential for the reliability of an analyzer in which the electrode is used as a detector.

Acknowledgements

This work was supported in part by the Central Programme for Basic Research CPBP No 01.17.

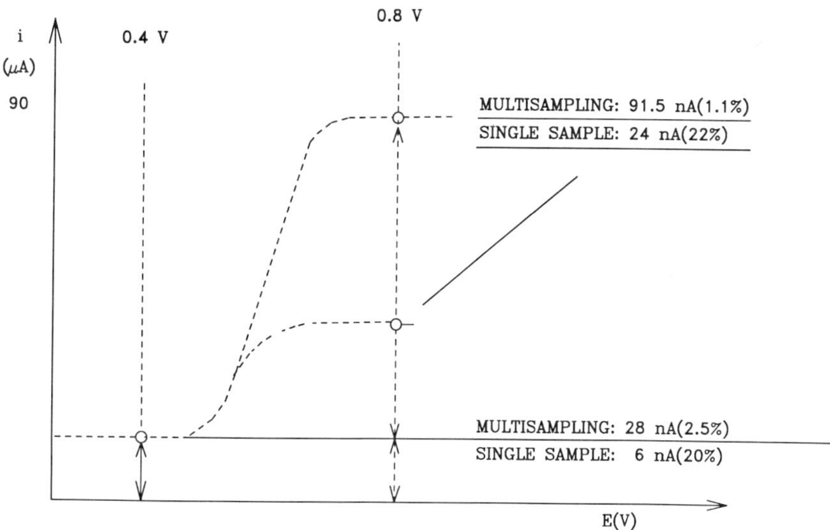

Fig.4. Graphic illustration of the advantages of multiple sampling using a CGMDE and a comparison with single sampling mode, Cd^{2+}, 10^{-6}M in 0.1 M KNO_3 (Numbers in brackets indicate the range of reproducibility for 7 values).

References

1. I.M. Kolthoff and Y. Okinaka, in: "Progress in Polarography", Vol. **II**, p. 352, Interscience Publishers, New York, London, 1962.
2. M.W. Peterson, Am. Lab., **69** (1979) 11.
3. L. Novotny, in: "Proceedings of the J. Heyrovsky Memorial Congress on Polarography", Prague, August 25-29, 1980, Vol. **II**, p. 129.
4. A.M. Bond and R.W. Jones, Anal. Chim. Acta, **121** (1980) 1.
5. Z. Kowalski, K.K. Wong, R.A. Osteryoung and J. Osteryoung, Anal. Chem., **59** (1987) 2216.
6. Z. Kowalski, Analyst, **113** (1988) 15.
7. J. Kuta and J. Smoler, in: "Progress in Polarography", Vol. **I**, p. 43, Interscience Publishers, New York, London, 1962.
8. Ch.N. Reilley and W. Stumm, ibid, p. 81.
9. B.H. Vassos, Anal. Chem., **45** (1973) 1293.
10. M. Lovrie, R.A. Osteryong, J.J. O'Dea and J. Osteryoung, unpublished work.

ELECTROCHEMICAL SENSORS

SOLID STATE POTENTIOMETRIC SENSORS

Jiří Janata

Center for Sensor Technology
University of Utah
Salt Lake City, Utah, 84112, U.S.A.

Introduction

It is more that fifteen years ago since the first integrated solid state electrochemical sensors have been described.[1,2] These were ion-sensitive field effect transistors (ISFET) which belong to the group of electrochemical sensors, specifically potentiometric sensors. Since then other solid state electrochemical and non-electrochemical sensors have appeared and the ISFETs themselves have reached the state of maturity which now brings some of them close commercialization. This is quite remarkable if we realize that only a relatively few researchers have been working in this area, certainly as compared to the related ion-selective electrodes.   The obstacles which had to be overcome have been both of the conceptual and practical nature originating from the fact that these devices are the product of integration of two seemingly diverse technologies: solid state device physics and electrochemistry. Soon after the initial explorations of the ISFETs transistors utilizing enzymatic selectivity[3] and potentiometric solid state sensors based on chemical modulation of electron work function[4] have appeared. The purpose of this paper is to review the present state of developement of these devices both from theoretical and practical point of view.

Ion Sensors

The ion selective membrane can be used in two basic configurations (Fig.1). If the solution is placed on either side of the membrane such arrangement is called symmetrical It is found in conventional ion-selective electrodes in which the internal electrical contact is provided through the solution in which an internal reference electrode is immersed. In the non-symmetrical arrangement one side of the membrane is contacted by the sample (usually aqueous) while the other side is contacted by some solid material. An example of this type are coated wire electrodes and ion sensitive field effect transistors.

In a conventional ion-selective electrode the membrane is placed between the sample and the internal reference solution. The composition of the internal solution can be optimized with respect to the membrane and the sample solution. In the interest of symmetry it is advisable to use the same solvent inside the electrode as is in the sample. This solution also contains the determinand ion in the concentration which is usually in the middle of the dynamic range of the response of the membrane. The ohmic contact with the internal reference electrode is provided by adding the salt which contains the appropriate ion to form a fast reversible couple with the solid conductor. In recent designs gel-forming polymers have been added into the internal compartment. They do not significantly alter the electrochemistry but add some convenience in handling.

The obvious advantage of the symmetrical arrangement is that the processes at all internal interfaces can be well defined and that most non-idealities at the membrane/solution interface tend to cancel out. Because the volume of the internal reference compartment is typically a few milliliters the electrode does not suffer from exposure to electrically neutral

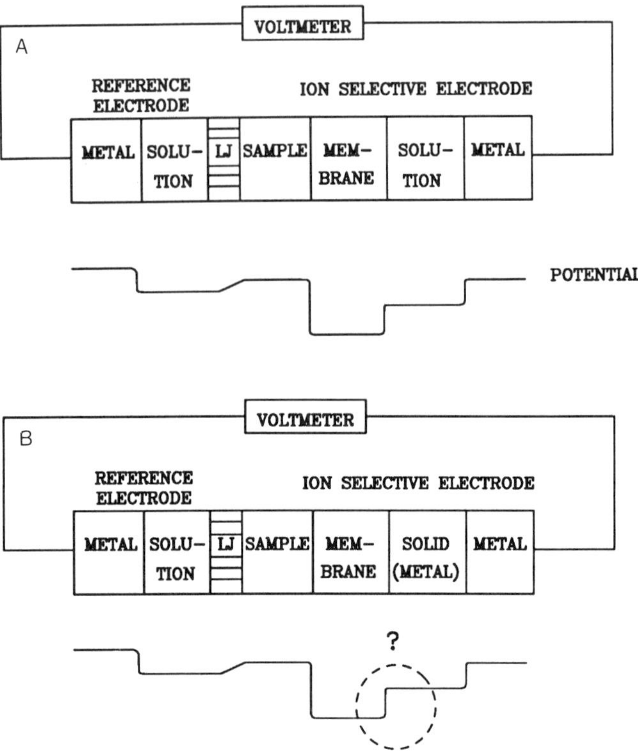

Fig.1. Schematic representation of the (A) symmetrical and (B) asymmetrical arrangement of ion selective membrane.

compounds which would penetrate the membrane and change the composition of this solution. This type of potentiometric ion sensor has been used in majority of basic studies of ion-selective electrodes and most commercial ion selective electrodes are also of this type. The drawbacks of this arrangement are also related to the presence of the internal solution and to its volume. Mainly for this reason it is not conveniently possible to miniaturize it and to integrate it into a multi-sensor package.

The general tendency to miniaturize chemical sensors and to make them more convenient has led to the developement of potentiometric sensors with solid internal contact. These sensors include coated wire electrodes (CWE), hybrid sensors and ion sensitive field-effect transistors (ISFET). The contact can be a conductor, semiconductor or even an insulator. The price which has to be paid for the convenience of these sensors is in the more restrictive design rules which have to be followed in order to obtain a sensor with the performance comparable to the conventional symmetrical ion-selective electrode.

The key issue in these sensors is the interface between the ion selective membrane and the contact. The most convenient way to present this problem is in the form of the equivalent electrical circuit in which the resistances and capacitances have their usual electrochemical meaning (Fig.2). It is necessary to include the electrometer (or at least its input stage) in the analysis of these sensors. In most modern instruments the amplifier is an insulated gate field-effect transistor (IGFET) which has the input dc resistance of greater than $10^{14}\Omega$ and the input capacitance on the order of picofarads.

The internal contact between the membrane and the preamplifier represents a parasitic impedance. Let us consider two limiting cases concerning the values of the equivalent resistances in Fig.2. A good ion-selective electrode has high exchange current density and therefore low value of the charge transfer resistance R_1 ($i_o = 10^{-3}$ A/cm^2, i.e., $R_{ct} = 25\ \Omega$

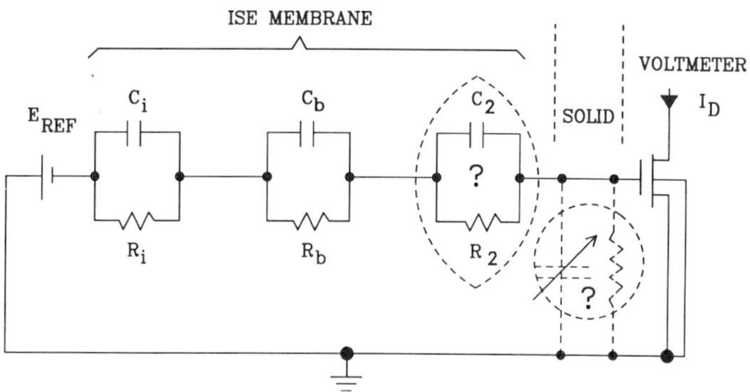

Fig.2. Equivalent electrical circuit of a coated wire electrode and an ISFET.

cm^2). This means that at least one charged species can transfer easily between the sample and the membrane and establish the potential difference according to one of the mechanisms discussed above. The bulk membrane resistance R_b can be as high as 10^6 Ω cm^2. However, because no net current passes through the membrane the potential in the bulk of the membrane is uniform, i.e., there is no electric field inside the membrane. In a conventional symmetrical ISE arrangement the composition of the internal solution can always be cosen in such a way that the interfacial charge transfer resistance R_2 is comparable to R_1. Thus, a well established potential profile exists throughout this structure. On the other hand if the R_2 is high the input capacitor together with the interfacial capacitance C_2 and the (variable) parasitic capacitance form a (variable) capacitive divider. In such a case the voltage which appears at the input capacitor of the electrometer depends not only on the electrostatic potential of the membrane but also on the undefined parasitic impedance of the connector and the electrode exhibits unstable characteristics.

The simplest example of a potentiometric ion sensor with solid internal contact is a coated wire electrode (CWE) in which the internal wire is covered with an ion selective polymeric membrane thus forming a compact and inexpensive ion sensor. Noble metal internal wires were used in the early designs of the CWE, mainly for the reasons of their chemical stability. Such electrodes often exhibited unacceptable drift. The high value of the internal charge transfer resistance at the membrane/wire interface has been soon recognized as being the cause of this problem.[6] It is not surprising if we realize that the typical electrode reactions at noble metals are of the redox type, i e , the charge transferring species is the electron. On the other hand in most ion selective membranes the charge conduction is done by ions. Thus, a mismatch between the charge transfer carriers can exist at the noble metal/membrane interface. This is particularly true for polymer based membranes which are invariably ionic conductors. On the other hand solid state membranes which exhibit mixed ionic and electronic conductivity such as chalcogenide glasses, petrovskites and silver halides form good contact with noble metals.

Several solutions of the problem of the conducting internal contact have been suggested. The most direct approach involves the interposition of a thin layer of aqueous gel which contains the fixed concentration of the salt of the primary ion. This approach, which has met with only a limited success, can be seen as the miniaturization of the conventional ISE structure. The main reason for its failure is the fact that water can permeate through the membrane and establish its own activity inside the sensor structure according to its chemical potential in each individual phase. In other words, the water permeating through the membrane reaches its osmotic equilibrium according to the concentration of the solutes present in the gel. This can lead not only to the significant change of the internal activity of the primary salt inside the gel but often to a catastrophic failure of the whole structure when the osmotic pressure inside the gel exceeds the limits of the mechanical stability of

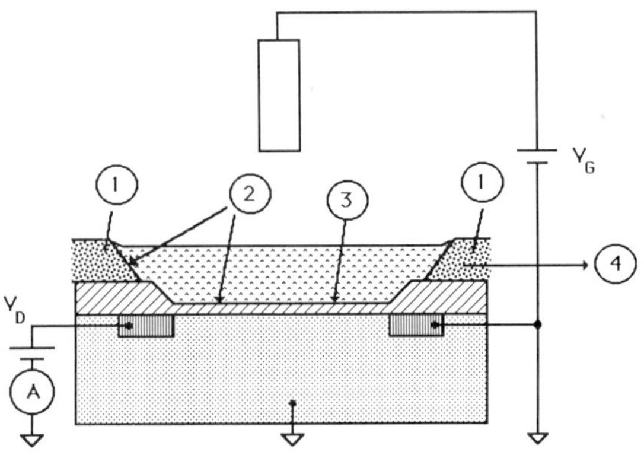

Fig.3. Problem areas of ISFET: (1) Encapsulation; (2) Membrane adhesion; (3) Acidobasic interference; (4) Crosstalk.

the membrane. This problem cannot be avoided because the osmolality of the sample is not known a priory and cannot be controlled during the experiment. The second problem relates to the Severinghaus effect (vide infra). In principle this problem can be avoided by establishing an electrochemical reaction with as high exchange current density as possible at the internal interface, by interposing a thin layer of Ag/AgCl between the membrane and the insulator[7] with the hope that the chloride ion exchange will dominate the internal interfacial potential. Using this strategy ISFETs without the Severinghaus interference have been prepared. In principle the same effect can cause problems in conventional ISE, however, mainly due to the relatively large volume (e.g. 3–5 ml) of the internal solution it is not observed in practice. In summary, it is possible to design well behaved CWEs and to expect that their electrochemical performance will be comparable to the conventional ISEs. However, the design of the membrane/solid interface has to be done with the understanding of the electrochemical processes which take place at that interface. If the final resistance is high then the problems such as drift become directly proportional to the length of the internal contact, specifically to the magnitude of the parasitic capacitance and resistance. Consequently, these problems can be minimized by decreasing the distance between the ion selective membrane and the amplifier. The logical stage in progression in the evolution of these sensors are then a hybrid ion sensor[8] and ion sensitive field effect transistor.[9]

The close integration of the ion selective layer with the amplifier, as it occurs in ISFET, offers some unique sensing possibilities but also creates its own problems. The problem areas of the ISFETs with polymeric membranes are shown in Fig.3. They have been recognized as the encapsulation (1), adhesion (2) of the membrane to the chip and to the encapsulation[9] the acidobasic interference (3) at the membrane/transistor interface.[10] Moreover, the early attempts to deposit several different membranes in a close proximity of each other is difficult due to the cross-contamination (4) of these membranes.[11] Most of these problems have now been solved.

It has be en found[10] that in some situations electrically neutral species can penetrate through the polymeric membrane of the ISFET and change the value of the interfacial potential at the insulator/membrane interface. This applies especially to the acidobasic species such as CO_2, lower carboxylic acids, ammonia etc. when the pH of the solution is far from their pK and they exist essentially in undissociated form. In other words, the ion selective membrane of the ISFET cannot reject electrically neutral species.

There have been several attempts to solve the problem of the adhesion of the membrane to the chip surface and to the surrounding encapsulation. The simplest approach has been to silanize the chip prior to the encapsulation step.[9] However, that improves only the adhesion of the encapsulant and of the membrane to the chip surface but does not prevent the development of the electrical shunt between the membrane and the encapsulant. A much more elaborate solution has been the construction of a suspended polymeric mesh which holds the membrane down.[12] Although it has improved the performance of the individual ISFET it has not solved the problem of preparation of a multiple-membrane ISFET sensor.

A major step forward in encapsulation and multiple membrane deposition has been made recently by using ionophore doping of the generic membrane[13] in combination with some other, previously reported techniques. It is known[14] that plasticized PVC can be made ion selective by doping it with ionophores. It is also known that the optimum performance is obtained from the ISFET if the ion-selective membrane forms a continuous envelope around the encapsulated chip.[15] This eliminates the interface between the encapsulant and the membrane which causes the electrical shunt. Because the distances between the individual ion-selective gates of a multisensing chip are short it is essential that the lateral diffusion of the electroactive ingredients of the individual membranes is minimized. The new plasticizer, bis-(2-ethylhexyl) sebacate used in these membranes[16] decreases the lateral diffusion 100 fold as compared with the conventional plasticizers. The final contribution to this developement came from the use of photolithographically produced encapsulation most of which is done at the wafer level.[17] The result is a dual-gate ISFET in which the long term drift is < 0.05 mV h^{-1} and with the usable lifetime of approximately 2 months. Clearly there is a room for further improvement. Thus, the encapsulation process will further benefit from the recently developed electrochemical encapsulation.[18] The primary adhesion of the membrane to the chip can be further improved by using carboxylated PVC.[19] The high concentration of carboxylic group has also beneficial effect on the electrochemical properties of the membrane, particularly on the rejection of the anions due to the Donnan failure.[20] Therefore, the optimum configuration is an ISFET with a continuous carboxylated PVC membrane which is individually doped with a suitable ionophore over the respective gate areas. The gates themselves contain a thin layer of Ag/AgCl which eliminates the interference from the neutral species.

Finally a few words about pH ISFET. We must remember that field effect transistor is basically a charge-sensing device. From the very beginning[1,2] of the research of this device it has been known that the FET with bare gate insulator exposed to the solution responds to changes of pH. Because pH is such an important parameter to measure a very considerable amount of research and speculation has been devoted to the explanation of the mechanism of this response. Clearly, there is a pH-dependent charge at the solution/transistor interface. The slope of this response has been found to vary between 50 % of theoretical for SiO_2 to 92 % for silicon nitride. Other materials such as Al_2O_3 or Ta_2O_5 and other oxides have been reported to yield nearly theoretical response. It is necessary to pause here and to realize that all these materials are very good bulk insulators. Therefore, they could not be used as pH-sensing membranes in a conventional ISE configuration. However, in an ISFET configuration these materials are the part of the input gate capacitor, and their use is possible.

It is quite adequate to accept the fact that the pH-dependent charge resides somewhere at the insulator "surface" and that the corresponding image charge in the transistor channel affects the transconductance. Let us now invoke the basic electrochemical characteristics of the interface, the distinction between the nonpolarized (resistive) and polarized (capacitive) interface. The argument about the location of the pH-dependent charge has revolved around this distinction: one school of thought has been that the charge is located in one plane, at the surface of the insulator and that the interface behaves as a capacitor at which the charge is generated by the de-protonation/protonation of the surface bound sites. This is the basis of so called "site-binding theory" (SBT). In this theory it is assumed that there are ionizable binding sites present at the surface of the insulator which determine the distribution of the compensating charge in the adjacent layer of solution. This distribution is the result of the interplay between the thermal forces and electrostatic forces originating in the fixed pH-dependent charge at the surface as governed by the Poisson- Boltzmann distribution. It leads to the Guy-Chapmann model of the diffuse layer which is formally identical with other models of space charge distribution. The potential decays exponentially from its surface value Ψ_0

$$\Psi_x = \Psi_0 \exp(-\kappa x) \tag{1}$$

with the effective thickness κ^{-1} of the space charge region being defined as

$$(8\pi C_0 z^2 e_0^2 \varepsilon kT)^{1/2} = \kappa \tag{2}$$

where ε is the dielectric constant, C_0 is the bulk concentration and e_0 is the charge of the electron.

The other model postulates the existence of the hydrated layer of finite thickness within which the de-protonated/protonated sites are situated. Clearly, in the limit of zero thickness of the hydrated layer the two theories merge. It is obvious that the SBT models the interface as a capacitor with one plate located at the surface and the other at the average distance κ^{-1} in solution. In other words the interface is considered to be ideally polarized. On the other hand the hydration layer model allows penetration of at least one type of ion through the interface and thus considers this interface to be non-polarized. How much non-polarized depends on the value of the exchange current density of the communicating ion(s). The equivalent electrical model is, of course, a parallel capacitor/resistor combination. From the point of view of the response it is obvious that the interface described by the capacitive model (SBT) would respond to adsorption of any charged species inside the space charge region. On the other hand adsorption would have little or no effect on the potential difference at the interface described by the hydration layer model because that potential difference is uniquely and unequivocally given by the Nernst equation. Of course the complicating factor in a real situation would be the existence of the mixed potential at the interfaces which have low total exhange current density.

The common ground to accommodate both models has been proposed by Sandifer[21] who has shown that the sub-Nernstian response of some materials and the presence of troublesome adsorption and stirring effects can be rationalized by considering the number of available binding sites and the thickness of the hydrater layer. The magnitude of the exchange current density depends on the concentration of the binding sites inside the hydrated layer. For glass electrode this is estimated to be 3.2 M.[22] As the number of binding sites decreases from 3 to 0.1 M the interfacial potential of a 10 Å thick layer is increasingly affected by adsorption (Fig.4) and the response deviates more from theoretical. On the other hand both the adsorption effects and the sub-Nernstian behavior vanish if the thickness of the hydrated layer is allowed to exceeds the thickness of the space charge layer inside the membrane Fig.5. This model goes a long way towards explaining some experimental results reported in the literature for ISFETs with oxide or nitride surfaces. Unfortunately, the properties of these materials prepared in different laboratories differ substantially. It is known that silicon nitride forms an oxygen-rich layer at the surface whose thickness depends on the deposition conditions. This "passivation" layer seems to form rapidly but is very stable even under continuous exposure to aqueous electrolyte solutions.[9] The hydration of this layer seems to fit the requirements and predicted behavior of the Sandifer model. The ISFETs which have been exposed to solution for more than one hour show no adsorption effects as would be expected from the SBT model.

Potentiometric Enzymatic Sensors

Enzymatic reactions combine substrate specificity with the high amplification factor. From that viewpoint they are ideal selective layers for chemical sensors, including potentiometric sensors. However, they are not specifically part of the information acquisition/processing scheme in the Nature. Their exclusive role is to lower, highly selectively, the activation energy barrier of certain reactions, thus acting as regulators. Because of this, their use in various sensing schemes may suffer from the problems related to the mode of coupling with the transducer.

Schematic diagram of an enzymatically coupled potentiometric sensor is shown in Fig.6. Its basic operating principle is simple: an enzyme (a catalyst) is immobilized inside a layer into which the substrate(s) diffuse. As it does it reacts according to the Michaelis-Menten mechanism and the product(s) diffuse out of the layer, into the solution. Any other species which participate in the reaction must also diffuse in and out of the layer. Because of the combined mass transport and chemical reaction this problem is often referred to as diffusion-reaction mechanism. It is quite common in electrochemical reactions where the electroactive

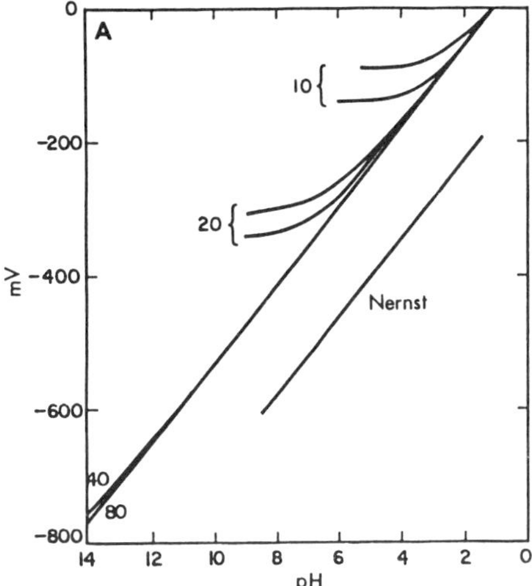

Fig.4. Effect of adsorption on ISFET surface. The thickness of the hydrated layer (in angstroms) is shown as parameter. The upper curve in each bracketed set corresponds to the absence of adsorption, the lower curve shows the adsorption of 100 mM adsorbate (reprinted from[21] with permission of J. Sandifer).

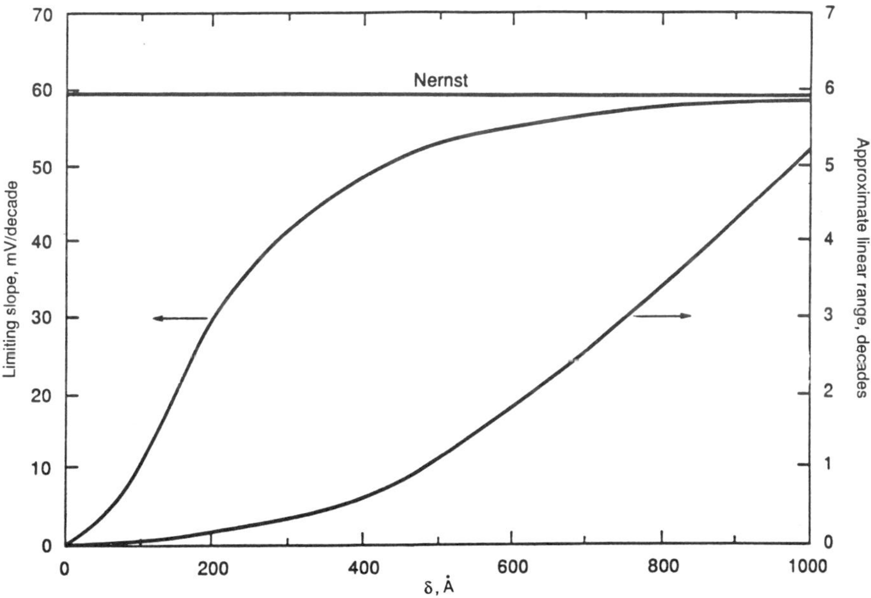

Fig.5. Effect of thickness of the hydrated layer on slope and dynamic range for the assumed concentration of binding sites Co = 0.1 mM (reprinted from[21] with permission of J. Sandifer).

species which undergo chemical transformation at the electrode, participates in some coupled chemical reactions. Mathematically this case is described by a set of second-order partial differential equations which are usually solved numerically. Because enzymes operate only in

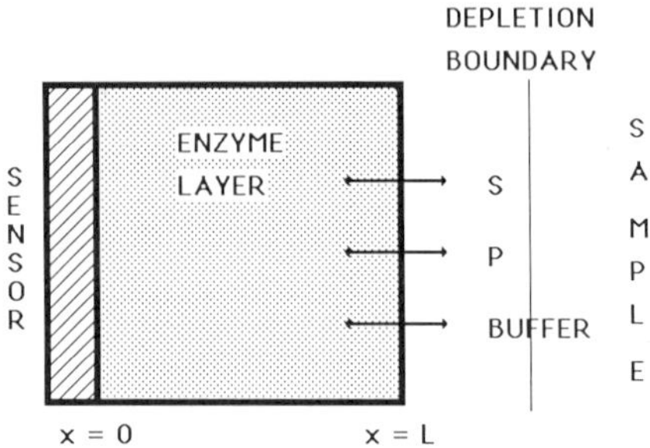

Fig.6. Potentiometric enzyme sensor.

aqueous environment the immobilization matrices are gels, specifically hydrogels.

The general diffusion-reaction equation for species i in one direction (x) is

$$\partial C_i/\partial t = +D_i(\partial^2 C_i/\partial x^2) \pm R(C_i) \tag{3}$$

When the pH-dependent Michaelis-Menten equation is substituted for the reaction term R (C_i) we obtain for substrate S

$$\frac{\partial C_s}{\partial t} = D_s\frac{\partial^2 C_s}{\partial x^2} - \frac{V_m C_s}{\Re(pH)(C_s + K_m)} \tag{4}$$

It is convenient to normalize the variables in Eq.4 with respect to the gel layer thickness, bulk concentrations and characteristic diffusional time. The normalized diffusion-reaction equation is then

$$\frac{\partial C_s}{\partial t} = D_s\frac{\partial^2 C_s}{\partial x^2} - \frac{\Phi^2 C_s}{1 + C_s} \tag{5}$$

where Φ is so called Thiele modulus.

$$\Phi = L[V_m/K_m D_S \Re(pH)]^{1/2} \tag{6}$$

This transformation is done for all variables. For $\Phi > 10$ the mechanism is diffusion controlled. Conversely, for small value of the Thiele modulus the process is reaction controlled, which means that the diffusional fluxes of all species participating in the reaction (Eq.5) are greater than the reaction term.

The actual solution of this problem depends on the initial and boundary conditions. These, in turn, depend on the approximations chosen for the model. Let us now review briefly the approximations which have been made for the enzymatically coupled potentiometric sensors and rank them in approximate order of seriousness.

(1) linear diffusion gradient inside the enzyme layer
(2) no partitioning of reactants and products between the gel and the sample
(3) no depletion layer at the gel/sample boundary
(4) no effect of mobile buffer capacity
(5) no effect of fixed buffer capacity
(6) no pH dependence of K_m
(7) no Donnan potential at the gel/sample boundary
(8) rates of protonation reactions are fast

Approximation (1) is a bad one despite the fact that it leads to simple mathematical solution: The concentration profiles are not linear. The partitioning of species between the gel and the sample (2) is also related to the existence of the Donnan potential (7) but it is a problem even for electrically neutral species (e.g. oxygen). If the solution is stirred the effect of the depletion layer at the gel/membrane interface is negligible (3). However, it could be a problem in stationary solutions. Approximations (4) through (6) would be the most serious for enzymatic sensors in which the output is related to the change of pH, because then the buffer capacity would have to be low. However, for sensors which use some other reactants/products but hydrogen ion a large excess of buffer would make the effects due to these assumptions negligible. Finally, rates of (almost) all protonation reactions (8) are extremely fast and they can be safely assumed to be instantaneous on the time-scale of all the other processes.

With these assumptions in mind we will now complete the outline of the solution of the diffusion-reaction problem as it applies to the most difficult case, the pH based enzymatic sensors. We assume only that there is no depletion layer at the gel/solution boundary (3), there is no fixed buffer capacity (5) and that the protonation reactions are very fast (8). The objective of this exercise is to find out the optimum thickness of the gel layer which is critically important for all zero-flux-boundary sensors.

As a rule hydrogen ion is involved not only in the pH-dependency of the reaction term (Thiele modulus) but also as the actively participating species involved in the acid-base equilibrium of the substrates, reaction intermediates, and products. Furthermore, enzymatic reactions are always carried out in the presence of a mobile buffer. By mobile we mean a weak acid or a weak base which can move in and out of the reaction layer, as opposed to immobile buffer represented by the gel (and the protein) itself. Thus, we have to include the normalized diffusion-reaction equation for hydrogen ion and for the buffer species. These equations have to be coupled with the buffer dissociation equilibrium.

Next we have to define the boundary and the initial conditions. For so called "zero flux" sensors there is no transport of any of the participating species across the sensor/enzyme layer boundary. Such condition would apply to, e.g., optical, thermal or potentiometric enzyme sensors. In that case the first space derivatives of all variables at point x are zero. On the other hand amperometric sensors would fall into the category of "non-zero-flux sensors" by this definition and the flux of at least one of the species (product or substrate) would be given by the current through the electrode.

The lack of the depletion layer at the gel/solution-boundary ($x=L$) is guaranteed by stirring of the sample. Under those conditions the concentration of all species at that boundary are equal to the bulk values and the initial conditions correspond to the situation in which all species except the substrate are initially present inside the enzyme layer.

This is a system of stiff, second-order partial differential equations which can be solved numerically to yield both transient, and steady state concentration profiles within the layer. Comparison of the experimental calibration curves and of the time response curves with the calculated ones provides the verification of the proposed model from which it is possible to determine the optimum thickness of the enzyme layer. Because the Thiele modulus is the controlling parameter in the diffusion-reaction equation it is obvious from Eq.6 that the optimum thickness L will depend on the other constants and functions included in the Thiele modulus. Because of this the optimum thickness will vary from one kinetic scheme to another.

Another inportant observation is related to the detection limit, dynamic range and sensitivity. For the expected values of the diffusion coefficient (in the gel) of approximately 10^{-6} cm^2 s^{-1} and substrate molecular weights about 300 the detection limit is approximately 10^{-4} M. This is due to the fact that the product of the enzymatic reaction is being removed from the membrane by diffusion with approximately the same rate as it is being supplied. The dynamic range of the sensor depends on the value of the K_m and V_m (which depends on the enzyme loading). Generally speaking, higher loading should extend the dynamic range at the top concentration range. It is often stated incorrectly that "the sensor has close to theoretical dependence" or a "nernstian response" which means that a one-decade change of the bulk concentration of the substrate is expected to yield a one-decade change of the surface ($x=0$) concenration. In the case of potentiometric enzyme sensors it would yield the slope of approximately 60 mV/decade. It is not intuitively obvious but clearly evident from the comparison of the experimental and calculated response curves that there is no "general theoretical" slope, each enzymatic sensor having its own depending on the mechanism and on the conditions under which it operates. We must remember, that decade/decade slope would occur only if a constant fraction of the product would reach the $x=0$ interface. The upper limit of the dynamic range depends on the value of the ratio $V_m/K_m D_s \Re(pH)$ in the

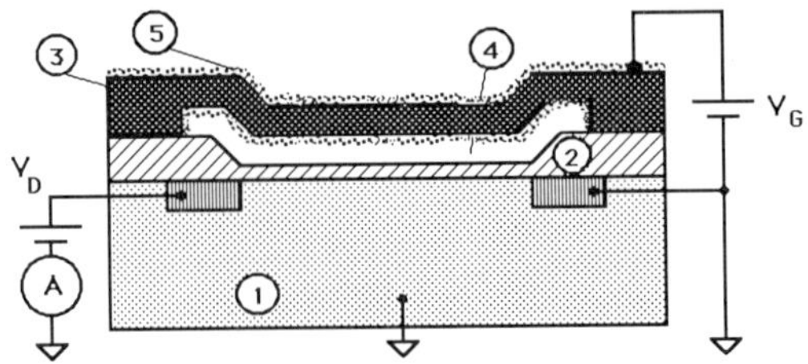

Fig.7. Diagram of the suspended gate field-effect transistor. 1 – substrate; 2 – insulator; 3 – suspended metal gate; 4 – gap; 5 – chemically selective layer.

Thiele modulus. It can be increased by enzyme loading but, obviously, only up to a point. Normally the dynamic range is approximately between 10^{-4} and 10^{-1}M.

The verification of the proposed model has been done on enzymatic field effect transistors for diffusion-reaction mechanism involving the oxidation of β-D-glucose catalyzed by glucose oxidase (GOD)/catalase system[23]

$$\text{GLUCOSE} + \text{E}_\text{o} \quad \Longleftrightarrow \quad [\text{E}_\text{o}\text{–GLU}] \quad \Longrightarrow \quad \text{E}_\text{R} + \text{GLUCONATE} + \text{H}^+$$

$$+$$

$$\text{O}_2$$

$$\Updownarrow$$

$$\text{E}_\text{o} + \text{H}_2\text{O}_2 \quad \Longleftarrow \quad \text{E}_\text{R} - \text{O}_2$$

$$\Updownarrow \quad \text{CATALASE}$$

$$\text{H}_2\text{O} + 1/2\,\text{O}_2$$

and for hydrolysis of penicillin catalyzed by β-lactamase.[24] In both cases hydrogen ion is the species detected by the transistor surface. As has been pointed out earlier the agreement between the predicted and experimental responses verifies the validity of the model and allows us to plot the concentration profiles inside the gel layer. The required thickness has been found to be approximately 150 μm for the glucose sensor and 20 μm for the penicillin. From these profiles we can see that the thinner membranes or directly immobilized enzymes would produce a reduced sensitivity. On the other hand the thicker the enzyme layer is the slower the response is. Thus, we can trade the time response for the signal-to-noise ratio by adjusting the thickness of the enzyme containing layer.

Other parameters which affect the response characteristics these sensors are the partitioning coefficients and the diffusion coefficients of all species, and the parameters related to the enzyme activity itself, such as the enzyme concentration, K_m and V_m. They are in turn affected by the preparation of the enzyme layer, e.g., by the degree of crosslinking, by the degree of the inactivation of the enzyme etc.. It is therefore not surprising to find that these devices have generally widely diverse life-time, time response, detection limits and sensitivity. The value of the experimentally verified model is then mainly in establishing the design parameters not only for potentiometric but also for other zero-flux boundary enzymatic sensors. It should be mentioned here that the above analysis of the potentiometric enzyme sensors has been done for ENFET but that it is equally applicable for other types of enzymatic potentiometric sensors.

Because each enzyme sensor has it own unique response it is necessary to construct the calibration curve for each sensor separately. The obvious way to reduce interferences is to use two sensors in differential mode. It is possible to prepare two identical enzyme sensors and either omit or deactivate the enzyme in one of them. This sensor then acts as a reference device. if the calibration curve is constructed by plotting the difference of the two outputs as the function of concentration of the substrate the effects of variations in the ambient composition of the sample as well as temperature and light sensitivity can substantially be reduced.

Work Function Sensors

The macroscopic device analogous to a solid state potentiometric work function sensor is the vibrating capacitor so called Kelvin probe.[25] The chemical modulation of the electron work function is a relatively new principle applied to chemical sensing. It relies on the fact that the two principal components of the electron work function, the Fermi level and the surface potential of an organic layer, change when a chemical species is adsorbed on or absorbed in such a layer. In order to make use of this effect we have developed in suspended gate field-effect transistor (SGFET) into which the organic polymer is introduced by electrochemical deposition (Fig.7). An advantage of this type of sensor is that a wide variety of organic semiconductors with different selectivity can be applied and used in a well controlled manner. This is particulary important for the fabrication of multisensors.[26,27] Furthermore, because this is a potentiometric (equilibrium) sensor there is no special requirement on the conductivity of the organic layer. The only pre-requisite is that it makes an electronic contact with the suspended metal gate, the condition which is easily satisfied for electrodeposited layers.

The origin of the chemical signal can be expressed in thermodynamic terms. At equilibrium the number of moles n of all species and their chemical potentials μ in a phase (e.g. in a chemically selective layer) are related through the Gibbs-Duhem equation which says that if a new species enters the organic layer the chemical potentials of all species in that layer must change. These include the change of the electrochemical potential of the electron - the Fermi level and therefore the electron work function.

The graphical representation of the situation in the gate of SGFET is shown in Fig.8 where the energy band diagram of the chemically selective layer is shown. In this figure the material is considered to be a p-type semiconductor. The position of the energy level for the dopant and thus the position of the Fermi level in the whole phase depends on the position of the intrinsic Fermi level of the pure material, E_{Fl}, on the electron donor/acceptor properties of the dopant and, if the dopant is charged, on the occupational density of the donor states E_D. Therefore, for a n-type material the dopant energy level (donor) would be located close to the conduction band edge, E_c, and the Fermi level would be closer to that edge accordingly. In Fig.8 the acceptor molecules (i.e. p-type semiconductor) are considered to be charged and therefore their distribution depends on their occupancy. This fact is shown by the symbol for a distribution ($\rangle$) in Fig.8.

The electrochemical potential of an electron can be expressed as the sum of the electrostatic energy and of the chemical potential

$$\tilde{\mu} = \mu - e\Phi_G \tag{7}$$

where Φ_G is the bulk (Galvani) energy of the phase. Because this energy is referenced against vacuum level it consists of the energy contributions resulting from the bulk potential ($\Psi = \Phi_B + E_{g/2}$) and from the surface (dipole) potential χ. Thus,

$$\tilde{\mu}' = \mu' + e\chi + e\Psi \tag{8}$$

The work function of the chemically sensitive layer Φ_L is then

$$-\Phi_L = \mu_L - e\chi_L \tag{9}$$

From Eq.9 we see that the chemical modulation of the work function can originate from two effects: action of the guest molecule on the energy state distribution in the bulk of the phase, i.e., by the absorption term μ_L in Eq.9 or by modulation of the surface potential χ_L, i.e., by adsorption. These two terms have different dependence on the activity of the guest molecule. The chemical potential follows the logarithmic law while the surface concentration depends on the type of the applicable adsorption isotherm. This may, in fact, create some problems

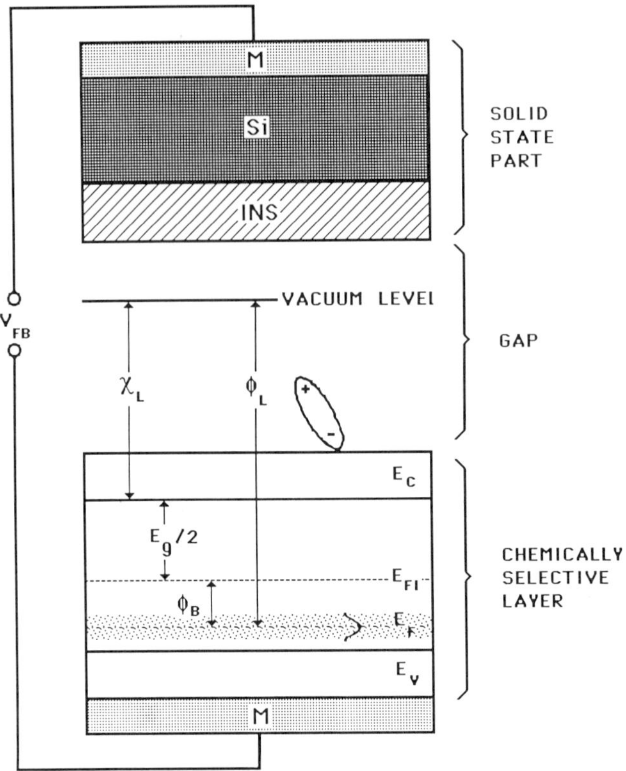

Fig.8. Energy band diagram of the gate of SGFET.

in resolving the two types of contributions to the overall change of the work function because the relative degree of their contribution is not known a priory.

An example of organic layer which changes its work function due to chemical modulation is electrochemically deposited polypyrrole. Because of its strong hydrogen bonding properties it binds water, alcohols, acetonitrile etc.[26,27] It can be deposited under variety of conditions such as different deposition potential, different electrolyte, different solvents and additives. The result is a spectrum of materials which show different affinity to different chemical species.

Acknowledgements

The financial assistance for presentation of this paper provided by the Humboldt Stiftung is gratefully acknowledged.

References

1. P. Bergveld, IEEE-Trans., BME-19, (1970) 70.
2. T. Matsuo, M. Esashi and K. Inuma, Digest of Joint Meeting of Tohoku, Sections of IEEE (1971).
3. S. Caras and J. Janata, Anal. Chem., **52** (1980) 1935.
4. I. Lundström and C. Svensson, in: "Solid State Chemical Sensors", J. Janata and R.J. Huber (Eds), Acad. Press, 1985.
5. H. Freiser, CW-ISE, Chapter 2, in: "Ion-Selective Electrodes in Analytical Chemistry", H. Freiser (Ed.), Vol. **2**, Plenum Press, N.Y., 1980.
6. M. Trojanowicz, Z. Augustowska, W. Matuszevski, G. Moraczewska and A. Hulanicky, Talanta, **29** (1982) 113.
7. X. Li, M.J. Verpoorte and D.J. Harrison, Anal. Chem., **60** (1988) 493.
8. T.A. Fjeldy and K. Nagy, J. Electrochem. Soc., **127** (1980) 1299.

9. J. Janata and R.J. Huber, in: "Ion-Selective Electrodes in Analytical Chemistry", H. Freiser (Ed.), Vol. **2**, 1983.

10. E.J. Fogt, D.F. Untereker, M.S. Norenberg and M.E. Meyerhoff, Anal. Chem., **57** (1985) 1995.

11. R.B. Brown, R.J. Huber and J. Janata, Proc. Transducers, Philadelphia, June 1985.

12. G.F. Blackburn and J. Janata, J. Electrochem. Soc., **129** (1982) 2580.

13. K. Bezegh, A. Bezegh, J. Janata, U. Oesch, A. Xu and W. Simon, Anal. Chem., **59** (1987) 2846.

14. E.J. Fogt, P.T. Calahan, A. Jevne and M.A. Schwinghammer, Anal. Chem., **57** (1985) 1155.

15. P.T. McBride, J. Janata, P.A. Comte, S.D. Moss and C.C. Johnsson, Anal. Chim. Acta, **101** (1978) 239.

16. U. Oesch, A. Xu, Z. Brzozka, G. Suter and W. Simon, Chimia, **40** (1986) 351.

17. N.J. Ho, J. Kratochvil, G.F. Blackburn and J. Janata, Sensors and Actuators, 4 (1983) 413.

18. K. Potje-Kamloth, M. Josowicz and J. Janata, Sensors and Actuators, 1989, in print.

19. T. Satchwill, and D.J. Harrison, J. Electroanal. Chem., **202** (1986) 75.

20. E. Lindner, E. Graf, Z. Niegreisz, K. Toth, E. Pungor and R.P. Buck, Anal. Chem., **60** (1988) 295.

21. J.R. Sandifer, Anal. Chem., **60** (1988) 1553.

22. G. Eisenmann, in: "Glass Electrodes for Hydrogen and Other Cations", Marcel Dekker, New York, 1967.

23. S.D. Caras, D. Petelenz and J. Janata, Anal. Chem., **57** (1985) 1920.

24. S.D. Caras and J. Janata, Anal. Chem., **57** (1985) 1924.

25. M. Josowicz and J. Janata, in: "Chemical Sensor Technology", Vol. **1**, T. Seiyama (Ed.), Kodansha Ltd., 1988.

26. M. Josowicz and J. Janata, Anal. Chem., **58** (1986) 514.

27. M. Josowicz, J. Janata, K. Ashley and S. Pons, Anal. Chem., **59** (1987) 253.

BIOSENSING BASED ON GAS SENSITIVE SEMICONDUCTOR DEVICES

I. Lundström and F. Winquist

Laboratory of Applied Physics
Linköping Institute of Technology
S-581 83 Linköping, Sweden

Introduction

The combination of catalytic metals and semiconductor devices has led to chemical sensors which appear to have both technical and medical applications.[1-5] More particularly, metal insulator semiconductor structures with the metal gate consisting of thin, discontinuous, iridium or platinum layers, have a large sensitivity to molecular ammonia.[6-9] These structures are so-called field effect devices which can be constructed, e.g., in the form of capacitors (Fig.1(a)) or transistors (Fig.1(b)). It was observed several years ago that hydrogen gas could shift the electrical characteristics of such devices along the voltage axis if the metal gate was made of a catalytic metal, namely palladium.[10] For these devices where the metal gate was thick enough to be continuous and non-porous, the voltage shift is due to hydrogen atoms adsorbed at the metal-insulator interface where they give rise to a dipole layer changing the work function of the metal at the metal-insulator interface (see Fig.2(a)). It was found, however, that this type of device was only to a very small extent sensitive to ammonia, although ammonia molecules can be dehydrogenated on a number of catalytic metal surfaces. Since ammonia molecules (or rather ammonium ions) are produced in a large number of biochemical reactions, we found it of interest to develop a field effect structure sensitive to ammonia. It was discovered that gates of catalytic metal films thin enough (of the order of 10 nm) to be discontinuous gave the field effect devices a large ammonia sensitivity. The details behind the ammonia sensitivity are not fully understood yet. The experimental observations suggest, however, that the voltage shift in this case is due to surface potential changes of the catalytic metal film which are capacitively coupled to the semiconductor surface through the discontinuities of the metal film (see illustration in Fig. 2(b)). The surface potential changes are caused by the adsorption of ammonia and its reaction with oxygen on the catalytic metal surface. It has been found that metal layer thicknesses between 3 and 30 nm in general yield a large ammonia sensitivity.[8] For a device (with a thin Pt-gate) operating at 150°C in air, the detection limit is lower than 1 ppm NH_3. Furthermore, devices with a thin Ir-gate operating at 30-50°C in air behind a gas permeable membrane can detect about 0.2 μM ammonium in a buffer at pH 8.5 on the other side of the membrane. The last example illustrates that the devices can be used also to monitor ammonium ions by detecting ammonia molecules in equilibrium with the ions.

Ammonia gas sensing is therefore an alternative to the use of ammonia selective electrodes, which may offer certain advantages in many cases. Gas sensors can be used for biosensing in several ways as illustrated in Fig.3. We have earlier demonstrated that urea, creatinine and a number of amino acids can be assayed with a flow through system and a reaction column containing the corresponding, ammonia liberating, immobilized enzyme.[11,12] The main purpose of the present paper is to describe some of the recent bioanalytical applications of the ammonia sensitive field effect structures. We have, e.g., used the devices in combination with urease to detect small amounts of mercury. Simple assay methods were developed using either solution chemistry in a small sample cell or dry reagent chemistry in a solid carrier.[13]

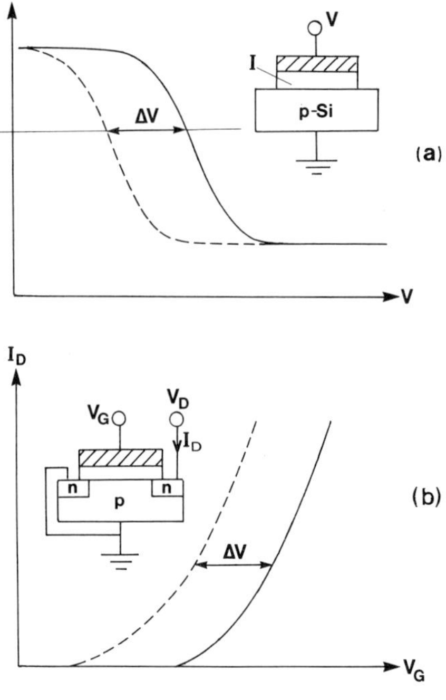

Fig.1. Two different field effect structures. (a) Schematic drawing of a metal-insulator-semiconductor capacitor. The insulator is often SiO_2 and the semiconductor Si. The C(V)-curve, the small signal capacitance of the MIS-structure as a function of the applied dc-voltage, V, is also shown. The C(V)-curve is the so-called high frequency curve normally measured at a frequency of 1 MHz. The dashed C(V)-curve illustrates the shift of the C(V)-curve when a gas sensitive MIS structure is exposed to a molecule it is sensitive to. The thin horizontal line illustrates the constant capacitance set by the feedback circuit used to measure ΔV. (b) Schematic drawing of a metal-insulator-semiconductor field effect transistor. In this case, gas exposure results in a shift of the so-called drain current (I_D)-gate voltage (V_G) characteristics along the voltage axis.

The assay is based on the inhibitory effect of mercury on the enzyme urease and hence on its rate of production of ammonia from urea. We also describe the assay of creatinine using dry reagent strips.

The test structures used were iridium - silicon dioxide-silicon capacitors with rather large area. Available commercial hydrogen and ammonia sensors consist, e.g., of a sensing element in the form of a transistor, a heater (a diffused resistor) and a temperature sensors (a pn-junction) on the same chip with dimensions smaller than about 1×1 mm^2.[14] Research and development work is, however, in several cases most easily done on metal-insulator semiconductor capacitors mounted on a thermostatted sample holder. The description of the test structures, their fabrication and physics, given below is, however, very short. More details can be found in several of the references, e.g., ref. [1 − 3, 8].

Monitoring of Ammonia

The sensors used were Ir-SiO_2-Si structures made on p-type silicon with 100 nm thermally grown silicon dioxide. Iridium was evaporated with an electron-gun through a metal mask having a circular hole of 2 mm diameter. The nominal thickness of the iridium film

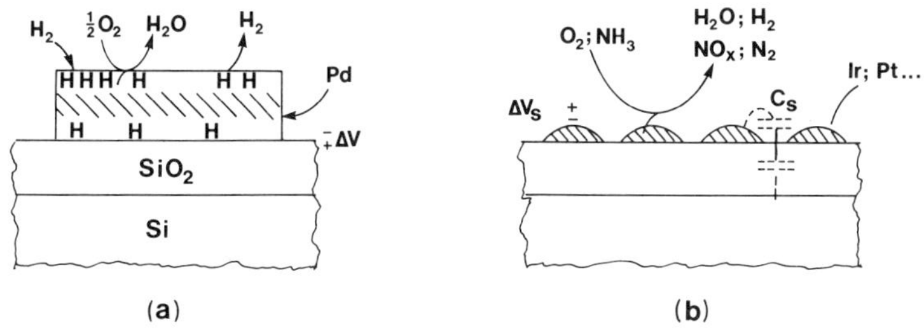

Fig.2. Two different detection mechanisms. (a) For a thick, non-porous, catalytic metal film (like Pd), hydrogen atoms give rise to a dipole layer at the metal-insulator interface. The hydrogen atoms emanate from catalytic reactions on the metal surface. The voltage drop, ΔV, across the dipole layer causes a shift of the $C(V)$ or $I_D V_G$ curve along the voltage axis. (b) For a thin, discontinuous catalytic metal film, also surface potential changes of the metal islands, ΔV_s, due to reaction intermediates and/or adsorbates will be felt by the semiconductor surface through a capacitive coupling over the stray capacitance, C_s. This causes a voltage shift ΔV of the electrical characteristics ($< \Delta V_s$).

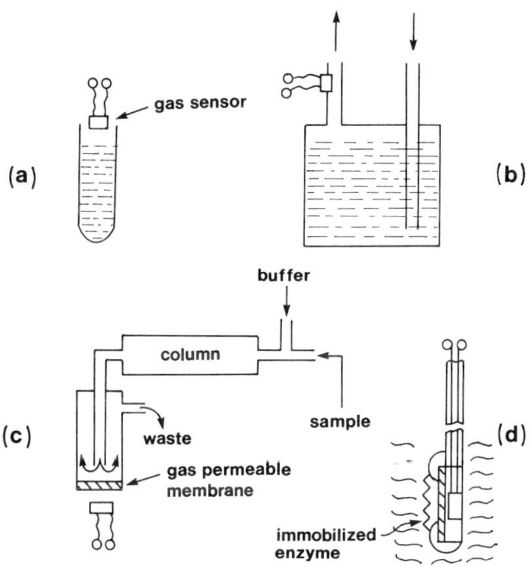

Fig.3. Schematic illustration of some different ways of using a gas sensitive device for biosensing or bioanalytical purposes:
(a) the device applied directly above a reaction vessel,
(b) the device applied in the exhaust of a reaction vessel (e.g., a fermenter),
(c) the device applied behind a gas permeable membrane in a flow through system,
(d) the device applied behind a gas permeable membrane in a probe for insertion in the solution, i.e. in a biosensor configuration.

was 3 nm and it was evaporated on top of the V-part of a previously evaporated Y-shaped thick (contact) layer of palladium. The back side of the semiconductor was covered with

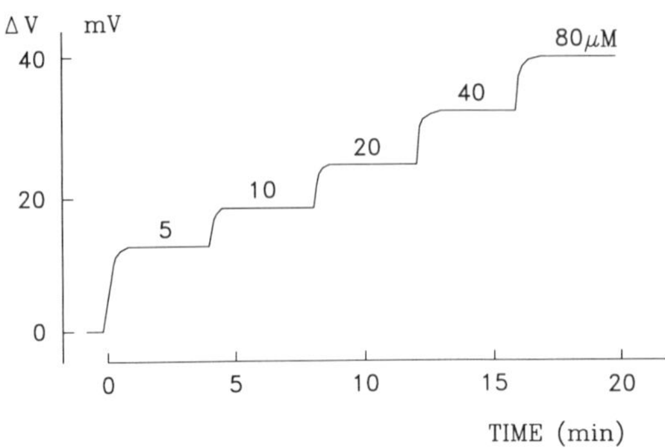

Fig.4. Response of a 3 nm IrTMOS to ammonium ions at different concentrations. The sensor was placed on a thermostatted sample holder (35°C) close to ($\approx$0.1 mm below) the gas permeable membrane of a "microcell".

aluminum. The thin iridium metal-oxide-semi-conductor capacitor (IrTMOS-) was placed on a thermostatted aluminum plate. The capacitance voltage curve of the IrTMOS-structure was monitored with a capacitance bridge and a feedback circuit making it possible to follow the voltage necessary to yield a given preset capacitance as illustrated by the horizontal line in Fig.1(a). In this way, the position of the C(V)-curve along the voltage axis can be followed by plotting this voltage on a strip chart recorder.

Fig.4 shows the response to ammonia of the IrTMOS behind a gas permeable membrane. The sample cell used was a microcell, later used for the determination of urease activity. The microcell was made of a 9.5 mm long teflon tube (inner diameter 2 mm) covered at the bottom end by a gas permeable (teflon) membrane.[13] The microcell was placed on top of the thermostatted IrTMOS structure with its membrane about 0.1 mm from the surface of the sensor. The sample consisted in this case of given amounts of ammonium chloride. The sensor normally reaches 90 % of its final response within 1 minute. The recovery time after removal of ammonium from the solution (not shown in Fig.4) is of the order of 2 minutes. It was found that, for the experimental setup in Fig.4 and with the 3 nm IrTMOS, the relationship between the ammonium concentration (in μM) and the steady state response, ΔV, of the device (in mV) is[13]

$$(\Delta V)^2 = 26.3[\mathrm{NH_4^+}] \tag{1}$$

Equation (1) is an empirical relationship which depends not only on the type of device used but also on the gas permeable membrane and the geometrical arrangements. The form of equation (1) is, however, consistent with the fact that for low temperature operation (30-50°C) and for sufficiently small concentrations of ammonia, the steady state response of IrTMOS devices to ammonia can be approximated by[15]

$$\Delta V = K[\mathrm{NH_3}]^{1/2} \tag{2}$$

where K is a constant.

In the experiments to be described, the rate of ammonia production is measured as a continuous change of ΔV. The response is therefore determined both by the sensitivity of the device as expressed by equation (1) and by the speed of response of the measurement system (device plus gas permeable membrane). If it is assumed that the rate of ammonia production is low enough, then $\Delta V^2 \sim [\mathrm{NH_4^+}(t)]$ at all times. This assumption is not important for the evaluation of the experiments since $\Delta V(t)$ is anyhow directly related to the enzymatic activity in the sample cell. It enables us, however, to estimate the concentration of ammonia at a given time, which is a more familiar parameter.

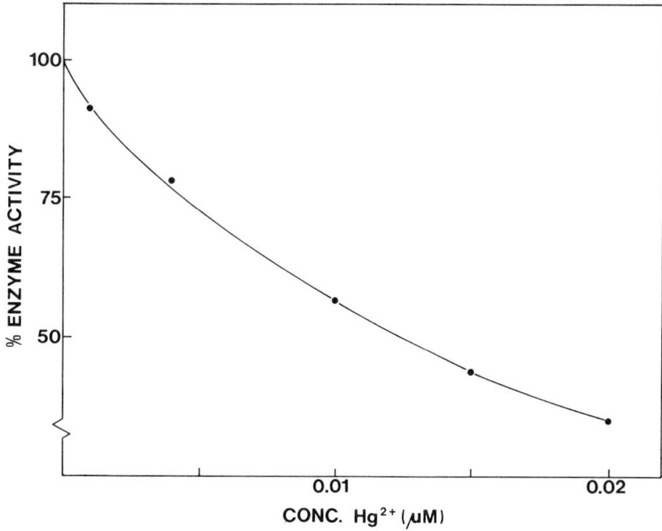

Fig.5. Relative urease activity in the microcell as a function of added mercury concentration. The activity was determined from the slope of ΔV^2 versus time plots obtained upon addition of 5 μl of 300 mM urea to the microcell containing 25 μl of urease with a specific activity of 3.9×10^{-3} U/ml. The microcell contained also different amounts of $HgCl_2$ as given on the abscissa.

Mercury Determined with a Microcell Arrangement

The inhibitory effect of Hg^{2+} on urease activity was studied by the addition of known standards of $HgCl_2$ into the microcell containing a specific activity of $3.9.10^{-3}$ U/ml of urease. The cell contained 25 μl of urease/$HgCl_2$ solution in 0.05 M Tris-HCl buffer at pH 8.3. 5 μl of the same buffer containing 300 mM urea was added to the microcell to initiate the enzyme catalyzed reaction. Details about the choice of enzyme and substrate concentration are found in ref. [13].

The production of ammonia from urea gives rise to a shift of the C(V)-curve along the voltage axis as discussed above. The inhibition of the urease activity by Hg^{2+} was estimated from the slope of ΔV^2 versus time plots, compared to the slope without added mercury(II) chloride. In Fig.5, we have shown the observed relative enzyme activity as a function of the Hg^{2+}-concentration. The detection limit appears to be about 0.001 μM Hg^{2+} in this measurement system. One sample could be processed within 5 minutes.[13]

Mercury Determined with Dry Reagent Carriers

Dry reagent chemistry offers several advantages over conventional analysis especially in terms of sample handling and safety. Several analytical methods have thus been developed based on the use of different types of sample carriers and mainly optical detection methods.[16] We therefore found it worthwhile to develop a dry reagent carrier for enzymatic activity measurements to be used together with an ammonia sensitive IrTMOS structure. The dry reagent carrier consists of two paper strips mounted on a polyethylene plate as illustrated in Fig.6. One carrier, the reagent carrier, was prepared by adding 20 μl of buffer (0.5 M Tris-HCl at pH 8.3) containing a specific urease activity of 5×10^{-3} U/ml to a piece of filter paper (Munktell, Sweden; area 100 mm^2; thickness 0.2 mm; void volume 80 %). The activator carrier was prepared by adding 10 μl of 50 mM urea in buffer to a filter paper of 50 mm^2 area. The prepared carriers were stored dried in a closed container at 4°C. Reagent and activator strips were cut into suitable sizes from the prepared papers (see Fig.6). Further details of the dry reagent carrier plate are found in ref. [13].

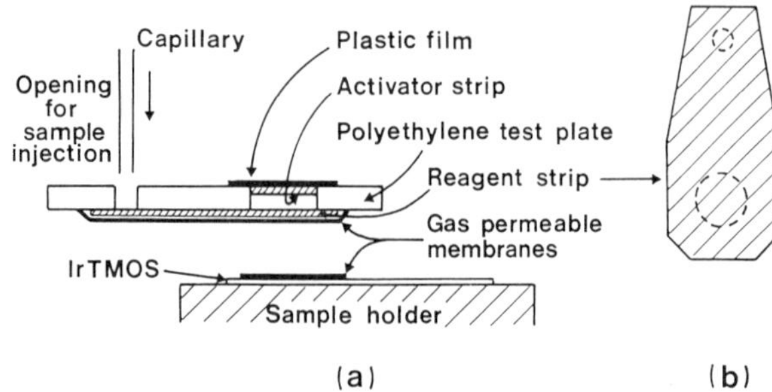

Fig.6. Schematic drawing of the solid sample and reagent carrier used to monitor enzymatic activity. A gas permeable membrane was applied also directly over the IrTMOS. The reagent strip was 5 mm long and 3 mm wide at its widest part. The holes in the polyethylene test plate were 1.2 mm (sample injection) and 2.5 mm (for the activator strip) in diameter, respectively. The activator strip had a diameter of 2.2 mm. The distance between the two strips was 0.8 mm. The completed test plate was placed on the gaspermeable membrane of the IrTMOS during the analysis.

A capillary is used for the sample collection (3 μl). The end of the capillary is pressed lightly against the reagent strip through the test hole. The reagent strip absorbs then 2 μl of the sample solution. When the activator strip is pressed against the reagent strip, the enzyme catalyzed reaction starts. The sensor is activated after each measurement by heating to 120°C.

Due to a small water vapor sensitivity of the IrTMOS structure, the signal from the sensor was allowed to stabilze for 3.5 minutes after the absorption of the sample before the activator strip was pressed down. Again the relative enzyme activity (in the reagent strip) was determined from slopes of ΔV^2 versus time plots. Some results are shown in Fig.7. The sensitivity of this solid carrier system is about 0.005 μM Hg^{2+}. One sample was analyzed in less than 8 minutes.[13]

Determination of Creatinine with a Dry Reagent Carrier

In the example above, we determined the enzymatic activity by using a given amount of substrate. The dry reagent carrier can, however, also be used to determine substrate concentrations where we have taken creatinine as an example. In this case, the reagent strip contained dried 0.1 M Tris-HCl buffer, pH 8.5 and the activator strip dried creatinine iminohydrolase (freeze dried enzyme dissolved in distilled water; total activity in the strip 6×10^{-3} U). A sample is absorbed in the reagent strip as before. The activator strip is pressed against the reagent strip and in this case, the reaction is allowed to run until a (roughly) constant sensor signal is obtained. This is thus a kind of "end point determination" which takes 3-4 minutes. Fig.8 shows some results obtained with creatinine standars (dissolved in distilled water). It is observed that $\Delta V^2 \sim$ [creatinine] in accordance with the discussion in the Introduction. These preliminary results are encouraging, although the spread between two different sets of experiments is still rather large.

Discussion

We have shown that the measurement of ammonia gas behind a gas permeable membrane can be an interesting alternative to the use of a pH electrode in solution. Ammonia sensors based on metal-oxide-semiconductor structures appear to be useful in this context; they are reasonably fast, stable and sensitive. It is furthermore demonstrated that these devices can be used in combination with dry reagent chemistry to provide simple assay methods. The determination of substrates which can be made to release ammonia in the presence of the

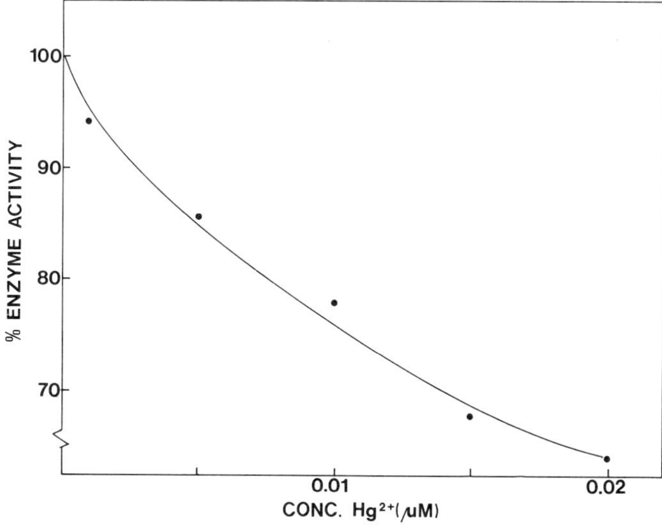

Fig.7. Relative urease activity in the reagent strip as a function of mercury concentration in the reagent strip. The activity was determined from slopes of ΔV^2 versus time plots after that the activator strip had been pressed against the reagent strip.

proper enzymes is straightforward as shown both in this paper and in earlier publications.[11,12]

Determination of traces of pollutants has become very important. The environment and living organisms are especially vulnerable to the toxic heavy metals. In many cases, a simple and fast method for the determination of heavy metals is desirable. Enzymes which are inhibited by the heavy metals offer an interesting possibility in this context, and several applications of this idea have been made earlier such as for the determination of small concentrations of mercury and copper.[17] We have shown that heavy metals (mercury) can also be detected, e.g., with the use of urease in combination with an IrTMOS ammonia sensor. Although several parameters can still be optimized, our results suggest that simple equipment for field use may be constructed around this type of sensor. The choice of enzyme (urease) was made out of convenience. There may be other enzymes which perform better. Furthermore the choice of enzyme will also determine the specificity of the enzyme-heavy metal system. It may be valuable to have both general heavy metal "sensors" as well as specific ones.

There are also possibilities to detect other species with the gas sensitive field effect structures. Palladium gate devices are, e.g., sensitive to H_2, H_2S and at elevated temperatures also to, e.g., alcohols.[18] Although IrTMOS structures, when operated up to 200°C, are highly selective to NH_3 (only H_2, low molecular amines and water vapor will interfere), it has been shown that PtTMOS structures operated at temperatures above 170°C are sensitive to a various degree also to alcohols, unsaturated hydrocarbones, ketones, and also some other organic compounds.[19,20] It is therefore possible to use the ideas presented in this paper also for other than ammonia producing enzyme-substrate pairs.

In the bioassay application, the specificity is guaranteed by the biochemical reaction used in combination with the sensor. It is of interest to note, however, that the sensitivity of the sensors for different species depends both on the nature of the catalytic metal and its operation temperature.[18-21] It is therefore possible to envisage arrays of sensors capable of giving real time information about the composition of a solution or gas stream which may lead to more versatile biosensors or bioanalytical systems. One of the advantages of the semiconductor technology is the possibility to make small devices in a reproducible manner. Furthermore, other functions than the sensing one can be integrated on the same semiconductor chip.

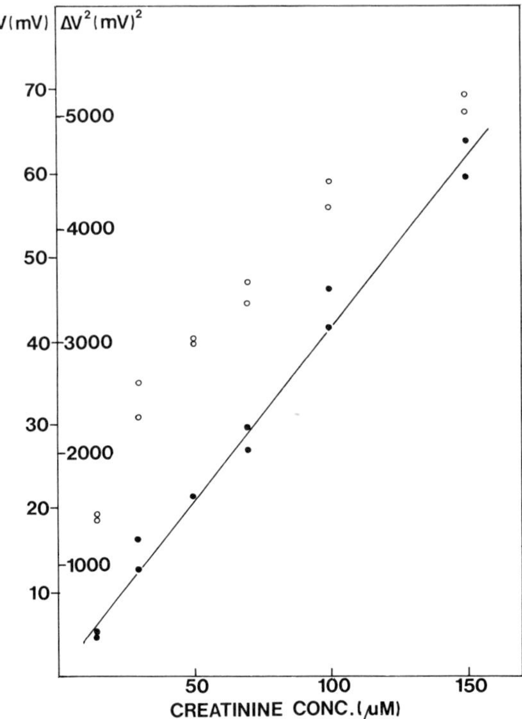

Fig.8. Observed "steady state" voltage shift of the IrTMOS structure caused by different concentrations of creatinine in the reagent strip (unfilled circles). The activator strip contained a given amount of creatinine iminohydrolase. The sample containing creatinine was adsorbed in the reagent strip and the enzyme catalyzed reaction was started by pressing the activator strip against the reagent strip. The voltage shift was read when the signal on the strip chart recorder became stable with time. The drawing shows also that ΔV^2 is approximately a linear function of the creatine concentration (black dots).

Acknowledgements

We like to thank Dr. Bengt Danielsson and Mrs. Anita Spetz for stimulating discussions on the present topic. Our work on chemical sensors is supported by a grant from the National Swedish Board for Technical Development.

References

1. I. Lundström and C. Svensson, in: "Solid State Chemical Sensors", J. Janata and R.J. Huber (Eds), Academic Press, New York, 1985, pp. 1-63.
2. I. Lundström, M. Armgarth, A. Spetz and F. Winquist, Sensors and Actuators, **10** (1986) 399.
3. A. Sibbald, Journal of Molecular Electronics, **2** (1986) 51.
4. F. Winquist, A. Spetz, M. Armgarth, I. Lundström and B. Danielsson, Sensors and Actuators, **8** (1985) 91.
5. I. Lundström, A. Spetz and F. Winquist, Phil. Trans. R. Soc. Lond., B 361, (1987) 47.
6. A. Spetz, F. Winquist, C. Nylander and I. Lundström, Proc. Int. Meeting Chemical Sensors, Fukuoka, 1983 p. 479.
7. F. Winquist, A. Spetz, M. Armgarth, C. Nylander and I. Lundström, Appl. Phys. Lett., **43** (1983) 839.

8. A. Spetz, M. Armgarth and I. Lundström, Sensors and Actuators, **11** (1987) 349.

9. J.F. Ross, I. Robins and B.C. Webb, Sensors and Actuators, **11** (1987) 73.

10. I. Lundström, M.S. Shivaraman, C.M. Svensson and L. Lundkvist, Appl. Phys. Lett. **26** (1975) 55.

11. F. Winquist, A. Spetz, I. Lundström and B. Danielsson, Anal. Chim. Acta, **163** (1984) 143.

12. F. Winquist, I. Lundström and B. Danielsson, Anal. Chem., **58** (1986) 145.

13. F. Winquist, I. Lundström and B. Danielsson, Anal. Lett., **21** (1988) 1801.

14. Sensistor AB, P.O. Box 76, S-581 01, Linköping, Sweden.

15. F. Winquist, A. Spetz, I. Lundström and B. Danielsson, Anal. Chim. Acta, **164** (1984) 127.

16. B. Walter, Anal. Chem., **55** (1983) 498A.

17. B. Mattiasson, B. Danielsson, C. Hermansson and K. Mosbach, FEBS Lett., **85** (1978) 203.

18. U. Ackelid, M. Armgarth, A. Spetz and I. Lundström, IEEE Electron Device Lett. EDL-7, (1986) 353.

19. F. Winquist and I. Lundström, Sensors and Actuators, **12** (1987) 255.

20. U. Ackelid, F. Winquist and I. Lundström, Proc. 2nd Int. Meeting Chemical Sensors, Bordeaux, 1986, p. 395.

21. M. Armgarth, U. Ackelid and I. Lundström, Digest of Technical Papers, Transducers' 87, Tokyo, 1987, p. 640.

CHEMICALLY MODIFIED ELECTRODES FOR THE ELECTROCATALYTIC OXIDATION OF NADH

L. Gorton[1], B. Persson[1], M. Polasek[2] and G. Johansson[1]

1. *Dept. of Analytical Chemistry*
   University of Lund
   P.O.Box 124 S-221 00 Lund, Sweden
2. *Analytical Chemistry Dept.*
   Faculty of Pharmacy
   CS-50165 Hradec Kralove, Czechoslovakia

Introduction

The development of chemically modified electrodes, CMEs, has now reached a level of maturity which allows the inclusion of such electrochemical cells into various sensing devices.[1] Special interest has been focused on the electron transfer or rather the hindrances for electron transfer during bioelectrochemical reactions. Studies of the oxidation of the reduced form of the nicotinamide coenzymes NADH and NADPH are attractive because a single electrochemical transducer reaction can be combined with any of the great number of dehydrogenases to give a sensor with the desired selectivity.

Direct electrochemical oxidation of coenzymes NADH and NADPH is possible at a high overvoltage, but it may result in electrode fouling.[2,3] Immobilization of mediating functionalities (catalysts) on an electrode surface can reduce the overvoltage and overcome difficulties encountered with unmodified electrodes.[4] We have found that mediators of the phenoxazine type, Fig.1, incorporating a paraphenylene-diimine moiety are particularly attractive because they seem to be selective for the NADH-catalysis.[5] Graphite and a number of other carbon electrode materials are easily modified with phenoxazine dyes by just dipping the electrode into a solution containing the mediator. The resulting coverage depends on the concentration of the dissolved mediator and the time allowed for adsorption. Very large coverages ($\Gamma > 10^{-8}$ mol cm^{-2}) can be obtained within a few minutes.

The stability of the modified layer towards desorption is mainly governed by the number of conjugated aromatic rings, so that a mediator containing a large number of rings makes a more stable CME than one with a low number of rings. The strong interaction between the carboneous surface and the phenoxazine (π-electron overlapping) results in a very fast charge transfer between the electrode proper and the modifier. This is demonstrated by the small peak separation, ΔE_p, between the peaks of the oxidation and reduction waves in cyclic voltammetry.[6,7] For coverages below 10^{-9} mole cm^{-2} the ΔE_p usually takes a value of about 5–20 mV, for scan rates up to about 400 mV s^{-1}. At higher coverages and for faster scan rates larger ΔE_p-values are obtained.[6] The oxidation and reduction waves are almost mirror images allowing the formal potential, $E^{0\prime}$, of the adsorbed species to be evaluated as the mean value of the peak potentials of the oxidation and reduction waves.[8]

Protons are invovled in the redox conversion of the mediator and the $E^{0\prime}$, will therefore move with a change of pH in the contacting solution. Phenoxazine dyes (see Fig.2) in solution undergo a 2e$^-$ redox conversion. Depending on the number of protons also taking part in the redox process, the $E^{0\prime}$ will move with 90 mV/pH (3H$^+$), 60 mV/pH (2H$^+$) or 30 mV/pH (1H$^+$). The number of electrons, n, taking part in the redox conversion for a mediator immobilized on the electrode, is reflected by the width of the voltammetric wave at half peak

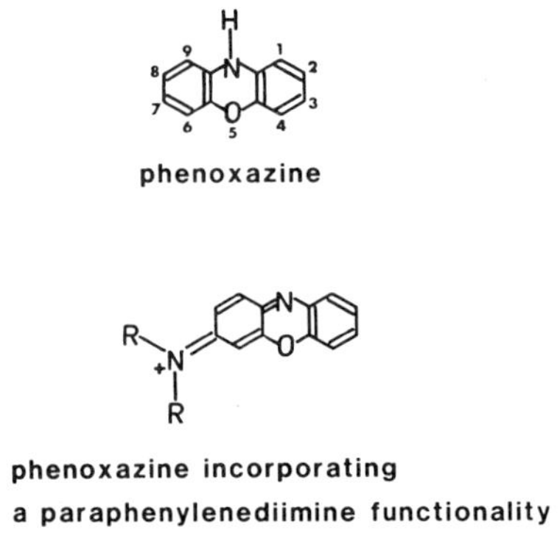

Fig.1. Structural formulae of phenoxazine and of a phenoxazine with a paraphenylene diimino functionality.

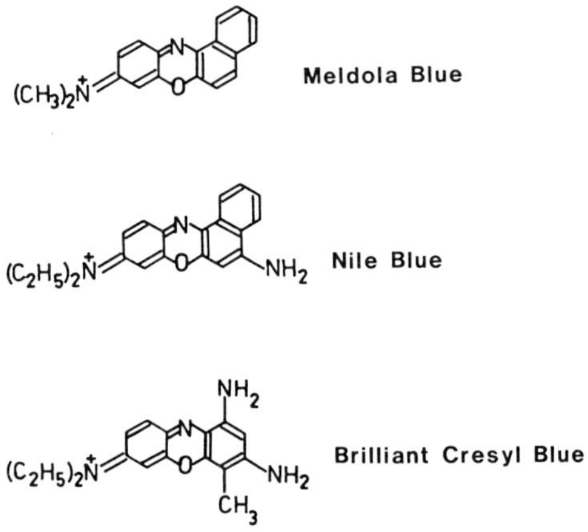

Fig.2. Structural formulae of Meldola Blue, Nile Blue, and Brilliant Cresyl Blue.

height $\delta_{0.5}$, which theoretically should equal $90.5/n$ mV.[8] The number n can also be obtained from the straight line obtained when plotting the peak current, i_p, versus the scan rate,

$$i_p = (n^2 F^2 / 4RT) A \Gamma v \tag{1}$$

Most adsorbed phenoxazine derivatives give slightly broader peaks than what is theoretically stipulated for a Langmuirian adsorption process. The number n, whether evaluated from the $\delta_{0.5}$ or from eqn.(1), is usually somewhat lower than two, reflecting interactions between the adsorbed molecules.[8] The immobilization process may strongly influence the properties of the adsorbed species, which is clearly revealed for Nile Blue in Fig.3. Both the energy level, as well as the various pK_a-values of the oxidized and the reduced forms, are changed on adsorption.[9-11]

Mediators

The phenoxazines can mediate the electron transfer from NADH in solution to the electrode. The reaction sequence can be summarized according to the following:

$$NADH + M_{ox} \xrightarrow{k_{obs}} NAD^+ + M_{red} \tag{2}$$

In the first step NADH reduces the oxidized form of the mediator, M_{ox}, whereby the reduced form, M_{red}, and NAD^+ are produced. In the next step M_{red} is reoxidized to form

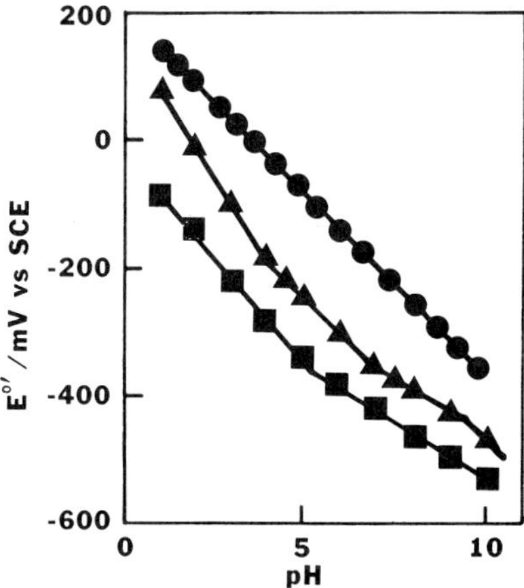

Fig.3. Variation of $E^{0\prime}$ with pH for Nile Blue adsorbed on graphite
(–□–□–), Nile Blue dissolved in aqueous solution (–△–△–), bis (benzophenoxazinyl) derivative of terephthaloic acid adsorbed on graphite
(–○–○–).

the active mediator. When a potential, E_{appl}, more positive than the $E^{0\prime}$, is applied to the
CME, the following reaction will take place.

$$M_{red} \xrightarrow{E_{appl} > E^{0\prime}} M_{ox} + H^+ + 2e^- \qquad (3)$$

The actual number of protons taking part in the reduction and oxidation of the mediator
depends on the structural elements of the phenoxazine.[4,11]

In a previous work we found that the reaction rate k_{obs} between NADH in solution
and the adsorbed phenoxazine mediator depends on a number of factors. To obtain kinetic
data the phenoxazine-CME was mounted in a rotating device and experiments were run and
evaluated according to Levich and Koutecky.[12]

The highest reaction rates were obtained for the mediators with the most oxidative
$E^{0\prime}$-values. A linear correlation was found when plotting the logarithm of the rate coefficients
versus the $E^{0\prime}$-values.[4] The commercially available phenoxazine Meldola Blue, with an $E^{0\prime}$-value at pH 7.0 of -185 mV vs SCE was found to be the most efficient one having a rate
constant of $3 \times 10^4 M^{-1} s^{-1}$ at this pH.[4]

The various structural elements of the mediator are also of great importance. Introducing other redox active groups, may drastically lower the reaction rate expected from the
linear log k_{obs} vs $E^{0\prime}$-relationship stated above.[4] Exchanging a charged iminogroup, Fig.1, for
an uncharged or a ketogroup has been stated to decrease the reaction rate with NADH.[13,14]

Different values of the rate coefficient, k_{obs}, were obtained when the NADH-concentration was varied.[2,6,7] This observation was found for all the various phenoxazine mediators
investigated. A formation of a complex between NADH and the adsorbed mediator could explain this behaviour. The reaction can be described (rewriting eq.(2)) to take place according
to the following:

$$\text{NADH} + M_{ox} \underset{k_{-1}}{\overset{k_{+1}}{\rightleftharpoons}} complex \overset{k_{+2}}{\rightarrow} \text{NAD}^+ + M_{red} \qquad (3)$$

The overal reaction rate, k_{obs}, can thus be described as:

$$k_{obs} = k_{+2} / (K_M + [NADH]) \qquad (4)$$

where

$$K_M = (k_{-1} + k_{+2}) / k_{+1} \qquad (5)$$

185

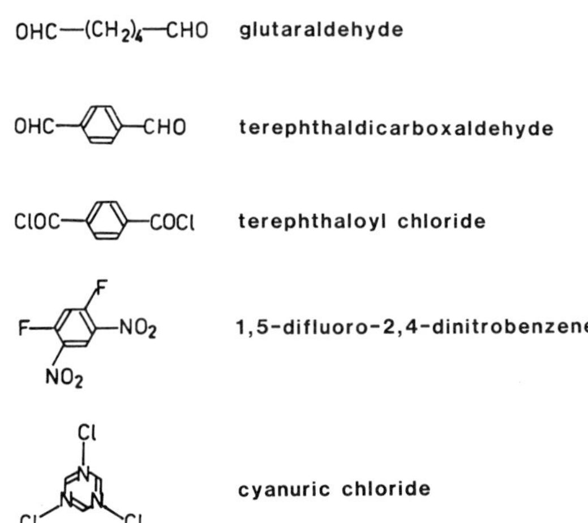

Fig. 4. Structural formulae of some bi- and trifunctional reagents.

The various rate coefficients, k_{obs}, k_{+2} and K_M can be evaluated from rotating disk electrode experiments.[4,6,7]

The reaction mechanism, eq.(3) will cause the current response to a constant NADH-concentration to increase linearly with surface coverage for small values of Γ and to level off towards a constant plateau for high values of Γ.[6] Thus once the coverage has reached a certain minimal value, the current response will virtually be unaffected by the coverage.[5]

The pH of the contacting solution has a profound effect on the reaction rate of eqn.(3). Under otherwise constant conditions, k_{obs} will decrease with an increase in pH. This effect seems to be common to all phenoxazine mediators.[4,7,15,16] For Meldola Blue, k_{obs} reached almost a value of 10^5 M^{-1} s^{-1} at pH 6.0.[4] A virtually mass controlled current could thus be obtained at this pH with this CME working as an NADH sensor in a FIA-setup.[5]

Straight calibration curves for NADH over at least three orders of magnitude were obtained between pH 6 and 9 with phenoxazine CMEs as sensors in flow systems despite the decreasing reaction rate. As long as the effective NADH-concentration at the electrode-solution interface is well below the K_M-value a linear response characteristic is expected. K_M-values in the millimolar level have been obtained for most phenoxazines.[4] When using phenoxazine-CME as sensors in FIA-systems with normal dispersion factors[17] of about 10 the upper linear response range is expected around 10–30 mM.

A high pH can also have a detrimental effect on the mediator. Phenoxazines, like Meldola Blue, which are derivatized in position 3 or 7, will decompose at pH-values higher than 7.[18,19] However, if they are derivatized in both these positions they will become alkaline-stable. The stability towards desorption of a phenoxazine-CME depends largely on the number of aromatic rings of the phenoxazine structure. The drawbacks of the restricted stability of electrodes modified with Meldola Blue, having only 4 rings and being alkaline instable, were partly overcome by reacting the amine function in position 3 of Nile Blue, whereby the aromatic ring system could be increased with either a naphthalene[10] or a pyrene moiety.[15]

Mediators with Several Aromatic Centra

Polyfunctional molecules including more than one catalytically active phenoxazine part can be synthesized by coupling the original phenoxazine dye with bi- or trifunctional reagents, see Fig.4. Such a coupling can overcome several problems encountered with simple phenoxazines. The desorption of the polyfunctional mediators, see e.g. Fig.5, seems to be very small due to the large number of aromatic rings. A Nile Blue derivative of terephthaloyl chloride, Fig.5, can be produced if the bifunctional reagent is reacted either with the reduced form of Nile Blue or with the imino form of oxidized Nile Blue, see Fig.6. Electrodes modified with this derivative could be used for a month, even in flowing solution, without noticeable

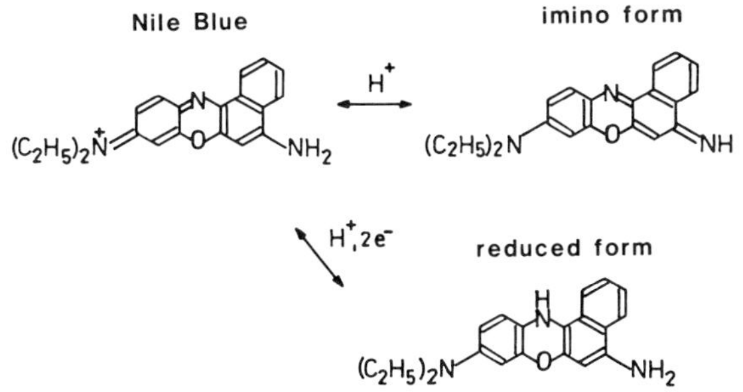

Fig.5. Structural formulae of Nile Blue, its imino and reduced forms.

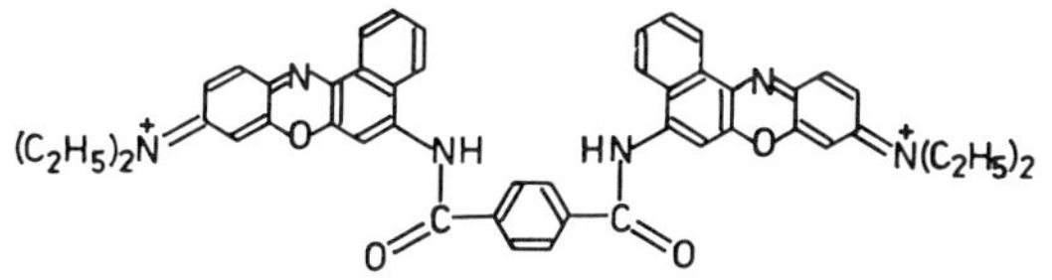

Fig.6. Proposed formula of a bis(benzophenoxazinyl) derivative obtained when reacting Nile Blue with terephthaloyl chloride.

indications of desorption, when followed by surface coverage measurements. The coupling increases the $E^{0\prime}$-value of adsorbed Nile Blue from −430 to −200 mV vs SCE at pH 7.0, see Fig.3. The alkaline stability of the compound permits the CME to be used at pH 9.0, which is about 2 pH-units higher than the upper pH-range for Meldola Blue.[6] The somewhat lower $E^{0\prime}$-value compared to Meldola Blue results in lower rate coefficients for the NADH oxidation, c.f. eqn.(3). preliminary values, $k_{+2} = 40\ s^{-1}$ and $K_M = 2 \times 10^{-3}$ M (pH 7.0) with a resulting $k_{obs(NADH=O)} = 2 \times 10^4\ M^{-1}\ s^{-1}$ (pH 7.0) were evaluated by rotating disk electrode measurements at various NADH-concentrations. The lower reaction rates compared to Meldola Blue implies that the electrode response is under partial kinetic rather than under full mass transfer control.

Fig.7 shows a series of cyclic voltammograms obtained at various pHs for one electrode modified with this Nile Blue derivative (Fig.6). It can be seen that the shapes of the waves will vary both with pH and with buffer constituents.[6] The variation of $E^{0\prime}$ with pH for this compound is depicted in Fig.3 and shows a single linear pH dependence, 60 mV/pH, through the whole pH range investigated, pH 1–9. This indicates that equal numbers of protons and electrons take part in the redox conversion of the adsorbed species.

An increased value of $E^{0\prime}$ of the adsorbed phenoxazine is expected to increase the reaction rate with NADH[4]. If it is large enough ($k_{obs} > 10^5\ M^{-1}\ s^{-1}$ at all pHs) it should be possible to produce a virtually pH insensitive NADH sensor. A number of other phenoxazines (e.g. Brilliant Cresyl Blue, Fig.2) and other coupling reagents (Fig.4) are therefore under study at present.

Two examples may be given, by reacting the reduced form of Nile Blue with 1,5-dichloro-2,5-dinitrobenzene, one compound could be obtained with an $E^{0\prime}$-value of −135 mV vs SCE (pH 7.0), and by reacting the iminoform of Brilliant Cresyl Blue with terephthaloyl chloride a compound could be obtained with an $E^{0\prime}$-value of −55 mV vs SCE (pH 7.0). They were both found to be catalytically active for NADH-oxidation. No kinetic data are available yet, however. Purification and identification of both the commercially available phenoxazines (usually with a purity of 50–90 %) and of the reaction products are far from straightforward.

The basic understanding of phenoxazines as mediators for NADH-oxidation and as modifiers for preparation of stable CMEs are demonstrated by the work discussed above. The prospects seem therefore to be good to make optimal phenoxazine mediators with high

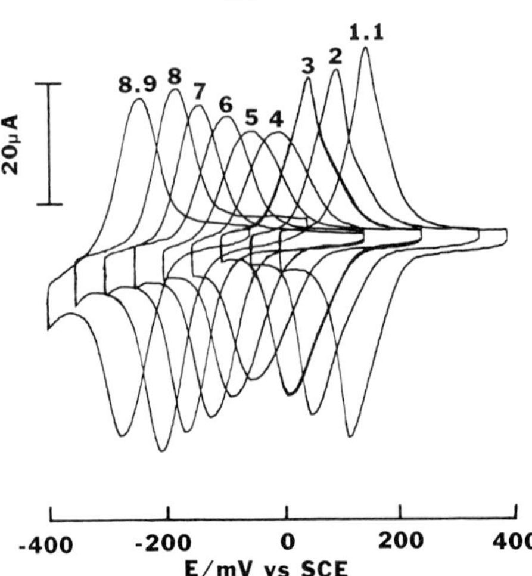

Fig.7. Cyclic voltammograms of a graphite electrode modified with the substance depicted in Fig.6. The surface coverage was 2.7×10^{-9} mole cm^{-2} and the scan rate was 50 mV s^{-1}. The buffers used were 0.1 M HCl (pH 1.1), 0.25 M phosphate buffer (pH 2–8) and 0.15 M pyrophosphate (pH 8.9).

reaction rates at all pHs, alkaline stability and desirable adsorption properties.

Acknowledgements

This work was granted by supports of the Swedish National Reserach Council (NFR), The Swedish Board for Tecnical Development (STUF) and the National Energy Research Board (STEV).

References

1. R.W. Murray, A.G. Ewing and R.A. Durst, Anal. Chem., **59** (1987) 379A.
2. Z. Samec and P.J. Elving, J. Electroanal. Chem., **144** (1983) 217.
3. J. Moiroux and P.J. Elving, J. Am. Chem. Soc., **102** (1980) 6533.
4. L. Gorton, J. Chem. Soc., Faraday Trans., I, **82** (1986) 1245.
5. R. Appelqvist, G. Marko-Varga, L. Gorton, A. Torstensson and G. Johansson, Anal. Chim. Acta, **169** (1985) 237.
6. L. Gorton, A. Torstensson, H. Jaegfeldt and G. Johansson, J. Electroanal. Chem., **161** (1984) 103.
7. L. Gorton, G. Johansson and A. Torstensson, J. Electroanal. Chem., **196** (1985) 81.
8. E. Laviron, in: "Electroanalytical Chemistry", A.J. Bard (Ed.), Vol. **12**, Marcel Dekker, New York, pp. 53-157 (1982).
9. H. Huck, Ber. Bunsenges. Phys. Chem., **87** (1983) 945.
10. H. Huck, Fresenius' Z. Anal. Chem., **313** (1982) 548.
11. J.M. Ottaway, in: "Indicators", E. Bishop (Ed.), Pergamon Press, Oxford, pp. 498-503 (1972).
12. A.J. Bard and L.R. Faulkner, in: "Electrochemical Methods, Fundamentals and Applications", Wiley, New York, 1980.
13. A. Kitani and L.L. Miller, J. Am. Chem. Soc., **103** (1981) 3595.
14. A. Kitani, Y.H. So and L.L. Miller, J. Am. Chem. Soc., **103** (1981) 7636.
15. B. Persson, L. Gorton and G. Johansson, in: "Proc. 2nd Int. Meeting on Chemical Sensors", J-L. Aucouturier, J.-S. Cauhapé, M. Destriau, P. Hagenmuller, C. Lucat, F.

Menil, J. Portier and J. Salardenne (Eds), Imprimerie Biscaye, Bordeaux, pp. 584-587 (1986).
16. M. Polasek, L. Gorton, R. Appelqvist, G. Marko-Varga and G. Johansson, in: "Thesis", R. Appelqvist, Lund University, 1987.
17. J. Růžička and E. Hansen, in: "Flow Injection Analysis", Wiley, New York, 1981.
18. M. Kotouček, J. Tomášová and S. Durčáková, Collect. Czech. Chem. Commun., **34** (1969) 212.
19. M. Kotouček and J. Zavadilová, Collect. Czech. Chem. Commun., **37** (1972) 3212.

SOLID POLYMER ELECTROLYTES FOR GAS SENSING ELECTRODES

Lionel S. Goldring*

Biomedical Sensors Unit, Chemistry Department
University College of Swansea
Swansea SA2 8PP, UK
and
Novametrix Medical Systems, Inc.
Wallingford, CT 06492, U.S.A.

Introduction

In 1954 Leland Clark demonstrated that a platinum cathode would measure the oxygen concentration of blood when it and a reference electrode were covered by an oxygen permeable membrane. Later in that same year Stow and Severinghaus showed that carbon dioxide could be estimated in blood with a glass electrode fitted with a gas permeable membrane. In the seventies the Huchs demonstrated that mechanical adaptations of these devices could be utilized to provide transcutaneous (non-invasive) measurement of arterial blood gas concentration if the skin area surrounding the sensor was heated to 44 - 45°C.

This work resulted from an effort to increase the convenience of transcutaneous electrode measurements by the development of a nonaqueous, solid electrolyte which would reduce electrode preparation problems and be capable of an extended storage life.

History of Ionic Conductivity in Solid Polymer Electrolytes

Ionic conductivity in polymer-based, water-containing, solid systems has been with us for a long time. A review of that history, which is intimately associated with the history of analytical electrochemistry and the physical chemistry of electrolyte solutions, will help to put the present work into perspective.

A. Aqueous Gels

Agar gels have been used for many decades; although it may seem strange to consider these systems as examples of solid polymer electrolytes, we will see that they share many properties with most polymer-based, ionically conducting systems. A variety of other polymers, both synthetic and natural product derivatives, will gel aqueous solutions.

B. Ion Exchange Membranes

1. Membranes in the 20'. Collodion-based, ion exchange membranes were prepared in the twenties; although their ion exchange capacity and conductivity were too low for most uses, it was demonstrated that most of the behavior predicted by Donnan was in fact observed. The theory of concentration potentials was developed and experimentally verified.

2. Synthetic Cross-linked, Polymer-based Materials. Synthetic, ion exchange resins based upon phenol-formaldehyde were prepared in the late thirties; this was followed by the

---

* Present Address: Ionetics, Inc., Costa Mesa, CA 92626, U.S.A.

Contemporary Electroanalytical Chemistry, Edited by A. Ivaska *et al.*
Plenum Press, New York, 1990

synthesis of polystyrene-based materials in the early forties. These materials were characterized as "homogeneous gels," implying that their internal structure, on a molecular scale, was as uniform as aqueous solutions.

By 1950 a number of workers were preparing "heterogeneous" membranes by mixing ion exchange materials with inert binders. These were utilized in electrochemical research, functioning as solid conductors, and some were commercially available for a few years.

In the early fifties a number of workers demonstrated the synthesis of "homogeneous", high capacity, high conductivity synthetic polymer membranes prepared both by condensation and addition polymerization. Both cation and anion exchange membranes based upon cross-linked polystyrene, reinforced by a fabric mesh became commercially available.

a) Electrodialysis for Water Purification. These materials had both the chemical and physical properties required for the electrical desalting of water and other ion separation processes. Their ion exchange capacity was sufficiently high so that near theoretical concentration potentials could be obtained for solutions that were several tenths molar.

3. Polymer Grafts/Blend — Chen AMF. Later in the fifties Chen and others demonstrated that ion exchange membranes could be prepared by "grafting" styrene and other vinyl monomers into polymer films such as PE and a modified ClTFE. After grafting, these were converted to cation exchange membranes by chlorosulfonation and into anion exchange membranes by chloromethylation and amination. These materials, which did not require reinforcement, were stronger and more flexible than the homogeneous membranes. Because the aromatic ring content of these membranes was lower than the poly(styrene)-based membranes, the ion concentration was lower, and, consequently, the resistivity of these materials was inferior to the homogeneous membranes. The improved mechanical properties made much thinner membranes possible resulting in lower overall membrane resistance.

a) Various Chemical Separations. Commercial electrodialysis equipment was developed for water purification, desalting of whey, removal of acid from food products, and a variety of waste treatment processes; most of these had only limited commercial success. In all of these processes solid polymer conductivity was of economic importance.

4. GE Membrane-based Fuel Cell. Around 1960 a group at General Electric invented the ion exchange membrane-based fuel cell. They were, I believe, the first to use the phrase: "solid polymer electrolyte," to describe the ionic conductor which it contained. Initially these cells utilized the graft-copolymer membranes described above, although 1 kW scale fuel cells were built for the Gemini space program, sustained operation at high current density showed a steady increase in internal resistance. GE initially believed that the polystyrene sulfonic acid was leaching out and efforts were made to increase cross-linking. Subsequently I showed that this was due to a loss in ion exchange capacity caused by free radical attack on the tertiary hydrogen of the polystyrene and subsequent oxidative chain cleavage. The short chain fragments which eventually resulted were water soluble and appeared in the product water of the fuel cell.

5. duPont Nafion. In parallel with the GE fuel cell development duPont developed a family of perfluorosulfonic and carboxylic acid membranes which have no sensitivity to free radical attack. These materials were rapidly adopted for the Gemini fuel-cell program. In the 70's it was demonstrated that these materials make effective separators for chloralkali cells and they have been commercially utilized for this on a significant scale.

C. Ion-Selective Electrode Membranes

Metal electrodes (and the hydrogen electrode) have been utilized for concentration measurement almost since the beginning of electrochemistry late in 19th century. Since early in this century chemists have been fascinated by the knowledge that a selectively permeable membrane would allow concentration measurement in a similar fashion. Before 1910 the remarkable selectivity of the glass electrode for hydrogen ion had been discovered and by the 30's it was commercially available and the ion exchange theory of its operation had been presented.

Although a wide range of synthetic and natural, organic and inorganic materials showing some selectivity and ionic conductivity were described from time to time, none was widely utilized until Ross described a "liquid" membrane sensor for calcium in 1967. This was based upon a calcium complexing agent dissolved in a low-volatility, hydrophobic solvent. It was captured within a porous hydrophobic polymer matrix which allowed it to function as a membrane. *1. PVC Based mtls from the Late Sixties.* Soon thereafter, several groups, nearly simultaneously, showed that similar performance could be achieved by utilizing the same or similar complexing agents dissolved in low-volatility solvents compatible with and capable of plasticizing poly(vinyl chloride). Membranes were prepared from these materials by casting from a volatile solvent or by conventional plastisol technology.

D. Battery Separator Materials

Some battery separators are swollen by battery electrolyte solutions and function as a solid polymer matrix with mobile conducting ions. In the last few years a number of alkali metal containing polymers, usually based upon poly(ethylene oxide) or derivatives thereof, have been described for use with non-aqueous Li based batteries.

Heterogeneous Nature of the Above Structures

All of the systems described above are solid, they contain polymer, and they are electrolytes but they all contain water. They differ enormously in their conductivity, ion concentration, mobility of the conducting ions, and even the nature of the structure which contains (retains?) the ions. In trying to classify these materials I have been unable to find clear demarcations between any of them. Depending upon the polymer, aqueous gels can be prepared with 1 - 10 % polymer but plasticized polymers are readily obtainable with polymer levels as low as 10 %. Most ion exchange resins are cross-linked but the fluorosulfonic acid ones are not; in terms of chemical structure they must be thought of as swollen gels, in which intermolecular bonding is not covalent. Some of these materials have covalently bonded ionic groups, but liquid membrane electrodes and their close relatives, the PVC-based ion-selective membranes, both of which function by ion exchange, do not.

In all of the water containing systems, it is possible to substitute high dielectric constant, non-agueous solvents so they cannot be distinguished from a plasticized polymer on this basis.

In all of the systems at least some of the ions are not covalently bonded and these account from the conductivity. At one extreme — the agar gels — the mobile ions may well be removable by water extraction. Clearly part of the salt can be removed, by water washing, from ion exchange membranes operated with high external electrolyte concentrations as is often the case. Charged or uncharged ion carriers of the PVC-based ion-selective electrodes may be leached from the plasticized polymer matrix; in practice the electrodes give satisfactory performance for a long enough time to be commercially acceptable. These membranes are quite permeable to water which plays a significant role in their performance.

A. Goldring, Mid 60's

On a molecular scale all of these polymer-based systems consist of two "phases:" a continuous, polymer network and a continuous solvent network. I consider them "heterogeneous" on a molecular scale because the dielectric constant in the solvent phase varies locally. I was one of the first to argue (in 1966) that the socalled "homogeneous gel" model of ion exchange resins was not possible. The dielectric constant in the vicinity of a sulfonate group cannot be the same as it is in the vicinity of a hydrocarbon (or fluorocarbon) backbone.

B. Exp. Demo. of Heterogeneity in the 80's

In the late 70's and early 80's several workers (Yeo and Eisenberg, Lee, and others) provided convincing experimental evidence of the molecular scale heterogeneity of some of these materials; especially the perfluorosulfonic acid membranes.

C. Conductivity via Ion Mobility in the Fluid Phase

All of these materials conduct by virtue of ionic mobility in a continuous fluid phase and they are solid by virtue of the continuous polymer phase.

Limitations of Above Materials

All of the solid polymer electrolyte systems described above have been considered for application to gas-sensing electrodes. All have one or more limitations which seemed to preclude successful application.

A. Gels

1. Solvent Volatility. Water is too volatile for effective use in gas sensors which may be expected to operate without attention for hours and even days and to be stable for a shelf-life which must exceed several months. While it is possible to gel non-aqueous, low-volatility solvents with suitable polymers, a number of problems prevented the successful developement of a commercially acceptable system.

2. Purification Problems. Many of the gelling agents are natural products or derivatives thereof and there can be severe impurity problems with them, because, trace impurities can produce profound effects with electrochemical sensors.

B. Ion Exchange Membranes

Ion exchange membranes have several limitations.

1. Thickness. The commercially available materials are too rigid and much too thick. In principle it might be possible to produce custom materials in the proper thickness but they lack the ability to conform to the microscopic contours of electrodes.

2. Lack of Solubility. With one notable exception (the fluorosulfonic acid membranes) they lack solubility so they are difficult (or impossible) to coat onto electrodes or membranes in the thin layers required for gas-sensing electrodes.

3. Reference Electrode Demands. More fundamental than the above is the problem of providing an exchangeable ion for the reference electrode. Cation exchange materials exclude the halide ions needed for silver halide reference electrodes.

4. Deleterious Effect of Quaternary Amines. Anion exchange membranes are all amine based and amines, even quaternary, may not be compatible with silver halide electrodes because they accelerate silver deposition on the cathode.

C. Ion-Selective Membranes

1. Conductivity. The PVC-based materials used for ion selective membranes have very low conductivities.

2. Polymers (and plasticizers) Lack Props for Conductivity. Furthermore, the properties of these materials (their low polarity and the resulting low dielectric constant) seem to preclude high conductivity. This may not be important for potentiometric sensors but it is crucial for voltammetric sensors.

System Requirements

It is complex to attempt to offer general specifications for a solid polymer electrolyte which will be useful under all conditions. However, a few generalizations are possible. To achieve a reasonable transient response — say of the order of a minute or less — the electrolyte should be of the order of 25 μm in thickness. Since the response time goes as the square of the thickness it is clear that a two or three-fold increase is the most that is likely to be tolerable.

Furthermore, the thickness should be small compared to the distance to the edge of the electrode else edge effects can disturb readings. These requirements place some limitations upon the materials themselves and the techniques for applying them to electrodes.

The conductivity in the electrolyte is essentially governed by the iR drop that can be tolerated; the current is, of cource, controlled by the cathode size, the geometric and diffusion properties of the electrolyte, the concentration of the analyte, and, perhaps the permeability of the membrane.

A. General Transcutaneous

In addition to the general problems of gas-sensing electrodes, transcutaneous electrodes have some special problems.

1. Electrolyte Thickness. A response time of the order of 10 to 20 seconds (T99) is desirable. This implies a solid polymer electrolyte thickness less than 25 μm. This is more or less the same for both oxygen and carbon dioxide electrodes.

2. Electrolyte Conductivity. For oxygen electrodes the electrolyte conductivity must be high enough so that iR drop does not lead to excessive non-linearity at high oxygen concentrations.

3. Diffusion and Sensitivity. The diffusion rate of the gases within the solid polymer electrolyte must be high enough to give an adequate reduction current; this is commonly expressed in terms of the electrode sensitivity, i.e. the current resulting from a given gas partial pressure.

4. Electrode Response Stability. For transcutaneous electrodes response stability of the order of 1 % per hour of operation is required; recalibration every four hours is the most that is clinically acceptable. Liquid electrolyte oxygen electrodes typically show stability to well under 10 % per week. Continuous electrode operation for the order of a week is expected.

For the carbon dioxide electrode this translates into a stability of the order of 0.3 to 0.4 mV per hour.

a) Silver Deposition Problems. Silver deposition on the cathode of oxygen electrodes can be a problem because it may alter the electrode shape and sensitivity. This is true even though the silver ion reduction current is negligible compared to the oxygen current because the electrode is operated for hundreds of hours. Hence, transcutaneous electrodes may be polished routinely.

With liquid electrolytes, it is possible to utilize a chloride ion concentration that will minimize silver solubility. This is desirable with solid polymer electrolytes but more difficult to achieve because of an absence of data. However, it is worth noting that it is reported that silver chloride shows higher solubility in most non-aqueous electrolytes.

Solid Polymer Electrolyte

To develop a suitable solid polymer electrolyte several chemical conditions must be satisfied simultaneously. For salts to ionize the solvent (plasticizer) should have a moderately high dielectric constant.

A. Plasticizers with Low Volatility and High Dielectric Constant.

For effective use as a plasticizer a liquid should have a low vapor pressure; this more or less implies a boiling point over 250°C.

B. Compatible Polymers

Compatibility of polymer and plasticizer implies solubility but there should also be a strong enough chemical interaction to insure longterm stability.

C. Salts with Suitable Properties

Soluble salts are needed and one must have an ion suitable for stabilizing the anode reaction; practically, this means a halide because the silver halides are the only practical reference electrodes that have been identified. Others may be possible theoretically but I'm not aware of any successful utilization.

D. Mutual Volatile Solvent

A volatile solvent in which polymer, plasticizer, and salts are soluble is needed to provide a vehicle for the preparation of suitably thin solid polymer electrolyte films.

Performance of Electrodes

In this work we first demonstrated that it was possible to gel a nonvolatile solvent with a polymer. Solid electrolyte coated membranes were prepared which could be shipped. Polymer impurities caused rapid silver deposition and this approach was abandoned.

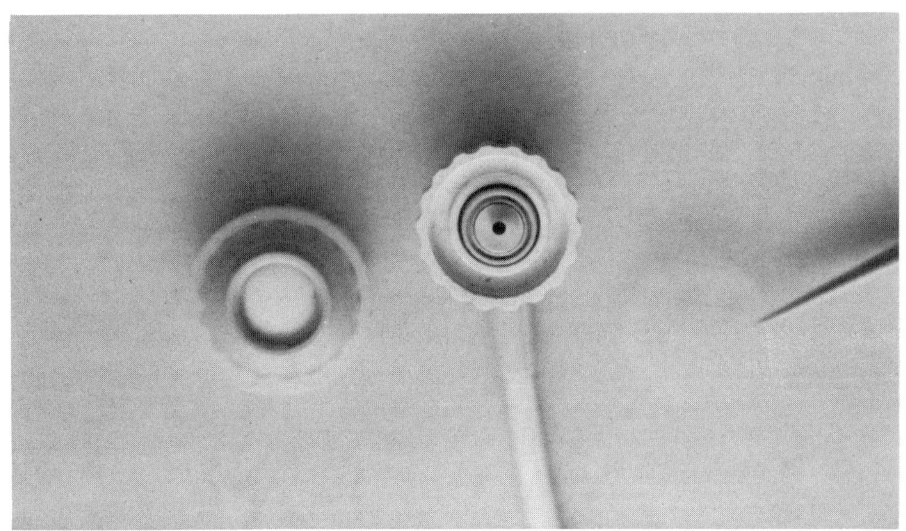

Phot.1. A gas sensing electrode.

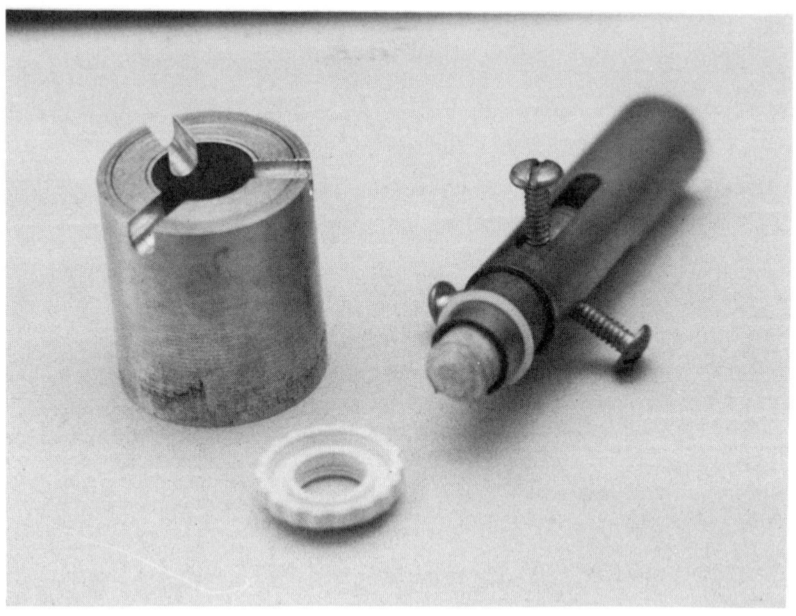

Phot.2. Membraning tool for a gas electrode.

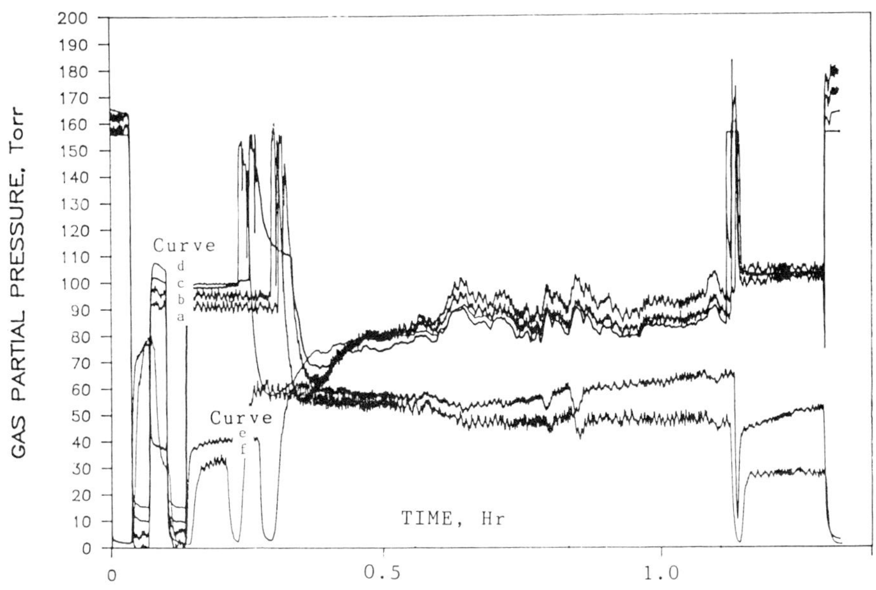

Fig.1. *In vivo*: transcutaneous O_2/CO_2 electrodes. Solid polymer electrolyte vs liquid electrolyte.
Curves:
    a, b, c, d: oxygen
        e, f: carbon dioxide
    a, b, f: solid polymer electrolyte
    c, d, e: conventional liquid electrolyte
    b, c, d: 5, 10, 15 torr displacement, respectively

Solid polymer electrolytes have been prepared from highly plasticized polymers. Early success with an oxygen electrode was demonstrated with several of the first systems developed. Again high silver deposition rates and excessive volatility prevented practical application of most of these. Finally, a polymer-plasticizer system showing adequate sensitivity and stability for application in some gas-sensing applications was developed.

A gas sensing electrode and the membraining tool can bee seen in photos 1 and 2.

Fig.1 compares the in vivo performance of oxygen and carbon dioxide electrodes with a non-aqueous, solid polymer electrolyte with electrodes containing a conventional liquid electrolytes. These electrodes have been operating for several weeks with virtually unaltered sensitivity.

The measurements were made with an Analog Devices AD 815 data aquisition board installed in an IBM PC/AT. They were recorded under the control of a software package: Laboratory Technologies Notebook. This also controlled the solenoids, by means of the AD 815 digital output channels, which changed the gases used in calibration. The readings obtained during calibration with 10 % CO_2 — 90 % nitrogen and 5 % CO_2 — 12 % oxygen — 83 % nitrogen were averaged and used to normalize both CO_2 and oxygen measurements. This was done by means of programs written in Lotus 1-2-3 macro language which also generated the plots.

Acknowledgements

The British Technology Group contributed to the support of this work.

The assistance of Dr. Rory Gowland and Mr. John Rowley in preparing and testing literally hundreds of formulations is gratefully acknowledged.

CARBON FIBER MICROELECTRODES

Karin Potje-Kamloth, Petr Janata* and Mira Josowicz
Institut für Physik, Fakultät für Elektrotechnik
Universität der Bundeswehr München
Werner Heisenberg Weg 39, D-8014 Neubiberg, F.R.G.

Introduction

Electrodes of the smallest characteristic dimensions less than few micrometers have unique properties which make them very attractive for use in many electrochemical and biomedical applications. Because of the response of these small electrodes it is possible to change their properties drastically just by variation of the electrode diameter. Depending on their size it is possible to decrease their sensitivity to the effects of solution resistance as well as also to increase their temporal electrochemical resolution.

Untill now one of the limitations in fabrication of these electrodes has been the encapsulation. The most common encapsulation procedure using glass capillary is not quite suitable if disk microelectrodes with the overall diameter less than a few μm have to be prepared. For these reasons we have studied new, generic encapsulation technique which can be applied to different microelectrodes regardless of their dimensions. The technique consists of electrochemical generation of either an organic or inorganic conducting precursor followed by thermal curing to form an insulating layer at elevated temperature.

For inorganic encapsulation the possibility of electrochemically depositing silicon[1,2] and then converting it irreversibly to silicon dioxide has been investigated. The primary reaction is the electrochemical reduction of trichlorosilane to silicon. The reaction is done in tetrahydrofuran under inert atmosphere.

For organic encapsulation we have used the electrochemical synthesis of poly(oxyphenylene) as originally presented by Mengoli et al.[3,4,5] It is done by the electrochemical oxidation of 2-allylphenol in water/methanol/butylcellosolve mixture to yield poly(oxyphenylene). This electrooxidation is done in the presence of allylamine in order to minimize the competing passivation of the substrate and to crosslink the linear polymer during the curing step.

Highly oriented carbon fibers ("graphite fibers") are a very convenient starting material for preparation of microelectrodes, namely because of their mechanical stability, high electronic conductivity and good biological compatibility. Preparation of encapsulated microdisc electrodes with diameter up to 1 μm will be discussed.

Experimental

For this study, single carbon fibers (Celion GY 70, Celanese Corp., U.S.A.) of 8 μm in diameter were clamped in platinum plates to form electrical contact. The electrodeposition of silicon and poly(oxyphenylene) on carbon fibers was done in a one-compartment cell at room temperature. For the electrodeposition of silicon the carbon fibers were used as a cathode. They were rinsed in THF immediately before the electrolysis. The silicon deposition was done in reaction vessel purged with inert gas (N_2 or Ar) containing solution of 0.03 M

---

* on leave from Reed College, Portland, Oregon, U.S.A.

Contemporary Electroanalytical Chemistry, Edited by A. Ivaska *et al.*
Plenum Press, New York, 1990

tetrabutylammonium bromide (TBAB) and 0.2 M $SiHCl_3$ (Merck) in freshly distilled THF. A platinum coil with area of 2 cm^2 surrounding the cathode was used as auxiliary electrode and a Ag/AgCl, sat.LiCl, THF// 0.03 M LiCl//reference electrode was used. The potentiostatic deposition was carried out in two modes: by applying a constant potential within the range of -1.85 V and -1.9 V (according to[2] deposition potential of Si) or by scanning the potential between -1.0 V and -3.0 V with 50 mV/s. The reduction potential was applied to the carbon fiber (working electrode) from 273 PAR potentiostat until the total charge of 31 mQ passed (in the first case). The film was then rinsed with THF and oxidized at 380°C for 30 minutes in an open-ended tubular furnace.

The poly(oxyphenylene) was electrodeposited from freshly prepared solution containing 0.23 M of 2-allylphenol, 0.4 M of allylamine, 0.2 M of butylcellosolve (ethyleneglycol monobutylether) in water/methanol mixture (1:1 by volume) by applying a constant potential of 4 V from constant voltage power-supply (Zentro-Elektrik, Type LA 15/156 B) between the cathode (a platinum coil of 1.5 cm^2 area) and the substrate (anode). The current was monitored with a Keithley 177 microvoltmeter and recorded with a Metrawatt, Model SE 780 recorder. The electrodeposited films of poly(oxyphenylene) formed within half of an hour (aprox. 1.5 μm thick) were rinsed with distilled water and cured at 150°C for 30 min.

The pointing of the carbon fibers was done by flame either in a Bunsen burner or in hydrogen flame of a micro-burner. The diameter of the flame was comparable with the diameter of the carbon fiber.

The carbon fiber "disk ultramicroelectrodes" were prepared from pointed fibers which were encapsulated in poly(oxyphenylene) by applying the same procedure as described above.

After the poly(oxyphenylene) encapsulation both type of electrodes were cut at the end with a scalpel in order to create a disk surrounded by the insulation.

The thickness of poly(oxyphenylene) layer was measured on Pt-glass substrates with profilometer (Sloan Dektak II).

The resistance of the deposited insulators was calculated from the slope of the current-voltage line measured at zero current crossing in the solution of 2.5 mM ruthenium hexamine chloride $Ru(NH_3)_6Cl_3$, (Aldrich) in 1 M KCl. These plots were obtained by applying a slow triangular voltage sweep (150 mV/s) while monitoring the current by the same way as is done in an ordinary cyclic voltammetric experiment except that a high current sensitivity was used. Either Wenking (Model VSG 72) or a EG&G PAR Potentiostat Model 273 and a Ag/AgCl,(3 M KCl) reference electrode with 1 M KCl bridge were used. The auxiliary electrode was a Pt coil of 1.5 cm^2 area.

Results

In order to get into the range of "disk ultramicroelectrodes" the carbon fiber of 8 μm used in our experiments was reduced in diameter to approx. 1 μm. It is possible to "point" the carbon fiber microelectrodes in the Bunsen burner or in the hydrogen flame to a tip less than 1 μm. However, the tips of the electrodes pointed in the hydrogen flame are sharper than the tips of the carbon fibers which were pointed in the Bunsen burner (Fig.1). Furthermore, the end of the electrode which was "pointed" in the Bunsen burner is split (Fig.1a). This can be explained by the existence of a nonuniform temperature gradient within the flame. Sharply pointed electrodes were obtained when using the hydrogen flame. This result is not surprising because the diameter of the flame is much smaller and has more uniform and steeper temperature gradient.

The encapsulation was tested for a leakage current as well as for the break down voltage. The testing for pinholes was done by applying -2 V from a battery to the fiber immersed in 1 M KCl and observing under a microscope H_2 bubbles which evolve when the auxiliary electrode touches the solution. Microscopically smoother and more continuous layers have been prepared on several carbon fibers if the deposition was done by scanning the potential between -1.0 V and -3.0 V. In order to eliminate the possibility of the artifact due to the high charge transfer resistance near zero current density the leakage tests were also carried out in the presence of a fast redox couple (ruthenium hexamine). The current voltage curves in the same solution were recorded for not encapsulated carbon fiber and also for encapsulated carbon fiber as shown in the Fig.2. No electrochemical activity of ruthenium hexamine could

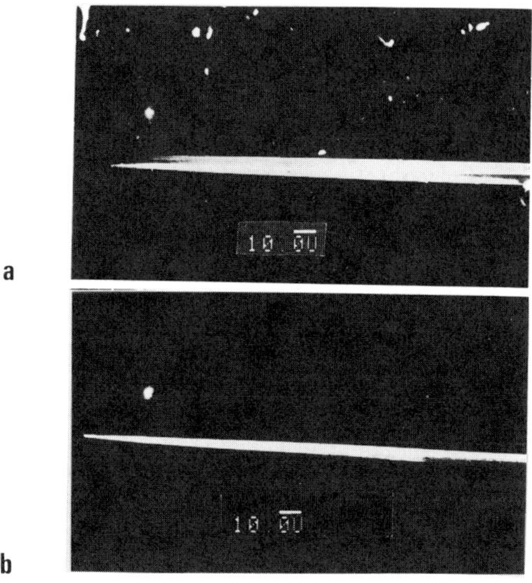

Fig.1. The SEM photograph of pointed oriented carbon fiber in the flame. a) from Bunsen burner, b) hydrogen micro-burner.

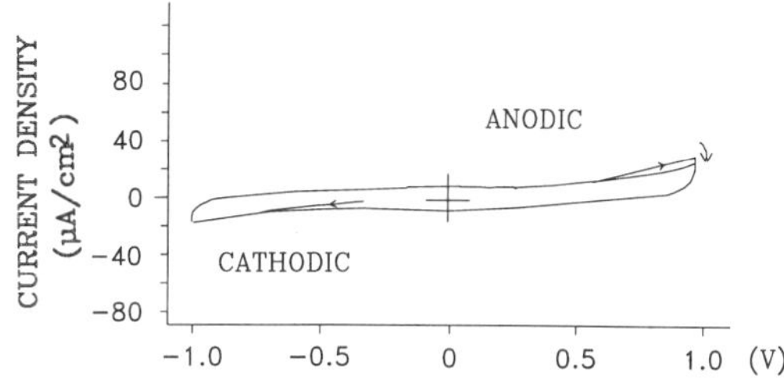

Fig.2. The blockage of the electrochemistry of ruthenium hexamine by the insulation film of Si/SiO_2. The solution was 2.5 mM $Ru(NH_3)_6Cl_3$ in 1 M KCl.

be detected which means that most of the electroactive area has been blocked by the silicon layer converted to the SiO_2 film, Fig.2. This test was carried out on a carbon fiber coated with approx. 0.2 μm thick film of silicon which was converted to silicon dioxide by applying the temperature step. However, none of the fibers coated with SiO_2 at constant potential within the range of -1.8 to -1.9 V were entirely pinhole or crack free.

The thickness of the encapsulating layer has been taken from Tannenberger[2] to be 0.2 μm. The resistance of the insulating layers were calculated from the recorded voltammograms obtained between 1.0 V and -1.0 V in the solution immediately after the thermal curing step and then at various intervals during continuous immersion in 0.1 M NaCl. The resistivity was calculated using the determined value of the insulator thickness. The initial resistivity was $1.2 \times 10^{11} \Omega$ cm and after the period of 10 days was decayed to $0.3 \times 10^{11} \Omega$ cm.

The thickness of poly(oxyphenylene) layer on carbon fiber was estimated from the deposition curve of poly(oxyphenylene) on Pt-glass substrate shown in Fig.3 and from the SEM-photograph of the encapsulated carbon fiber shown in Fig.4. The initial resistivity for the poly(oxyphenylene) encapsulation was $4.5 \times 10^{12} \Omega$ cm and after the period of 9 days was decayed to $7.3 \times 10^{11} \Omega$ cm.

The 8 μm diameter non-encapsulated carbon fibers used in our experiments behave as cylindrical microelectrodes.[6] The cyclic voltammograms are shown in Fig.5 for different scan rates. As the scan rate decreases the shape of the voltammogram evolves from the peak-shaped to the sigmoidal shape. Cyclic voltammograms recorded for this encapsulated

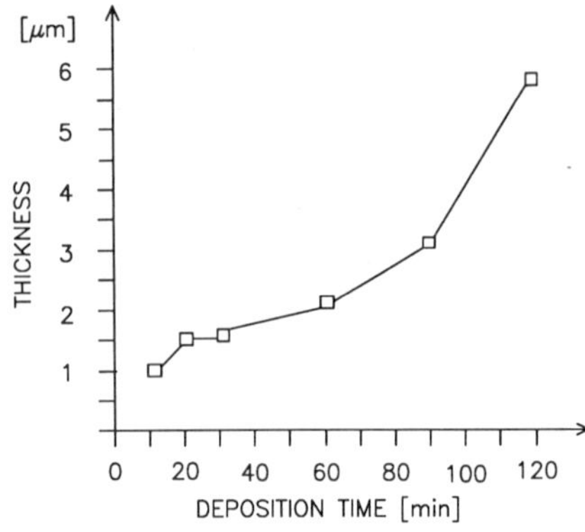

Fig.3. Dependence of the poly(oxyphenylene) film thickness on the deposition time after curing at 150°C for 1 h.

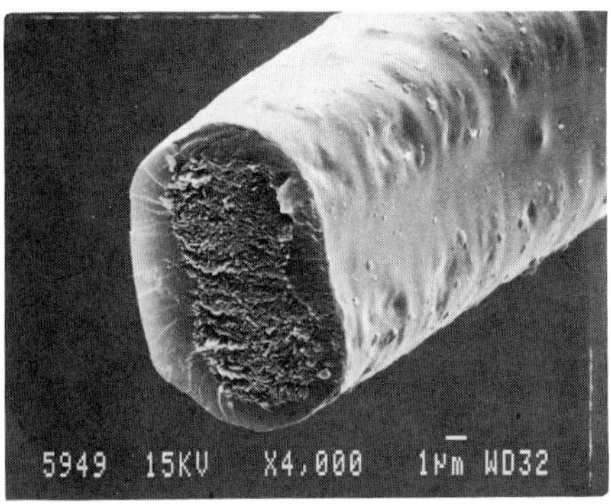

Fig.4. SEM photograph of 8 μm carbon fiber encapsulated with 1.5 μm thick poly(oxyphenylene).

microdisk carbon fiber electrode shown in the Fig.6 show a steady-state behavior of the limiting current. Similar shape of the curve was obtained for a disk of approx. 1 μm also shown in Fig.6b.

Discussion

Our experimental results show that the carbon fiber microelectrodes can be encapsulated by electrochemically generated insulator. The encapsulation with SiO_2 is experimentally demanding and does not yield reliable results. This is presumably due to the fact that the thin layer of silicon is extremely reactive and traces of residual oxygen, present at the electroactive (i.e. to be encapsulated) surface cause the formation of the insulating film which inhibits further deposition of the precursor. Therefore, only very thin layers which were obtained at very slow scan rates within the potential range of -1 V up to -3 V are pinhole free. The inconsistency of the quality of the silicon deposits has been observed also by Tannenberger.[2]

The organic encapsulation has been found to work on carbon fiber without any problems. The long term continuous exposure to aqueous electrolytes degrades the overall resistance somewhat but the final value is still sufficiently high.

The new encapsulation procedure is particularly useful for the preparation of carbon fiber disk electrodes of different disk diameters. Furthermore, the new encapsulation technique give a possibility to insulate microelectrodes more reproducibly and to automatize this process if necessary.

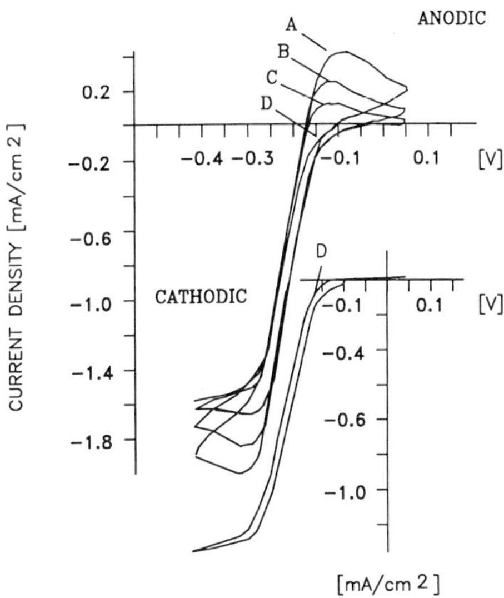

Fig.5. Cyclic voltammograms of cylindrical carbon fiber of 8 μm in diameter and 3 mm length : a) 200 mV/s, b) 100 mV/s, c) 50 mV/s, d) 10 mV/s.

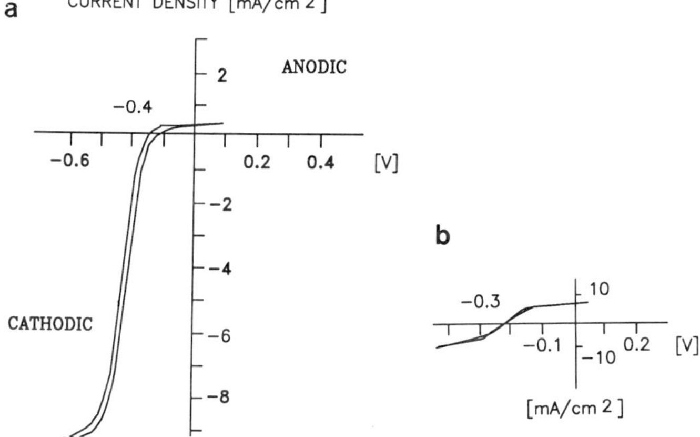

Fig.6. Cyclic voltammograms of encapsulated carbon fiber disc microelectrode. Scan rate is 150 mV/s: a) 8 μm diameter of the disk, b) approximately 1 μm diameter of the disk.

Acknowledgements

This work was supported by the DFG Grant Number Jo 157/2-1. The authors express their thanks to Prof. Dr. Jiri Janata for valuable discussions and encouragement.

References

1. A.K. Agrawal and A. Austin, J. Electrochem. Soc., **128** (1981) 2292.
2. J. Gobet and H. Tannenberger, J. Electrochem. Soc., **135** (1988) 109.
3. G. Mengoli, P. Bianco, S. Daolio, M.T. Munari, J. Electrochem. Soc., **128** (1981) 2276.
4. G. Mengoli, S. Daolio and M.M. Musiani, J. Appl. Electrochem., **10** (1980) 459.
5. G. Mengoli, and M.M. Musiani, J. Electrochem. Soc., **134** (1987) 647C.
6. M.R.M. Wightman and D.O. Wipf, Ultramicroelectrodes, in: "Electroanalytical Chemistry", A.J. Bard (Ed.), Vol. **15**, Marcel Dekker, New York, 1988.

VOLTAMMETRIC DETERMINATION OF ORGANIC COMPOUNDS USING CLAY MODIFIED CARBON PASTE ELECTRODES

Lucas Hernandez, Pedro Hernandez and Encarna Lorenzo

Department of Chemistry
Autonoma University
28049 Madrid, Spain

Introduction

The use of modified electrodes for the voltammetric determination in the trace analysis is a field of great progress.[1] They allow an easy preconcentration and direct separation from some matrices, especially when modified carbon paste electrodes are used.[2-6]

Several groups have reported on the use of clays[7,8] and zeolite[9,10] film electrodes. Both these materials are able to adsorb and incorporate electroactive species for the direct determination of the analyte.

In this paper we describe the use of several clays (sepiolite, bentonite and hectorite) as modifiers of carbon paste electrodes in the determination of the following organic compounds: flunitrazepam, nitrobenzene, aniline, phenol, linuron, dinocap and endosulfan.

The clays allow preconcentration of these compounds on the electrode surface in some matrices. Then the selected molecules are oxidized or reduced using the linear or differential pulse scan techniques.

Experimental

Apparatus and Reagents

The carbon paste electrodes were constructed from spectroscopic graphite (particle size <0.45 mm), which was mixed with nujol at a concentration of 1 g/ml and the necasary amount of moist clay to obtain a mass proportion of 10 %. This mixture was placed in a polyethylene tube with a geometric surface area of 1.6 mm^2. The connection was made with copper wire.

Two cells were employed troughout the study: one in which the solution to be studied was preconcentrated, and the other containing a suitable supporting electrolyte.

The measurement was made using a saturated calomel electrode as reference electrode and platinum as counter electrode. A Metrohm Polarecord E-506 polarograph was used.

Stock solutions of the pure compounds were prepared by dissolving each compound in methanol. Working standards were prepared daily by dilution with deionized water (Milli-Q and Milli RO- Millipore systems).

Procedure

The carbon paste electrode was placed in the preconcentration cell (20 ml) containing the solution of the analyte. The solution was stirred and the circuit kept open for a predetermined time. The electrode was then removed from the preconcentration cell, briefly rinsed with deionized water and placed in the measurement cell, which contained normally KNO_3 as supporting electrolyte and the voltammograms were recorded. All studies were carried out at room temperature.

Contemporary Electroanalytical Chemistry, Edited by A. Ivaska *et al.*
Plenum Press, New York, 1990

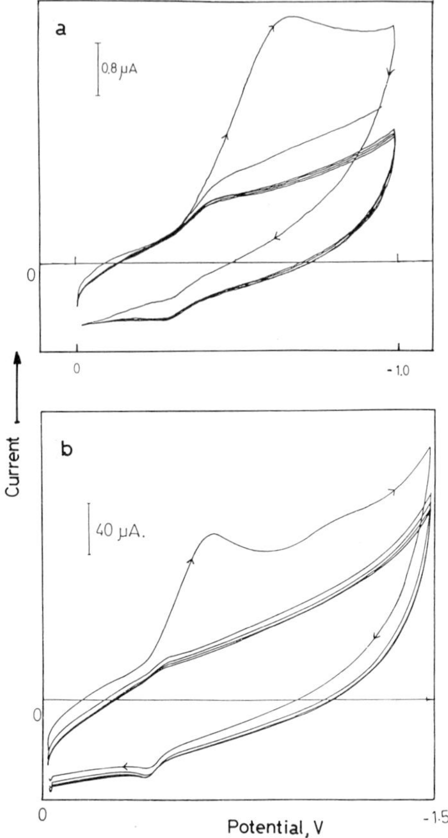

Fig.1. Cyclic voltammograms obtained with a sepiolite modified carbon paste electrode after 5 min residence time in (a) 0.5 μg/ml of nitrobenzene (pH=3.5) and (b) 4.0 μg/ml of dinocap (pH=4.0). Supporting electrolyte, 0.5 M KNO$_3$ (pH=5.3); scan rate, 250 mV/s.

After the scan, the potential was held at a high negative (-1.0; -1.2 V) or positive potential (1.0; 1.5 V) for about one minute to regenerate the electrode when the selected molecules were reduced or oxidized, respectively.

Results and Discussion

Accumulation and Stripping Behaviour

The reduction mechanism of nitro derivatives (nitrobenzene, dinocap, flunitrazepan and others) when a clay modified carbon paste electrode is used agrees well with that reported for the reduction of these compounds.[11] Figs 1 and 2 show the voltammograms obtained by cyclic and by differential pulse scan rsdifferential pulse voltammetryrespectively, when a sepiolite modified carbon paste electrode was kept for 5 minutes in a solution containing (a) nitrobenzene or (b) dinocap.

The behaviour observed for aniline and phenol when a clay modified carbon paste electrode is used, agrees well with the mechanisms put forwards by Lines[12] for the oxidation of aniline and with the oxidation of phenol to quinone.[13−15]

Optimun Conditions for the Determination

In order to establish the method of determination, a study was performed to find the controlling parameters in the preconcentration step, on the electrode and in the measurement step itself.

206

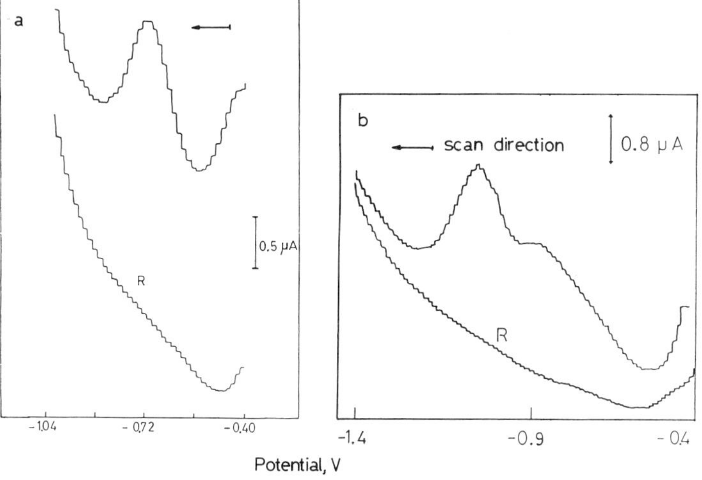

Fig.2. Differential pulse voltammograms obtained with a sepiolite modified carbon paste electrode after 5 min residence time in (a) 0.5 μg/ml of nitrobenzene (pH=3.5) and (b) 0.4 μg/ml of dinocap (pH=4.0). Supporting electrolyte, 0.5 M KNO_3 (pH=5.3). R, residual current. Scan rate: (a) 50 mV/s, (b) 30 mV/s. Pulse amplitude: (a) -100 mV, (b) -50 mV.

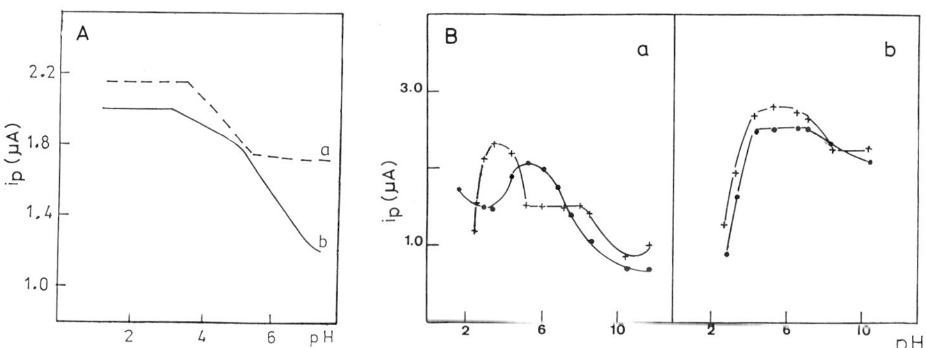

Fig.3. Influence of pH in (a) preconcentration and (b) measurement steps. Residence time, 5 min. A: phenol (5.0 μg/ml); scan rate 40 mV/s; pulse amplitude: 100 mV; supporting electrolyte, 0.02 M KNO_3. B: dinocap (0.4 μg/ml); scan rate: 30 mV/s; pulse amplitude: -50 mV; supporting electrolyte, 0.5 M KNO_3. ($\times$) sepiolite and ($\bullet$) hectorite modified electrode.

In view of the disadvantages found in the measurement of the voltammograms, which show a decrease in the peak intensity the use of buffered solutions in the preconcentration cell should be ruled out. This means that the pH should exclusively be adjusted either with HNO_3 or KOH according to the need to add one or the other reagent.

The results of the study indicate that, in general, the best electrolyte for analytical purposes is 0.01–0.5 M KNO_3 solution at pH between 1.0 and 2.0 or 4.0 and 6.0 for oxidation or reduction respectively (see Table 1).

Fig.3 shows the peak intensity values obtained for (A) phenol and (B) dinocap at different pH values.

The influence of the residence time of the electrode in the analyte solution was studied

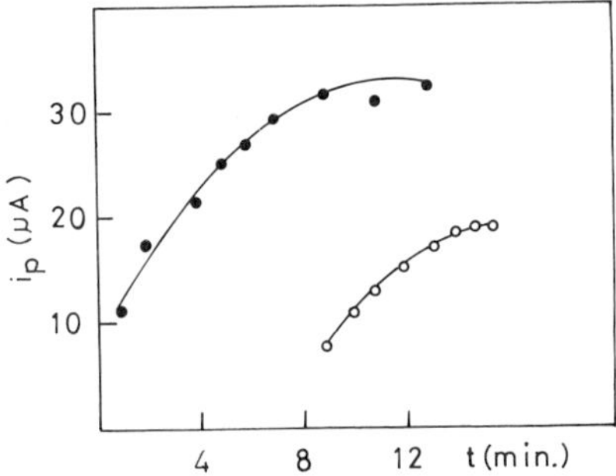

Fig.4. Influence of residence time of the electrode in a solution containing (o) 0.3 μg/ml or (•) 5.0 μg/ml of linuron (pH=2.0). Supporting electrolyte, 0.01 M KNO$_3$ (pH=1.7); scan rate: 30 mV/s; pulse amplitud 100 mV.

by increasing time until maximum peak intensity (i_p) was observed. The curves with linuron are given in Fig.4. The optimium residence times for every compound studied are summarized in Table 1. These times are usually between 5 and 15 minutes.

Table 1

| Compound | Preconcentration | | Measurement |
	t (min)	pH	Electrolyte
Aniline	5	6.5	0.02 M KNO$_3$ (pH=1.5)
Phenol	10	1.5	0.02 M KNO$_3$ (pH=2.0)
Nitrobenzene	5	3.0–4.0	0.5 M KNO$_3$ (pH=5.5)
Flunitrazepam	12	3.0–4.0	0.5 M KNO$_3$ (pH=3.9)
Dinocap	10	4.0	0.5 M KNO$_3$ (pH=5.3)
Linuron	15	2.0	0.01 M KNO$_3$ (pH=1.7)
Endosulfan	10	13.0	0.05 M KNO$_3$ (pH=13.0)

The stirring rate was not found to affect the response of the electrode.

In the measurement step, the change of the pulse amplitude, ΔE, gives a linear increase in i_p reaching a maxium value at $\Delta E = 100$ mV. This is received without any modification in the voltammograms. An increase in i_p was found with increase of the scan rate. In order to get an adequate analytical response, a scan rate between 30 and 50 mV/s was chosen in this work.

Fig.5 shows the influence of (a) scan rate and (b) pulse amplitude on i_p for aniline.

In cyclic voltammery, when the scan rate was varied with the other parameters fixed, the voltammograms showed peaks with all scan rates up to 600 mV/s. The residual current was also found to increase with the scan rate. Therefore the best results in intensity were obtained for 250 mV/s (see Fig.6).

The response both in cyclic and differential pulse voltammetry shows a linear relationship between the peak intensity and the analyte concentration.

The optimun conditions in analytical determinations are given in Table 2. In all the cases studied and the concentration ranges indicated above, the typical standard deviation is lower than 8 %.

Determination of the compounds given above was carried out in different matrices (see Table 2) without any previous separation. The results obtained were good. Fig.7, e.g., shows the voltammograms for aniline in differet beverages.

Based on our studies it is possible to conclude that clay modified carbon paste electrodes allow the efficient preconcentration of organic compounds for voltammetric measurement.

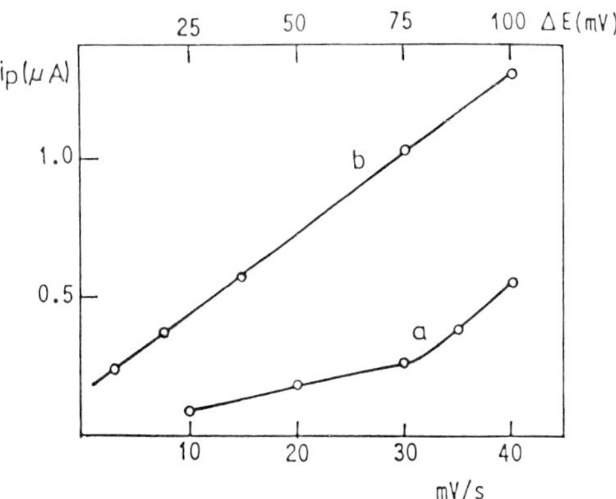

Fig.5. Influence of (a) scan rate and (b) pulse amplitude on i_p for a 1.0 μg/ml aniline solution at pH=6.5. Supporting electrolyte, 0.1 M KNO$_3$ (pH=1.5); residence time, 5 min. (a) pulse amplitude: 100 mV, (b) scan rate: 15 mV/s.

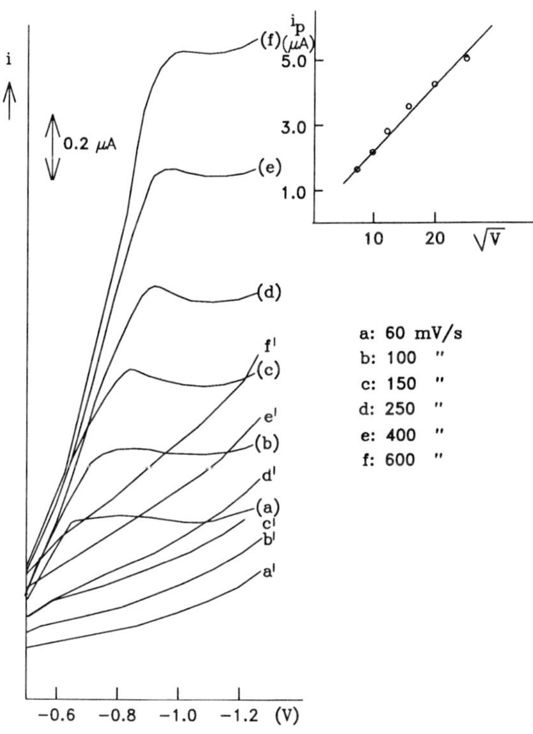

Fig.6. Effect of scan rate on peak current for a 0.5 μg/ml nitrobenzene solution (pH=3.5). Residence time, 5 min; supporting electrolyte, 0.5 M KNO$_3$ (pH=5.5). a) 60 mV/s, b) 100 mV/s, c) 150 mV/s, d) 250 mV/s, e) 400 mV/s, f) 600 mV/s.

The procedure is simple and it gives limits of determination that are lower or similar than the limits obtained by regular polarographic methods.

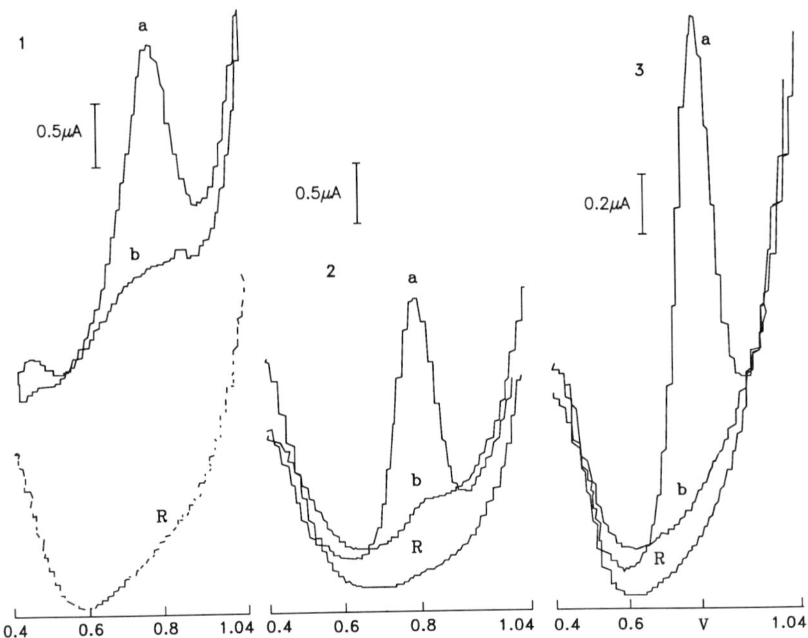

Fig.7. Determination of aniline in different samples: (1) red wine, (2) gin, (3) water. R, residual current. (b): blank without addition of aniline. (a): with addition of 0.5 μg/ml. Scan rate: 40 mV/s. Pulse amplitude: 100 mV.

Table II

Compounds	CPME	Anal.Technique	Linear response	Limit of detection ng/ml	Matrices
Aniline	Sepiolite	Diff.Pulse Voltammetry	20–500 ng/ml	15	vine, gin, milk, beer and water.
Phenol	Sepiolite	Diff.Pulse Voltammetry	0.2–5.0 μg/ml	100	soft drinks.
Nitrobenzene	Sepiolite	Cyclic Voltammetry	50–500 ng/ml	40	
		Diff.Pulse Voltammetry	50–500 ng/ml	30	wine, beer, cider.
Flunitrazepam	Bentonite	Diff.Pulse Voltammetry	0.2–4.0 μg/ml	130	human serum, urine.
Dinocap	Sepiolite	Diff.Pulse Voltammetry	2–14 ng/ml	1.2	river water.
	Hectorite	Diff.Pulse Voltammetry	18–72 ng/ml	12	
Linuron	Sepiolite	Diff.Pulse Voltammetry	100–500 ng/ml	75	river water.
Endosulfan.	Sepiolite	Diff.Pulse Voltammetry	0.1–5.0 ng/ml	0.1	river water.

Acknowledgements

The authors thank CITYT for financial support to this project: Grant number PA 86 - 0367.

References

1. R.W. Murray, A.G. Ewing and R.A. Durst, Anal. Chem., **59** (1987) 379A.
2. K.H. Lubert, M. Schurrbusch and A. Thomas, Anal. Chim. Acta, **144** (1982) 123.
3. K. Izutsu, T. Nakamura, R. Takizuva and H. Hanava, Anal. Chim. Acta, **149** (1983) 147.
4. J. Wang, B. Greene and C. Morgan, Anal. Chim. Acta, **158** (1984) 15.
5. R.P. Baldwin, J.K. Christensen and L. Kriger, Anal. Chem., **58** (1986) 1790.
6. J.G. Redepenming, Trends Anal. Chem., **6** (1987) 18.
7. E. Deniz, P.K. Ghosh, J.R. White, J.F. Equey and A.J. Bard, J. Am. Chem. Soc., **107** (1985) 5644.
8. J.R. White and A.J. Bard, J. Electroanal. Chem., **197** (1986) 233.
9. C.G. Murray, R.J. Novak and D.R. Rolison, J. Electroanal. Chem., **187** (1985) 205.
10. P. Hernandez, E. Alda and L. Hernandez, Fresenius' Z. Anal. Chem., **327** (1987) 676.
11. H. Lund, in: "Organic Electrochemistry", M.M. Baizer and H. Lund (Eds), Chapter 8, Marcel Dekker, New York (1983).
12. R. Lines, in: "Organic Electrochemistry", M.M. Baizer and H. Lund (Eds), Chapter 15, Marcel Dekker, New York (1983), p. 463.
13. O. Hammeric, in: "Organic Electrochemistry", M.M. Baizer and H. Lund (Eds), Chapter 16, Marcel Dekker, New York (1983), p. 485.
14. P.T. Kissinger, in: "Laboratory Techniques in Electroanalytical Chemistry", Marcel Dekker, New York (1984), p. 631.
15. M.S. Smyth and W.F. Smyth, Analyst, **103** (1978) 529.

THE ROLE OF SURFACE PROCESSES IN SIGNAL FORMATION WITH SOLID- STATE ION-SELECTIVE ELECTRODES – CHLORIDE INTERFERENCE ON COPPER ION-SELECTIVE ELECTRODE

A. Lewenstam,[1] A. Hulanicki[2] and E. Ghali[3]

1. Laboratory of Analytical Chemistry
   Åbo Akademi University
   SF-20500 Turku - Åbo, Finland
2. Department of Chemistry
   University of Warsaw
   Pl- 02 093 Warsaw, Poland
3. Department of Mining and Metallurgy
   Laval University
   Quebec, Canada, G1K 7P4

Introduction

Although there is no doubt that surface science is a discipline on its own, in practically oriented electroanalytical chemistry it is mostly an auxiliary force to establish the mechanism of electrode signal. A recent, rapid development of electroanalytical sensors provided the evidence that to get more incisive knowledge of surface-located signal forming processes it is necessary to set a new or to improve existing analytical procedures. Therefore the inspection of surface processes by suitable methods seems to be an indispensable feature of electroanalytical chemistry of today.

This paper gives the illustration of how surface techniques engaged in the interpretation of the electrode signal can help to introduce a new analytical procedure. For this purpose the long discussed case of chloride ion interference on solid-state copper ion-selective electrode (Cu-ISE)[1,2] will be exploited. However, the same methodology may be applied to any other similar case situations.

Theoretical

Solid-state copper ion-selective electrodes are usually equipped with the membranes containing a mixture of copper(II) and silver sulphide[3] or monocrystalline copper(I) sulphide.[4] These electrodes are sensitive towards copper(II) ions with Nernstian slope (29.6 mV/decade at 25 °C) in wide concentration range (down to 10^{-8} M). Their attractive properties, such as: good selectivity, short response time and long life time resulted in a variety of successful analytical applications.[5] However, with extensive analytical use of Cu-ISEs it became evident that in the presence of chlorides, in concentrations higher than ~0.1 M, these primarily divalent electrodes act with variable super-Nernstian slope, which in limiting cases can reach the value typical for monovalent electrodes.[6-8] This fact, hindering the application of copper electrode in the analysis of natural water,[9,10] body fluids[11] and plating baths,[12] focused the attention not only for its practical importance, but also because of insufficiencies of ISE theory, providing conflicting interpretations of chloride interference.[1,2] Two general mechanisms were proposed to explain the influence of chloride ions, considering either the redox reactions at the membrane surface:[11,14-19]

Contemporary Electroanalytical Chemistry, Edited by A. Ivaska *et al.*
Plenum Press, New York, 1990

$$I \qquad Cu^{2+} + CuS + 2nCl^- \rightleftharpoons 2CuCl_n^{1-n} + S \qquad (1)$$

$$2Cu^{2+} + Cu_2S + 4nCl^- \rightleftharpoons 4CuCl_n^{1-n} + S \qquad (2)$$

$$2Cu^{2+} + Ag_2S + 4nCl^- \rightleftharpoons 2CuCl_n^{1-n} + 2AgCl_n^{1-n} + S \qquad (3)$$

or ion-exchange processes, without the change of the oxidation state of participating species:[6,11,17,18,20−23]

$$II \qquad Cu^{2+} + Cu_2S + 2nCl^- \rightleftharpoons 2CuCl_n^{1-n} + CuS \qquad (4)$$

$$Cu^{2+} + Ag_2S + 2nCl^- \rightleftharpoons 2AgCl_n^{1-n} + CuS \qquad (5)$$

By both of these approaches it was possible to interpret the observed increase of slope of the characteristics of Cu-ISE in the presence of chlorides with fairly good agreement with the experimental data. However, the fact that these two essentially different mechanisms of surface reactions result in the same theoretical prediction, clearly shows that the real explanation of the mechanism of the signal is still missing. Therefore the crucial question about the type of membrane surface reaction should be answered. Such a possibility exists, though, and it can be seen by comparing the reaction schemes (I) and (II). Namely, the interference of chlorides on pure copper(II) sulphide can be developed only via redox type of reaction (1). In the case of mixed copper(II)/silver sulphide membrane this creates clearcut alternative: the negative result for CuS would mean that the chloride influence is resulting from the ion-exchange reaction (5), the positive will prove the real role of redox reaction (1). For this reason more thorough studies of pure copper(II) sulphide membranes were attempted.

Experimental

Compounds and Apparatus

Pure copper foil (99.999 %) supplied by Fisher and 99.999 % pure sulphur by Cerac were used. All other chemicals were of Baker analytical grade quality. The gases (H_2 and Ar) used were of zero grade quality supplied by Air Products. Double distilled water was used in all experiments. Taccusel PRT 20-2X potentiostat with PAR 175 programmer were used for current-voltage measurements. A graphite auxiliary electrode and saturated calomel reference electrode from Fisher were applied. A pHM64 potentiometer coupled with REC-61 recorder, both from Radiometer were used in the potentiometric experiments. Orion 94-29A copper and 94-17A chloride electrodes as well as home made copper(I) sulphide electrode were used for determination of copper and chloride ions. Perkin-Elmer 603 Atomic Absorption Spectrometer (AAS) was also used for determination of copper ions. Philips X-ray diffractometer was used for X-ray diffraction (XRD) analysis. Hitachi S 500 scanning electron microscope coupled with energy dispersion X-ray facility (SEM/EDAX) for the copper and sulphur surface analysis were used. The temperature in all electrochemical experiments was controlled by thermostat ($\pm$ 1 °C). The solutions were purged with appropriate gases for 1h before each experiment and then during the experiment, as described elsewhere.[16] Except of potentiostatic experiment solutions were not agitated.

Preparation of Copper(II) Sulphide Membranes

Copper(II) sulphide was obtained by direct synthesis from copper foils prereduced with hydrogen and resublimated sulphur taken in molar ratio 1.000:1.000. Both elements were put in quartz boats into evacuated ($< 10^{-4}$ torr) and sealed pyrex-quartz tube. Two independent furnaces were used to control the temperature during synthesis. Sulphur was allowed to distil over copper foils by using 450 °C in both parts of reaction tube for 48 h. Then the temperature in the copper part was decreased to 425 °C for 1h and both parts of tube were further on cooled at rate 25 °C per day. The product obtained was powdered and sieved. A pure copper(II) sulphide was identified as a product by XRD analysis.

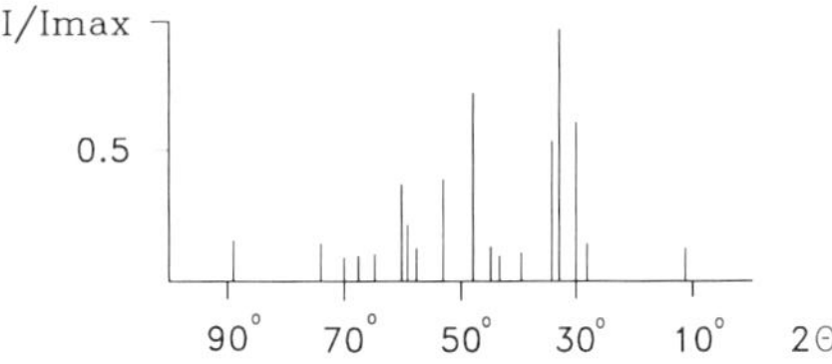

Fig.1. X-ray spectrum of freshly synthesized copper(II) sulphide.

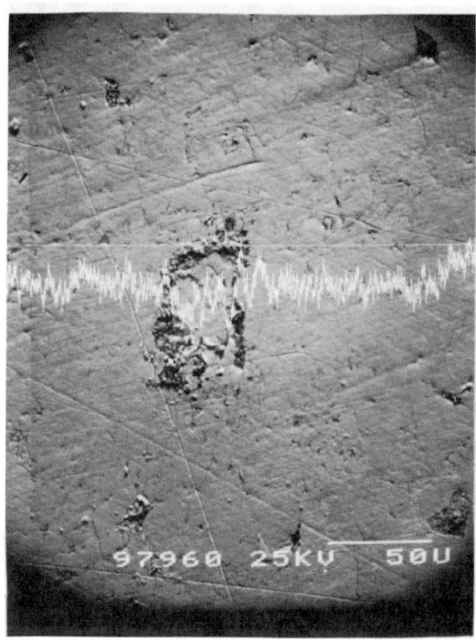

Fig.2. SEM picture (×400) of the surface of freshly synthesized copper(II) sulphide. EDAX performed with sulphur probe.

The membranes were pelletized using hydrostatic or mechanical compression techniques with highest pressure 20 T/cm^2. The highest density of the discs made was 96 % of the theoretical copper(II) sulphide density (4.67 g/cm^2). The discs were then equipped with graphite or silver electrical contacts by using graphite or silver-based conductive glues. Only one face of the electrode (surface 0.5 cm^2) was exposed to the solution, the remainder was embedded in epoxy resin. The membrane surface was polished successively on water-cooled emery papers (240 - 1000 grit) followed by final polishing on diamond paste of 1 μm. The electrodes were then thoroughly cleaned with distilled water and the membrane surfaces were checked with XRD (Fig.1) and SEM/EDAX techniques (Fig.2).

Pure copper(II) sulphide membranes were obtained on the way described above evidently show the superiority of this method compared to wet precipitation one as the latter leads to nonsoichiometric, contaminated, leaching, porous or occasionally inactive copper(II) membranes.[24,25]

Results and Discussion

Basic Characteristic of Copper(II) Sulphide Electrodes

The calibration curves obtained in copper(II) nitrate solutions of pH = 4.8, containing acetic buffer, are shown in Fig.3. (For comparison the calibration curve for a Cu$_2$S electrode is also presented). The calibration slopes are near to the expected divalent slope, i.e., 29.6 mV at 25 °C and the basic performance (response time, stability and repeatability of readouts) is similar to that observed with Orion Cu-ISE. Standard potentials observed are close to theoretical values,[13,26,27] for graphite contacted membranes standard potential observed is

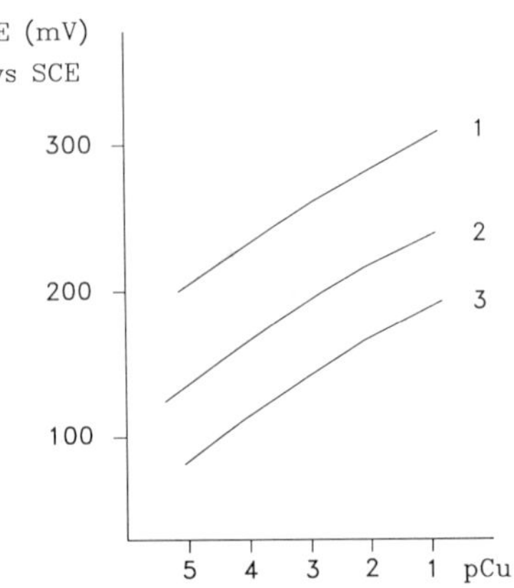

Fig.3. Calibration curves of copper sulphide based electrodes: 1 – CuS electrode with carbon contact, 2 – CuS electrode with silver contact, 3 – Cu_2S electrode with silver contact (pH of the solution 4.8).

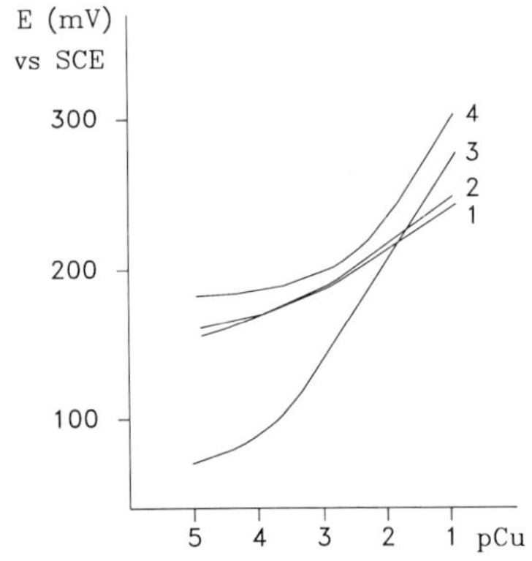

Fig.4. Calibration curves of copper(II) sulphide electrode in: (1) – 1×10^{-2} M HNO_3, (2) – 1×10^{-2} M HCl, (3) – 1×10^{-2} M HCl + 1 M KCl, (4) – 1 M HCl.

$E^0 = + 341$ mV and theoretically predicted: $E^0 = + 312$ mV, for silver contacted membranes $E^0 = + 271$ mV and $E^0 = + 281$ mV vs SCE, respectively.

The Influence of Chloride Ions on Copper(II) Sulphide Electrode Behaviour

The influence of chloride ions on the behaviour of copper(II) sulphide electrode was studied by using potentiometric and voltammetric methods in four different media: 1×10^{-2} M HNO_3, 1×10^{-2} M HCl, 1×10^{-2} M HCl + 1 M KCl and 1 M HCl. The potentiometric calibration curves obtained in diluted acids are similar to those reported above (Fig.4, curves 1,2), indicating no difference in the mechanism of copper(II) sulphide electrode action.

The linear shape of the voltammetric curve registered in the same conditions (Fig.6, curve 1) shows that only soluble anodic dissolution products are formed in diluted acids. The same conclusion was drawn from SEM/EDAX analysis of copper(II) sulphide surface after anodic dissolution at +600 mV (SCE) for 3 h in diluted acids (Fig.5). The results are supporting the statement that in diluted acids the concentrations of H^+ and Cl^- ions and the redox potential of $Cu^{2+}/CuCl_n^{1-n}$ pair are too small to disturb the fast and reversible electrochemical surface reaction:

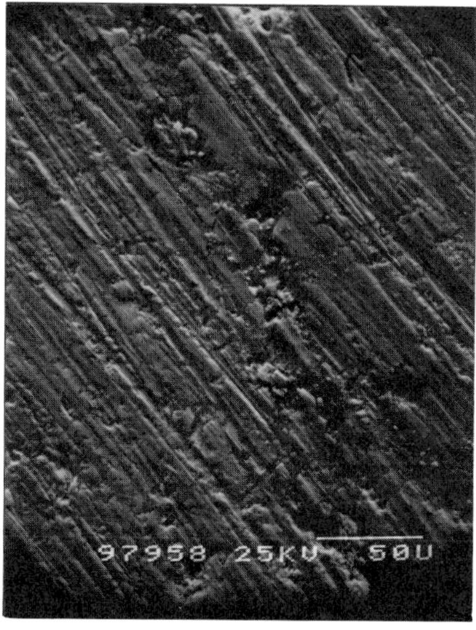

Fig.5. SEM picture ($\times$400) of the surface of copper(II) sulphide after anodic dissolution in 1×10^{-2} M HCl.

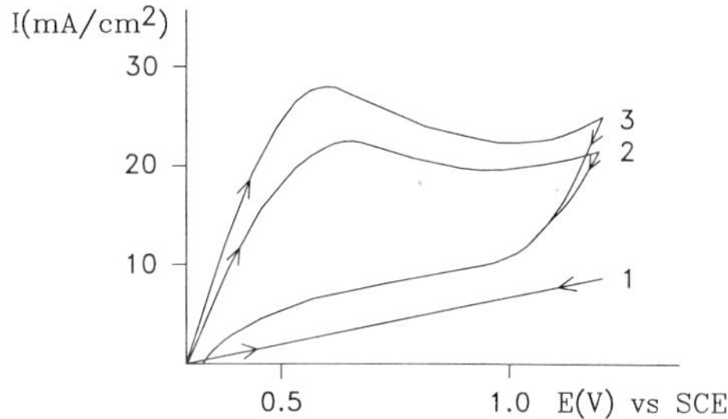

Fig.6. Current-potential first anodic cycle for copper(II) sulphide electrode in: (1) – 1×10^{-2} M HCl, (2) – 1×10^{-2} M HCl + 1 M KCl, (3) – 1 M HCl.

$$CuS \rightleftharpoons Cu^{2+} + S + 2e^-$$

(6)

which is responsible for the potential build-up and the divalent electrode response.[28] However, this situation changes dramatically when the chloride ion concentration is increased, as can be seen for 1×10^{-2} M HCl + 1 M KCl and 1 M HCl, solutions (Fig.4, curves 3,4 and Fig.6 curves 2,3). The slope of the potentiometric calibration curve is increased, but at the same time it was observed that the response of the electrode is slower and apparently also dependents on stirring rate as well as that the potential readouts are less reproducible ($\pm$ 5mV) with the tendency of slow drift towards higher potential values. The current-voltage behaviour is also basically changed. Instead of linear behaviour, as observed in diluted HCl, the current-voltage curve is parabolic indicating active - passive behaviour of the electrode. This can be attributed to the inhibition of mass transport in the membrane surface region due to the diffusion barrier imposed by insoluble products and/or pore saturation.[15,16] To get further insight on CuS properties under discussed conditions, Pourbaix diagrams were drawn in coordinates: potential vs pH and potential vs pCl at pH $= 0$ and pH $= 2$ using thermodynamic data from Latimer[29] (Fig.7 and 8).

The remaining conditions as in Fig.7. $CuCl_2^-$ was chosen as the representative form of $CuCl_n^{1-n}$ species.

Remembering that this type of analysis is always of limited value because of idealized thermodynamic assumption of total reversibility of all reactions considered, the following

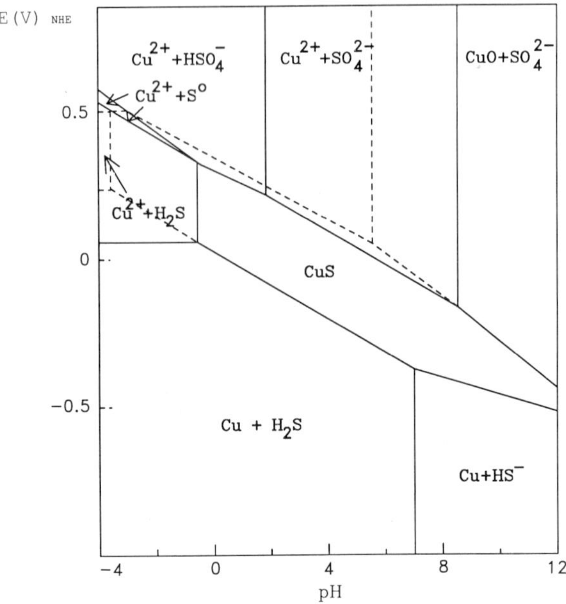

Fig.7. E-pH diagram for CuS – H$_2$O system at 25 °C.
___ [Cu^{2+}] = 10^{-9} M, [HSO$_4^-$]= [SO$_4^{2-}$] = [H$_2$S]= [HS$^-$]= 10^{-5} M
- - - [Cu^{2+}] = 10^{-3} M, [HSO$_4^-$]= [SO$_4^{2-}$] = [H$_2$S]= [HS$^-$]= 10^{-5} M

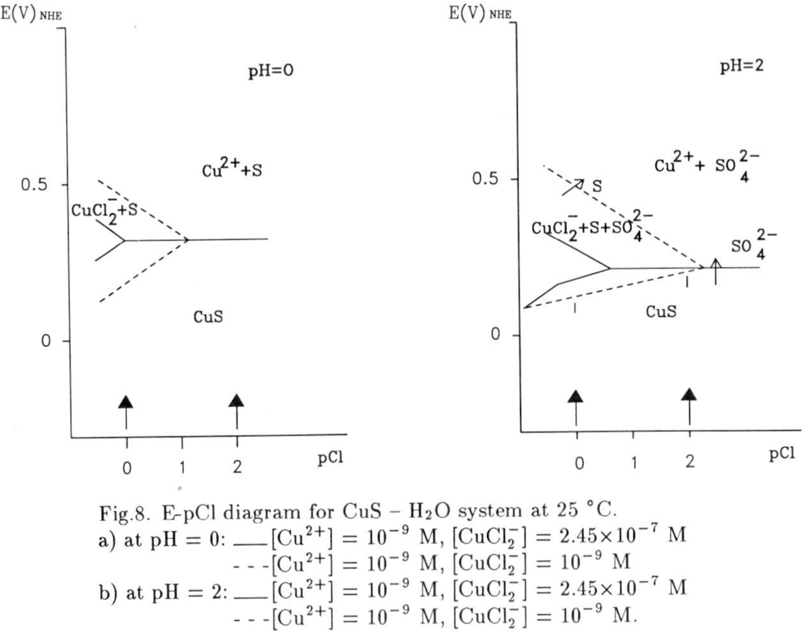

Fig.8. E-pCl diagram for CuS – H$_2$O system at 25 °C.
a) at pH = 0: ___[Cu^{2+}] = 10^{-9} M, [CuCl$_2^-$] = 2.45×10^{-7} M
         - - -[Cu^{2+}] = 10^{-9} M, [CuCl$_2^-$] = 10^{-9} M
b) at pH = 2: ___[Cu^{2+}] = 10^{-9} M, [CuCl$_2^-$] = 2.45×10^{-7} M
         - - -[Cu^{2+}] = 10^{-9} M, [CuCl$_2^-$] = 10^{-9} M.

conclusions can be drawn. The appearance of a new phase: orthorombic sulphur is possible in the case of anodic dissolution of CuS in 1 M HCl under condition of low copper(II) ion concentration, Fig.7. This condition can in fact be fulfilled because of efficient conversion of copper (II) ions in the presence of chlorides according to chemical reaction:[30]

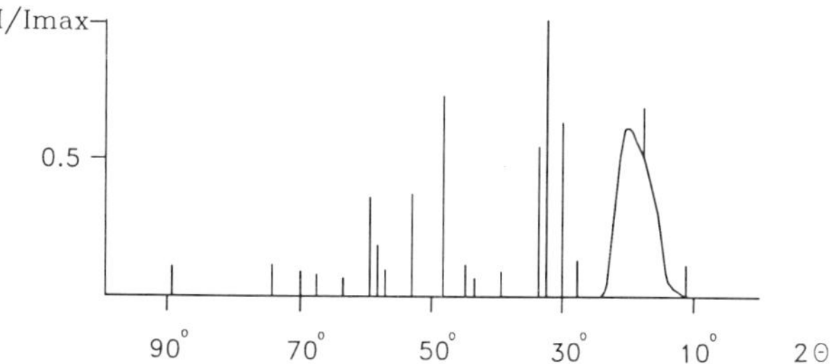

Fig.9. X-ray diffractogram of copper(II) sulphide electrode after anodic dissolution in 1×10^{-2} M + 1 M KCl.

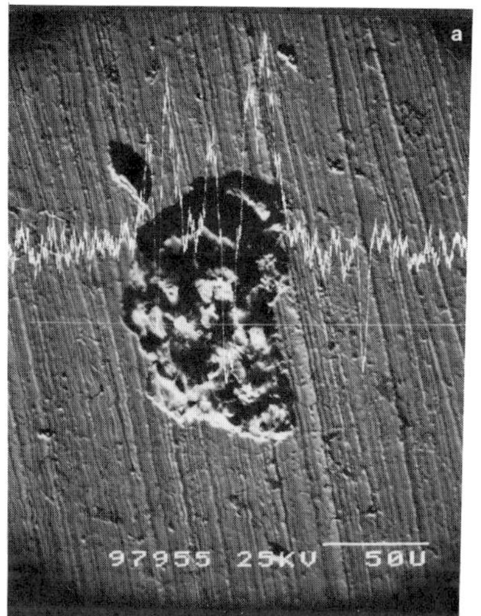

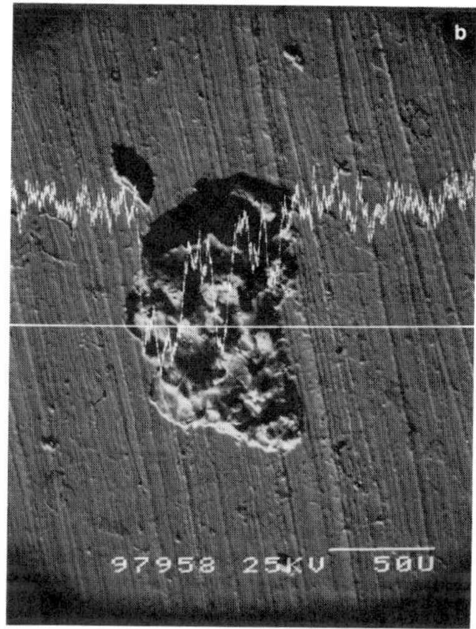

Fig.10. SEM picture ($\times400$) of copper(II) sulphide at the first anodic half-cycle in 1 M HCl. Sulphur (a) and copper (b) were identified with EDAX probes (CuS synthetized by Ventron was used in this experiment).

$$Cu^{2+} + CuS + 2nCl^- \rightarrow 2CuCl_n^{1-n} + S \tag{7}$$

The influence of reaction (7) on the overall anodic dissolution of CuS can clearly be predicted from E-pCl diagrams, for $pCl \cong 0$ at pH = 0 and pCl < 1 at pH = 2 (Fig.8). This means that one can expect the formation of elemental sulphur not only in 1 M HCl, but also in the solutions of higher pH, whenever chloride ion concentration is high enough; for instance: in the conditions of author's interest: 1×10^{-2} M HCl + 1 M KCl.

To test this prediction the surface of CuS membranes, dissolved anodically in the media discussed above, was analyzed for the presence of sulphur by using SEM/EDAX technique with sulphur probe. In 1 M HCl the sulphur presence can clearly be seen already after one anodic half-cycle (Fig.10). In 1×10^{-2} M HCl no sulphur could be detected. In 1×10^{-2} M HCl + 1 M KCl the result of the analysis was not definite, the surface analyzed by XRD, however, provided the strong indication of sulphur presence, because of the characteristic low-angle spectrum, Fig.9. To facilitate sulphur formation under these conditions potentiostatic dissolution at +600 mV (SCE) for 72 h was attempted. The presence of sulphur was quite evident (Fig.11) which is opposite to the observation of CuS membrane surface dissolved under the same conditions in 1×10^{-2} M HCl, (Fig.5). Finally, the surface of CuS disc left in

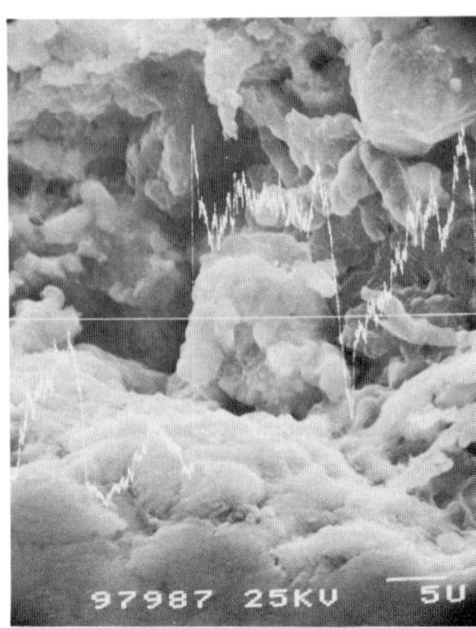

Fig.11. SEM picture ($\times 2000$) of copper(II) sulphide after anodic dissolution in 1×10^{-2} M HCl + 1 M KCl. EDAX performed with sulphur probe.

contact with 1 M $CuCl_2$ was inspected. In this case the presence of sulphur was indicated by SEM/EDAX technique and the XRD analysis also revealed the presence of copper(I) chloride.

Referring to these observations it is possible to conclude that the role of chloride ions in anodic dissolution of copper(II) sulphide is manifested by a synergic effect of the chemical reaction:

$$Cu^{2+} + CuS + 2nCl^- \rightleftharpoons 2CuCl_n^{1-n} + S \tag{8}$$

and the electrochemical reactions:

$$\text{at pH} = 0: \qquad CuS \rightleftharpoons Cu^{2+} + S + 2e^- \tag{9}$$

$$\text{at pH} = 2: \quad CuS + 4H_2O \rightleftharpoons Cu^{2+} + SO_4^{2-} + 8H^+ + 8e^- \tag{10}$$

Therefore the overall anodic reactions can be presented in the following form:

$$\text{at pH} = 0: \qquad CuS + nCl^- \rightleftharpoons CuCl_n^{1-n} + S + e^- \tag{11}$$

$$\text{at pH} = 2: \quad CuS + nCl^- + 2H_2O \rightleftharpoons CuCl_n^{1-n} + \frac{1}{2}SO_4^{2-} + \frac{1}{2}S + 4H^+ + 4e^- \tag{12}$$

Estimated current efficiencies obtained from potentiostatic experiments (+600 mV for 3h) after current-time integration coupled with the determination of total copper dissolved (by AAS) were: 100 ± 3 % for reaction (11) and 99 ± 3 % for reaction (12).[16]

Conclusions

Theoretical

The influence of chloride ions on electrochemical behaviour of copper(II) sulphide, both in open-circuit and voltammetric experiments, is a result of the electrochemical reactions (9) and (10) of the redox reaction (8). In the case of open-circuit measurements with the increase of chloride ion concentration the increase of electrode slope is observed and can be assigned to the change of potential determining reaction from (9) to (11). As a consequence two limiting slopes are observed: divalent (29.6 mV/decade) in absence of chloride ions and monovalent (59.2 mV/decade) in 1 M KCl. The voltammetric behaviour of CuS is also influenced by the same reaction, although with increase of pH reaction (12) instead of reaction (11), controls the anodic behaviour. It is at first look surprising that sulphate formation, which influences anodic dissolution at pH $= 2$ and pCl $= 0$, has no significant role in the open-circuit mechanism under the same conditions. The same behaviour, however, was previously observed with other sulphides (Ag_2S, PbS) and attributed to slow kinetics of aqueous sulphur oxidation.[31] Our results show that copper(II) sulphide is not different in this respect either.

It can be concluded in general, that electrochemical properties of copper(II) sulphide are influenced by chloride ions and consequently that copper(II) sulphide plays an active role in build up of chloride interference whenever this substance is present in the membrane. The redox reaction responsible for the interference discussed above shows the importance of consideration and inclusion of such mechanism in the general picture of interfering pathways on solid-state ion-selective electrodes. Explanation of the mechanism has so far dominated by traditional reasoning restricted only to simple ion-exchange reactions at the membrane surface.

Practical

Based on the theoretical results presented above it is also possible to draw conclusions of practical importance, i.e., by applying the mechanisms described to propose a new method for direct determination of copper in the matrices containing high concentrations of chloride ions.

As it was said already the chloride ion interference on copper (II) ISE is typically accompanied by a sluggish response which is undoubtedly unfavorable for the purpose of practical use of this sensor. On the other hand the apparent increase of the slope in presence of interfering ions (here: chlorides) is a potentially advantageous feature because of increased sensitivity in direct measurements. The conjunction of these facts encouraged us to find such conditions in which the chloride interference could be blocked, but the over-Nernstian slope preserved and - at the same time - response time shortened. It was presumed that this triple advantage can be achieved by adding a properly selected ligand to the analytical matrces. From different possible ligands thiosulfate was finally selected because of its complexing and redox properties ($\beta_{Cu(S_2O_3)_2^{3-}} = 10^{12.3}$, $\beta_{Cu(S_2O_3)_2^{2-}} = 10^{12.9}$, $E^0_{S_4O_6^{2-}/S_2O_3^{2-}} = 0.15\ V^{18}$). It was expected that thiosulfate will:

- eliminate the chloride interference by substitution of chloride ions in copper(I) complexes according to the reaction:

$$CuCl_n^{1-n} + 2S_2O_3^{2-} \rightleftharpoons Cu(S_2O_3)_2^{3-} + nCl^- \tag{13}$$

- moderate the corrosive properties of copper(II) ions because of the reaction:

$$Cu^{2+} + 2S_2O_3^{2-} \rightleftharpoons Cu(S_2O_3)_2^{2-} \tag{14}$$

which would result in better reversibility of the redox reaction at the membrane surface:

$$Cu(S_2O_3)_2^{2-} + CuS + 2S_2O_3^{2-} \rightleftharpoons 2Cu(S_2O_3)_2^{3-} + S \tag{15}$$

- and finally it was expected that the electrode potential will still be controlled advantageously by one-electron reaction:

$$CuS + 2S_2O_3^{2-} \rightleftharpoons Cu(S_2O_3)_2^{3-} + S + e^- \tag{16}$$

In the complete derivation of usable analytical equation the fact of difference between the surface and bulk concentrations was taken into account. Therefore the final equation, combining the potential of copper ISE with total (bulk) copper concentration was obtained by using diffusion-layer model principles[6] - in the form:[18]

$$E = E^0 + \frac{RT}{F} \ln \frac{C_{Cu^{2+}}}{(C_{S_2O_3^{2-}} - A \times C_{Cu^{2+}})^2} \tag{17}$$

where: E is the potential of ISE, $C_{Cu^{2+}}$ – total copper concentration $C_{S_2O_3^{2-}}$ – total thiosulfate concentration, A – the ratio of diffusion coefficients of copper(I) and copper(II) complexes.

Equation (17) enables the fast and direct determination of copper in the concentration range $10^{-2} - 10^{-5}$ M, in the presence of up to 3 M chlorides. The method described was successfully applied to the determination of copper in nickel plating baths and also proved to be a competitive method to the routinely used spectrophotometric one.[32]

The authors leave the reader with the hope that in the course of this presentation we were able to prove our primary statement on the importance of theoretical considerations which combined with the use of independent techniques can result in successful developement of good analytical methods.

Acknowledgements

The authors are indebted to A. Ivaska for his encouragement and valuable discussions as well as to B. Dandapani for his helpful advices.

References

1. J. Gulens, Ion-Sel. Electrode Rev., **2** (1980) 117.
2. J. Gulens, Ion-Sel. Electrode Rev., **9** (1987) 127.
3. J.W. Ross and M.S. Frant, Pittsburgh Symposium on Analytical Chemistry and Spectroscopy, Cleveland, Ohio, 1969, p. 60.
4. A. Hulanicki, M. Trojanowicz and M. Cichy, Talanta, **23** (1976) 47.
5. J. Koryta, Anal. Chim. Acta, **61** (1972) 329.
6. A. Hulanicki and A. Lewenstam, Talanta, **23** (1976) 661.
7. D. Midgley, Anal. Chim. Acta, **87** (1976) 19.
8. G.B. Oglesby, W.C. Duer and F.J. Millero, Anal. Chem., **49** (1977) 877.
9. R. Jasinski, I. Trachtenberg and A. Andrychuk, Anal. Chem., **46** (1974) 364.
10. J. Barica, J. Fish. Res. Board. Can., **35** (1978) 141.
11. A. Lewenstam, in: "Ion Measurements in Physiology and Medicine", M. Kessler, D.K. Harrison and J. Höper (Eds), Springer Verlag, Berlin, 1985, p. 24.
12. M.S. Frant, Plating, **58** (1971) 686.
13. D.M. Westall, F.M.M. Morel and D.N. Hume, Anal. Chem., **51** (1979) 1792.
14. E. Ghali and A. Lewenstam, in: "Ion-Selective Electrodes", E. Pungor (Ed.), Akademiai Kiado, Budapest, 1980, 235.
15. E. Ghali, B. Dandapani and A. Lewenstam , Surface Techn., **12** (1981) 265.
16. E. Ghali, B. Dandapani and A. Lewenstam, J. Appl. Electrochem., **12** (1982) 369.
17. A. Hulanicki, A Lewenstam and T. Sokalski, in: "Ion- Selective Electrodes", E. Pungor (Ed.), Akademiai Kiado, Budapest, 1985.
18. A. Lewenstam, T. Sokalski and A. Hulanicki, Talanta, **32** (1985) 531.
19. E.G. Harsanyi, K. Tòth, E. Pungor and M. Ebel, Microchim Acta, **II** (1987) 177.
20. J.W. Ross, in: "Ion-Selective Electrodes", R.A. Durst (Ed.), NBS Spec. Publ., **314** (1969) 82.
21. D.J. Crombie, G.J. Moody and J.D.R. Thomas, Talanta, **21** (1974) 1094.
22. P. Lanza, Anal. Chim. Acta, **105** (1979) 53.
23. T. Hepel, Anal. Chim. Acta, **123** (1981) 151.
24. G.J.M. Heijne, W.E. van der Linden and G. den Boef, Anal. Chim. Acta, **89** (1977) 287.
25. E. Pungor, K. Tòth, M.K. Papay, L. Polos, H. Malissa, M. Grasserbauer, E. Hoke, M.F. Ebel and K. Persy, Anal. Chim. Acta, **109** (1979) 279.
26. M. Koebel, Anal. Chem., **46** (1974) 1559.
27. R.P. Buck and V.R. Shepard, Anal. Chem., **46** (1974) 2097.
28. P. Ruetchi and R.F.J. Amlic, Electrochem. Soc., **112** (1965) 665.
29. W.M. Latimer, in: "Oxidation Potentials", Prentice-Hall, Englewood Cliffs, 1952.
30. G.W. McDonald and S.H. Langer, Metall. Trans. B., **14** (1983) 559.
31. G.W. Warren, B. Drouven and D.W. Price, Metall. Trans. B., **15** (1984) 235.
32. A. Hulanicki, T. Sokalski and A. Lewenstam, Chem. Anal. (Warsaw) **32** (1987) 601.

FTIR-ATR AND ION-CHROMATOGRAPHIC INVESTIGATIONS OF THE ION TRANSPORT THROUGH ION-SELECTIVE PVC-MEMBRANES

R. Kellner,[1] E. Zippel,[1] E. Pungor,[2] K. Toth[2] and E. Lindner[2]

1. *Institute for Analytical Chemistry*
   Technical University of Vienna
   A -1060 Vienna, Austria
2. *Institute for General and Inorganic Chemistry*
   Technical University of Budapest
   H-1111 Budapest, Hungary

Introduction

The overall response of an ion-selective membrane electrode, such as the BME-44 ionophore*-PVC-DOS system for potassium ions, described earlier[1-9] is the result of several consecutive and parallel processes of which the following four main ones are most frequently considered.[7]

1. The diffusion of K^+-ions from the bulk of the solution to the membrane;

2. The charge transfer process through the membrane solution interface (the heterogeneous step);

3. The formation of potassium ionophore complex in the membrane surface area and

4. The diffusion of the formed complex through the membrane bulk.

Steps 1-3 are known to be very fast and are called phase boundary processes. In contrast, in step 4, the rate of diffusion of species through the membrane is a rather slow process and it is controlled by the bulk membrane properties.

The aim of this report is to present the results of the experimental work done recently in our laboratories to study both the surface and the bulk mebrane processes to demonstrate the importance of one over the other in the potential response function. The short time behaviour of model membranes was studied by response time measurements, while the transport processes within the membranes were followed under *in situ* conditions by means of FTIR-ATR-spectrometry and ion-chromatography. "*In situ* conditions" means that the membranes remained in contact with aqueous K^+-containing solutions throughout the FTIR-ATR-measurements simulating normal operating conditions of the electrode membranes. Prior to

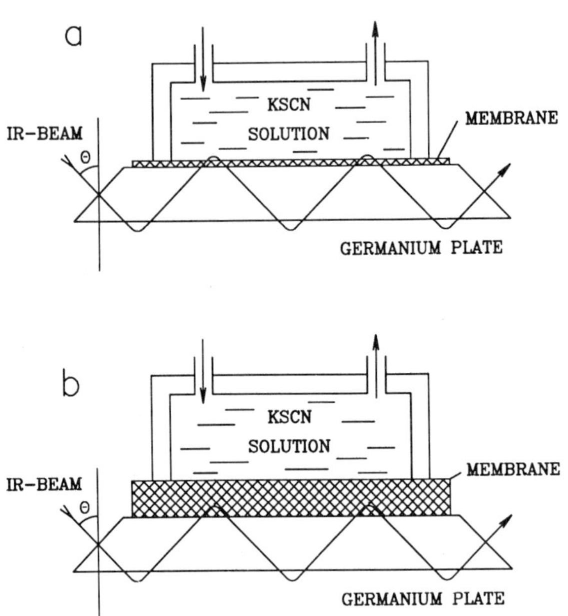

Fig.1. IR-ATR-cell for *in situ* measurements at thin films (a) and thick films (b) of PVC-DOS-Crown ether membranes

the *in situ* measurements the chemical nature of the membrane surfaces before and after contact with aqueous solutions of KSCN, KCl, NH_4SCN and NaCl was investigated by *ex situ* FTIR-ATR-spectroscopy.[1–4]

In Situ Measurements

The experimental set-up used in connection with FTIR-ATR-spectrometry enabled us to follow the diffusion of the components of interest through the membrane under realistic conditions for ISFETs . One side of the membrane is in contact with a solid state electrode the other side with the aqueous solution containing the ion to be determined. This means that the membrane being analysed remains in contact with the solution throughout the whole measurement rather than being separated and dried. This set-up allows an investigation of the diffusion process in model membranes and a direct comparison of the results with the electrochemically measured time dependent parameters of the corresponding electrodes.

Experimental

Fig.1 shows a schematic drawing of the ATR-cell used for thin (a) and thick (b) membrane layers as compared to the penetration depth d_p according to Harrick.[10] For our case this penetration depth in the ν-SCN band region (2050 cm^{-1}) is around 0.2 μm (Ge-reflection element, $n_1 = 4$, $\theta = 45°$, $n_2 = 1.5$).

In order to guarantee a good adhesion of the membrane to the reflection element in contact with water the Ge-surface was made hydrophobic by silanization prior to the dip coating procedure.[11] The silanization was carried out according to a previously described method.[12] Thus 1 % diethoxydimethylsilane (Merck 818817) was dissolved in 200 ml absolute toluene which was dried by molecular sieve (4Å) for three days. The reaction was carried out at 100°C for 75 h. This pretreatment delivered sufficently stable coating of PVC/DOS-membranes of sub-μm to 10 μm thickness depending on the concentration of the solution used for coating. (65 mg PVC and 120 mg DOS in 15-20 ml THF or 65 mg PVC and 120 mg DOS in 5-7 ml THF, resp.).

The technique of attenuated total reflection (ATR)[10] has been used for some time and recently for *in situ* studies by Fringeli,[16] Gendreau[17] and others in the investigation of protein

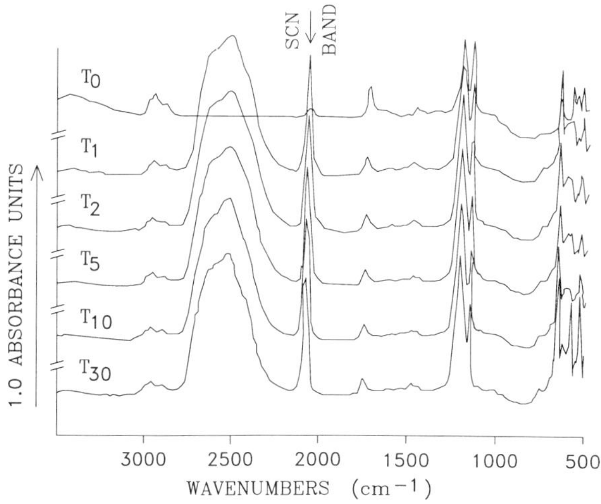

Fig.2. AFTIR-ATR-spectra of a thin PVC/DOS/BME-44-membrane
in contact with 3 mol/l KSCN solution in D_2O. T is the measurement
time in minutes after the solution contact.

adsorption at polymer surfaces and by Neckel et al.[18] for spectroelectrochemistry (Fig.1).

We made experiments with both crown ether containing and dummy membranes of
varying thickness, using KSCN solutions of different concentrations (0.1, 0.5, 3 mol/l) or 1
mol/l KCl solution. It was also necessary to change the solvent from distilled water (H_2O) to
heavy water (D_2O) due to the fact that normal water as solvent has vibrational bands in the
interesting spectral region of the crown ether complex (1500 – 1600 cm^{-1}). The sample cell
was filled by a medical syringe with 10 ml test solution (volume of the cell is approximately
6 ml). The measurements were carried out in static solutions without streaming or stirring
and the spectra were recorded at times 0, 1, 2, 5, 10 and 30 min. The cell was filled between
time 0 and 1 min.

Response Time Measurements

Parallel to the spectroscopic studies the dynamic response characteristics of the membranes were investigates with the activity step method[13] by means of a new switched wall jet
type experimental set-up.[14,15]

Results and Discussion

Thin Membranes (d < 0.2 μm). A thin membrane containing crown ether, Fig.1a, was
exposed in the cell to a 3 mol/l solution of KSCN in D_2O. Small complex bands at 1520
and 1580 cm^{-1} could be determined in the membrane after contact with the solution. The
increase of intensity as a function of time is not significant (Fig.2). Also the more significant
SCN-vibration at 2054 cm^{-1}, which clearly can be discerned, was of constant intensity. This
is explained by the fact that solution species and species adsorbed at the surface by a rapid
process made the greatest contribution to the ν-SCN band.

Thick Membranes (1 μm< d <10 μm). The thick film case (according to Fig.1b) is
markedly different from the case of thin film because the transport of SCN$^-$-ions with the
K-crown-ether complex can clearly be observed by the FTIR-ATR-technique. The increase
of the intensity of the ν-SCN band within 30 min after the first contact between the 0.5
mol/l solution of KSCN and the membrane can be measured from Fig.3 and is displayed in
Fig.4 (crown ether curve). A comparison of the analogous curve with the dummy membrane
shows that even a low content of crown ether in the membrane (2 %) significantly increases
the SCN-amount extracted into the membrane. When the concentration of KSCN in the
bathing solution was increased a further increase in the amount extracted was observed by

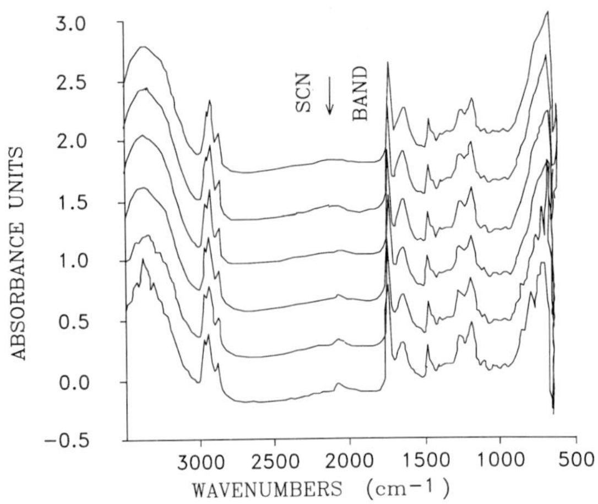

Fig.3. FTIR-ATR-spectra of a thick PVC/DOS/BME-44-membrane in contact with 0.5 mol/l KSCN solution in D_2O.

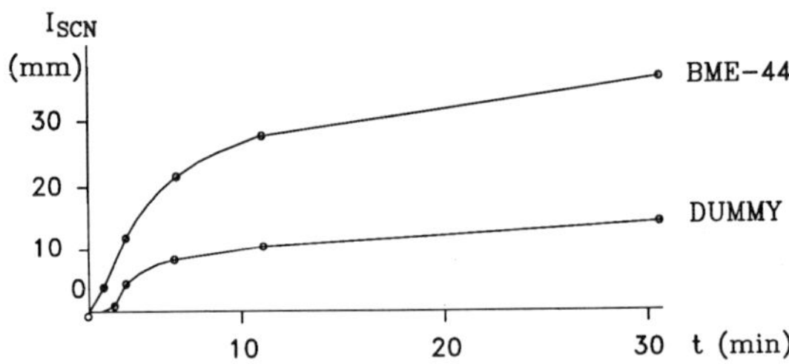

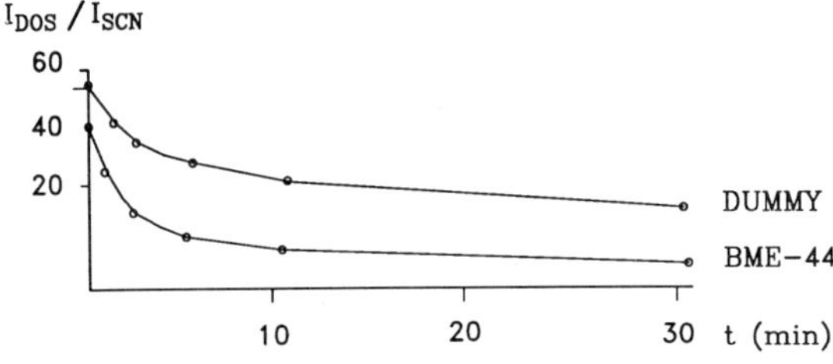

Fig.4. Plot of the intensity of the ν-SCN at 2054 cm^{-1} and the ratio of the ν-DOS ν-SCN versus time for crown ether and dummy membranes in contact with 0.5 mol/l KSCN solution in D_2O each.

ATR-spectroscopy. This is a direct proof for the increase of the site concentration in the liquid membrane electrode, which is in excellent agreement with the findings in another study.[20]

The experiments with KCl contacted liquid membranes showed that no measurable amount of K-BME-44 complex was found to be transported through the membrane as can be explained using the fixed site membrane model.[19,20]

Response Time Measurements. In order to check the effect of equilibration processes at membrane surfaces in contact with KCl and KSCN solutions of different concentrations dynamic response curves were recorded with the wall jet type apparatus as described in the litterature.[14,15] The effect of concentration steps from 10^{-3} to 10^{-2} mol/l KCl, and from

226

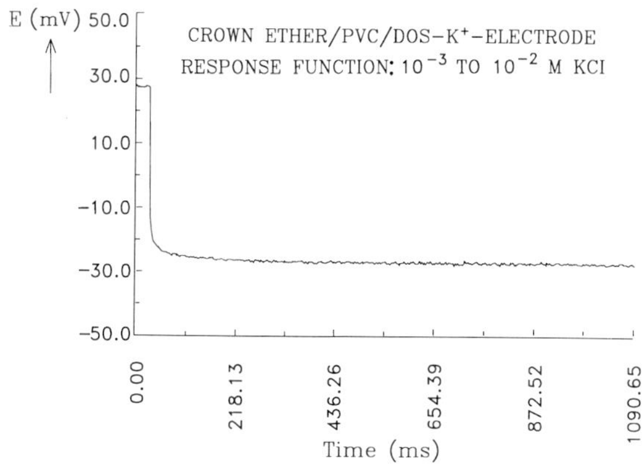

Fig.5. Example of a response time curve BME-44/DOS/PVC membrane conditioned in KCl, c = 10^{-3} mol/l for 2 days, concentration step from 10^{-3} to 10^{-2} mol/l KCl.

conditioned membrane (conditioned in 10^{-3} mol/l KCl or 10^{-1} mol/l KSCN for varying lengths of time.[7])

In all cases the response time values were of the same order of magnitude as those found for precipitate based electrodes[14] which means in the range of a few tenths of ms. Fig.5 gives a typical example of a response time curve for a BME-44 containing PVC/DOS-membrane 2 days conditioned in 10^{-3} mol/l KCl. The potential changes immediately after the concentration is stepped from 10^{-3} to 10^{-2} mol/l KCl and the electrode response is Nernstian.

An overall equilibration of the ion-selective membrane with the sample solution is not a precondition for Nernstian behaviour, i.e., the role of conditioning in the primary ion containing solution is not crucial.[21] However, if the chloride concentration level is dramatically increased in the bathing solution (in the concentration range, at which anion extraction cannot be neglected, i.e., in the range of the Donnan exclusion failure) then the extent of Cl$^-$ extraction into the membrane is no longer negligible . This is reflected by a smaller potential change, and by a somewhat decreased rate of change in the response time curves (activity step from 1 mol/l KCl to 0.1 mol/l KCl) and also by the appearance of BME-44 potassium complex bands in the FTIR-ATR spectra.

Similar conclusions can be drawn when the potential time functions recorded in high concentrations of KCl and KSCN solutions are compared with each other (Fig.6). The diverging direction of the potential changes are due to the fact that the applied activity step in potassium chloride solution corresponds to the cation response range of the electrode while that in KSCN solution is in the anion interfering section of the potential vs. log a_K response curve (see insert to Fig.6).

The rates of the potential response both in KCl and KSCN solutions are nevertheless by several orders of magnitude larger than the rate of the inner membrane diffusion as studied by the FTIR-ATR method in this work (Fig.3) and by other techniques presented elsewhere.[22] But in the long time range rate of the potential shift (E or asymmetry potential change) was found to be the same as that observed in the diffusion processes.

Ion-Chromatographic Measurements

Under normal experimental conditions, ion-selective electrode membranes are in general immersed at both sides in aqueous solutions and are operating under near zero current. The following experiment was done in addition to the ATR-diffusion measurements described in chapter 2, done at membranes placed on a solid state support, in order to compare the diffusion behaviour of KSCN and KCl through the membrane. The experimental set up is shown in Fig.7.

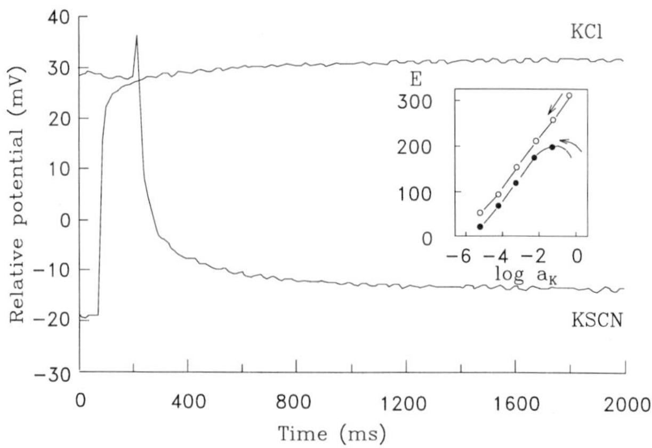

Fig.6. Comparison of the response time curves of BME-44/DOS based potassium selective membranes recorded in concentrated KCl and KSCN solutions. Inset: EMF-log a_K calibration plot of the BME-44/DOS based electrode in KCl (o) and KSCN (•) solutions. The response time curves were recorded in the concentration ranges marked with arrows. Activity step: from 0.1 mol/l KCl to 1.0 mol/l KCl or to 0.1 mol/l KSCN. KSCN: flow rate: 115 ml/min.

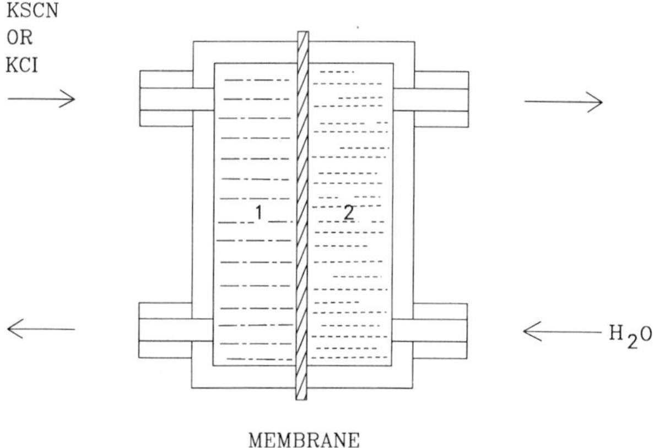

Fig.7. Flow-trough-cell for diffusion experiments with ion selective membranes conditioned on one side with 3 mol/l KSCN solutions, or with 2 mol/l KCl solutions, and with distilled H_2O on the other side.

A PVC/DOS-membrane containing the crown ether was prepared as described before,[8] placed in the flow through cell as shown in Fig.7 and conditioned with distilled H_2O on both sides for approx. 20 hrs. After that time the water on side 1 was replaced by an aqueous solution of KSCN (3 mol/l). From time to time samples were taken from the water reservoir on side 2 and analyzed by ion-chromatography for K^+. The water was slowly circulated through side 2. The data shown in Table 1 indicate a transfer of K^+ from side 1 to side 2 through the membrane.

In a similar way a second membrane was treated with a 2 mol/l KCl-solution. In that case, however, no K^+ was detected during the experiment in the reservoir on side 2 (Tab.1). This data is supported by the fact that K^+ could be found by electron probe microanalysis

Table 1. Ion-chromatographic measurements
of K$^+$ in samples from side 2

Sampling times (hrs)	3M KSCN (ppm)	2M KCl (ppm)
0 (blank)	*	*
1.5	-	*
5	-	*
15	8.5	*
48	1.5	-
72	1.0	-
96	0.8	-

* not detectable
- not measured

only at the surface in contact with KCl-solution (side 1) and not at all on the surface of side 2. These data in combination with the earlier results[1] described in this survey clearly support the validity of the space charge effect theory.

Conclusions

The results presented in this paper suggest that the diffusion of complex species through the membrane is by far the slowest process among those listed in the introduction. On the other hand, the dynamic response measurements proved that the rate of potential response is in the ms range; thus one must conclude that the potential determining process cannot solely be explained by the inner membrane diffusion, but rather by phaseboundary processes, i.e. by the building up of the space charge.

According to the space charge membrane model potassium ions are solubilized by the ligand in the boundary region of the membrane, while hydrophilic anions such as chloride ions are remaining in the solution phase. The concentration of the BME-44 potassium complex within the membrane bulk is determined by the low concentration of the fixed sites[20] (to fulfil electroneutrality), which are evenly distributed in the membrane bulk. This fact causes a steep concentration profile for the BME-44 potassium complex in the subsurface membrane region which could be followed with the FTIR-ATR-technique.[1]

As know, the FTIR-ATR-technique with variable angles of incidence, provides additional information of the concentration in regions of different thicknesses. Unfortunately, the minimum thickness of the observed layer is about one order of magnitude bigger than that of the space-charge region. In spite of this, the small but significant variation of the integral concentration in depth certainly shows an enrichment of the BME-44 potassium complex in areas close to the membrane surface.

When, however, the BME-44 ionophore based membrane is treated with relatively concentrated KSCN solution both the anions and cations are extracted into the membrane in the form of an electroneutral ion pair. Thus, under this condition not only the concentration of the complex is higher but the concentration profile for the complex in the subsurface membrane layer is less steep, too. This high concentration of the complex could be detected in depth by FTIR-ATR spectrometry and due to the more even distribution of the complex in the membrane it could not be washed out easily.[1]

In the light of response time experiments the basic potential determining process in both KCl and KSCN solutions seems to be the ion transfer rates at the membrane-solution interface. These transfer rates are characterized by the exchange current densities and were found to be high for ionophore based membranes. The slow inner membrane processes, however, do not influence the short-term potentiometric behaviour of the solvent polymeric membranes.

Selective transport of primary ions does take place in the membrane bulk as found in the electrodialysis experiments. It is, however, not of any greater importance in the potential forming process in the membrane because the rate of the ion transport through the membrane is several orders of magnitude less than the rate by which the membrane potential is established after an activity step.

References

1. R. Kellner, G. Fischböck, G. Götzinger, E. Pungor, K. Tóth, L. Pólos and E. Lindner, Fresenius' Z. Anal. Chem., **322** (1985) 151.
2. R. Kellner, G. Götzinger, E. Pungor, K. Tóth and L. Pólos, Fresenius' Z. Anal. Chem., **319** (1984) 839.
3. R. Kellner, G. Götzinger, E. Pungor, K. Tóth and L. Pólos, in: "Proceedings of the 4th Scientific Sessions on Ion-Selective Electrodes", E. Pungor (Ed.), Matrafüred, Hungary Akademiai Kiado, 1985.
4. R. Kellner, E. Zippel, E. Pungor, K. Tóth and E. Lindner, Fresenius' Z. Anal. Chem., **328** (1987) 464.
5. E. Lindner, K. Tóth, M. Hórvath, E. Pungor, B. Agai, I. Bitter, L. Töke and Z. Hell, Fresenius' Z. Anal. Chem., **322** (1985) 157.
6. P. Läuger, R. Benz, G. Stark, E. Bamberg, P.C. Jordan, A. Fahr and W. Brock, Q. Rev. Biophys., **14** (1981) 513.
7. K. Tóth, E. Lindner, E. Pungor, E. Zippel and R. Kellner, Fresenius' Z. Anal. Chem., in press (1988).
8. G.J. Moody, R.B. Oke and J.D.R. Thomas, Analyst, **95** (1970) 910.
9. H. Freiser, in: "Ion-Selective Electrodes in Analytical Chemistry", H. Freiser (Ed.), Vol. 1, Plenum Press, New York, p. 246 (1978).
10. N.J. Harrick, in: "Internal reflectance spectroscopy", Interscience New York (1967).
11. R. Kellner and G. Götzinger, Mikrochim. Acta, **II** (1984) 61.
12. P. Hofer and U.P. Fringeli, Biophys. Struct. Mech., **6** (1979) 67.
13. E. Lindner, K. Tóth and E. Pungor, Pure Appl. Chem., **58** (1986) 469.
14. E. Lindner, K. Tóth and E. Pungor, Anal. Chem., **48** (1976) 1071.
15. E. Lindner, K. Tóth, E. Pungor, T.R. Berube and R.P. Buck, Anal. Chem., 59 (1987) 2213.
16. a) U.P. Fringeli, Z. Naturforsch., **320** (1977) 20
    b) U.P. Fringeli, Günthayrd HH., Infrared membrane spectroscopy, in: "Membrane spectroscopy", F. Grell (Ed.), Springer, New York (1981).
17. R.M. Gendreau, Biological and biomedical applications of FTIR-Spectrometry, International Conference on Fourier and Computerized Infraded Spectroscopy, J.G. Grasselli and D.G. Cameron (Eds), Proc. SPIE, **553** (1985) 4.
18. R. Kellner, H. Neugebauer, G. Nauer and A. Neckel, In situ-spectroelectrochemistry via infrared techniques, International Conference on Fourier and Computerized Infrared Spectroscopy, J.G. Grasselli and D.B. Cameron (Eds), Proc. SPIE, **553** (1985) 12.
19. G. Horvai, E. Gráf, K. Tóth, E. Pungor and R.P. Buck, Anal. Chem., **58** (1986) 2735.
20. R.P. Buck, K. Tóth, E. Gráf, G. Horvai and E. Pungor, J. Electroanal. Chem., **223** (1987) 51.
21. W.E. Morf and W. Simon, Helv. Chim. Acta, **69** (1986) 1120.
22. U. Oesch and W. Simon, Anal. Chem., **52** (1980) 692.

ON THE ELECTROCHEMICAL APPROACH TO SOLID-STATE ION SELECTIVE MEMBRANE PREPARATION

M. Neshkova

Institute of General and Inorganic Chemistry
Bulgarian Academy of Sciences
1040 Sofia, Bulgaria

Introduction

The classical approach to the preparation of solid-state membrane electrodes is restricted to a limited number of sensing materials which allow formation of compact membrane discs mainly by pressing. For materials of poor plasticity the technology is further complicated by the introduction of additional procedures like hot pressing, sintering, cutting of the membrane discs from melts or sometimes even growing of monocrystal membranes. When membranes prepared according to the above technology are used in flow systems or applied in microconstructions, other limitations of this approach become evident, too. The search for new techniques of membrane preparation is also important from a theoretical point of view. The basic potential generating processes are known to take place at the membrane/solution interface, hence the way of building up the membrane surface proved to be decisive for revealing fully the abilities of a particular sensitive material. The aim of the present report is to outline the possibilities offered by the electrochemical approach to the preparation of solid-state chalcogenide membranes and its role in the elucidation of their response mechanism. Copper ion-selective electrodes (ISE) are used as an example.

Electrochemical Preparation of Chalcogenide Membranes

The essence of this technological approach consists in electrochemical *in situ* deposition of the active chalcogenide material onto an inert conducting substrate (most often of Pt) pre-shaped and sized appropriately as an electrode.[1] In the presence of heavy metal ions, Se(IV), Te(IV) and As(III) are known to be electroreduced to selenides, tellurides and arsenides respectively. This fact can be used for the preparation of membrane coatings with a thickness of a few micrometers. This idea is best illustrated in Fig.1. A growing interest to this approach has been noted recently in relation to the investigations on solar energy conversion by photovoltaic or photoelectrochemical cells.[2-4] Unfortunately, the idea has not yet gained its due position among the techniques for ion-selective electrode preparation.

The electrochemically deposited chalcogenide coatings can be used as ion-selective membranes provided the following requirements are met: *(i)* A definite composition of the coating must be obtained to ensure maximum sensitivity of the electrodes. Codeposition of either the metal or non-metal must be avoided. Very often the electrochemical reaction may result in the formation of more than one compound and then a proper choice of the chemical and electrochemical conditions of the electrolytic bath has to be made.[5] *(ii)* The coatings must be of good adherence, mechanically strong and chemically stable to garantee a long-term usage of the electrodes. From the afore said it is obvious that complementary electrochemicals studies (under potentiostatic control) of each particular system are needed in order to define the optimal conditions for membrane deposition.

Contemporary Electroanalytical Chemistry, Edited by A. Ivaska *et al.*
Plenum Press, New York, 1990

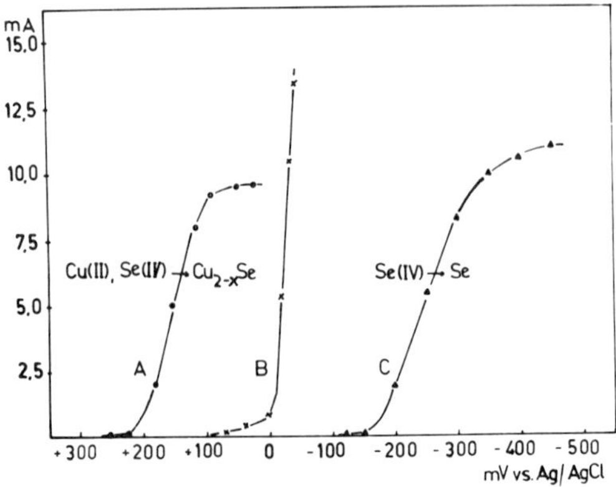

Fig.1. Voltammetric curves at a Pt-electrode in 0.5 M H_2SO_4: Curve A – 0.3 M $CuSO_4$ + 3 × 10^{-2} M Na_2SeO_3; Curve B – 0.3 M $CuSO_4$; Curve C – 3 × 10^{-2} M Se(IV).

In 1976[1] and in a later publication of ours[5] we pointed out the possibilities of the electrochemical deposition of chalcogenide membranes as a promising approach to solid-state ion-selective membrane preparation. In a series of papers[5-9] the electrode performance of Cu-ISEs based on electrochemically prepared membranes of the general formula $Cu_{2-x}Se$ has been examined and compared to that of the pressed-pellet selenide membranes. Recently, using the same approach, Cu-ISEs with a thin ternary CuAgSe-and Cu_3As- membranes have been developed and investigated, too.[10] The Ag-ISE with electrodeposited Ag_2Se membrane proved to be very sensitive and stable.[11] Since the possibilities of this approach are not confined only to the preparation of the above mentioned ISEs, we would like to emphasize some of its advantages and limitations:

(1) Homogeneous membranes of a definite structure and composition can easily be prepared from all chalcogenides, even from such of poor plasticity which cannot be shaped by pressing.

(2) Electroplating is a very cheap technique for *in situ* membrane preparation.

(3) The membrane thickness is reduced to a few micrometers. This loweres the membrane considerably.

(4) Electrode constructions of various shape and size can easily be designed.

(5) Electrocrystallization, in general, ensures coatings of very fine structural defects,[12] something that cannot be obtained by other techniques. When compared to the corresponding single crystals, the electrodeposited films ensure unique surface morphology, which gives rise to characteristic electrophysical properties.[13]

(6) The composition and structure of the coating membranes can easily be modified through control of the chemical and electrochemical parameters of the electrolytic bath. This allows model studies of the response mechanism.

(7) These type of membranes have a considerably shorter life time as compared to the corresponding pressed-pellet ones. Their life times varies from 5 months to one year for the different membranes. This drawback is compensated by the possibility to redeposit a new membrane to place the worn-out one.

Electrochemical Approach in Elucidating the Response Mechanism of Selenide-Based Copper Ion-Selective Electrodes

In some of our earlier investigations[6-9] we demonstrated that, contrary to the statement of Hirata,[14] Cu-ISEs with very good performance characteristics can be prepared on selenide basis. Suggestions were also made concerning the response mechanism on the basis of the experimentally established fact that the electrodes are primary for Cu(I), and the good agreement observed between the experimentally determined selectivity coefficients, $K_{Cu^+,Cu^{2+}}$, and those calculated according to the theoretical concepts of Buck[15] and Pungor and Toth.[16]

The assumption was made that the Cu(II) response of the selenide sensors is governed by the following metathesis reaction proceeding at the electrode surface:

$$Cu^{2+} + Cu_2Se \rightarrow CuSe + 2Cu^+ \tag{1}$$

This model appeared to be adequate to explane the electrode lower limit of the Nernstian response of the electrode in Cu(II) buffer and non-buffer solutions as well as of some "anomalous" behaviours of the electrodes.[8-9] It did not, however, give an exact answer to the question what are the great differences in the behaviour of the $Cu_{2-x}Se$ electrochemical thin membrane we proposed and the $Cu_{1.8}Se$ monocrystal one.[17] It was not clear enough how the new CuSe phase, formed as the result of reaction (1), was attached onto the electrode surface. In order to give answers to the questions presented above, an electrochemical study was started to investigate the interrelation betveen the chemical and phase composition of the membrane, and the electrode function for Cu(II).

Mathieu et al.[18] demonstrated that composition of the binary sulphide, Cu_xS, evaporated as a thin layer could successfully be varied by constant current flow in the following galvanic cell:

$$+ \; Cu_xS/Cu^{2+}(sol.)/Pt- \tag{I}$$

An alternative approach was used by us. The nonstoichiometric membrane, denoted as $Cu_{2-x}Se$, was rendered stoichiometric, Cu_2Se, by shortcircuiting the galvanic cell:

$$Cu/Cu^{2+}(sol.)/Cu^{2+}(sol.)//Cu_{2-x}Se/Pt/Cu' \tag{II}$$

until an e.m.f. zero was reached. In the above galvanic cell a 10^{-2}M $CuSO_4$ with pH=2 (H_2SO_4) was used as the electrolyte. A thermostated cell deaerated with Ar was used with separate cathodic and anodic compartments.

This experiment allowed us to determine the exact value of x for the nonstoichiometric membrane. For this purpose, first, Q_{dep}, the quantity of electricity consumed for membrane deposition was registered, and then, Q_{red}, that for membrane reduction in the galvanic cell (II), according to the reaction:

$$CuSe + Cu^{2+} + 2e^- \rightarrow Cu_2Se \tag{2}$$

The value of x can be calculated following Faraday's law by the formula:

$$\frac{Q_{dep} + Q_{red}}{8} = \frac{Q_{dep}}{8 - x}$$

Three parallel experiments gave the value $x = 0.67$. The X-ray diffraction analysis of thin electrode coatings is strongly hampered by their fine texture and by the interference of the Pt substrate. It was therefore necessary to analyze a greater number of electrode samples so that a higher reliability of results could be attained. It has been established that with thin evaporated $Cu_{2-x}Se$ films the homogeneous phase is preserved up to $x = 0.6$.[20] The experimentally determined coefficient ($x = 0.67$) for our non-stoichiometric membranes turned out to correspond to the co-existence of two phases: a predominating one of Cu_3Se_2 and an admixture of CuSe. The membrane surface was further characterized by XPS technique, too. The photoelectron spectrum of Cu $2p_{3/2}$ implies the presence of Cu(I) and Cu(II) at the membrane surface as well (Fig.2).

The galvanic cell (II) was used to vary the coating composition in the range $0.67 \geq x \geq 0$. Fig.3 represents a typical dependence of the membrane potential on its composition at open circuit. The plateaus registered on the curve correspond to the region of x values for which two phases coexist in the membrane as can be predicted from thermodynamic considerations.[18,19] For $x \leq 0.4$ of the plateau region the coexistence of Cu_3Se_2 and $Cu_{2-x}Se$ ($x = 0.2$) phases was proved, while for $x < 0.2$ Cu_2Se and $Cu_{2-x}Se$ ($x = 0.2$) were found.

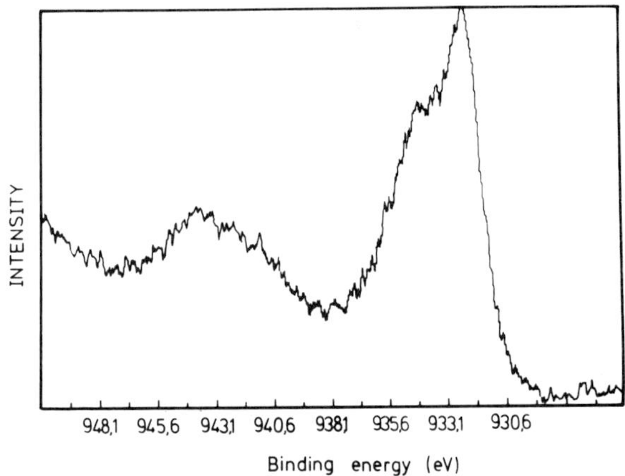

Fig.2. Cu $2p_{3/2}$ photoelectron spectrum of the $Cu_{2-x}Se$ – membrane ($x = 0.67$).

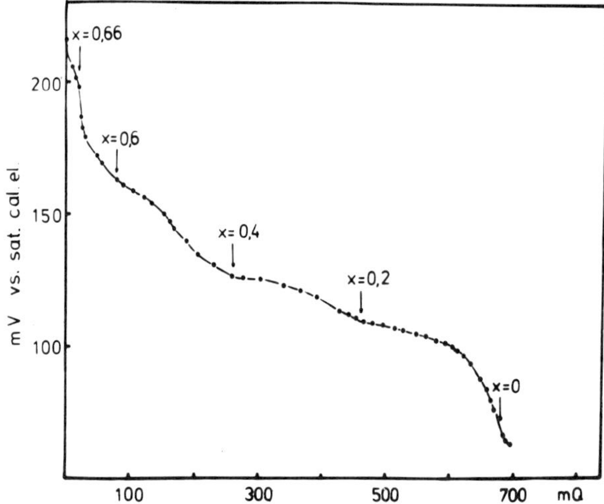

Fig.3. $Cu_{2-x}Se$ – membrane potential dependence on the value of x resp. on the mQ_{red} consumed during its step by step reduction in galvanic cell II.

The electrode behaviour of a stoichiometric Cu_2Se membrane that was obtained as a result of a complete reduction of the initial membrane in galvanic cell (II) was investigated first. The change in structure and composition of the membrane resulted also in a change in the colour: from ink blue ($x = 0.67$) to light grey ($x = 0$). Although the stoichiometric membrane senses Cu(II) in solution, its potential proved to be unstable with time. Curve C in Fig.4 represents the initial potential values. The potential changes exponentially with time going towards a steady-state value. The time for reaching a stable potential depends on the concentration of the Cu(II) standards (Fig.5). While shifting to its steady-state potential value the membrane regains its blue colour and the copper solution becomes opalescent. It was established by X-ray diffraction that membranes which had reached equilibrium potential had undergone some structural changes, too. In place of the single Cu_2Se phase, Cu_3Se_2 and $Cu_{2-x}Se$ ($x = 0.2$) were also detected being characteristic for the plateau region $x = 0.4$ (Fig.3). The membranes thus restructured exhibit stable electrode performance but they have higher limit of Nernstian response as compared to the initial membranes ($x = 0.67$). A mixture of Cu(I) and Cu(II) was found by XPS in the precipitate obtained after centrifuging

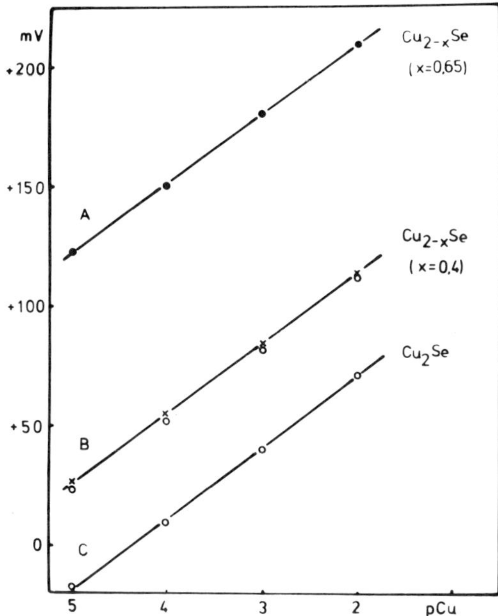

Fig.4. Calibration curves for membranes of different composition:
Curve A – (●) with the initially prepared membrane ($x = 0.67$); Curve
B – (×) with a partially reduced membrane ($x = 0.4$); (o) with the
restructured stoichiometric membrane; Curve C – (o) with the stoi-
choimetric Cu_2Se – membrane.

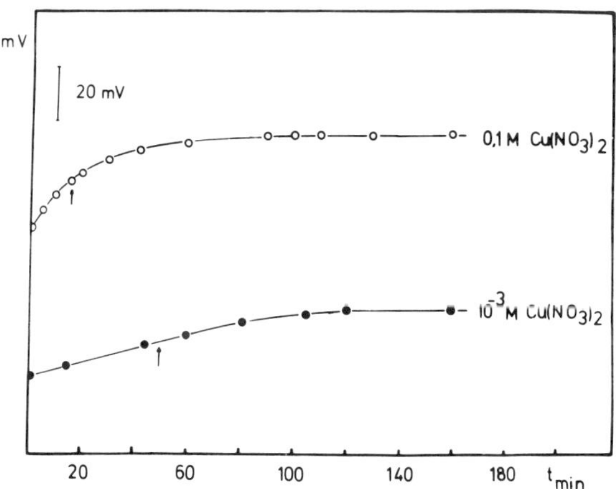

Fig. 5. Potential – time transient for the stoichiometric Cu_2Se – membrane.

the copper solution.

By partial reduction of the initial membrane ($x = 0.67$) a composition corresponding to
$x = 0.4$ was obtained. Such membranes showed a stable potential in Cu(II) solution from the
very beginning and a standard potential identical to that of the stoichiometric membranes
after reaching a steady-state (curve B in Fig.3).

The above described phenomena strongly support the suggestion that reaction (1) is
responsible to the electrode response to Cu(II). To exhibit a stable response in Cu(II) solutions
the phase composition of the membrane must obviously be adequate to accept the new CuSe
phase generated according to reaction (1) so that no changes in the Cu activity does occure

in the coating. This can only be ensured providing that there is also a measurable quantity of CuSe in an appropriate structural arrangement in the membrane. That is however, not the case with the stoichiometric membrane. The observed potential change could be due to the following reaction:

$$2Cu_2Se + Cu^{2+} \rightarrow Cu_3Se_2 + 2Cu^+ \tag{3}$$

This reaction gives a new equilibrium between the two phases. When the initial phase composition of the membrane implies coexistance of Cu_2Se and $CuSe$ (for compositions in the range $0.7 > x > 0.25$), no activity changes of Cu in the coating will be provoked by reaction (1), and the potential will be determined only by the copper activity in the solution.

The electrode sensitivity, on the other hand, differs for membranes with different phase composition and exhibits the highest value for x ranging from 0.6 to 0.7. That is probably due to changes in the conductivity as a function of the phase composition as was demonstrated by Young.[21] This kind of investigations are presenty under way in our laboratory.

The above discussed facts demonstrate unequivocally that the electrochemical approach to ion-selective membrane preparation is of a definite interest both from technological point of view, and as an experimental technique for model response studies.

Acknowledgements

This study was supported by a grant from the Bulgarian National Committee for Science (No 465).

References

1. M. Neshkova, Bulgarian Patent, No. 33078 (1976).
2. G.F. Fulop and R.M. Taylor, Am. Rev. Mater. Sci., **15** (1985) 197.
3. D. Elwell and R.S. Feigelson, Sol. Energy Mater., **6** (1982) 123.
4. Y.W. Chen, J.A. Turner and R. Noufi, Appl. Phys. Comm., 4 (1984-85) 241.
5. M. Neshkova and H. Sheytanov, J. Electroanal. Chem., **102** (1979) 189.
6. Idem, Talanta, **32** (1985) 654.
7. Idem, Ibid., **32** (1985) 937.
8. Idem, Mikrochimica Acta, **II** (1985) 161.
9. M. Neshkova, Conference Materials, Analytiktreffen 1985, Leipzig, p. 35.
10. M. Neshkova, Compt Rend. Acad. Bulg. Sci., in press.
11. M. Neshkova, Ibid., in press.
12. K.M. Gorbunova and Yu.M. Polukarov, in: "Elektrokhimia", (1964) 1 Moscow (1968), p. 59.
13. M. Tomkiewicz and I. Ling, W.S. Parsons, J. Electrochem. Soc., **129** (1982) 2016.
14. H. Hirata and K. Higashiyama, Talanta, **19** (1972) 391.
15. R.P. Buck, Anal. Chem., **40** (1972) 1432.
16. E. Pungor and K. Toth, Analyst, **95** (1970) 1625.
17. J. Vesely, Collect. Czech. Chem. Commun., **36** (1971) 3364.
18. H.J. Mathieu and H. Rickert, Z. für Physik. Chemie, **79** (1972) 315.
19. M. Sato, Electrochim. Acta, **11** (1966) 361.
20. R.N. Kordumova, Krystallographia, **13** (1968) 796.
21. V. Young, Solid State Ionics, **25** (1987) 9 and 21.

SENSOR TECHNIQUE FOR MONITORING CHANGES IN pH, Ca^{2+}, p_{O_2}, p_{CO_2} AND ELECTRICAL CONDUCTIVITY IN MILK DURING FERMENTATION

Pekka O. Lehtonen*, Hanna Laitinen, Tuomo Tupasela and Matti Antila

University of Helsinki
Department of Dairy Science
SF-00710 Helsinki, Finland

Introduction

Many dairy products are made through the process of fermentation. The quality of such products depends strongly on the cultures used with fermentation. The functional micro-organisms involved in these processes may have very different properties, but they are often lactic acid bacteria. Those cultures are referred to as "starters".

Sensors offer the possibility to realtime analysis, which is particularly important for the rapid analysis of samples. We have used sensors in a study to develop cheese starters. The action of the starter bacteria is central to the production of cheese. Their primary function is the production of the lactic acid from lactose which ensures the desired pH at all stages throughout the process.

Lb.helveticus is used as a bacterial starter in Emmental cheese. *Lb.casei* occurs normally in cheese milk and cheesemaking equipment and is a part of the natural flora. It seems, however, to have a benefical role in cheese ripening. We compared how these two bacteria differ when growing in milk, investigating two different strains of each bacterium. *Lb.casei* strains were isolated from cheese and *Lb.helveticus* strains from pure cultures intended for Emmental cheesemaking.

The objectives of the work were
- to get new physical/chemical information for the study to develop cheese starters
- to evaluate the relationship between metabolic activity and changes in pH, p_{O_2}, p_{CO_2}, Ca^{2+} and electrical conductivity
- to compare how two bacteria differ when growing in milk, investigating two different strains of each bacterium.

Materials and Methods

Strains and Media

Two strains of *Lb.helveticus* and of *Lb.casei* were investigated when growing in reconstituted skim milk in a fermenter. The strains of *Lb.helveticus* were isolated from freeze dried pure cultures of Chr. Hansen's Laboratory: *Lb. helveticus* CH 1 and *Lb.helveticus* LH 7. *Lb.casei* strains RG 2 and G 2 were isolated from cheese.

Before culturing in a fermenter the bacteria were first grown in MRS-broth medium at 37°C for 24 h (1 % inocolumn) and then in sterilized (95°C, 55 min) skim milk at 39°C until coagulated: *Lb.helveticus* 11-12 h and *Lb.casei* 17-18 h.

---

* Present address: University of Helsinki, Department of Chemistry, SF-00100 Helsinki, Finland

Contemporary Electroanalytical Chemistry, Edited by A. Ivaska *et al.*
Plenum Press, New York, 1990

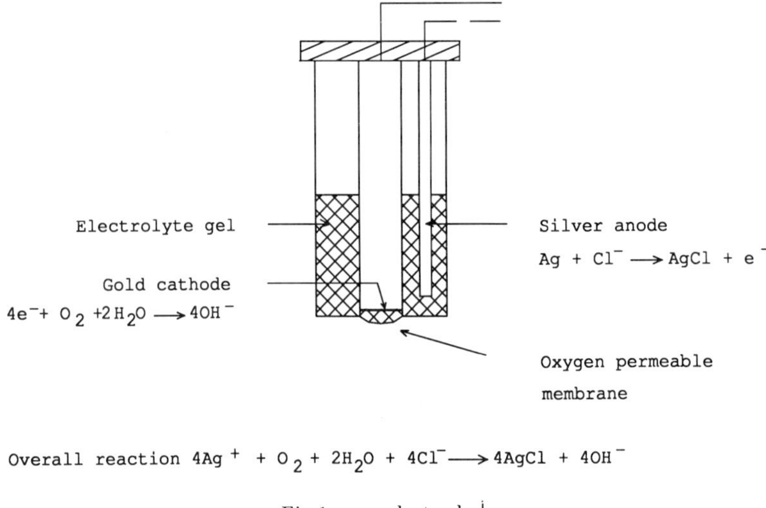

Fig.1. p_{O_2}-electrode.[i]

The inoculumn used in the fermenter was 5 % (200 ml/4 l milk). Milk was sterilized in an autoclave at 95°C for 55 min and it was put into the fermenter just before each start. The temperature was set to 39°C and when it was attained, cultured skim milk was inoculated. Fermentation was done aerobically.

Assays

During the fermentations pH, dissolved oxygen and dissolved carbon dioxide were measured continuously with electrodes which were in the fermenter. The first sample for calcium activity and electrical conductivity measurement was taken just after inoculumn. Thereafter samples for calcium activity and electrical conductivity measurements were collected at hourly intervals.

Equipment

Fermenter. Fermentations were carried out in a BIOSTAT E fermenter (B Braun Melsungen) with a working volume of 10 l.

pH-electrode. Measurement of the pH-value in the culture medium was carried out with a sterilizable combined-glass electrode manufactured by INGOLD.

Oxygen electrode. Dissolved oxygen was measued with a sterilizable pO_2-electrode with measurement range 0 – 100 % produced by INGOLD. The construction of pO_2-electrode is shown in Fig.1. The amperometrically operating pO_2-electrode consists of a Ag-anode and Pt-cathode which are separated from the measurement solution by a gas-permeable polymeric membrane (Clark principle).[1] Anode and cathode are conductively connected to each other by an electrolyte. The electrolyte builds a small definable coating between the membrane and the electrode. If the polarization voltage is suitable the oxygen which diffused through the membrane is completely reduced at the cathode. This reaction results in an electrical current in the nA range, which is directly proportional to the oxygen partial pressure, the actual value to be measured, in the measurement solution.

Electrode reactions:

Cathode reaction: $O_2 + 2H_2O + 4e^- \rightarrow 4\,OH^-$

Anode reaction: $4\,Ag + 4\,Cl^- \rightarrow 4\,AgCl + 4e^-$

As the permeability of the membrane depends on temperature, an exponential increase in electrode current (3 %/°C) with rising temperatures can be observed.

Carbon dioxide electrode. The CO_2-proportion in the culture medium is measured as CO_2-partial pressure, p_{CO_2}, in the measurement range 1 – 100 mbar p_{CO_2} (Fig.2). For p_{CO_2}-measurement in the culture medium there is an INGOLD-p_{CO_2}-electrode with the corresponding installation fitting. It is a potentiometric electrode which operates according to

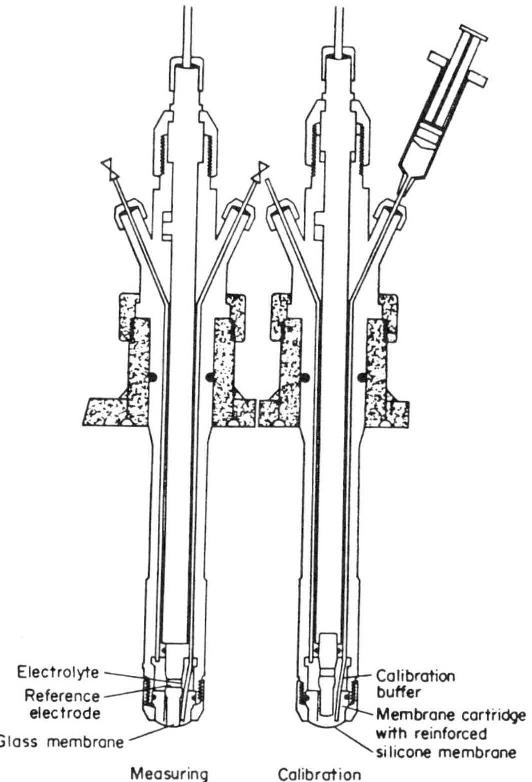

Fig.2. Basic design of p_{CO_2}-electrode.[2]

the Severinghouse-principle.[3] In that construction CO_2 diffuses from the culture medium through a gaspermeable membrane of the electrode into a bicarbonate-electrolyte until the same CO_2-partial pressure is attained on both sides of the membrane, i.e., an equilibrium condition is obtained. Because pH of solutions of pure water and bicarbonate depend on the concentration of dissolved CO_2, the pH of the internal buffer solution will be dependent on the partial pressure of carbon dioxide, p_{CO_2}, in the sample solution.

$$pH = pk - \log p_{CO_2}$$

Where k is the first dissociation constant of carbonic acid. An internal pH-electrode is used to register the pH-value of this (internal) buffer. The pH-value then directly correlates with the CO_2-partial pressure in the culture medium.

Calcium electrode Ca^{2+} activities were determined with a Radiometer ION 85 digital ion meter equipped with an Orion 93-20-01 Ca^{2+} ion-selective electrode and an Orion 90-01 single junction reference electrode with a filling solution with 4 M KCl and saturated with Ag^+. The ion-selective electrode was calibrated against an aqueous standard solutions containing $CaCl_2$ and adjusted with KCl to an ionic strength of 0.08 M. The slope of the electrode response was determined for the Ca^{2+} activity in the interval 10 to 0.1 mmol/l before each series of measurements. The Orion calcium ion-selective electrode is a liquid membrane probe that is used to measure calcium ion activity in aqueous solutions and biological samples. The electrode consists of a replaceable module with a solid plastic membrane containing a calcium-selective ion exchanger. When this membrane is in contact with a solution containing calcium ions, a potential, dependent upon the level of calcium ion activity in the solution, develops across the membrane. The potential is measured against a reference electrode. Response times at Ca^{2+}-activities ranging from 0.10 to 10 mmol/l (I=0.08 mol/l) were less than 5 s. In reconstituted skim milk the electrode response time was considerably longer.

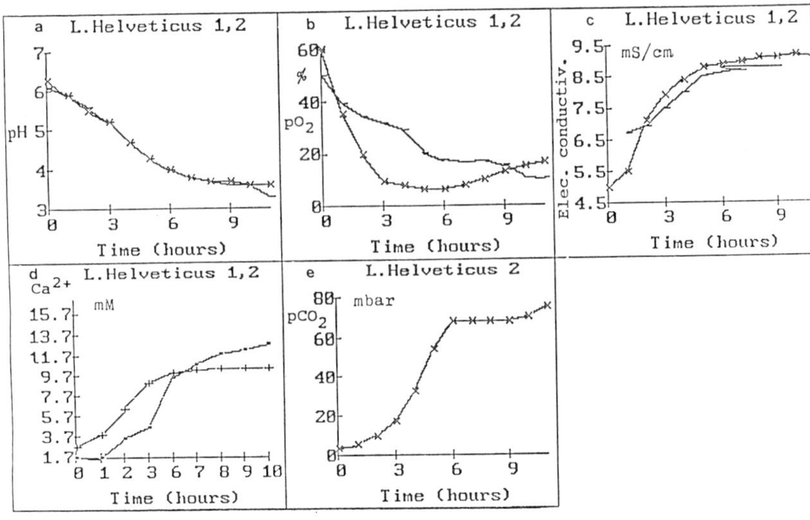

Fig.3a-3e. Typical fermentation profile of *Lactobacillus helveticus* (culture I and culture II) in a 10 l laboratory fermenter. Temperature 39°C.

(3a) pH as a function of time.

(3b) p_{O_2} as a function of time.

(3c) Electrical conductivity as a function of time.

(3d) Ca^{2+} activitiy as a function of time.

(3e) p_{CO_2} as a function of time.

The time necessary for a stable reading was less than 3 min.

Electrical conductivity cell. Impedance is a measure of the total opposition to the flow of a sinusoidal alternating current in a circuit containing resistance, inductance, and capacitance. Inductance and capacitance together are called the reactive part of the circuit. The changes in impedance that occur in a microbial culture can be measured by placing two metal electrodes into the culture medium and introducing an alternating potential into the circuit.

A conductivity cell consists of two electrodes which may be two parallel sheets of platinum fixed in position by sealing the connecting tubes into the sides of the measuring cell.

We measured electrical conductivity with a Radiometer CDM 83 conductivity meter and conductivity cell.

Experimental Design

During the fermentations the stirring speed was kept constant at 100 rpm. The lactobasilli bacteria was grown at 39°C.

Biochemistry of Starters

Homofermentative lactobasilli split lactose to glucose and galactose. The glucose is then converted to pyruvate and then to lactic acid with trace amounts of acetic acid and carbon dioxide (Collins, 1977).[4] Certain species and strains of homofermentative organisms may produce aroma and flavor-enhancing intermediates in addition to lactic acid.

Results and Discussion

Diagrams of different cultivations are presented in Figs 3a-3e and 4a-4e.

In the course of the microbe-induced fermentation the pH decreased from 6.7 to 3.3 during bacterial growth of *Lb.helveticus* growing on reconstituted milk.

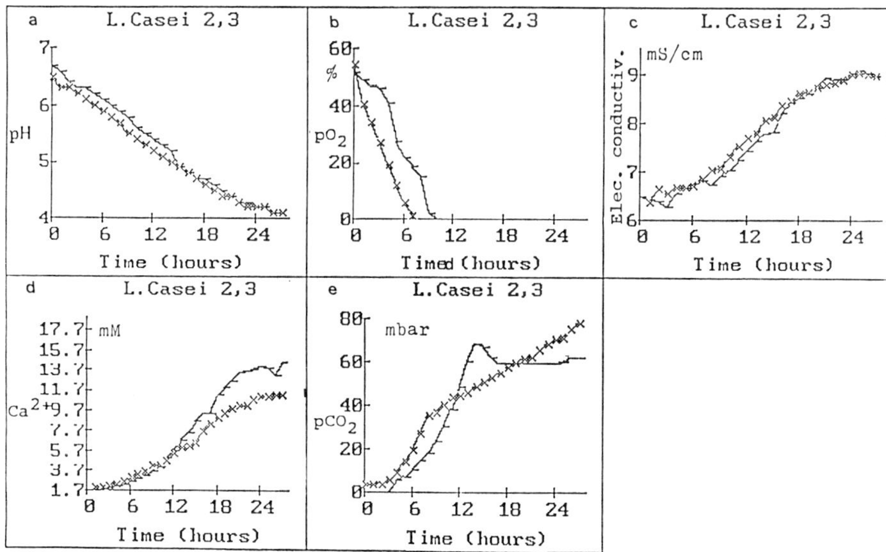

Fig.4a-4e. Typical fermentation profile of *Lactobacillus casei* (culture I and culture II) in a 10 l laboratory fermenter. Temperature 39°C.
(4a) pH as a function of time.
(4b) p_{O_2} as a function of time.
(4c) Electrical conductivity as a function of time.
(4d) Ca^{2+} activitiy as a function of time.
(4e) p_{CO_2} as a function of time.

The pH was found to change from 6.7 to 4.0 during bacterial growth of *Lb.casei* growing on reconstituted skim milk.

An increase in the electrical conductivity of a culture occurs when actively metabolizing bacteria utilize large molecules in a culture medium and form smaller ion pairs. In this breakdown are formed low molecular nitrogen compounds which starter bacteria need for growing.

The electrical conductivity increased from 5.0 to 9.0 mScm^{-1} during the growth of *Lb.helveticus*.

Calcium plays a significant role in the physico-chemical properties of milk. Though its exact role in milk stability is unknown. The calcium exists in protein-bound, complexed and ionized forms, all of them being in equilibrium. Even though ionized calcium represents only about 8 % of the total calcium, it is of special interest because it is free to react and therefore change the total calcium equilibrium.

In our work during the growth of *Lb.helveticus*, calcium ion activity increased from about 1.8 to 11.7 mM. This was due to the production of lactic acid by bacteria and followed by decrease in pH.

It can be seen that there are no significant differences between different strains of the two species. *Lb.helveticus* strains are faster in growing and so all changes are faster with *Lb.helveticus* than with *Lb.casei*.

Conclusion

Cheese starter research is continuing at an accelerating pace. This study provides new chemical and physical information for cheese starter research. This new information obtained here by using sensors makes it possible both to monitor the fermenter and to improve the cheese starter.

Despite the many studies characterizing the biochemical nature of starter growth, there is a dearth of information on the nature and importance of the various cheese making techniques, and yet these are clearly essential for the definition of desirable starter properties.

By using sensors and on-line analytical methods we can obtain a better understanding of cheese starters and improve the cheesemaking process.

References

1. L.C. Clark, R. Wolf, D. Granger and Z. Taylor, J. Applied Physiol., **6** (1958) 189.
2. E. Puhar, A. Einsele, H. Buhler and W. Ingold, Biotechnol. Bioeng., **22** (1980) 2411.
3. J.W. Severinghaus, Ann. N. Y. Acad. Sci., **18** (1968) 115.
4. E.B. Collins, J. Dairy Sci., **60** (1977) 779.

ELECTROCHEMICAL FLOW ANALYSIS

EXPLOITATION OF ELECTROCHEMICAL TECHNIQUES BY FLOW INJECTION ANALYSIS

Elo Harald Hansen

Chemistry Department A
Technical University of Denmark
Building 207, DK-2800 Lyngby, Denmark

Since its introduction in 1975, almost 2200 papers have been published on Flow Injection Analysis (FIA), more than 25 % of these dealing with procedures comprising electrochemical detection.[1] In fact, the very first FIA experiment was executed with an electrochemical sensor, that is, an ammonia gas sensing probe,[2,3] which demonstrated the extreme reproducibility with which the samples repeatedly were being presented to the detecting device by this approach, thereby allowing attainment of steady-state conditions to be abandoned. Giving the impetus to further research, it was subsequently established that FIA is based on a combination of sample injection, controlled dispersion of the injected sample zone during its transport from the point of injection to the point of detection, and reproducible timing of all events within each assay cycle. Thus, FIA should prove an ideal vehicle for systems incorporating electrochemical transducers, the performance of which is diffusion controlled (Table 1).

Table 1. Electrochemical Detection Procedures Used in FIA.

Amperometry
Conductometry
Coulometry
Piezoelectricity
Polarography
Potentiometry
Chronopotentiometry/ potentiometric stripping
Ion-selective electrodes/ISFETs
Voltammetry

Besides, since any FIA curve respresents a continuum of concentrations – between zero and that corresponding to the peak maximum, C^{max} (cf. Fig.3, left) – and that there is no single element of fluid that has the same concentration as the neighbouring one, the analyst has at his disposal, in reality, an infinite number of elements of different concentration ratios of sample and reagent which each might be exploited for analytical purposes. Yet, going through the literature, it is apparent, however, that most users of the technique still maintain to exploit solely the peak maximum of the FIA gradient, thus in fact limiting themselves to applying FIA as a convenient sample transport system replacing the batch mode approach, rather than as a tool to enhance instrument performance. It is the purpose of this communication, by selected examples, to demonstrate why it might be attractive to execute electrochemical measurements by FIA and which advantages it might entail, and to show that this approach implies some unique applications.

Contemporary Electroanalytical Chemistry, Edited by A. Ivaska *et al.*
Plenum Press, New York, 1990

FIA Potentiometry

Electrochemical measurements are either performed on the sample itself or following suitable pretreatment. Whatever the approach, these manipulations are readily performed in a FIA system operated under either limited or medium dispersion.* In the former instance the FIA system serves merely as a means of rigorus and precise transport of the sample material to the flow cell in undiluted form or as a means of reproducible sample introduction. Procedures of medium dispersion, however, allows the samples to be mixed with one or several reagents in order to form an electroactive species, or to condition the individual sample solutions before they are presented to the electrochemical sensor in order to measure them under optimal conditions. Thus, for instance in potentiometry, advantage is taken of the constant and reproducible time during which the sample is exposed to the electrode, allowing the recorded transient signal to be related to the concentration of analyte[1,4] (Fig.1). Besides, due to the high wash-to-sample ratio, the detector is subjected to the sample solution for a limited period of time only, permitting the sensor to return to its baseline level between each measurement. Thus this level serves as a continuous check of the detector performance, which is a feature of particular importance in on-line monitoring applications.[1,5] A typical measuring cycle is less than 30 seconds, while sample and carrier stream consumption is a fraction of a millilitre. Therefore the system can be calibrated by known standards whenever necessary, which allows corrections to be made for possible sensor drift or change of response characteristics. The short contact time between sample and detector furthermore reduces any risks of adsorption problems and probe deterioration; and in addition clearing the flow system, the continuously flowing carrier stream can be exploited for possible regeneration of the sensor,[6,7] i.e., by addition of species which recondition the sensing surface – their effect being judged from the way by which the baseline is being restored between individual injections of sample or standard solutions. Indeed, it is well established that when an ion-selective electrode is operated in the FIA mode and if the sample contains high levels of an interfering ion, a negative dip is observed in the baseline immediately after the zone has cleared the ion-selective surface, the extent of the negative deviation being proportional to the concentration of interferent while its duration corresponds to the time necessary to restore the composition of the electrode surface by washing with the carrier stream.[8] Properties of CHEMFETs[9] and of bilayer lipid membranes have been investigated in a similar manner.[10]

Operated in a dynamic mode many ion-selective electrode do also exhibit kinetic discrimination towards the ion under investigation and interfering species, that is, during the short duration of sample exposure the response time of the sensor to the different species might vary considerably, which in turn might be exploited to increase the selectivity and the detection limit of the sensor.[11,12,13] Since many electrochemical devices such as ion-selective electrodes can be made very small, they are perfectly suitable to be incorporated into FIA-microconduits, where the detector can be placed exactly in that position of the flow path where the dispersion for the particular application is optimal[1,14] - cf. Fig.2.

FIA Voltammetry

Generally speaking, detectors can be divided into two groups, static and dynamic. The former simply yields an electrical signal which is either linearly or logarithmically related to the concentration of species in the analyte. In order to obtain analytical information on several species simultaneously, potentiometric detectors must be placed sequentially[4,15,16] or grouped into a multidetector.[17] Voltammetric or polarographic detectors are qualitatively different and belong to the dynamic group. Not only can they detect the concentration of an electroactive

---

* While dispersion is defined as the physical dilution process of the injected sample zone in the FIA system, the dispersion coefficient D is a numerical expression of the dispersion process in a particular element of fluid of the dispersed sample zone, being defined as the ratio of the concentrations before and after the dispersion process has taken place in that element of fluid that yields the analytical readout, that is, $D = C^0/C$, which for $C = C^{max}$ yields $D = C^0/C^{max}$, where Co is the original concentration of the injected sample solution, and C is the concentration of any element of fluid along the gradient of the dispersed sample zone (cf. Fig.3), which element especially may be that one corresponding to the peak maximum (C^{max}).

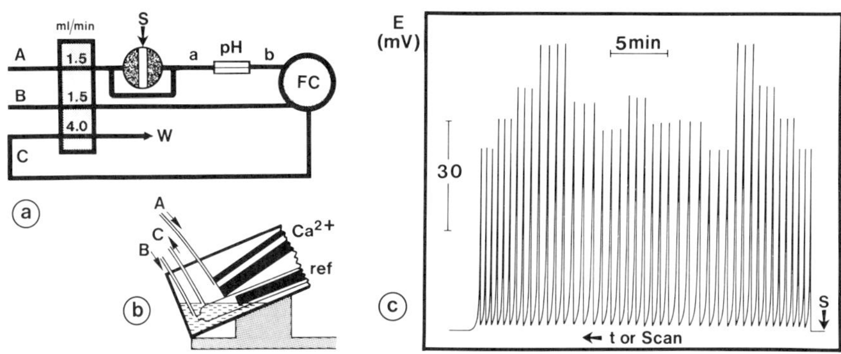

Fig.1. (a) FIA manifold for simultaneous determinations of pH and pCa in serum samples as operated under conditions of limited dispersion. S, point of injection (30 μl); pH, flow through capillary glass electrode; FC, flow through cell containing a PVC-membrane-based calcium selective electrode and the common reference electrode, details of which are shown below (b). The carrier solution (A and B) is TRIS buffer of pH 7.4, the connecting tubes (a and b) being made as short as possible. The carrier solution supplied via line B to the tip of the reference electrode is included in order to stabilize the reference electrode junction potential for sera measurements. In (c) is shown the potentiometric determination of the ionized calcium content in 6 serum samples, bracketed by two calibration runs of aqueous calcium standards (0.5 to 5 mM), all assays made in triplicate.[1,4]

species at a given potential,[18-21] but they can also scan their own function on a solution of a given composition, which capability makes them powerful tools in instrumental analysis. Combining this feature with the unique ability of the FIA technique to create a perfectly reproducible concentration gradient from an injected zone does, however, offer entirely new possibilities for conducting electrochemical experiments. Relying on the principle that each and every element of fluid along the gradient corresponds to a fixed dispersion value – i.e., a certain concentration ratio of sample and reagent – which in turn can be translated into a fixed delay time elapsed from the moment of injection,[1,22,23] it was Janata and Růžička[24] who originally suggested the FIA gradient scanning technique where a physical parameter is measured continuously within a certain range – such as current vs. potential (Fig.3, left) – along the dispersed sample zone. Provided that the rate of the scanning is much faster than the movement of the carrier stream, the dispersed zone may be advanced during the scan and still appear as to be stationary. If this is not the case, or if the chemical reaction occuring is too slow, then selected sections of the zone can be stopped consecutively within the detector. The resulting record will have a multidimensional form, like the three-dimensional set of voltammograms depicted in Fig.3, right.

The combination of rapid scanning instruments (both electrochemical and optical) with the FIA gradient approach is natural. The tedious manual handling necessary to prepare solutions for batchwise measurement by means of these advanced instruments is not flexible and does not exploit their full potential. Seen from a purely analytical viewpoint, the advantages of gradient scanning, as a convenient means of studying the chemical and physical phenomena which take place when solutions are mixed, should become even more apparent when multicomponent analysis will be performed routinely on unknown samples, resolution of the spectra being facilitated by means of chemometric procedures.

Cañete et al.[25] have recently elaborated on this approach by implementation of normal and derivative cyclic voltammetry. Pointing out that these types of three-dimensional analytical information is very powerful indeed (being similar to that provided by image detectors such as diode arrays), offering, besides theoretical studies, interesting practical possibilities especially in the multidetection of electroactive compounds, the authors used their cyclic

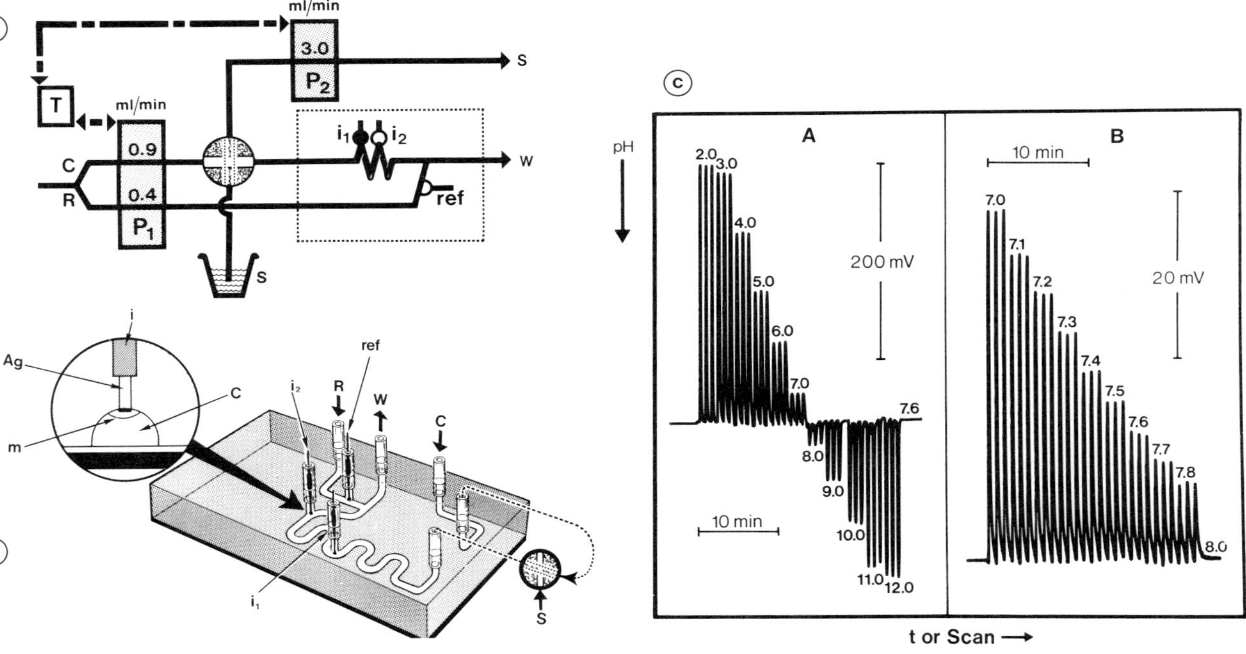

Fig.2. (a) Manifold and (b) FIA microconduit unit for potentiometric determination of pH, the components within the boxed area of the manifold being those contained in the microconduit. Pumps P1 and P2 are operated sequentially, controlled by timer T, so that sample (volume 30 μl) is first aspirated by P2 into the injection valve (during which period P1 is stopped) and subsequently propelled forward by carrier stream C (diluted buffer) towards indicator electrode i_1 (or i_2), the construction details of which are shown in the insert (m is the pH-sensitive membrane). The reference electrode (ref) is placed in a side channel fed by stream R, and the combined streams are ultimately led to waste, W. (c) pH-response curves obtained with the microconduit system shown to the left. (A) A series of pH standards in the range 2.0 to 12.0; the carrier solution being 0.01 M phosphate buffer containing 0.1 M NaCl, at pH 7.6. (B) Performance of the system in the physiologically important pH range, the carrier solution here being adjusted to pH 8.0.[1,14]

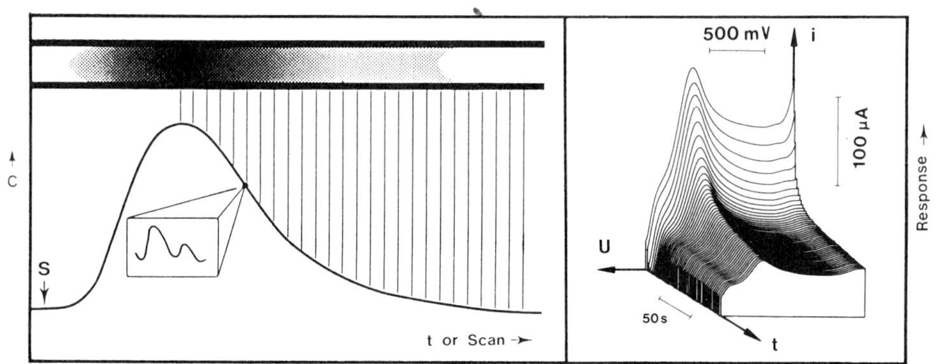

Fig.3. The principle of the FIA Gradient scanning technique, allowing a series of spectra to be recorded on a dispersed sample zone (left); S, point of sample injection, and t, time. The three-dimensional recording (right) shows an electrochemical spectrum, that is, a voltammogram of a 0.01 M solution of copper(II)amine complex being reduced on a mercury electrode. The peak (forming a ridge) is located at a potential (U) at which maximum current intensity (i) is observed due to reduction of Cu(II). Fifty scans (each lasting 2 sec.) were recorded on the tailing section of the dispersed sample zone.[1,24]

voltammetric procedures for the determination of binary and ternary mixtures of phenol derivatives in different types of water.

Enzymatic Assays and Enzyme Electrodes

Taking advantage of the selectivity of enzymes towards specific substrates, these catalytic reagents have gained increased significance in several areas of analytical chemistry such as clinical, pharmaceutical, agricultural and industrial applications, and most recently process monitoring. Because of its rapidity and versatility and its ability to handle microvolumes, FIA has proven itself an ideal vehicle for enzymatic assays. Aimed either at sample pretreatment or matrix modification, enzymatic methods thus belong to the FIA conversion techniques, which in their broadest sense may be defined as procedures by which a nondetectable species is converted into a detectable substance through controlled kinetic chemical reaction.[26] Of all the FIA conversion methods used, systems with immobilized enzymes are by far the most common.[1] They offer the selectivity of enzymatic reactions and the economy and stability gained by immobilizing the often costly catalysts, and applied in the FIA mode they secure strict repetitiousness and hence that a fixed degree of turnover from cycle to cycle is maintained. Besides, by getting a high concentration of enzyme immobilized within a small volume, the ensuing high activity will facilitate an extensive and rapid conversion of substrate at minimum dilution of the sample, the small dispersion coefficient in turn promoting a lower limit of detection. The enzymes needed for a particular assay might either be incorporated into a single or into sequentially placed packed reactors, or integrated into the detection device itself as encountered in the enzyme sensors. For oxidases, which give rise to the generation of hydrogen peroxide, as well as for dehydrogenases, which require the presence of suitable coenzymes such as NADH or NADPH, the oxidized or reduced forms of which are monitored, optical as well as electrochemical detection can be used. Thus hydrogen peroxide has been detected optically by coupling to appropriate chromogenic agents, by chemiluminescence via reaction with luminol, and electrochemically by amperometry, while the coenzymes usually are detected by spectrophotometry or fluorometry or by various electrochemical techniques including (pulsed) amperometry and chemically modified electrodes. For other enzymes, such as the iminohydrolases, which degrade substrates with the ultimate formation of ammonium ammonia, selective determination might be achieved potentiometrically by an ion-selective electrode or via gas diffusion followed by optical or electrochemical detection.

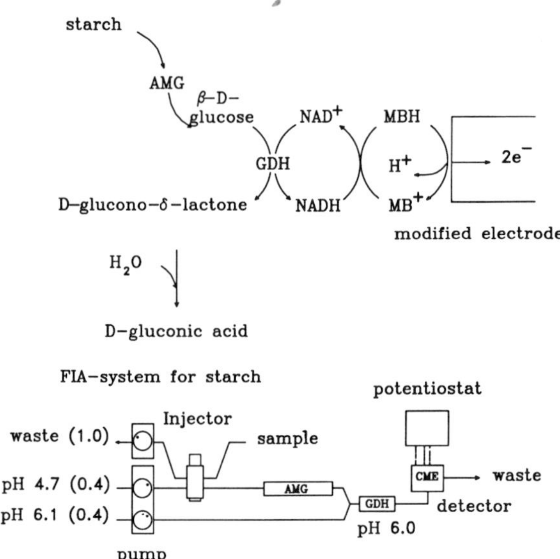

Fig.4. (top) Reaction scheme for the enzymatic determination of starch as executed in the FIA manifold shown below, where AMG refers to amyloglucosidase and GDH to glucose dehdrogenase, both of which enzymes are immobilized on porous glass in separate packed-bed reactors. The detection of the generated NADH is facilitated amperometrically by means of a modified electrode at 0 mV vs. Ag/AgCl, 0.1 M KCl. Flow rates are given in ml/min; injected volume 30 μl.[18]

Focusing on electrochemical detection procedures, the works of a Swedish research group in Lund should be emphasized. Having investigated a number of enzymatic assays, primarily aimed at determinations of various carbohydrates, they have described several very elegant approaches.[18,27−30] As an example might serve the determination of starch according to the reaction scheme depicted in Fig.4 (top), the actual assay being performed in the FIA manifold shown in Fig.4 (bottom).[18] The starch is first hydrolyzed to β-D-glucose units by amyloglucosidase in the AMG reactor at pH 4.7. After mixing with a second buffer stream containing soluble NAD$^+$, the generated glucose is further oxidized in the glucose dehydogenase (GDH) reactor at pH of 6.0, whereby an equimolar amount of NADH is formed (the two different pH values used were selected to obtain optimal performance of the individual enzyme preparations). The detection is performed by mediated oxidation of NADH at a modified electrode, mounted in a confined wall-jet flow through cell. The mediator is N,N-dimethyl-7-amino-1,2-benzo-phenoxazine (Meldola Blue, MB$^+$) which is adsorbed on the surface of a graphite rod. NADH and MB$^+$ form a complex which decomposes in a rate-limiting step to NAD$^+$ and the reduced mediator. The latter, MBH, is reoxidized amperometrically, the current drawn being directly related to the amount of starch in the sample. Instead of employing sequentially placed reactors, the enzymes might coveniently be combined into a single reactor, which furthermore reduces any problems due to back pressure of the system. In fact, the same Swedish group which did the initial starch work has recently tested this approach (including in their immobilized enzyme reactor also mutarotase in order to convert any generated α-glucose to β-glucose thereby increasing the ultimate yield).[30]

Within the last few years there has been an almost explosive surge in the development and applications of electrochemical and optical enzyme sensors. In enzyme electrodes a membrane containing one or several immobilized enzymes is placed in front of the active surface of an electrochemical detector. The analyte is transported by diffusion into the mem-

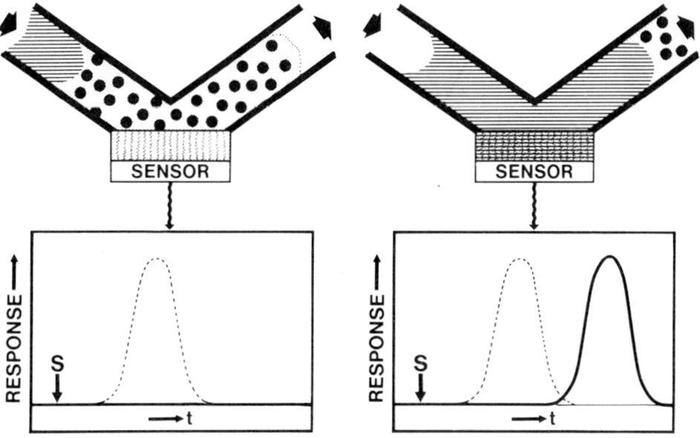

Fig.5. Operation of a latent image sensor during the passage of a heterogeneous sample zone, showing on the left the penetration of the interstitial liquid into the enzyme containing sensor structure so that a latent image is formed as a result of enzymatic degradation of substrate to a nonelectroactive species (dotted line), and on the right the development of the latent image with the aid of the passing reagent zone, resulting in the formation of an electroactive component and the ensuing recorded response (solid line).[26,34]

brane and here degraded enzymatically forming a product which subsequently can be sensed electrochemically. A condition for obtaining a linear relationship between analyte and signal is, however, that pseudo first-order reaction conditions are fulfilled, i.e, the concentration of converted analyte reaching the detector surface must be much smaller than the Michaelis-Menten constant. Since the Michaelis-Menten constant for most enzyme systems is of the order of 10^{-3} M, and many sample matrices (e.g., sera) contain much higher concentrations, the use of enzyme sensors in static systems often calls for complicated use of additional membrane layers aimed at restricting diffusion of analyte to the underlying enzyme layer, or alternatively for chemically modified electrodes where the electron transfer is governed by appropriate mediators (one such example is the glucose sensor based on ferrocene/ glucose oxidase as suggested by Cass et al.,[31] yet it has a very limited lifetime). However, in FIA the degree of conversion might simply be adjusted by the time the analyte is exposed to the enzyme layer, and therefore the amount of converted analyte can be regulated directly by adjusting the flow rate of the FIA system.[1] Additionally, the time of exposure might possibly be exploited for kinetic discrimination, advantage being taken of differences in diffusion rates of analyte and interfering species within the membrane layer of the electrode. Thus Petersson, when using an amperometric glucose sensor based on glucose oxidase, achieved a total spatial resolution of the signals of glucose and paracetamol, even when the interfering agent was added at excessive levels.[32] Another example of instrument enhancement using the kinetic approach is encountered in voltammetry. Contrary to normal practice, where the sample has to be deaerated prior to analysis, the oxygen in the test solution does not have to be removed if the differential pulse technique is performed in the FIA mode.[33] This is due to the irreversibility of oxygen reduction so that kinetic advantage can be taken by comparing the readouts of two electrodes sequentially monitoring the same sample zone.

When heterogeneous sample solutions, such as undiluted blood, are to be assayed by means of an enzyme sensor incorporated into a FIA system, it is a prerequisite that the individual samples, at the time they are exposed to the enzyme, are undiluted by the carrier stream,[26,34] i.e., the dispersion coefficient of the sample is equal to 1. If this condition is not fulfilled, the analytical readout will be severely influenced by the composition of the

matrix of the sample (i.e., in the case of blood, the hematocrit level, the pH value and
the buffering capacity). If the enzymatic degradation of the substrate leads to a product
which can be directly sensed by the detector, this does not present any problem. Thus,
Petersson recently described the determination of glucose and lactic acid in whole blood by
means of enzyme detectors containing glucose oxidase and lactate oxidase, respectively, the
generated hydrogen peroxide being quantified amperometrically via oxidation to oxygen.[32] If
the product, on the other hand, requires an ensuing chemical reaction, facilitated by a reagent
in the carrier stream, in order to be sensed by the detector, a paradox appears to exists,
because at $D = 1$ no physical mixing of the carrier stream and the sample solution takes place,
and therefore the composition of the carrier stream should have no influence. It is, however,
important to realize that a perfectly controlled spatial and kinetic discrimination prevails in
the sandwich membrane operated under FIA conditions, so that the sensing process, in fact,
occurs in two steps (Fig.5): In the first step, where the injected sample is in contact with
the enzyme layer, the sample leaves an "imprint" promoting the degradation of substrate
to a nonelectroactive species. In the next step, when the undiluted sample zone is swept
away by the reagent containing carrier stream, the generated and entrapped species is then
being converted to an electroactive component which subsequently is transferred to the active
surface of the sensing element. In fact, the procedure might be compared to a photographic
process: when the picture is taken, its imprint is indeed on the film, yet latently, and it only
becomes visible to the spectator after its has been developed via chemical processing. This
approach was recently successfully implemented for the assay of urea in undiluted blood
samples.[26,34,35]

Impulse-response FIA

The recent analytical literature abounds with reports on the development of electrochemical, fiber optic, piezoelectric and other sensors. Novel detection principles are frequently
announced, and attainment of improved sensor quality in terms of selectivity, sensitivity and
lifetime are the goal of many established as well as newly formed research groups. The need
for continuous monitoring of critical parameters in clinical chemistry, biotechnology, pharmaceutical, chemical and nuclear industry, chemical warfare or environmental control is the
driving force behind the sensor research.

While the majority of such research activities has novel aspects, sound theoretical basis
and uses high technology to fabricate the sensing devices, the Achilles Heel of many chemical
sensors is the lack of their time stability as reflected in deterioration of their response – up
to the ultimate lack of it. Surprizingly, most of the research papers published to date do
not contain the vital data on the performance of these sensors, such as speed of response,
sensitivity and detection limit, as these parameters necessarily undergo changes during the
lifetime of each device. Information on poisoning, or recovery, of sensors as well as on typical lifetimes at various operating conditions are even more scarce. Whether this is due to
proprietory information or lack of data is difficult to judge.

Introduction of an automated, well reproducible rigorous testing procedure which would
allow critical testing of chemical sensors should become a powerful tool for sensor development. There is increasing experimental evidence that FIA is suitable for sensor and material
testing by means of the impulse-response technique.[36]

Impulse-response FIA is based on repetitive action of a well defined zone of a selected
chemical species (Reagent) on a target (T) situated in a continuous unsegmented carrier
stream (C) consisting of an inert fluid (Fig.6). The target may be a sensor, the carrier stream
may be an aqueous solution to which the sensor does not respond (or responds at the detection
limit - thus establishing a continuous baseline). The injected species is the one to which
the sensor is to respond. By continuously monitoring the sensor output, while repeatedly
injecting, at a fixed frequency, the same species at the same concentration level in the form of
a well defined zone, a series of peaks will be obtained, which would reflect the response of the
sensor. Thus the series of peaks will show the response to either remain unchanged (Fig.6b
(A)) or to be deteriorating (Fig.6b (B)) over a selected test period. Since the procedure
is automated it is not prone to subjective judgement or manual operational errors as are
the generally used batchwise or "steady-state" test procedures. Indeed, instead of injecting

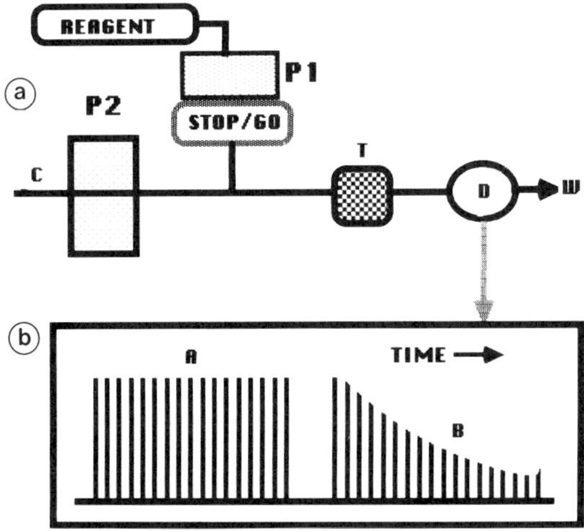

Fig.6. (a) Impulse response FIA system comprizing a target material T situated within carrier stream (C) propelled by pump P2. At preselected time intervals a metered volume of test liquid (reagent), aspirated by pump P1 by operating it for a fixed period of time (GO), is introduced into the carrier stream, the effect of the exposure being monitored by detector D. (b) In **A** is shown a run where the target is unaffected by the repeated bombardments of test solution, while **B** represents a situation where the target is gradually deteriorating due to the repeated exposure.[1,36]

repeatedly the same solution, a series of solutions, R1, R2, R3, of different concentrations, may be injected in preprogrammed sequences. Similarly, interfering or regenerative species may be injected repeatedly or singly.

The FIA impulse-response technique should prove most suitable for testing and quantitative evaluation of the lifetime and change of response characteristiques of sensors (electrochemical, fiber optic). Since target materials can be tested automatically with high reproducibility under prolonged periods, the FIA impulse response technique might become, once its potential has become generally recognized and exploited, the most important modification of the flow injection method, perhaps even surpassing its applications as a tool for serial assays.[1]

Besides examining a tendency of a series of response peaks, as illustrated in Fig.6, a change in individual peak shape will be indicative of e.g., adsorption-desorption phenomena at the surface of a sensor. Since most chemical sensors utilize a chain of chemical reactions taking place in the sensing area, it is possible to examine each set of sensing reactions separately and thus localize the weakest link of the sensing process.[1]

Acknowledgements

The author expresses his gratitude to the Danish Council for Scientific and Industrial Research for financial support of the current FIA research programme. Thanks are also due to Jaromir Růžička of the University of Washington/Chemistry Department A for continuous stimulating discussions.

References

1 J. Růžička and E.H. Hansen, in: "Flow Injection Analysis. 2nd Edition", John Wiley, New York, 1988.
2 J. Růžička and E.H. Hansen, Anal. Chim. Acta, **78** (1975) 145.
3 E.H. Hansen, In Focus, **6** (1983) 12.

4 E.H. Hansen, J. Růžička and A.K. Ghose, Anal. Chim. Acta, **100** (1978) 151.

5 W. Frenzel, Fresenius' Z. Anal. Chem., **329** (1988) 668.

6 W. Frenzel, Fresenius' Z. Anal. Chem., **329** (1988) 698.

7 M. Trojanowics and W. Frensel, Fresenius' Z. Anal. Chem., **328** (1987) 653.

8 E.H. Hansen, A.K. Ghose and J. Růžička, Analyst, **102** (1977) 705.

9 A.U. Ramsing, J. Janata, J. Růžička and M. Levy, Anal. Chim. Acta, **118** (1980) 45.

10 M. Thompson and U.J. Krull, Anal. Chim. Acta, **142** (1982) 207.

11 E.H. Hansen, J. Růžička and A.K. Ghose, STI/PUB/**535** (IAEA, Vienna) (1980) 77.

12 M. Trojanowicz and W. Matuszewski, Anal. Chim. Acta, **151** (1983) 77.

13 L. Ilcheva and K. Cammann, Fresenius' Z. Anal. Chem., **322** (1985) 323.

14 J. Růžička and E.H. Hansen, Anal. Chim. Acta, **161** (1984) 1.

15 J. Růžička, E.H. Hansen and E.A.G. Zagatto, Anal. Chim. Acta, **88** (1977) 1.

16 R. Virtanen, in: "Ion-Selective Electrodes, 3. Proceedings of the Third Symposium, Mátrafüred", Hungary, E. Pungor (Ed.), Anal. Chem. Symposia Series, **8** Elsevier, 1981, p. 375.

17 J. Janata and R.J. Huber, Ion-Selective Electrode Rev., **1** (1979) 31.

18 J. Emnéus, R. Appelqvist, G. Marko-Varga, L. Gorton and G. Johansson, Anal. Chim. Acta, **180** (1986) 3.

19 A.G. Fogg, R.M. Alonso and M.A. Fernández-Arciniega, Analyst, **111** (1986) 249.

20 T.Z. Polta and D.C. Johnson, J. Electroanal. Chem., **209** (1986) 159.

21 J. Wang, P.A.M. Farias and J.S. Mahmoud, Analyst, **111** (1986) 837.

22 J. Růžička and E.H. Hansen, Anal. Chim. Acta, **145** (1983) 1.

23 E.H. Hansen, Fresenius' Z. Anal. Chem., **329** (1988) 656.

24 J. Janata and J. Růžička, Anal. Chim. Acta, **139** (1982) 105.

25 F. Cañete, A. Rios, M.D. Luque de Castro and M. Valcarcel, Anal. Chim. Acta, **214** (1988) 375.

26 J. Růžička and E.H. Hansen, Anal. Chim. Acta, **214** (1988) 1.

27 G. Johansson, L. Ögren and B. Olsson, Anal. Chim. Acta, **145** (1983) 71.

28 B. Olsson, B. Stålbom and G. Johansson, Anal. Chim. Acta, **179** (1986) 203.

29 B. Olsson, G. Marko-Varga, L. Gorton, R. Appelqvist and G. Johansson, Anal. Chim. Acta, **206** (1988) 49.

30 G. Marko-Varga, in: "Carbohydrate Determinations in LC and FIA Using Immobilized Enzymes and Electrochemical Detection - Extension of Selectivity and Sensitivity", Ph.D. Thesis, Univ. of Lund, Sweden (1988).

31 A.E.G. Cass, G. Davis, G.D. Francis, H.A.O. Hill, W.J. Aston, I.J. Higgins, E.V. Plotkin, L.D.L. Scott and A.P.F. Turner, Anal. Chem., **56** (1984) 667.

32 B.A. Petersson, Anal. Chim. Acta, **209** (1988) 231.

33 D.C. Johnson, S.G. Weber, A.M. Bond, R.M. Wightman, R.E. Shoup and I.S. Krull, Anal. Chim. Acta, **180** (1986) 187.

34 B.A. Petersson, H.B. Andersen and E.H. Hansen, Anal. Lett., **20** (1987) 1977.

35 B.A. Petersson, Anal. Chim. Acta, **209** (1988) 239.

36 J. Růžička and E.H. Hansen, Anal. Chim. Acta, **179** (1986) 1.

POTENTIOMETRIC DETECTION IN HIGH-PERFORMANCE ION-CHROMATOGRAPHY

Marek Trojanowicz

Department of Chemistry, University of Warsaw
PL 02-093 Warsaw, Poland

A development of high-performance ion-chromatography can be dated to the middle of seventies. Although liquid chromatographic methods for separation of metals as metal-organic compounds or metal chelates were investigated much earlier,[1] there was a lack of fast, multicomponent methods for determination of anions. They are required for environmental science, food and clinical analysis and many other areas of chemical analysis. A development of high performance ion-chromatography was initiated by the pioneer work by Small et al.,[2] who introduced specially developed pellicular resin to effect the separation and a second ion-exchange column to reduce the background conductivity of the eluent. This procedure allowed to apply a conductivity detection. Several years later Fritz et al.[3,4] developed macroporous ion-exchangers of much lower capacity, which enabled the use of eluents of very small ionic strength and thus conductivity detection can be used without the need of a suppressor column.

At the moment, after more than ten years of the development in this field, several other detection methods can successfully be used in ion-chromatography.[5,6] Conductivity detection is still most frequently used as well in single column systems as in those involving a suppressor column. In both systems a similar detectability is reported. Among its disadvantages, a high temperature coefficient and insensitivity to anions of very weak acids has to be mentioned. The latter drawback can be eliminated in indirect conductivity detection with use of potassium hydroxide as eluent, however sensitivity of such procedure is limited.[7] An interesting feature of the conductivity detection is its possibility for simultaneous detection of cations and anions.[8,9] UV detection, very common in HPLC, can also be utilized for ion-chromatography of inorganic species, however, many common ions as fluoride, phosphate, carbonate or sulphate do not absorb UV radiation. An increase of sensitivity for UV detection in ion-chromatography can be obtained in measurements below 200 nm[10] and in indirect vacancy mode non-absorbing anions[11] and cations[12,13] can be detected. Observed detectabilities are poorer than in conductivity detection. For electroactive anions and cations very low detection limits can be obtained using amperometric detection with silver[14-16] or mercury electrodes.[17] A choice of detected species in that case is rather limited.

Modern potentiometry with variety of classical and membrane sensing electrodes has, so far, rather occasionally been used in ion-chromatography . The first applications of potentiometric detection appeared in mid-seventies, when it was utilized for very sensitive detection of halide with silver/silver chloride electrode,[18] and nitrate and nitrite with liquid-state membrane nitrate electrode.[19] In later works number of indicating electrodes used for potentiometric detection has significantly increased. A list of reported applications is shown in Table 1.

One essential aspect in the use of potentiometric detection for chromatographic applications is the design of flow through cell, providing small dead-volume, rapid, reproducible response with stable base-line. In flow-cell used by Hershcovitz et al.[22] for detection of halide, a silver wire as an electrode of the second kind was used with pH glass electrode as the reference electrode in a cell of relatively large volume (Fig.1A). Another design of flow-cell with a wire indicating electrode, used in measurements with metallic and of the second kind electrodes has much smaller dead-volume (Fig.1B).[33-38] Typical construction of a thin-layer flow

Table 1. Potentiometric detection in high-performance ion-chromatography.

Sensing electrode	Detected species	Reference
Electrodes of the second kind		
Ag/AgCl	halide	18
Ag/AgCl	halide, SCN^- *)	20
Ag/AgCl	halide, SCN^-, $S_2O_3^{2-}$	21
Ag/Ag Sal	halide, SCN^-	22
Membrane electrodes		
Solid state membrane		
F - ISE	F^-	23
Cl - ISE	halide, SCN^-	24, 25
Br - ISE	halide, SCN^-	23, 25
I - ISE	halide, SCN^-, $S_2O_3^{2-}$	26
S - ISE	halide, SCN^-, CN^-	25, 27
pH glass membrane	acetate, oxalate formate, CO_3^{2-}	28
Liquid-state membrane		
NO_3 - ISE	NO_3^-, NO_2^-, phtalate	19
Cl, NO_3, ClO_4 - ISE	NO_3^-, halate	29, 30
K - ISE	Na, K, Rb, Cs	31
PVC membrane		
K - ISE	alkaline metal ions, NH_4^+	32
Cu metallic electrode	Mg, Ca, Sr, Ba	33
	Cd, Co, Cr, Cu,	34, 35
	Fe, Mn, Ni, Pb, UO_2^{2+}	
	halide, halate, pseudohalide	36
	HCO_3^-, NO_2^-, NO_3^-, SO_4^{2-}	

*) replaced on-line by chloride ions

cell for detection with solid-state membrane electrodes is shown in Fig.1C.[26] For detection of alkali metal ions in open-tubular column liquid chromatography an ion-selective microelectrode was inserted directly into the end of the open-tubular column, giving configuration with extremely small dead volume (Fig.1D).[31]

Indicator Electrodes of the Second Kind

The response of silver electrodes of the second kind, reported in chromatographic detection by several authors, is based on the shift of the ionic equilibrium between the electrode coating and the determined anions in effluent. In determination of halide, thiocyanate and thiosulphate most often silver/silver chloride electrode was used.[18,20,21] An improvement in the base-line stability in halide determination was observed, when silver coating contained the same anion as used in eluent.[22] The magnitude of the observed potential changes for various electrodes of the second kind depends on the thermodynamic stability of species formed by detected anions with silver ions and on the kinetics of the ion-exchange equilibria at the electrode surface in each case. Both those factors may influence also the base-line stability. This can be illustrated by the results of flow-injection measurements without separation on the column, performed with silver electrodes of the second kind, obtained by anodic coating of silver wire (Table 2).[38] Measurements were carried out using flow through cell shown in Fig.1B with 1 mM sodium perchlorate as the carrier solution. For twelve investigated anions the largest potential changes were exhibited by Ag/AgBr electrode, except iodide, for which largest but tailing response was observed with Ag/AgI electrode. Tailing for iodide was much smaller than with Ag/AgBr electrode. Some examples of recorded flow-injection peaks for

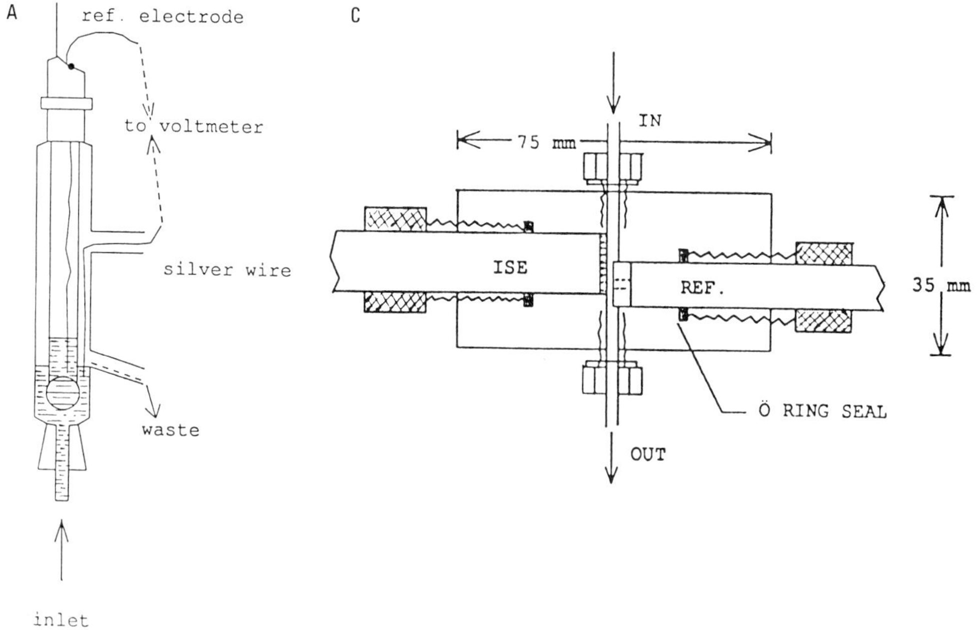

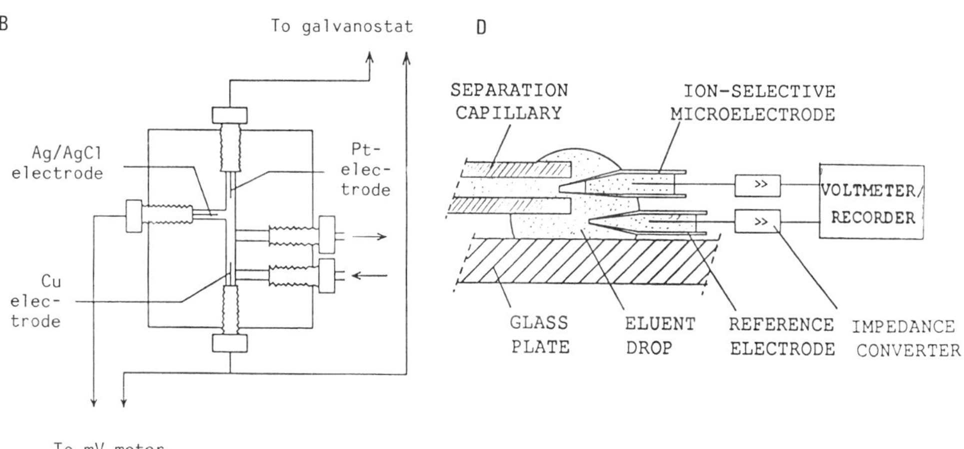

Fig.1. Flow through cells used for potentiometric detection in ion-chromatography: A – acc. to Hershcovitz et al.,[22] B – acc. to Alexander et al.[49] C – acc. to Butler and Gershey,[26] D – acc. to Manz and Simon.[31]

four different anions examined are given in Fig.2.

Excluding chromate, detection limits for the injected samples are much below 1 nmol. For anions providing smaller potential changes, the main positive peaks were accompanied with negative dips in falling part of the peak. Their magnitude have increased with increase of concentration of the samples injected. Similar observations were reported for Ag/AgCl electrode[21] and membrane iodide electrode[26] in chromatographic applications. Observed phenomena are consistent with theoretical considerations about transient response for anions, for which the bulk electrode membrane shows poor selectivity.[39] For Ag/AgBr electrode most stable base-line was also observed. Studies on chromatographic application of that indicating electrode for detection of anions are in progress. The application of less sensitive Ag/AgCl electrode for ion-chromatography was reported recently by

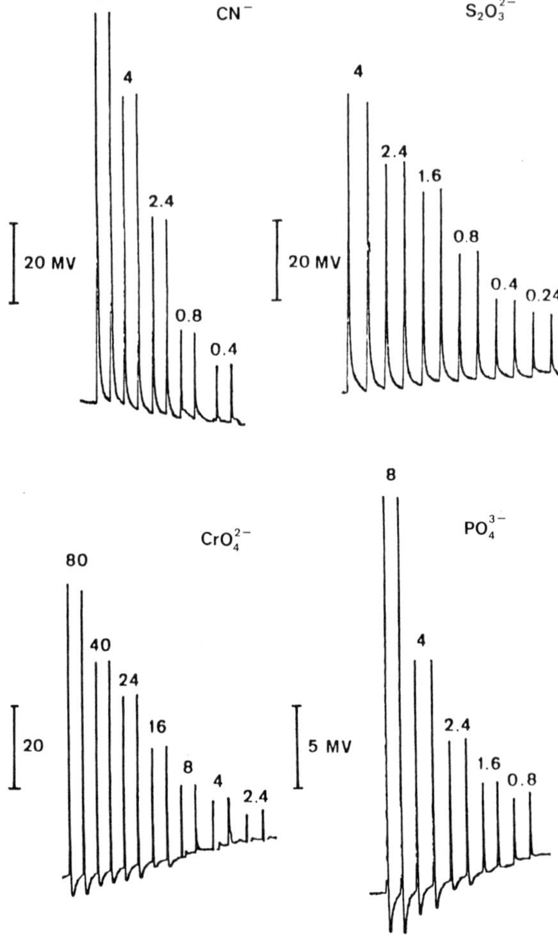

Fig.2. Flow-injection response of silver/silver bromide wire electrode. Injected amounts are shown in nmol. Carrier: 10 mM sodium perchlorate. Sample volume 80 μl. Flow-rate 7.8 ml/min.[38]

Lockridge et al.[21] Using sodium perchlorate solution of constant or increasing concentration as eluent, the isocratic and gradient chromatography of halide, thiocyanate and thiosulphate with very stable base-line was demonstrated (Fig.3).

The injected amount 20 nmol of each species, seem to be much above the detection limit. Authors admit, that similar response was observed with Ag/AgSCN electrode and that in all the investigated electrodes of the second kind coatings with smaller particle size provide faster electrode response.

Membrane ISEs as the Indicator Electrodes

The application of potentiometric detection in ion-chromatography is favoured by the progress in the field of membrane ion-selective electrodes (ISE). The electrodes with solid-state membranes were mostly employed for determination of halides, pseudohalides and some other anions binding silver ions.[23−27] The use of fluoride electrode in multidetector, chromatographic system offers very low detection limit of 1.2 ng fluoride in injected samples.[23] Application of bromide electrode in the same system provides even five-fold better detectability. The same level of detectability was reported by Butler and Gershey for iodide with iodide ISE.[26] In the system with preconcentration step the detectability can be lowered by an order

258

Table 2. Response of anodically coated Ag/AgX to various anionic species in flowinjection measurements. Carrier: 1 mM NaClO$_4$; injected amount: 0.8 nmole of each species.

Injected ion	Response of different coating, mV			
	AgCl	AgBr	AgI	Ag$_2$S
Cl$^-$	7.2	13.4	1.5	0
Br$^-$	8.5	38	35	6.0
I$^-$	8.0	10.5	60	8.0
CN$^-$	6.7	16.2	17.7	13.3
SCN$^-$	8.5	22	6.0	5.0
S^{2-}	7.0	20.8	20	27.0
SO$_3^{2-}$	1.8	7.0	1.2	2.2
S$_2$O$_3^{2-}$	7.0	31	11	5.5
HCO$_3^-$	0.7	2.2	0.5	0.8
CrO$_4^{2-}$	0.8	2.7	0.7	0.7
C$_2$O$_4^{2-}$	0.7	1.0	0	0
PO$_4^{3-}$	0.6	3.2	2.0	1.2

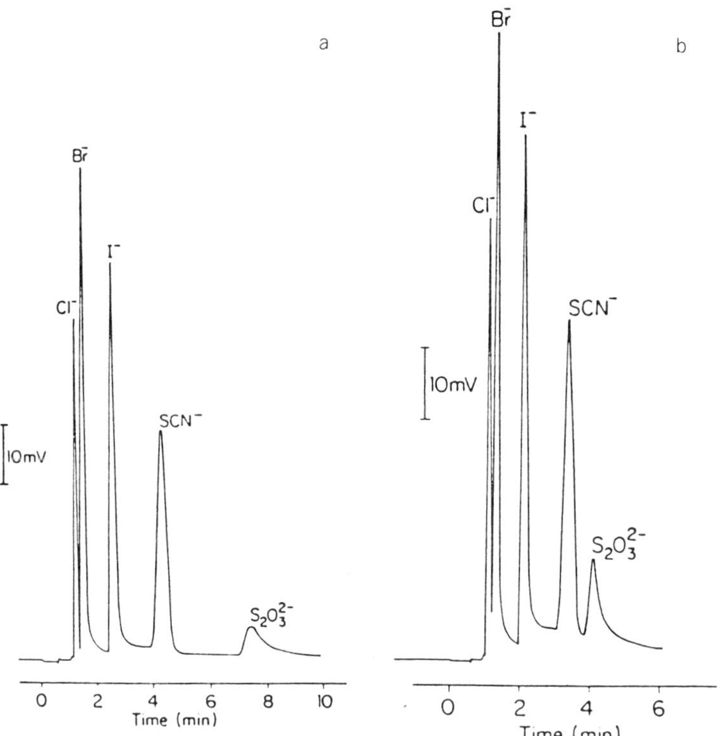

Fig.3. Chromatograms obtained for isocratic (A) and gradient (B) elution with potentiometric detection using Ag/AgCl electrode. Sample volume 20 μl. Injected 20 nmol of each species.[21]

of magnitude. For the optimum electrode performance, a constant iodide background of about 0.5 μM was reported in the column effluent passing through the detector. In comparative studies of solid-state membrane electrodes used as chromatographic detectors for determination of halide and thiocyanate among chloride, bromide, iodide and sulphide electrodes,

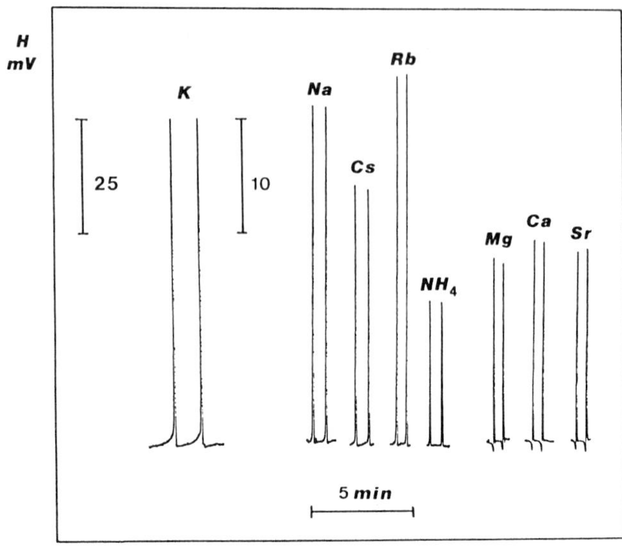

Fig.4. Flow-injection response of PVC electrode containing in the membrane dibenzo-18-crown-6 and dioctyl adipate conditioned in potassium chloride solution. Injected 80 nmol of each species. Carrier: 1 mM LiCl. Flow-rate 6.8 ml/min. Wall-jet flow-cell arrangement.

the last one has shown the best sensitivity.[25] In the investigation of iodide and sulphide ISE response to cyanide, faster and more sensitive response was observed for the latter electrode.[27]

Sporadic use in ion-chromatography was reported for pH glass electrode[28] and micro-type membrane cell with commercial ion-exchange membranes, separating two compartments of the flow-cell.[40]

Many possible applications can be expected for the use of electrodes with liquid or plasticized membranes. As mentioned above, liquid-state electrodes were introduced as chromatographic detectors by Schultz and Mathis.[19] Later on they were also employed for detection of halate[29,30] and as microelectrodes for detection of alkali metal ions.[31] PVC membrane electrodes sensitive to alkali metal ions and ammonium ion were investigated in ion-chromatography by Suzuki et al.[32] A substantial advantage of liquid and plasticized membrane ISE, known from fundamental studies on ISE, is a possibility to adjust the selectivity and detectability of their response by choice of respective ionophore, solvent and plasticizer in case of PVC membrane. For instance, in the mentioned application of PVC membrane electrodes for chromatography of alkali metal ions, four different ionophores valinomycin, nonactin, tetranactin and benzo-15-crown-5 were examined.[32] Most similar response with regard to magnitude of potential change for various sensed cations was obtained with tetranactin. Similar results can be obtained using common ionophore dibenzo-18-crown-6 in PVC membrane and changing the selectivity of response to alkali and alkaline earth metal ions by addition of different plasticizer. The examination of the selectivity in non-flowing conditions for four different plasticizers, namely, dipenthyl and dioctyl phthalates, dioctyl adipate and di(ethylhexyl) sebacate has shown, that the best detectability and the smallest differences in response to various cations can be expected for dioctyl adipate membrane.[38] It was confirmed in flow-injection measurements using lithium chloride as carrier and wall-jet arrangement of sensing PVC membrane electrode with internal reference solution (Fig.4). The flow-injection response obtained, allows to expect a detection limit below 10 nmol. Study on application of that sensor for ion-chromatography is in progress.

A comparable response for different alkali metal ions in chromatographic detection was achieved using a liquid microelectrode with a classical cation-exchanger (potassium tetra-p-chlorophenylborate in 2,3-dimethylnitrobenzene).[31]

The sensitivity of response of liquid-state or plasticized membrane anion electrodes can

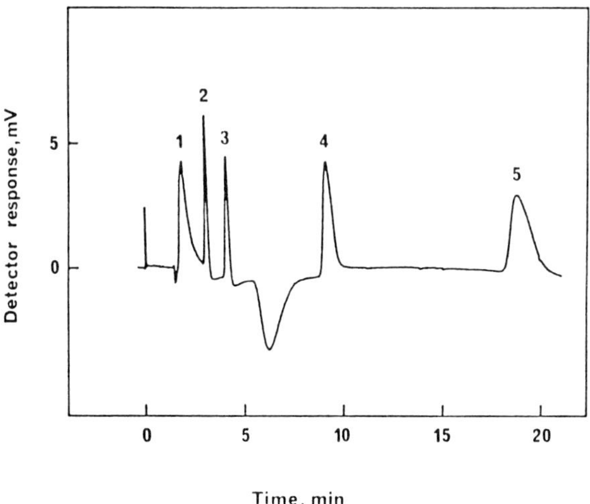

Fig.5. Chromatogram of mixture of cyanide (1), chloride (2), bromide (3), iodide (4) and thiocyanate (5) obtained with a metallic copper electrode and 40 mM sodium tartrate (pH 3.2) as eluent. Vydac 302 IC column. Injected amounts: CN – 0.8, Cl – 80, Br – 10, I – 0.4 and SCN – 0.4 nmol.[36]

be modified by change of the anion incorporated in the membrane phase. In flow-injection measurements of ten different inorganic anions with PVC membrane electrode with Aliquat 336S in chloride, nitrate and perchlorate form, the most suitable properties for chromatographic application were found for the nitrate form.[38] Analogous studies were carried out with oleophilic anion-exchange resin liquid membranes for flow-injection measurements and chromatography of oxyacid anions.[29,30]

Table 3. Comparison of available data on absolute detection limit (ng in injected sample) for ion-chromatography employing different detection methods.

Detected ions	Potentiometry with mebr. ISE	Amperom.	Conductivity		UV	
			Suppr.	SC	Direct	Indirect
F^-	1.2 [23]	–	5 [42]	0.4 [44]	–	25 [11]
Cl^-	4.0 [18]	NE	0.4 [43]	0.5 [44]	20 [45]	2 [46]
Br^-	0.25 [23]	1.0 [14)]	1.0 [43]	NE	10 [10]	100 [43]
I^-	0.24 [26]	0.14 [15]	NE	NE	15 [10]	500 [11]
NO_3^-	6.2 [19]	–	0.8 [43]	NE	10 [10]	140 [43]
NO_2^-	15 [19]	–	62.5 [42]	0.9 [44]	10 [10]	200 [43]
CN^-	NE	0.2 [14]	–	10*) [7]	–	–
HS^-	NE	0.01 [16]	–	10*) [7]	–	–
NH_4^+	3.6 [32]	–	4.0 [12]	NE	–	29 [12]
Na^+	NE	–	0.6 [12]	3.8 [13]	–	7.5 [13]
K^+	NE	–	2.3 [12]	25 [13]	–	11.7 [13]
Ca^{2+}	NE	–	NE	5 [13]	–	301 [13]

*) reversed, NE - not estimated

Application of a suitable potentiometric membrane electrode for detection in ion-chromatography gives a detectability similar and in some cases even better than reported for other more common detectors. This can be concluded from data presented in Table 3, where figures collected for detection limits of different methods are the lowest of those reported in the literature.

Detection with Metal Electrodes

Modern analytical potentiometry is predominated with membrane electrodes, however, especially interesting example of potentiometric detection in ion-chromatography is the application of metal electrodes. Very intensive studies were carried out for a metallic copper electrode. It was utilized for the detection of anions binding cuprous or cupric ions, for anions of oxyacids oxidizing a metallic copper and for vacancy indirect detection of anions practically non-binding copper ions.[36] With the use of moderately copper complexing ligands in eluent an indirect chromatographic detection of alkaline earth metal ions[41] or transition metal ions[34] is possible. A list of species detected with obtained detection limit is shown in Table 4. The data presented indicate significantly better detectability found in direct detection mode, than in indirect vacancy or subtractive detection. Example of an obtained chromatogram is shown on Fig.5.

Table 4. Detection limit estimated for ion-chromatography with metallic copper electrode.

Detector response	Determined species	Detection limit [ng]	Detection limit [ppm][*)	Ref.
Direct	Cl^-	70	0.28	36
binding	Br^-	20	0.08	36
of Cu(I)	I^-	3	0.012	36
or Cu(II)	CN^-	1	0.004	36
	SCN^-	3	0.012	36
Cu^0 oxidation	ClO_3^-	750	3	36
	BrO_3^-	6	0.024	36
	IO_3^-	50	0.20	36
Indirect	HCO_3^-	1600	6.4	36
vacancy	NO_2^-	230	0.92	36
	NO_3^-	3200	13	36
	SO_4^{2-}	9200	37	36
Indirect	Ca^{2+}	160	0.64	37
subtractive	Ni^{2+}	17	0.068	34
	Mn^{2+}	100	0.40	34
	UO_2^{2+}	240	0.96	34
Direct	Cu^{2+}	0.4	0.0016	34

[*) for a sample volume of 250 μl

In the interpretations of the mechanism of detection, as presented up to date the equilibrium between metallic copper and more or less complexed copper ions in solution was considered. A voltammetric study with a metallic copper disc electrode in neutral phosphate buffer solution, used frequently as eluent in chromatographic detection, shows formation of an oxide layer on the copper surface.[47] Its thickness and stoichiometric composition distinctly depends on the electrode pretreatment.

Other metallic electrodes has not been, so far, examined as detectors for ion-chromatography. Flow-injection measurements for different anions with potentiometric detection using mercury film on a metal silver electrode indicate satisfactory detectability and dynamic properties (except iodide) for such a detection (Fig.6).[38]

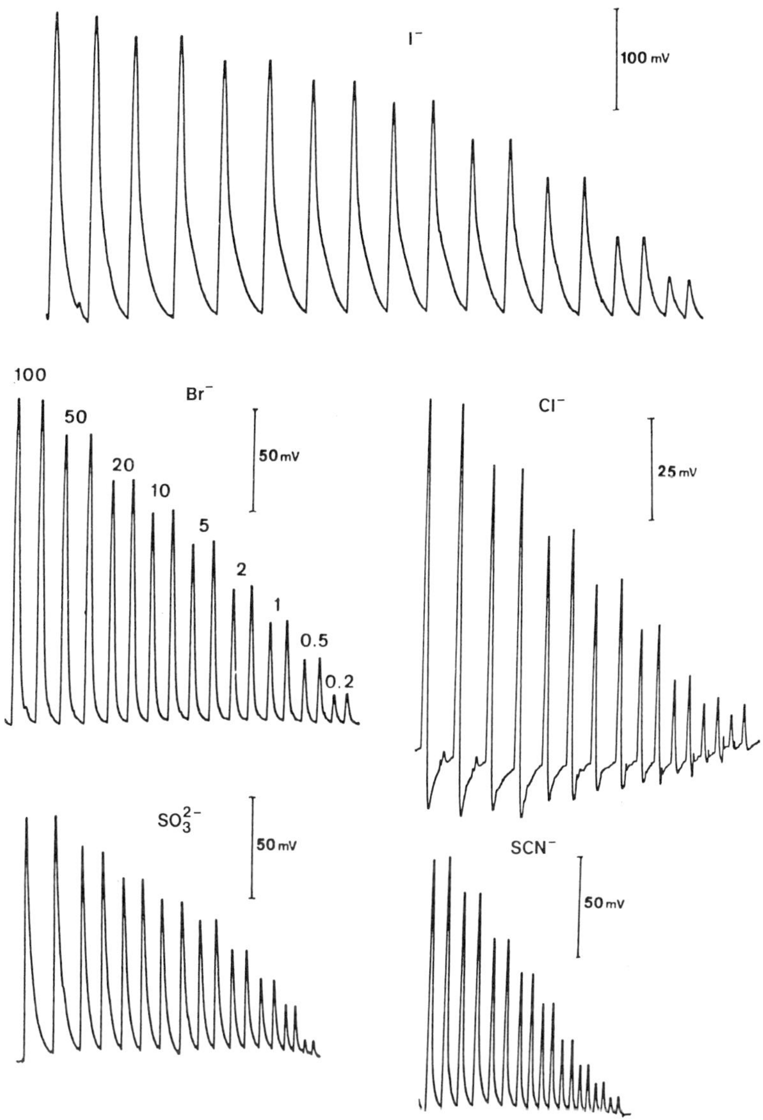

Fig.6. Flow-injection potentiometric response of mercury film electrode on silver wire support. Carrier: 0.1 M acetate buffer pH 4.6. Flow-rate 1.8 ml/min. Injected amounts shown in nmol (the same for all anions).[38]

Linear Calibration of Potentiometric Detectors

It is a very common opinion, that a serious disadvantage of potentiometric detection in chromatography is its Nernstian, logarithmic relationship between the signal magnitude (peak height or area of the peak) and the concentration of sensed species. As it was elucidated by several authors theoretically and confirmed experimentally in, e.g., flow-injection measurements with silver/silver chloride electrode.[48] in the range of small potential changes, peak height linearly depends on concentration in injected solutions. Similar results were produced in flow-injection measurements of various anions with other silver electrodes of the second kind electrodes (Fig.7 and 8).[38]

Linear relation between peak height and concentration was also found in chromatographic detection with Ag/AgCl electrode,[18,20] with sulphide ISE used for cyanide

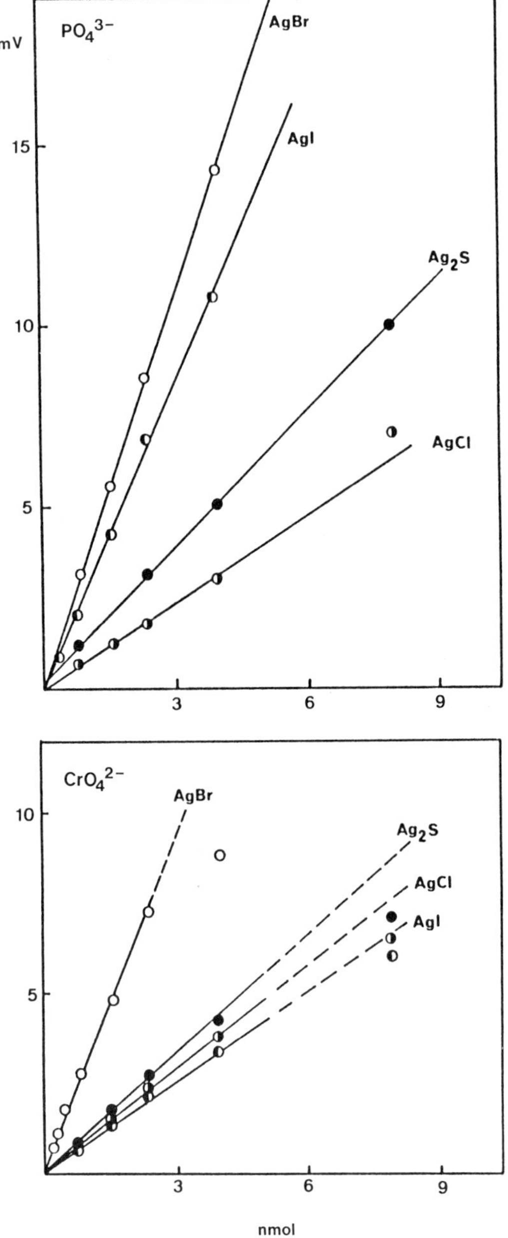

Fig.7. Calibration plots of phosphate and chromate ions for flow-injection response of wire electrodes of the second kind. Carrier: 10 mM sodium perchlorate. Sample volume 80 μl. Flow-rate 7.8 ml/min.

detection,[27] with PVC membrane electrode sensitive to alkali metal ions[32] and with a metallic copper electrode in direct detection. In chromatographic detection with liquid-state membrane nitrate electrode[19] and indirect chromatographic detection with a metallic copper electrode a wider range of linear relationship was observed between peak area and concentration of sensed species.

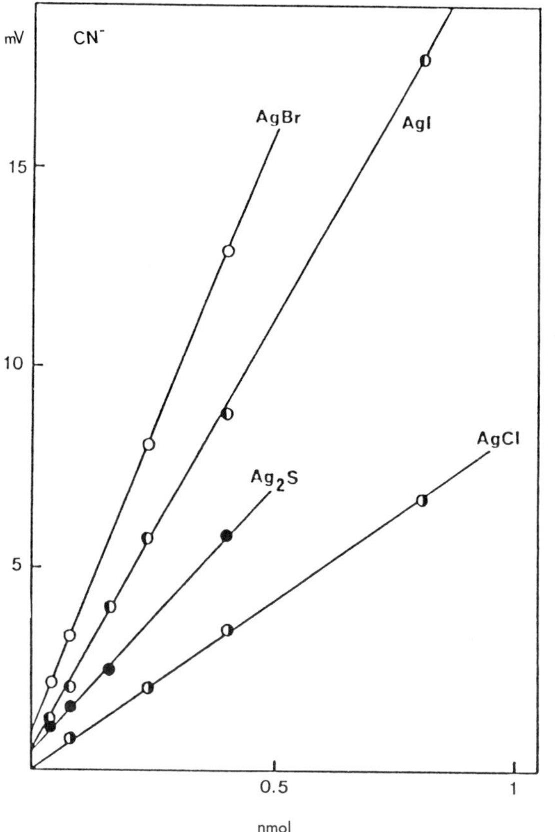

Fig.8. Calibration plots for cyanide for flow-injection response of wire electrodes of the second kind. Carrier: 10 mM sodium perchlorate. Sample volume 80 μl. Flow-rate 7.8 ml/min.

Final Remarks

The application of potentiometric detection for ion- chromatography needs certainly further studies. Flow-injection measurements without separation column as the preliminary test of detection method for chromatographic applications yield numerous promising results. They indicate also, that in most examples examined a dynamic behaviour of detectors was observed beeing satisfactory in chromatographic applications. A linear dependence of peak height on concentration facilitates interpretation of the results. Easy preparation of flow through potentiometric detector and common availability of signal transducers make this detection method very attractive.

Acknowledgements

The author wishes to thank Dr. P.W. Alexander and Dr. P.R. Haddad of University of New South Wales, Sydney, Australia for fruitful cooperation and Dr. T. Krawczynski vel Krawczyk, Dr. W. Matuszewski, Mrs. E. Pobozy and Mr. B. Szostek of University of Warsaw, Poland for their experimental contribution to this paper.

References

1. G. Schwedt, in: "Chromatographic methods in inorganic analysis", Dr. A. Huthig Verlag, Heidelberg, 1981.
2. H. Small, T.S. Stevens and W.C. Bauman, Anal. Chem., **47** (1975) 1801.

3. D.T. Gjerde, J.S. Fritz and G. Schmuckler, J. Chromatogr., **186** (1979) 509.

4. D.T. Gjerde, G. Schmuckler and J.S. Fritz, J. Chromatogr., **187** (1980) 35.

5. J.S. Fritz, Anal. Chem., **59** (1987) 335 A.

6. P.R. Haddad and P. Jandik, in: "Ion chromatography", J.G. Tarter (Ed.), Marcel Dekker, New York, 1987.

7. T. Okada and T. Kuwamoto, Anal. Chem., **57** (1985) 829.

8. M. Yamamoto, H. Yamamoto, Y. Yamamoto, S. Matsushita, N. Baba and T. Ikushige, Anal. Chem., **56** (1984) 832.

9. D.J. Pietrzyk and D.M. Brown, Anal. Chem., **58** (1986) 2554.

10. R.J. Williams, Anal. Chem., **59** (1987) 541.

11. H.J. Cortes and T.S. Stevens, J. Chromatogr., **295** (1984) 269.

12. D.L. McAleese, Anal. Chem., **59** (1987) 541.

13. H. Shintani, K. Tsuji and T. Oba, Bunseki Kagaku, **24** (1985) 109.

14. R.D. Rocklin and E.L. Johnson, Anal. Chem., **55** (1983) 4.

15. P. Jandik, D. Cox and D. Wong, Am. Lab., March 1986, 114.

16. K. Han and W. Koch, Anal. Chem., **59** (1987) 1016.

17. T. Hsi and D.C. Johnson, Anal. Chim. Acta, **175** (1985) 23.

18. M.C. Franks and D.J. Pullen, Analyst, **99** (1974) 503.

19. F.A. Schultz and D.E. Mathis, Anal. Chem., **46** (1974) 2253.

20. V.V. Bardin, Yu.M. Ivanov and O.F. Shartukov, Zh. Analit. Khim., **33** (1978) 1732.

21. J.E. Lockridge, N.E. Fortier, G. Schmuckler and J.S. Fritz, Anal. Chim. Acta, **192** (1987) 41.

22. H. Hershcovitz, Ch. Yarnitzky and G. Schmuckler, J. Chromatogr., **252** (1982) 113.

23. J. Slanina, F.P. Bakker, P.A.C. Jongejan, L. van Lamoen, and J.J. Mols, Anal. Chim. Acta, **130** (1981) 1.

24. A. Jyo, K. Mori and N. Ishibashi, Bull. Chem. Soc. Jpn., **56** (1983) 3507.

25. H. Muller and R. Scholz, in: "Ion-selective Electrodes, 4. Proceedings of the 4th Symposium, Mátra füred, Hungary", E. Pungor and I. Buzás (Eds), Anal. Chem. Symposia Series, **22** Elsevier, 1985, p. 553.

26. E.C.V. Butler and R.M. Gershey, Anal. Chim. Acta, **164** (1984) 153.

27. Wang-nang Wang, Yeong-jgi Chen and Mou-tai Wu, Analyst, **109** (1984) 281.

28. S. Egasira, J. Chromatogr., **202** (1980) 37.

29. N. Ishibashi, T. Imato, M. Yamauchi, M. Katahira and A. Jyo, Anal. Chem. Symp. Ser., **22** (1985) 57.

30. S. Koizumi, T. Imato and N. Ishibashi, Anal. Sci., **3** (1987) 319.

31. A. Manz and W. Simon, J. Chromatogr., **21** (1983) 326.

32. K. Suzuki, H. Aruga and T. Shirai, Anal. Chem., **55** (1983) 2011.

33. P.R. Haddad, P.W. Alexander and M. Trojanowicz, J. Chromatogr., **294** (1984) 397.

34. P.R. Haddad, P.W. Alexander and M. Trojanowicz, J. Chromatogr., **324** (1985) 319.

35. P.R. Haddad, P.W. Alexander and M. Trojanowicz, Anal. Chim. Acta, **177** (1985) 183.

36. P.R. Haddad, P.W. Alexander and M. Trojanowicz, J. Chromatogr., **321** (1985) 363.

37. P.W. Alexander, P.R. Haddad and M. Trojanowicz, Chromatographia, **20** (1985) 179.

38. M. Trojanowicz, T. Krawczynski, W. Matuszewski, E. Pobozy and B. Szostek, in preparation.

39. W.E. Morf, Anal. Chem., **55** (1983) 1165.

40. R.S. Deelder, H.A.J. Linssen, J.G. Koen and A.J.B. Beeren, J. Chromatogr., **203** (1981) 153.

41. P.R. Haddad, P.W. Alexander and M. Trojanowicz, J. Chromatogr., **294** (1984) 397.

42. J.A. Mosko, Anal. Chem., **56** (1984) 629.

43. B. Rossner and G. Schwedt, Fresenius Z. Anal. Chem., **320** (1985) 566.

44. J.S. Fritz, D.L. DuVal and R.E. Barron, Anal. Chem., **56** (1984) 1177.

45. D.R. Jenke, P.K. Mitchel and G.K. Pagenkopf, Anal. Chim. Acta, **155** (1983) 279.

46. P.J. Naish, Analyst, **109** (1984) 809.

47. M.L. Hitchman and M. Trojanowicz, unpublished results.

48. M. Trojanowicz and W. Matuszewski, Anal. Chim. Acta, **151** (1983) 77.

49. P.W. Alexander, P.R. Haddad and M. Trojanowicz, Anal. Lett., **17** (A4) (1984) 309.

STABLE MODIFIED ELECTRODES FOR USE IN AMPEROMETRIC DETECTORS IN FLOW SYSTEMS

James A. Cox and Thomas J. Gray

Department of Chemistry
Miami University
Oxford, OH 45056, U.S.A.

Practical analytical applications of modified electrodes, particularly examples that are based on mediated reactions, have been slow to emerge. Only a few attempts, other than our work which is described herein, to apply such electrodes to flow systems have been reported. Baldwin and co-workers used an electrode that was prepared by mixing cobalt phthalacyanine with carbon paste for the determination of hydrazine[1] and some sulfhydryl compounds[2] by high performance liquid chromatography with amperometric detection. Although the results were promising, the day-to-day stability without mechanically renewing the surface was somewhat limited.

Johnson and Larochelle[3] developed a coulometric detector based on a platinum surface that was modified by iodine adsorption for the chromatographic determination of Cr(VI). They did not report the stability of the surface. Along with many others, we found that in static solution this modified electrode is stable;[4] however, in a flow system, it lacked long term stability, at least for the oxidation of nitrite[5] and thiocyanate.[6]

Reim and VanEffen[7] generated an active nickel oxide that permitted amperometric detection of certain carbohydrates after chromatographic separation, but the working curves were nonlinear. Active oxides have also been used by other groups to promote a variety of electrocatalytic oxidations.

There are two general problems with the use of electrocatalytic modified electrodes in flow systems. In order for the method to have a wide linear dynamic range, the rate of electrolysis must be controlled by mass transport of the analyte from the bulk solution to the electrode surface. Second, the modified electrode must be stable during both experimental operations and storage.

The current for a mediated electrolysis will be limited by mass transport if the modified electrode is highly conductive, the mediator concentration is high and the rate constant of the mediating reaction is favorable. In 1984, we demonstrated that modification of a glassy carbon electrode by cyclic voltammetry of a fresh solution of $RuCl_3$, $K_4Ru(CN)_6$ and NaCl at pH 2 yielded a surface at which As(III) was oxidized at a rate limited only by mass transport.[8] For example, with a 2 mM As(III) solution in 0.5 M NaCl at pH 2, linear scan voltammetry over the range 5 – 500 mV/s yielded peak current functions, $i_p v^{-1/2}$, which were quite constant; they varied in the range 6.3 – 6.2 $(\mu A)s^{1/2}(mV)^{-1/2}$. Moreover, the limiting current in a rotating disk voltammetry experiment was directly proportional to the square root of the rate of rotation over the range 400 – 3600 rpm.

The results of the hydrodynamic voltammetry suggested that the mixed-valent ruthenium film on glassy carbon, mvRu, was suitable for an amperometric detector in a flow system. This hypothesis was verified.[5] A working curve prepared by flow injection amperometry at a mvRu electrode was linear over the range 5 – 200 μM As(III). The slope was 7.4 $\pm$ 0.4 $(\mu A)l/mmol$ (7 points) at a 0.071 cm^2 (geometric area) electrode in a Bioanalytical Systems TL5A cell; the correlation coefficient was 0.9999; and the intercept was −1.5 nA. The detection limit using the criterion of a signal equal to three-times the standard deviation of blank injections was 200 pg As in a 20 μl injection. The half-width of the peaks was 22 s

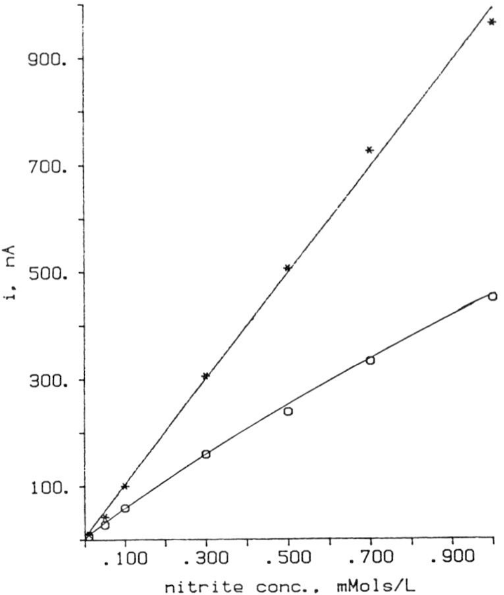

Fig.1. Flow injection amperometry of nitrite at glassy carbon (*) and metalated polymer (hexachloroiridate in quaternized poly(4-vinyl)pyridine) electrodes. Applied potential, 0.85 V vs Ag/AgCl; carrier, 0.2 M phosphate buffer at pH 4.6; flow rate 1 ml/min; injection volume, 20 μl; electrode area, 7.1 mm^2.

at a flow rate of 1.0 ml/min, so the calculated sample throughput was 64 injections per hour with baseline resolution. The latter calculation assumed a model of six standard deviations between gaussian peaks for baseline resolution.

This electrode was also very stable. In voltammetric experiments on static solutions, the electrode was stable for 10 weeks,[8] and in the flow system, reproducible working curves were obtained over a period of several days.[5] It is important to note that As(III) is not electroactive at bare glassy carbon, so these observed stabilities demonstrate that the film remained intact.

In marked contrast, electrodes modified by incorporating the mediator into a quaternized poly(4-vinyl)pyridine, qPVP, film did not exhibit day-to-day stability in a flow system.[5] Moreover, as shown in Fig.1, working curves for flow injection determinations at such electrodes are nonlinear and have a lower slope than at the bare electrode. The illustrated case is for the determination of nitrite. The electrode for the experiment in Fig.1 was glassy carbon which was dip-coated with qPVP and metalated with hexachloroiridate; the experimental conditions were following:
- applied potential, 0.85 V vs Ag/AgCl
- carrier solution, pH 4.6 phosphate buffer
- flow rate, 1 ml/min
- injection volume, 20 μl
- geometric area of the electrode, 7.1 mm^2

The same general result was observed with As(III) as the analyte, qPVP as the ionomer and hexacyanoruthenate as the mediator.

The flow injection peaks at the modified electrode of Fig.1 are quite broad. From the peak half-widths, 35 s, a sample throughput of 40 injections per hour is calculated using the assumptions and the model presented previously. Thus, the throughput at the electrode modified with a metalated qPVP film is less than at the surface which provides a current that is limited by mass transport in solution.

A greater peak width is expected if the current is significantly governed by penetration of the analyte into the film. By rotating disk voltammetry, the rate of oxidation of nitrite at an electrode modified by a hexachloroiridate-containing film of qPVP was demonstrated to

268

be limited indeed by a combination of the rate of the cross-exchange reaction between NO_2^- and Ir(IV) and the diffusion of nitrite into the film, but not to the base metal,[9] using the theoretical treatment of Saveant et al.[10-12]

Our work to date has demonstrated that both of the above types of modified electrode can alleviate the problem of electrode passivation by accumulation of electrolysis products on the surface. With nitrite as the analyte, flow injection data at a bare Pt electrode showed a continuous loss of sensitivity with consecutive injections;[5] the current for 40 μM nitrite introduced into a pH 4.6 phosphate buffer flowed at 1.0 ml/min dropped from 60 nA to 51 nA after five 20 μl injections. At an Ir(IV)-containing qPVP film, the peak current at 0.85 V vs Ag/AgCl was constant (0.0 % relative standard deviation of 5 trials with a peak current of 49 nA under the same conditions).

The oxidation of thiocyanate is also well-known to cause electrode passivation.[12] Although it is usually attributed to accumulation of electrolysis products, it may be just deactivation of the surface by consumption of the oxide. The latter occurs, for example, with the oxidation of As(III) at a Pt (oxide-coated) electrode.[14] We have found that with an electrode modified with the ruthenium-containing inorganic film, mvRu, the problem is alleviated.[6] The general characteristics of flow injection determinations of thiocyanate with an amperometric detector based on the mvRu electrode are summarized in Table 1. The sample throughput was calculated as described above.

The linear dynamic range which extends up to 4 μM thiocyanate is not surprising considering that the peak current for the oxidation of thiocyanate by linear scan voltammetry at the mvRu electrode is limited by diffusion of the analyte in the bulk solution. As shown in Table 2, the peak current function, $i_{pa}v^{-1/2}$, is independent of scan rate in the range 10 – 300 mV/s.

Table 1. Flow injection amperometric determination of thiocyanate at a glassy carbon electrode modified with a Ruthenium-containing inorganic film.

characteristic	result
calibration curve sensitivity[a,c]	52.6 ± 0.8 nA/μM
linear dynamic range[a,c]	$0.2 - 4.0$ μM
sample throughput[b,c]	120 injections/h
reproducibility of peak current for consecutive injections (50)[b]	46.7 ± 0.7 nA

[a] 50 μl injections in pH 4.6 buffer flowed at 2 ml/min; applied potential, 0.85 V vs Ag/AgCl.
[b] same as the above except 20 μl injections of 1.8 μM thiocyanate.
[c] these data are reported in.[6]

Table 2. Effect of scan rate on the peak current for the oxidation of SCN^- at the glassy carbon electrode modified with Ru-containing film.

v, mV/s	$i/(mV)^{1/2}$, μA(s)$^{1/2}$(mV)$^{-1/2}$
10	1.2
50	1.2
100	1.1
300	1.1

Solution: 0.6 mM SCN^- in 0.2 M $Ca(H_2PO_4)_2$, KH_2PO_4 at pH 2. Electrode area: 0.8 mm^2.

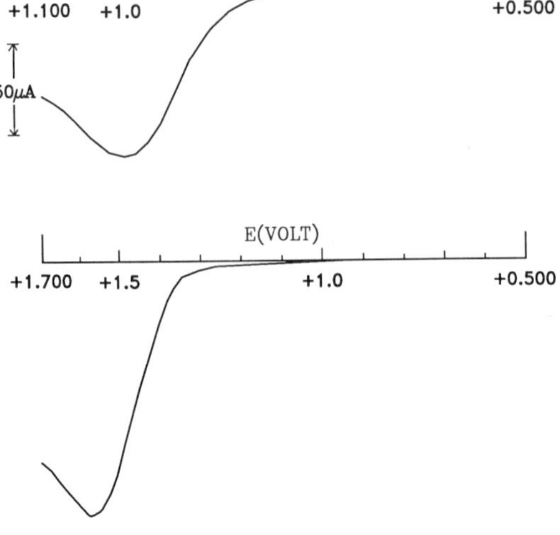

Fig.2. Linear scan voltammetry of cystine at glassy carbon (lower curve) and mvRu (upper curve) electrodes. Cystine, 4.4 mM; electrolyte, 0.2 M K_2SO_4 at pH 2; scan rate, 50 mV/s; area, 0.07 cm^2. The reference electrode is Ag/AgCl (1.0 M chloride).

Perhaps as important as the data in Table 1 is the demonstration of long term stability. A single preparation of the electrode was used for well over 1000 injections of a variety of organic and inorganic analytes.[6] The sensitivity remained virtually constant for a 3-month period. This stability was achieved by using a preparation procedure that was a modification of our previous method[5,8] which provided a thicker, more active film. The procedure is still not optimized and the exact chemical nature of the electrode is still under investigation, so the procedure is not specified herein. The method that has already been published[8] provides virtually the same results except for the long term stability in a flow system.[5]

A potentially important application of the method is the determination of thiocyanate in biological fluids. For example, the level of this ion in blood, urine and saliva provides an indicator of smoking behaviour, a monitor of dosages of certain drugs and a means of diagnosing chronic exposure to cyanide. Flow injection amperometry is not sufficiently selective to use directly on these samples since some response to uric and ascorbic acid, which are at much higher concentrations in these samples (except saliva), is also observed. With 20 μl samples and a 2.8 ml/min flow rate of a 0.2 M K_2SO_4 carrier at pH 2, the sensitivities of thiocyanate, uric acid and ascorbic acid are 32, 18 and 14 nA/μM respectively.[6] This modified electrode is therefore better suited as a detector for high performance liquid chromatography than for flow injection analysis.

The mvRu electrode also mediates the oxidation of organic sulphur compounds. Fig.2 contains a comparison of the linear scan voltammograms for the oxidation of cystine, a disulfide, at bare glassy carbon and at the mvRu electrode. Typically, we observe that the oxidation occurs at about 0.9 V vs Ag/AgCl at the latter and at about 1.5 V vs Ag/AgCl at glassy carbon. These data suggest that the modified surface is suitable for inclusion in the detector of a flow injection amperometry system.

As in the previous work, the flow injection system consisted of a Cole Parmer Masterflex peristaltic pump, Rheodyne 7125 injector, Bioanalytical System (BAS) TL-5A flow through electrochemical cell and either an IBM EC/230 or a BAS CV 37 potentiostat. These instruments respectively are designed for use with an electrochemical detector for liquid chromatography and for low current measurements such as with ultramicroelectrodes. A pulse dampener was also included in the flow system.

Fig.3 illustrates a sequence of peaks that are developed in the flow injection amperometric determination of cystine at a concentration near the detection limit. From the half-widths

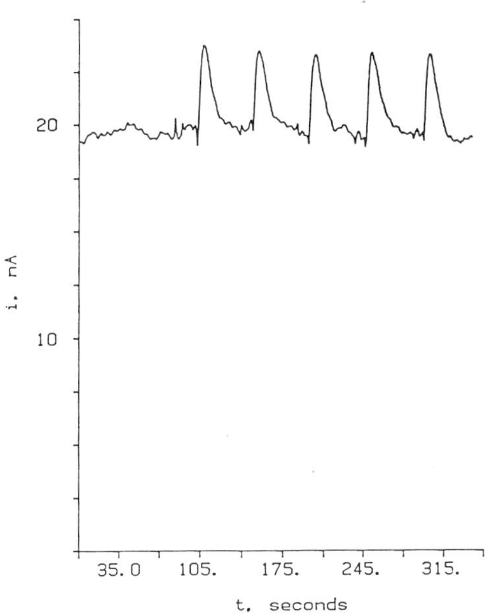

Fig.3. Flow injection amperometric peaks at the mvRu electrode near
the limit of detection for cystine. Applied potential, 0.92 V vs Ag/AgCl;
carrier, 0.2 M K_2SO_4 at pH 2; flow rate, 1.9 ml/min; injection volume,
20 μl; cystine concentration, 0.2 μM.

of the peaks for a 1 μM cystine sample in a 20 μl injection, 11.6 s, a throughput of 122
samples is calculated.

With 20 μl injections, a 0.2 M K_2SO_4 carrier at pH 2, a flow rate of 1.9 ml/min, and an
applied potential of 0.9 V vs Ag/AgCl (1.0 M Cl^-), the flow injection amperometric working
curve is linear from 0.4 to 6.4 μM at the mvRu electrode. The data which are shown in Fig.4
have the following linear least squares characteristics: slope, 5.6 $\pm$ 0.1 nA/μM; y-intercept,
4.3 $\pm$ 0.4 nA, correlation coefficient; 0.9985.

The applied potential in the above experiment was one suggested to yield acceptable
signal-to-background ratios by linear scan voltammograms for cystine and a blank with the
carrier solution as the supporting electrolyte. In order to optimize the sensitivity, point-
by-point hydrodynamic voltammograms were obtained for cystine and the blank. Here, the
applied potential of the modified electrode in the flow cell was varied, and both the background current and the flow injection peak current were measured at each potential under
potentiostatic conditions. The difference between the currents is plotted versus potential.
The data in Fig.5 suggest that a potential of 0.92 V vs Ag/AgCl, should be used. A linear
least squares fit of a 5-point working curve over the range 0.2 to 3.4 μM cystine under the
Fig.4 conditions except at 0.92 V vs Ag/AgCl yielded the following: slope, 11.0 $\pm$ 0.2 nA/μM;
correlation coefficient, 0.9993; y-intercept, 1.8 $\pm$ 0.4 nA.

Hence, both the sensitivity and the y-intercept were improved by optimizing the applied
potential.

The hydrodynamic voltammogram does not yield a plateau current. A possible reason
is that a second faradaic process begins to add to the backgound current; this is better seen
by cyclic voltammetry, Fig.6. This process will cause a decrease in the current due to the
mediated oxidation of cystine in the event that the further-oxidized product of the background
process is a less effective mediator than the parent species.

As suggested by the previously-cited applications to the determinations of As(III),
thiocyanate and cystine and by a report on its use for the oxidation of methanol,[11] the mvRu
electrode is not selective. Cystine is often determined in mixtures with glutathione and
cysteine. As shown in Fig.7, the last two species also undergo a mediated oxidation at the

271

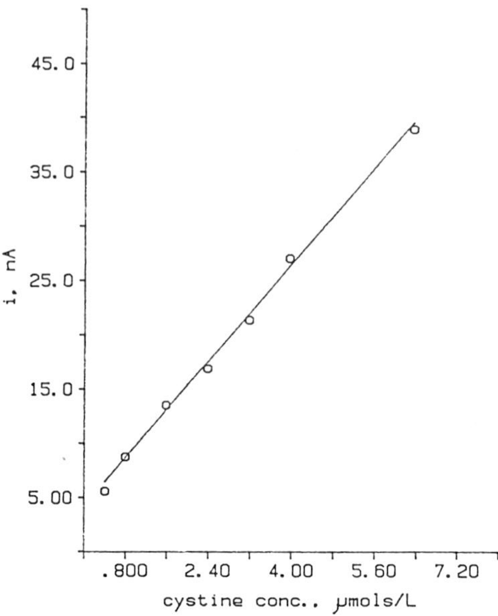

Fig.4. Flow injection amperometry of cystine at the mvRu electrode. Applied potential, 0.90 V vs Ag/AgCl; carrier, 0.2 M K_2SO_4 at pH 2; flow rate, 1.9 ml/min; injection volume, 20 μl.

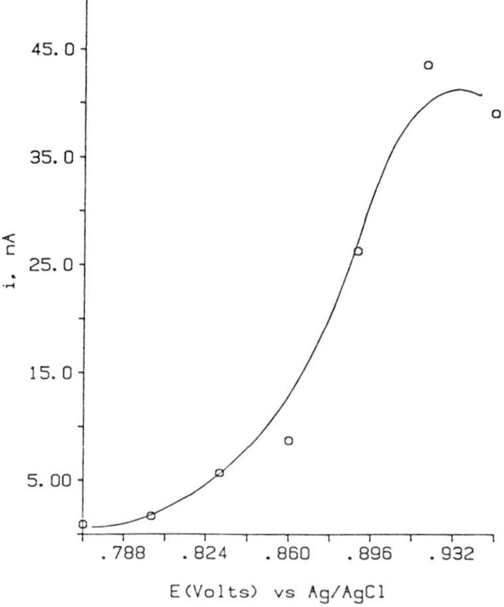

Fig.5. Hydrodynamic voltammogram for the oxidation of cystine at the mvRu electrode. Conditions are those in Fig.4 except the applied potential of the flow injection experiment was varied. Cystine concentration, 3.4 μM.

mvRu electrode at 0.9 V vs Ag/AgCl. At bare glassy carbon electrodes, a distinct oxidation of these species was not observed at pH 2, even when the glassy carbon electrode was pretreated by anodization. Hence, the mvRu surface appears to be well-suited to the determination of glutathione and cysteine by flow injection amperometry and to the determination of mixtures of these compounds after separation by high performance liquid chromatography.

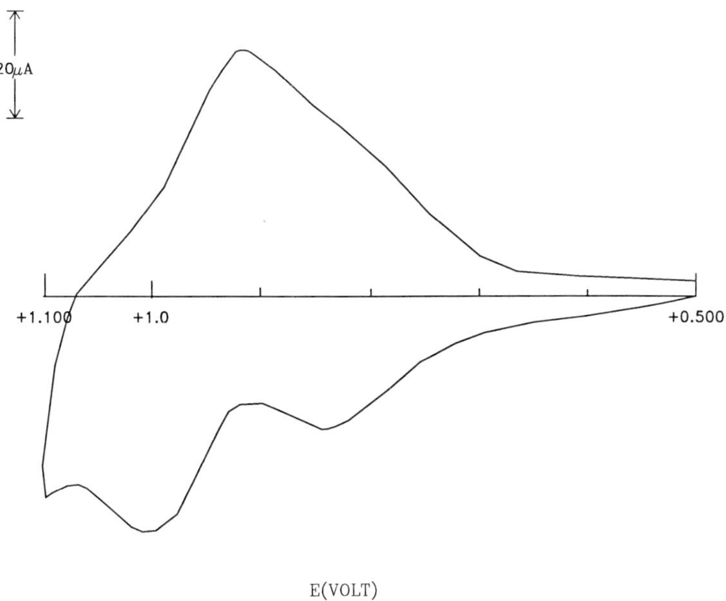

Fig.6. Cyclic voltammetry of the mvRu electrode. Scan rate, 50 mV/s; electrolyte, 0.2 M K_2SO_4 at pH 2; reference electrode, Ag/AgCl.

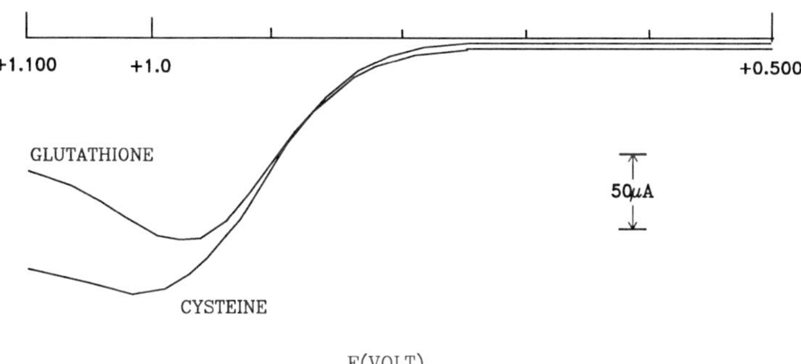

Fig.7. Linear scan voltammetry of glutathione and cysteine at the mvRu electrode. Glutathione, 4.6 mM; cysteine, 5.5 mM; all other conditions are the same as those in Fig.2.

Fig.8 contains the flow injection amperometry working curve for glutathione. Linear least squares analysis of given data points in range 0.2 – 3.2 μM yielded the following: slope, 14.7 ± 0.2; y-intercept, 1.2 ± 0.3 nA; correlation coefficient, 0.9997.

From the half-width of the peaks at 1.9 ml/min, 13.2 s, a sample throughput of 107 injections per hour with baseline resolution is calculated.

The behaviour of another thiol, cysteine, is similar to that of glutathione. The plot shown in Fig.8 (0.2 to 3 μM) has a slope of 15.5 ± 0.5 nA/μM and a correlation coefficient of 0.9982. When data at a higher concentration are also considered (6.1 μM yields 66.5 nA), it is apparent that there is a negative deviation from linearity; therefore, the result at 3 μM cysteine can be deleted. Over the range 0.2 to 1.5 μM, the correlation coefficient improves to 0.9999; the slope in this range is 17.4 ± 0.1 nA/μM and the y-intercept is 0.7 ± 0.1 nA. The calculated sample throughput at 2.0 ml/min is 150 injections per hour.

These results demonstrate that electrodes which are modified with a mediator-containing inorganic film have the combination of stability, conductivity and kinetic characteristics toward important analytes to serve as practical amperometric sensors in flow systems. As

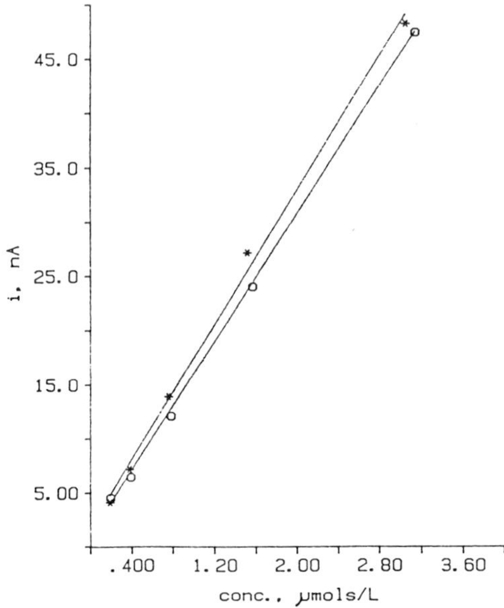

Fig.8. Flow injection amperometry of glutathione (o) and cysteine (*)
at the mvRu electrode. Applied potential, 0.90 V vs Ag/AgCl; carrier,
0.2 M K_2SO_4 at pH 2; flow rate, 1.9 ml/min; injection volume, 20 μl.

yet, we have not found a means of making these films selective, except for an isolated example where a film of Ag(I) and $Mo(CN)_8^{4-}$ only developed a current in the presence of potassium ion,[15] so these sensors may be more useful in conjunction with high performance liquid chromatography than for flow injection analysis.

References

1. K.M. Korfhage, K. Ravichandran and R.P. Baldwin, Anal. Chem., **55** (1983) 1608.
2. M.K. Halbert and R.P. Baldwin, Anal. Chem., **57** (1985) 1536.
3. D.C. Johnson and J. Larochelle, Talanta, **20** (1973) 959.
4. J.A. Cox, P.J. Kulesza and M.A. Mbugwa, Anal. Chem., **54** (1987) 787.
5. J.A. Cox and K.R. Kulkarni, Talanta, **33** (1986) 911.
6. J.A. Cox, T.J. Gray and K.R. Kulkarni, Anal. Chem., **60** (1988) 1710.
7. R.E. Reim and R.M. Van Effen, Anal. Chem., **58** (1986) 3203.
8. J.A. Cox and P.J. Kulesza, Anal. Chem., **56** (1984) 1021.
9. J.A. Cox and P.J. Kulesza, J. Electroanal. Chem., **175** (1984) 105.
10. C.P. Andrieux, J.M. Dumas-Bouchait and J.M. Saveant, J. Electroanal. Chem., **131** (1982) 1.
11. C.P. Andrieux and J.M. Saveant, J. Elcetroanal. Chem., **134** (1982) 163.
12. C.P. Andrieux and J.M. Saveant, J. Electroanal. Chem., **142** (1982) 1.
13. D.S. Austin, J.A. Polta, T.Z. Polta, A.P-C. Tang, T.D. Cabelka and D.C. Johnson, J. Electroanal. Chem., **168** (1984) 227.
14. W.-H. Kuo and T. Kuwana, J. Electroanal. Chem., **169** (1984) 167.
15. J.A. Cox and B.K. Das, Anal. Chem., **57** (1985) 2739.

FLOW STREAM DETECTORS BASED ON THE ELECTROCATALYTIC OXIDATION OF POLYHYDROXY COMPOUNDS AT SILVER OXIDE ELECTRODES

Terrence P. Tougas,[1] Edwin G.E. Jahngen[1] and
Michael Swartz[2]

1. Department of Chemistry, University of Lowell
   Lowell, Massachusetts 01854, U.S.A.
2. Waters Chromatography Division, Millipore Corporation
   Milford, Massachusetts 01854, U.S.A.

Many simple carbohydrates and other polyhydroxy compounds can be oxidized at a silver oxide surface. The oxidation is via an electrocatalytic mechanism involving a Ag(I) oxide. This forms the basis of a flow stream detector operated in an amperometric mode which may be used for either flow injection or high performance liquid chromatography (HPLC) applications. The title electrode has been applied to the detection of simple carbohydrates, triglycerides and nucleic acid components.

This paper will discuss the characteristics of this detector with respect to the above applications. Stable oxides are formed at potentials in excess of 200 mV vs. SCE. A chrono-coulorometric experiment has been used to follow the formation of the oxide layer, loss of silver due to the formation of soluble hydroxy species, and reduction of the oxide to metallic silver. With respect to carbohydrate oxidation, several factors are important including hydroxide concentration, temperature and potential. The long term stability has been considered and is analyte dependent. Many carbohydrates (i.e. glucose, fructose, galactose) do not cause any observable dimunition while some analytes (cytidine, glycerol) cause a noticable loss of sensitivity. In all cases the sensitivity can be restored by electrochemical treatment in which the oxide is reduced and regenerated.

Introduction

The detection of carbohydrates in chromatographic effluent has traditionally been a difficult problem due to the lack of a suitable chromophore or electroactive functionality.[1] As a result, detection has typically been based on derivatization to form a chromophore, refractive index or UV detection at low wavelengths. Each of these approaches has inherent drawbacks which renders them less than satisfactory where sensitivity is of concern. If one considers the thermodynamics of carbohydrate oxidation, one would predict they are relatively easy to oxidize. In fact, the direct heterogeneous oxidation at conventional electrode materials (i.e. carbon, platinum or gold) does not occur due to kinetic constraints.

One approach to facilitating carbohydrate oxidation is to introduce into the electro-chemical system a charge transfer mediator. The mediator oxidizes the carbohydrate and then the reduced form of the mediator is oxidized through direct electronation to regener-ate the mediator. Several cell arrangements are possible, but a surface bound mediator is most desirable for detection purposes. The analytical significance of this electroanalytic approach to redox chemistry is that the current required to re-electrolyze the surface mediator is directly related to the concentration of analyte.

Contemporary Electroanalytical Chemistry, Edited by A. Ivaska *et al.*
Plenum Press, New York, 1990

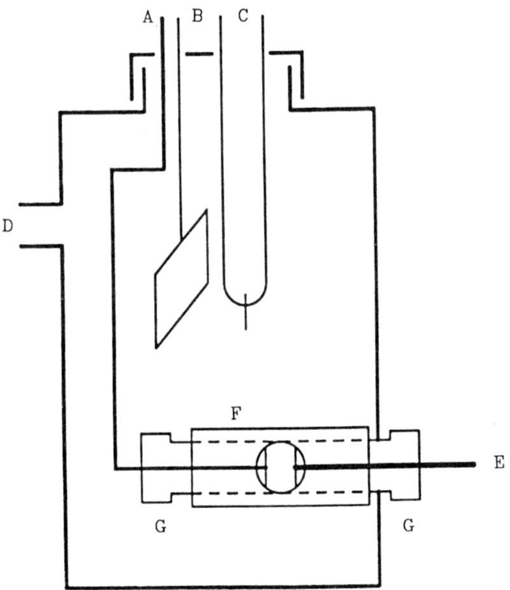

Fig.1. Electrochemical cell for electrocatalytic detection based on silver oxide electrode. A: Inlet tubing from flow injection system or HPLC, B: Platinum counter electrode, C: SCE, D: Overflow for polyethylene reservoir containing electrode assembly, E: Silver rod sealed in polyethylene fitting (2 mm diameter), F: Polyethylene union (Upchurch P-603), 6 mm diameter opening in two sides, G: Polyethylene fittings (Upchurch P-200).

Several metal oxides (platinum,[2,3] gold,[4] nickel,[5] copper,[6]) and cobalt phtalocyanine[7] have been employed as surface bound mediators for carbohydrate detection. In a dc amperometric mode of operation detectors based on these mediators exhibit a significant loss of response with time and/or exposure to analyte. Various potential pulse programs[2-4] have circumvented this stability problem, but at the expense of sensitivity and complexity of the instrumentation. Silver electrodes coated with electrogenerated silver oxide exhibit electrocatalytic activity with respect to carbohydrate oxidation. This paper describes our efforts to utilize an electrode as a carbohydrate detector in a dc amperometric mode.

Experimental Section

Reagents and Solutions

All reagents used were reagent grade unless otherwise stated. All were used without further purification. The distilled-deionized water (17.7 M$\Omega\times$cm) used was obtained from a Nanopure still and 4-cartridge purification system. Acetonitrile and methanol were HPLC grade obtained from Fisher Scientific. Carbonate free water was prepared by boiling water for 15 minutes. Solutions of base prepared from this water were stored and used under helium.

Equipment

The flow injection system used was a Control Equipment Corp. model MCA-103. The potentiostat used in conjunction with the flow injection work was either a PARC model 174 or an IBM 225. HPLC was performed on Waters Chromatography equipment consisting of model 510 pumps, model 680 gradient controller, model 460 electrochemical detector and U6K injector.

Two types of elctrochemical cells were used in this work. One was a thin layer geometry cell (from Water's model 460) equipped with a 3 mm diameter silver disk as working electrode.

276

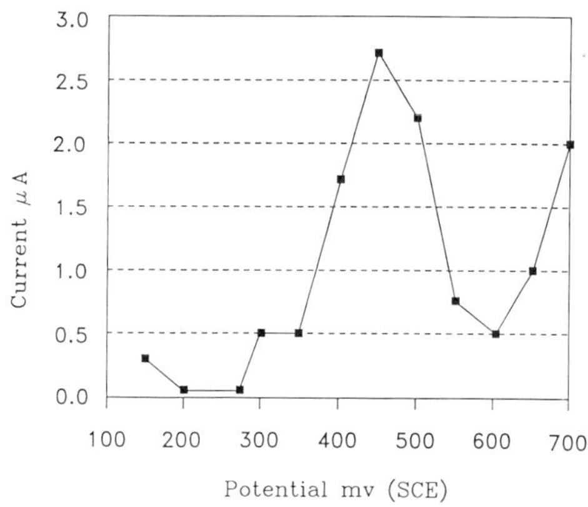

Fig.2. Hydrodynamic voltammogram of glucose. Constructed from the peak height of the flow injection response. The electrode was reconditioned via potential program (-200 to $+700$ to 450 mV, 15 minutes at each step) between each change of potential. Flow rate: 1 ml/min, injection volume: 65 μl, analyte: 1mM glucose in 1 M NaOH, carrier stream: 1 M NaOH.

The other was constructed from Upchurch P-200 series fittings as illustrated in Fig.1. The silver working electrode was a 2 mm diameter disk. The Water's cell contained a Ag/AgCl reference electrode, while the homemade cell used an SCE. Unless otherwise indicated all potentials are with respect to SCE.

Results and Discussion

Voltammetric Studies

In alkaline media silver exhibits 6 peaks (4 oxidations, 2 reductions) in the cyclic voltammetry experiment.[8] These have been assigned to oxidation and reduction of various silver oxides and hydroxides as shown in Table 1.

Table 1. Cyclic voltammetric data from silver in 0.4 M NaOH.

	E_p (mV)
$Ag^o \longrightarrow Ag(OH)_2{}^- + e^-$	175
$Ag^o \longrightarrow Ag(OH) + e^-$	260
$2Ag^o \longrightarrow Ag_2O + 2e^-$	385
$Ag_2O \longrightarrow 2AgO + 2e^-$	630
$2AgO + 2e^- \longrightarrow Ag_2O$	380
$Ag_2O + 2e^- \longrightarrow 2Ag^o$	75

Scan rate: 20 mV/s   Supporting electrolyte: 0.4 M NaOH

Hydrodynamic voltammograms for various carbohydrates have been obtained through the flow injection experiment. The result for glucose is shown in Fig.2. This experiment implicates the silver(I) oxide as the charge transfer mediator for the oxidation of glucose since the maximum response occurs at potentials where this species is the stable oxide. The peak shaped response is a typical of the hydrodynamic experiment, where a wave is expected for a mass transfer limited reaction. The loss of current at higher potentials corresponds with the formation of a silver(II) oxide surface. This peak shaped response presumably indicates that silver(II) is incapable of oxidizing glucose within the time scale of this experiment. Similar hydrodynamic voltammograms were obtained for the other carbohydrates examined.

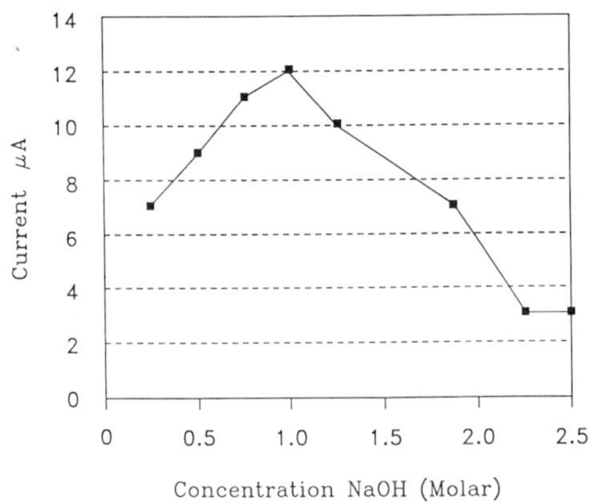

Fig.3. Dependency of glucose response on hydroxide concentration. Measured current represents peak height in flow injection experiment. Conditions same as in Fig.2 except: Detection potential: 450 mV, carrier stream and analyte prepared with the same base strength.

Not unexpectedly, the response and thus the rate of oxidation is a function of both temperature and hydroxide concentration. In previous work[9] it was shown that the rate of oxidation of glycerol at silver oxide electrodes increase with temperature and with base concentration up to about 2.5 M. Fig.3 shows similar results for glucose obtained in a flow injection experiment. In this case a maximum response was obtained at about 1 M hydroxide.

There seems to be no clear advantage to working at elevated temperatures. Even though the response increases with temperature, an increase in noise was also observed.

Response in a Flow Injection System

A number of carbohydrates and related compounds have been examined to ascertain whether they can be oxidized at these electrodes. These are summarized in Table 2. In addition the linearity of response as a function of concentration has been determined for a number of these compounds (Table 2).

The scope of response includes a variety of compounds all possessing multiple hydroxy functionalites. One notable exception is triolein. Response in this case is probably due to hydrolysis of this triglyceride in strong base to form the active glycerol. Simple alcohols (methanol) yield no observable signal. This is significant since the latter is a common organic modifier in chromatographic separations. The detection of nucleic acid components is through their sugar moiety, since the base alone (adenine) is not oxidized. Further, ribose is active while deoxyribose is not. As a result, detection is selective for RNA components over DNA components. Detection of carbohydrates and nucleic acid constituents in the effluent of a chromatographic system has been achieved by using post-column addition of base.

Stability of Electrodes and Response

Both the physical stability of the electrodes and the stability of response towards a particular analyte have been considered. The former concern arises out of the mechanism of silver oxidation which includes the formation of a soluble hydroxy species. To measure any loss of electrode upon potential conditioning a coulometric experiment was conducted. A series of double-step chronocoulometry experiments were performed and the limiting charge for the oxidative and reductive steps were compared as a function of potential. A loss of electrode is observed during the initial formation of silver(I) oxide. From the geometrical area of the electrode it is estimated that a loss of about 0.1 μm occurs upon conditioning the surface at 450 mV. The practical consequence is that electrodes need to be repolished daily.

Of perhaps more significance to the practical operation of this detector is the stability of the response of analytes. This was evaluated in two ways. First a flow injection experiment was performed, repeatedly injecting the same solution. Peak heght was monitored as a

Table 2. Response of various compounds at a silver oxide based detector.

A) Response of various compounds relative to ribose.

Compound	Response	Compound	Response
CARBOHYDRATES			
Ribose	100.0	Sucrose	35.6
Glucose	74.0	Galactose	+
Xylose	+	Fructose	+
2-Deoxyribose	−		
NUCLEIC ACID CONSTITUENTS			
Adenine	−	2'-deoxyuridine	−
Adenosine	+	Guanosine	+
Cytidine	+	Uridine	+
AMP, ADP, ATP	+	GMP, GDP, GTP	+
MISC.			
Inositol	+	Gentamicin	+
Inositol Phosphate	+	Glycerol	148.7
Methanol	−	1,2-Pentanediol	−
cis-1,2-Cyclohexanediol	−	trans-1,2-Cyclohexanediol	−
Triolein	+	1,2-Hexanediol	−

Response is relative to Ribose=100 and represents the relative peak height in the flow injection experiment, normalized for concentration. Flow Rate: 1 ml/min; Carrier Stream: 0.1 M NaOH; Potential: 450 mV(SCE); Spl. Volume: 60 μl. − indicates less than 2 % of the ribose response, while + indicates significant response, but as yet undetermined relative to ribose.

B) Calibration data from flow injection experiment.

Compound	Slope (nA/μM)	Intercept (nA)	r^2	n
Galactose*	5.50 ± 0.06	8 ± 3	0.9991	9
Xylose*	4.68 ± 0.06	n.s.	0.998	10
Ribose	30 ± 1	n.s.	0.995	5
Cytidine	1.41 ± 0.09	16 ± 6	0.981	5
Guanosie	5.5 ± 0.1	13 ± 5	0.997	9
UMP	0.69 ± 0.05	−10 ± 5	0.98	5
Inositol*	9.9 ± 0.2	56 ± 8	0.998	8

Flow injection conditions were same as in Part A. Concentration range was 1-100 μM.

* Water's cell used for determination.

n.s.: not significant (T-test).

function of time and number of injection. When the electrodes were conditioned by stepping from −200 mV to 450 mV(SCE) the response of all analytes examined decreased rapidly and eventually reached a fairly constant level (Fig.4A). In contrast, when electrodes were conditioned by stepping from −200 mV to +700 mV and then to +450 mV the response of many analytes although initially lower was constant over a long period of time (several hours) and was higher than the level reached by the other method of conditioning (Fig.4B). Some analytes (notably cytidine and glycerol) still exhibit a dimunition of signal with exposure of the electrode to analyte. In the case of cytidine the loss of response is a reversible process. If sufficient time is allowed between injections the response is stable and near its initial level (Fig.5).

Alternatively, solutions containing either just base or base and analyte were pumped continuously by the electrode. Measuring current continuously and switching via a valve from one stream to the other allowed continuous monitoring of the response vs. time. Results from

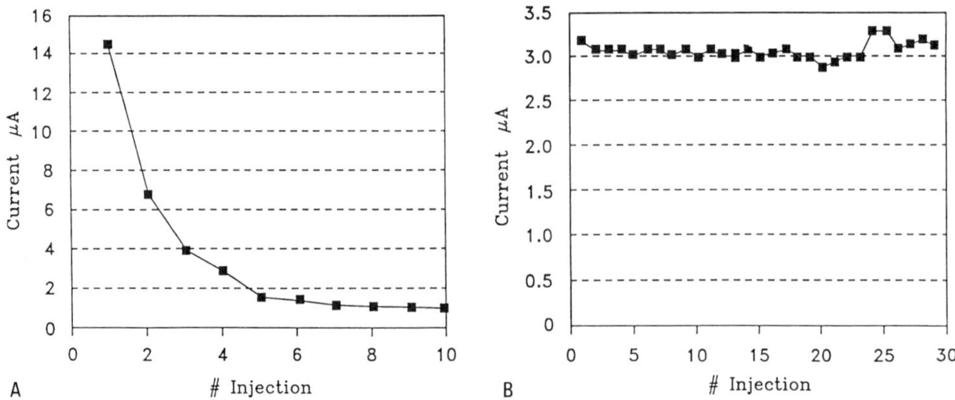

Fig.4. Dependency of response on potential conditioning. Current was obtained from peak heght in flow injection experiment. Analyte: 100μM Adenosine in 1 M NaOH, carrier stream: 1 M NaOH, other flow injection conditions same as in Fig.3. **A.** Conditioning from −200 mV to +450 mV. **B.** Conditioning from −200 mV to +700 mV to +450 mV. (Note: This represents a time period of about 2 hours.)

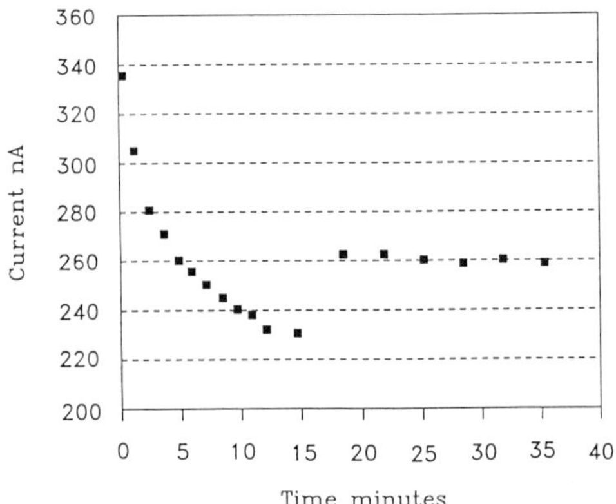

Fig.5. Flow Injection Response of Cytidine. Flow injection conditions same as in Fig.4B except: Analyte: 100μM cytidine in 1 M NaOH.

this experiment are consistent with the flow injection approach. All carbohydrates examined exhibited a fairly constant response, while cytidine response diminished with time. The initial cytidine response could be recovered if base alone was pumped through the cell for 3-4 minutes.

Functional Group Requirements for Detection

While we have not completed any bulk electro-synthetic reactions to ascertain what comprises the products of the oxidative process, we have arrived at some conclusions regarding the requisite structural features for electroactivity. Although the carbohydrates will exist in their open form under the basic conditions of the analysis, we could eliminate the requirement of a carbonyl function due to the activity of some of the alcohols such as glycerol and inositol. The requirement for a polyhydroxy functionality is evident form the inactivity of methanol and the 1,2-dihydroxy pentanes, hexanes and cis- and trans-cyclohexanes. The exception here would appear to be 2-deoxyribose, which is inactive, albite a polyhydroxy compound. Indeed, it would appear that regardless of what oxidative process is occuring, there is a requirement for three hydroxy groups in a 1,2,3-trihydroxy configuration. This is consistent with the results with the alcohols and the data concerning the 2-deoxyribose, which has three hydroxy groups but not in a 1,2,3-trihydroxy arrangement. The sole exception to this rule of thumb is gentamicin. This trisaccaride is composed of 2-deoxystreptamine, purpurosamine and garosamine, the latter two are amino sugars where the amine occupies one of the 1,2,3-trihydroxy positions.

Without an accurate determination of the products of the electrolytic process, we cannot state definitively the structural features of a molecule required for electrodetection. However, we have expanded the work into the area of nucleotides, nucleosides and bases, and predicted that the bases would be inactive due to their lack of a ribose moiety, while either purine or pyrimidine nucleosides and nucleotides would be electro-active. It was observed that adenine and uracil (a purine and pyrimidine base respectively), were not detected by this method, while the nucleotides and nucloesides were responsive. Due to the inactivity of 2-deoxyribose we predicted that the 2'-deoxynucleotides would not respond, and this lack of response was observed with 2'-deoxyuridine.

With respect to the purine and pyrimidine species, this detection methodology coupled with our previously reported data[10] using an amperometric detector with a Ag/AgCl electrode where we demonstrated that oxidation occurs on the purine ring, permits the selective monitoring of either purine nucleotides, nucleosides and bases, (Ag/AgCl), or the analysis of any ribonucleic acid species in the presence of the 2'-deoxyribonucleic acid species. These results suggest that this process will easily distinguish between ribonucleic acids (RNA), and 2'-deoxyribonucleic acids (DNA).

Acknowledgements

The authors wish to thank Mark DeBendetto and James DeMott for their technical assistance and useful discussions. We also acknowledge the financial assistance of the University of Lowell in the form of a Seed Money Grant.

References

1. T. Okada and T. Kuwamoto, Anal. Chem., **58** (1986) 1375.
2. S. Hughes and D.C. Johnson, Anal. Chim. Acta, **132** (1981) 11.
3. S. Hughes and D.C Johnson, Anal. Chim. Acta, **149** (1983) 1.
4. G.G. Neuburger and D.C. Johnson, Anal. Chem., **59** (1987) 150.
5. R.E. Reim and R.M. Van Effen, Anal. Chem., **58** (9186) 3203.
6. K.G. Schick and C.O. Huber, U.S. Patent 4, 183, 791 (9179).
7. L.M. Santos and R.P. Baldwin, Anal. Chem., **59** (1987) 1766.
8. J.M.M. Droog and F. Huisman, J. Electroanal. Chem., **115** (1980) 221.
9. D. Kyriacou and T.P. Tougas, J. Org. Chem., **52** (1987) 2318.
10. D.P. Malliaros, M.J. DeBenedetto, P.M. Guy, T.P. Tougas and E.G.E. Jahngen, Anal. Biochem., **169** (1988) 121.

APPLICATION OF MODIFIED ELECTRODES FOR ANALYSIS IN FLOWING SOLUTIONS

Gordon G. Wallace[1], Mary Meaney[2] and Malcolm R. Smyth[2]

1. Department of Chemistry, University of Wollongong
   Wollongong, N.S.W. 2500, Australia
2. School of Chemical Sciences, NIHE Dublin
   Glasnevin, Dublin 9, Ireland

Introduction

Analysis in flowing solutions, as performed in particular with high performance liquid chromatography (HPLC) and flow injection analysis, (FIA) has developed rapidly over the last decade and now plays an important function in most analytical laboratories throughout the world. There is little doubt, however, that even HPLC lacks the resolving power required to solve analytical problems in complex matrices with minimal sample preparation. Often, the resolving power of the detection method is called upon to assist in the solution of these problems. This is particularly true with electrochemical detection (ED) systems which offer a certain degree of selectivity based on differences in oxidation or reduction potentials of the species to be determined.[1] In recent years, the advent of chemically modified electrodes (CMEs) has provided a stimulus to further improve both the sensitivity and selectivity of ED systems used in HPLC and FIA.

Chemically Modified Electrodes in Quiescent Solution

Procedures suitable for the incorporation of chemical or biochemical substances or ion-exchange sites onto an electrode surface have been developed in various laboratories.[2-6] Such modified surfaces can then be used to perform a variety of functions. They have, for instance, been employed to preconcentrate analytes prior to voltammetric analysis and hence impove sensitivity. For example, dimethylglyoxime has been used to preconcentrate nickel[3] and EDTA to preconcentrate silver.[4] If this preconcentration can be achieved in an environment conducive to rapid electron transfer, then sensitivity will be enhanced even further. Alternatively electrocatalysts have been attached to electrode surfaces, and by speeding up what would otherwise be a sluggish electron transfer process, increased sensitivity has been attained.[5] However, this approach yields no added selectivity and a large background must usually be tolerated.

This is illustrated in Fig.1 for the determination of nitrite at a ruthenium polymer modified electrode where it can be seen that the oxidation response for nitrite is shifted to a less positive (and hence more analytically useful) potential coinciding with the Ru(II)/Ru(III) couple. Although there is an obvious improvement in the peak shape for nitrite oxidation, the response is set on top of a large background current.

Another approach has been to improve performance by covering conventional electrode surfaces with a protective membrane. By excluding species on the basis of size or charge, then increased selectivity has been obtained.[6] In some cases a protective membrane can indirectly enhance sensitivity by improving the electron transfer rate due to an absence of contaminants.

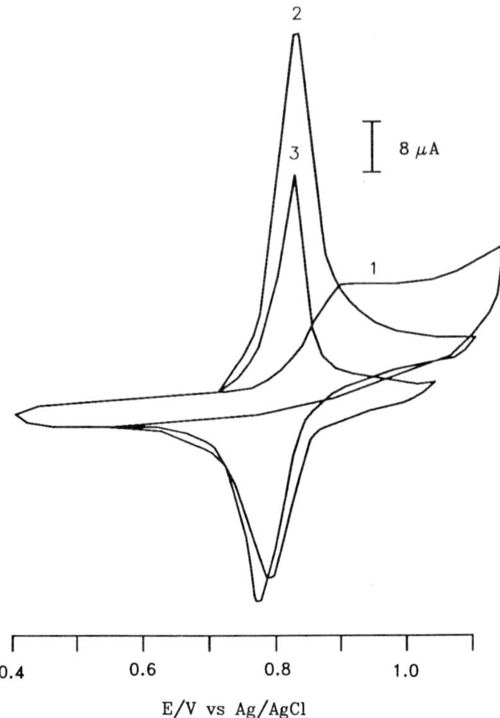

Fig.1. A comparison of cyclic voltammetric CV behaviour of nitrite obtained at modified and non-modified glassy carbon electrodes.
1. CV behaviour of 20 ppm nitrite at a bare glassy carbon electrode.
2. CV behaviour of 20 ppm nitrite at a $[Ru(bpy)_2(PVP)_5 Cl]Cl$ polymer modified glassy carbon electrode (details of preparation are given in reference 24).
3. Background response of $[Ru(bpy)_2(PVP)_5Cl]Cl$ polymer modified glassy carbon electrode.
Conditions: supporting electrolyte 1.0 M $NaNO_3$, scan rate 100 mV/s.

All of the above modes of operation serve to increase sensitivity, and to some degree selectivity. They do, however, have some disadvantages when compared to conventional electrodes. With CMEs, the electrode preparation stage is obviously more detailed and time-consuming.[2] Preparation of a conventional substrate is usually the first, and most important, step. For example, we have found that the analytical performance of polypyrrole coated on platinum substrate depends critically on the substrate preparation. Furthermore, the method of attaching the modifier, whether by chemisorption, covalent bonding or electropolymerisation,[2] must also be carefully controlled if reproducible electrode surfaces are to be obtained. There is no doubt, however, that these time consuming procedures can be tolerated if the electroanalytical chemist is provided with a surface which displays improved performance, reproducibility and stability over an extended time period.

In addition, there is no doubt that CMEs are more suspectible to matrix effects than conventional electrodes. This is quite understandable since the electrode surface is now, not only responsible for the electron transfer process, but also for the chemistry (e.g. complexation, ion-exchange, or a catalytic reaction). Obviously an ion-exchange process will be very dependent on the pH and the ionic strength of the sample, as will complexation or precipitation processes. Even more subtle, perhaps, is the susceptibility of modified electrodes to morphology changes in different matrices. Changes in surface area or porosity will influence analytical performance, and we have found that such changes do occur with chemically modified electrodes. If such morphology changes are irreversible then the problems are compounded.

Furthermore, most CMEs have a limited capacity/life time, and are overloaded easily, which influences performance and necessitates a regeneration process.

Applications of Modified Electrodes in HPLC and FIA

Electrochemical detection (ED) in high performance liquid chromatography (HPLC) and flow injection analysis (FIA) continues to be of increasing importance for the determination of trace quantities of compounds which are difficult to detect spectrophotometrically.

The electrode materials that have been used most in this regard are glassy carbon and carbon paste, which are most amenable for monitoring oxidation processes, e.g. of neutrotransmitters, phenols etc. The use of mercury-based detection systems have of course been widely used in quiescent solution for the determination of a wide variety of inorganic and organic species which undergo electrochemical reduction, but have seen only limited application in flowing streams, due to problems with cell design, mechanical stability and removal of oxygen from the eluants typically used in HPLC and FIA.

Various approaches have been used to achieve modification of the surface of carbon electrodes used in flowing streams. One approach has been based primarily on electrochemical pretreatment. Gregg[7] reported a method involving electrochemical pretreatment of the surface of a glassy carbon electrode at +1.32 V until a steady baseline was achieved, followed by disconnecting the reference electrode for two minutes. After re-equilibration, he reported that this pretreatment resulted in enhanced sensitivity for the HPLC determination of timolol and oxprenolol, but slightly decreased sensitivity for prenalterol. In a further publication, Gregg[8] also reported that preanodisation at 6.0 V for two minutes at a current ≤ 2 mA was required for the determination of oxprenolol in blood by HPLC-ED. Iwamoto et al.[9] also reported that the limit of detection for serum methionine by HPLC-ED could be lowered 360 times by holding the potential of the glassy carbon electrode at +1.9 V for two minutes before using this electrode at +1.7 V for detection purposes. In terms of stability of response, Wang and Peng[10] reported increased stability of glassy carbon electrodes for FIA determination of NADH and dopamine after polishing, rinsing, sonication and holding the potential constant at +1.5 V for thirty minutes. No significant decreases in response for these two compounds were noticed after 360 injections. Significant decreases were found, however, for both phenol and chlorophenol. The effect of polishing a glassy carbon electrode and subsequent performance as an electrochemical detector for HPLC has been studied by Hoogvliet et al.[11] They reported that extensive polishing can render a glassy carbon electrode more adsorptive and therefore more susceptible to blocking of the response. This can be overcome by polishing with alumina particles of size < 6 μm followed by 10–30 linear potential scans from 0.0 to +1.5 V. This was demonstrated to be important for the amperometric detection of epinephrine. Cathodic pretreatment has also been demonstrated to be important for the determination of amino acids (such as L-cystine and L-histidine) and proteins (such as albumins, globulins, ribonucleases and lysosomes) using Cu wire (1 mm diameter). Cathodic pretreatment resulted in the Cu electrode exhibiting rapid, reproducible and near Nernstian responses towards these biological compounds.[12] In addition, the application of a triangular wave from 0.0 to +1.8 V at 1 V/s for 30 s has been shown to be important for on-column electrochemical detection of various catechols using graphite fibres located inside the chromatographic column.[13]

Thus electrochemical pretreatment is seen to be an important means of modifying the surface of carbon electrodes prior to their use in electrochemical detectors for HPLC and FIA. This is mostly probably due to the formation of more surface oxygen-containing sites which are capable of promoting oxidation of various organic species.

A second approach to modification of electrode surfaces prior to their use in flowing streams has been to immobilize enzymes onto the surface. These techniques have long been used in the development of amperometric enzyme electrodes , but have only recently been applied to on-line analysis using (primarily) FIA. Some examples from the recent literature include the use of glucose oxidase immobilised onto either a Clark oxygen electrode[14] or nylon net[15] for the determination of glucose down to sub mM levels, and chymotrypsin covalently immobilised onto the surface of a IrO_2- coated Ti electrode for detection of N-benzoyl-L-tyrosine ethyl ester.[16]

A third approach has been to use chemically modified electrodes in flowing streams. Baldwin and co-workers[17,18] have reported on the use, for instance, of Co-phthalocyanine modified electrodes for the determination of mercaptopurines in plasma and oxalic acid and α-keto-[2-oxo-] acids in blood and urine using HPLC-ED, whereas Marko-Varga et al. used Meldola Blue (C.l. Basic Blue 6) on carbon rods to mediate the response of adsorbed. glucose dehydrogenase[19] The electrodes were mounted in a flow through cell for FIA. The reagent stream contained 2.5 ml NAD^+ in 0.1 M phosphate buffer, and the portion of NAD^+ reduced to NADH when glucose was injected into the system was re-oxidised amperometrically at 0.0 V. In this way, glucose could be determined between 5 μM to 2 mM concentration. In a recent publication, Kulesza et al.[20] have reported on the use of a mixed metal Ni-$Fe(CN)_6$ deposit on a vitreous carbon electrode for HPLC determination of Fe(III). The benefit of this approach is that there was no need to remove O_2 from the eluant.

Another early attempt using this approach was made by Wang and Hutchins[21] who coated a vitreous carbon electrode with cellulose acetate, then hydrolysed the film in OH^- to increase its porosity. By hydrolysing for different time periods, the authors could achieve films of different porosities. The coated electrodes showed high selectivity towards smaller analytes in both HPLC and FIA, and electrode poisoning due to protein adsorption was minimised. The method was applied to the determination of paracetamol in diluted urine. A reduction in surface poisoning by SCN^- in the FIA determination of NO_2^- at a vitreous carbon disk (7.5 mm^2) coated with poly-(4-vinyl-pyridine) incorporating $IrCl_6^{4-}$ as redox mediator was also reported by Cox and Kulkarni.[22] In this case, however, the response for NO_2^- was found to be halved when compared with a non-modified electrode surface, and the response was non-linear. The modified electrode, however, showed decreased interference from Pb(II), Mn(II) and Fe(II).

The use of polymer modified electrodes for the FIA determination of electroinactive anions has been reported by Ikariyama and Heineman.[23] This method is based on repetitive doping-undoping of the polypyrrole coating due to influx/regress of anions such as PO_4^{3-} and CO_3^{2-}, and could determine concentration of these ions between 10 μM to 1 mM. The coating was found to be stable for two weeks in an anaerobic atmosphere.

In these laboratories both chemisorbed and electrochemically generated polymers have been used in flowing streams. CMEs with electrocatalytic properties are most suited to operation under these conditions since the problems associated with high background levels and the lack of selectivity are alleviated. We have, for example, employed a ruthenium-containing polymer to enhance the detection of several analytes whose responses are known to be insensitive due to sluggish electron transfer.[24-26] By operating in a flowing system using the amperometric mode, the background current problems observed in a conventional stationary cell are alleviated, and the system is amenable to use with FIA following suitable stabilisation of the CME. Such an approach has also been used by Wang and co-workers[27] in the application of electrocatalytic CMEs.

References

1. P.T. Kissinger and W.R. Heineman, in: "Laboratory Techniques in Electroanalytical Chemistry", Marcel Dekker, New York, 1984.
2. G.G. Wallace, in: "Chemical Sensors", T.E. Edmonds (Ed.), Blackie and Son, Glasgow, 1988.
3. R.P. Baldwin, J.K. Christensen and L. Kryger, Anal. Chem., **58** (1986) 1790.
4. G.G. Wallace and L. Yuping, J. Electroanal. Chem., in press.
5. S.M. Geraty, D.W.M. Arrigan and J.G. Vos, in: "Electrochemistry, Sensors and Analysis", M.R. Smyth and J.G. Vos (Eds), Anal. Chem. Symposia Series, **25** Elsevier, Amsterdam, 1986, p. 303.
6. G. Batley, T.M. Florence and B. Hoyer, Anal. Chem., **59** (1987) 1608.
7. M.R. Gregg, Chromatographia, **20** (1985) 129.
8. M.R. Gregg, Chromatographia, **21** (1986) 705.
9. T. Iwamoto, M. Yoshiura and K. Iriyama, Bunseki Kagaku, **36** (1987) 98.
10. J. Wang and T. Peng, Anal. Chem., **58** (1986) 1787.

11. J.C. Hoogvliet, C.M.B. van der Beld, C.J. van der Poel and W.P. van Bennekom, J. Electroanal. Chem., **201** (1986) 11.
12. M.L. Hitchman and F.W.M. Nyasulu, J. Chem. Soc., Faraday Trans. I, **82** (1986) 1223.
13. R.L. St.Claire (III) and J.W. Jorgensen, J. Chromat. Sci., **23** (1985) 186.
14. L. Campanella, M. Tomasetti, B. Rappuoli and M.R. Bruni, Int. Clin. Prod. Rev., **4** (1985) 46.
15. G.J. Moody, G.S. Sanghera and J.D.R. Thomas, Analyst, **111** (1986) 665.
16. J.A. Osborn, A.M. Yacynych and D.C. Roberts, Anal. Chim. Acta, **183** (1986) 287.
17. M.K. Halbert and R.P. Baldwin, Anal. Chim. Acta, **187** (1986) 89.
18. L.M. Santos and R.P. Baldwin, J. Chromat., Biomed. Applic., **58** (1987) 161.
19. G. Marko-Varga, R. Appelqvist and L. Gorton, Anal. Chim. Acta, **179** (1986) 371.
20. P.J. Kulesza, K. Bratjer and E. Dabek-Zlotorzynska, Anal. Chem., **59** (1987) 2776.
21. J. Wang and L.D. Hutchins, Anal. Chem., **57** (1985) 1536.
22. J.A. Cox and K.R. Kulkarni, Analyst, **111** (1986) 1219.
23. Y. Ikariyama and W.R. Heineman, Anal. Chem., **58** (1986) 1803.
24. N. Barisci, G.G. Wallace, E.A. Wilke, M. Meaney, M.R. Smyth and J.G. Vos, Electroanalysis, in press, 1989.
25. G.G. Wallace, M. Meaney, M.R. Smyth and J.G. Vos, Electroanalysis, in press, 1989.
26. M. Meaney, J.G. Vos, M.R. Smyth and G.G. Wallace, Analytical Proceedings, **26** (1989) 15.
27. J. Wang, P. Tuzhi and T. Golden, Anal. Chem., **59** (1987) 740.

STUDIES OF THE MODULATED FLOW TECHNIQUE FOR FLOW POTENTIOMETRIC STRIPPING ANALYSIS

Gerhard Schulze, Raymond Sänger and Edzard Han

Institut für Anorganische und Analytische Chemie
Technische Universität Berlin
Straße des 17. Juni 135, D-1000 Berlin 12, F.R.G.

Electrochemical stripping analysis has become widely used for trace determinations of heavy metals. Anodic stripping voltammetry (ASV) is the most important stripping technique in environmental analysis. However, recently potentiometric stripping analysis (PSA) became an alternative to ASV.[1] PSA is a low cost system that doesn't need a complicated instrumentation. In PSA preconcentration of the respective element is achieved as in ASV by electrochemical deposition at a stationary electrode. Subsequently the element is redissolved in PSA by an oxidizing agent. The big advantage of PSA is that no interferences of oxygen[2] and less of nitro-organic compounds[3] are observed.

PSA in combination with a flow system has been investigated and a flow injection manifold[4] as well as a continuous flow configuration[5] have been described. The modulated flow technique was introduced for sensitivity enhancement of flow PSA.[6] The enhancement depends on the design of the flow cell . The stripping time t_s is given by:

$$t_s = K \times \left(\frac{\dot{V}_e}{\dot{V}_s}\right)^\alpha$$

with:

    K = constant
    $\dot{V}_e$ = volume flow rate during electrolysis
    $\dot{V}_s$ = volume flow rate during stripping
    α = exponent depending on cell design

A high volume flow rate during the electrolysis and a low flow rate during the stripping step lead to a sensitivity enhancement, which depends on the value of α as well. This paper describes a comparison of various types of cells which were tested for their suitability for the modulated flow technique.

Experimental

Apparatus and Reagents

Three flow cells and a rotating disc electrode in batch mode for comparison measurements were used. Two of the cells were wall-jet cells (WJC), type A and B, the third was a thin-layer cell. Type A was described earlier,[7] type B is shown in Fig.1a. The first is a cell with a small volume and small dispersion. The diameter of the inlet is 0.5 mm. Type B is able to accept working electrodes with diameters up to 7 mm. In this work two glassy-carbon working electrodes were used with diameters of 3 mm and 7 mm. The latter was drilled concavely (Fig.1b). Both electrodes were polished using diamond pastes down to 0.7 μm. Two different inlets with diameters of 0.5 mm and 0.05 mm, respectively, can be placed rectangular to the working electrode. The thin-layer cell (TLC) is made of PTFE and plexiglas (Fig.2).

Contemporary Electroanalytical Chemistry, Edited by A. Ivaska *et al.*
Plenum Press, New York, 1990

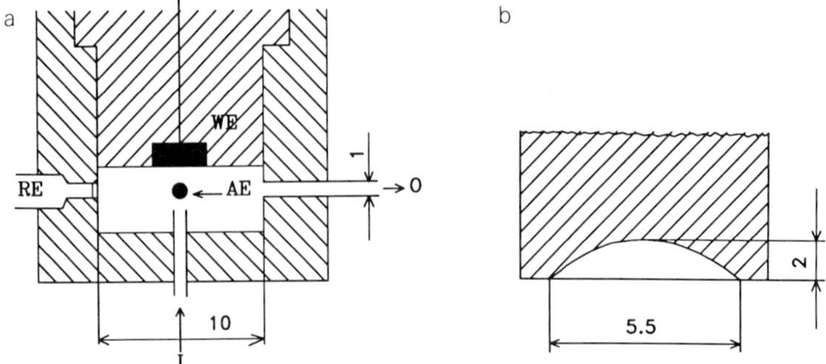

Fig.1. a) Wall-jet cell type B. WE: working electrode, AE: auxiliary electrode, RE: reference electrode, I: inlet, O: outlet.
b) Design of the concave electrode.

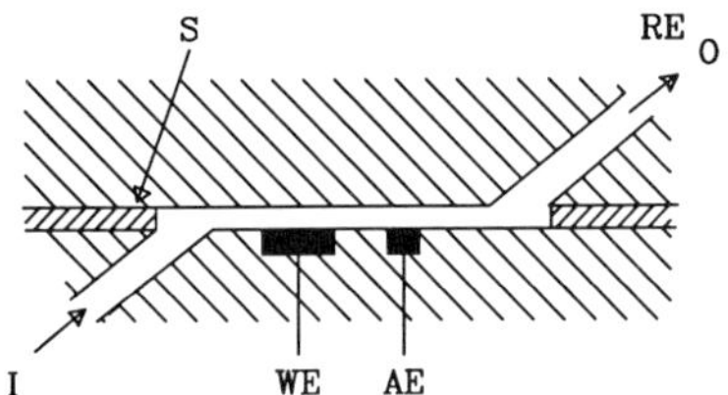

Fig.2. Thin-layer cell. WE: working electrode, AE: auxiliary electrode, RE: reference electrode, I: inlet, O: outlet, S: spacer.

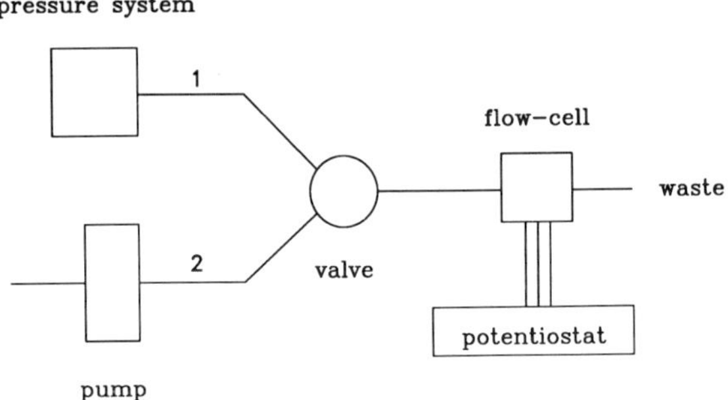

Fig.3. Experimental set-up.

The lower part made of PTFE is separated from the upper part made of plexiglas by a PTFE spacer (0.05 mm). From this spacer a part with a length of 12 mm and a width of 4 mm was cut out thus forming the cell. The working electrode and the auxiliary electrode were fitted in and the inlet was drilled into the PTFE block. The reference electrode was placed in the outlet which was drilled into the plexiglas. The rotating disc electrode (RDE) was the 663 VA Stand of Metrohm. In all cases the auxiliary electrode was made of glassy carbon, the reference electrode was a Ag/AgCl electrode.

The experimental set-up (Fig.3) includes a two-channel system. One channel (1) is equipped with a pressure system (FIAstar, Tecator) to adjust low flow rates and to avoid pulsation, in the other channel (2) a peristaltic pump is inserted which allows high flow rates. A valve is employed to switch between the two channels. For detection the cell is connected with a potentiostat (Striptec, Tecator). Ion stock solution containing 1 g/l lead and mercury, respectively, were prepared from Merck Titrisol. All other reagents were p.a. grade (Merck). All solutions were freshly prepared by dilution with double distilled water.

290

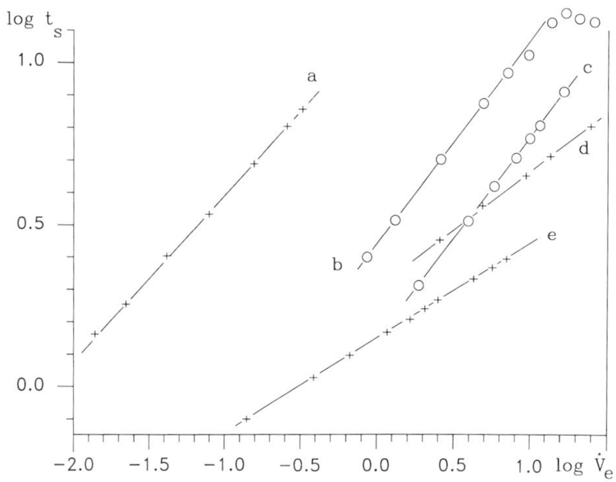

Fig.4. Determination of α-values. (a) WJC type B, 0.05-mm-nozzle; (b) WJC type B, 0.5-mm-nozzle; (c) WJC type A, 0.5-mm-nozzle; (d) WJC type B, 0.5-mm-nozzle, concave electrode; (e) Thin-layer cell.

Procedure

The investigations were performed by using lead as example. Thus the sample solutions contain $Pb(NO_3)_2$, $Hg(NO_3)_2$ and HNO_3. The working electrode was preplated with mercury by running sample solution through the cell for 2 minutes. The electrolysis potential was -1.0 V and the electrolysis time was set to 30 s. For modulated flow technique the valve was switched after 25 s from channel 1 with the flow rate $\dot{V}_1$ to channel 2 with the flow rate $\dot{V}_2$. Both channels contain the same solution. Using the rotating disc electrode the rotating rate was changed after 25 s.

Results and Discussion

Design of the Cell and of the Electrode

To compare the different flow cells the specific α-values were determined. A solution containing 1.5 mg/l Pb^{2+}, 10 mg/l Hg^{2+} and 0.1 M HNO_3 was used for electrolysis and stripping to obtain t_s in dependence on $\dot{V}_e$; $\dot{V}_s$ and t_e were kept constant at 0.1 ml/min and 30 s, respectively. The plots log t_s vs. log $\dot{V}_e$ gave $\alpha=0.52$, for the RDE (batch mode), $\alpha=0.28$ for the TLC and $\alpha=0.60$ for the WJC type A. For the WJC type B the values are $\alpha=0.50$ (0.05-mm-nozzle) and $\alpha=0.61$ (0.5-mm-nozzle). In addition, a new concavely shaped electrode was tested and resulted in $\alpha=0.35$ (Fig.4).

In the case of the RDE α-value measured is in good agreement with the theoretical prediction ($\alpha_{th}=0.50$). The deviations from the expected values for the other cells are about 15 % (TLC: $\alpha_{th} = 0.33$), 33 % (WJC type B with 0.05-mm-nozzle: $\alpha_{th} = 0.75$) and 20 % (WJC types A and B with 0.5-mm-nozzle: $\alpha_{th} = 0.75$).[8] Because of the missing theoretical basis for the concave electrode it is not possible to give a difference to the ideal α-value. To give an interpretation of the effects, it is useful to consider the flow pattern in the inner of the cell. Studies on the shape of the jet were made by pumping a dye solution through the flow-channel and observing the sequence of instants during the entry of the solution into the cell, made with WJC type A and 0.5-mm-nozzle and a high-speed camera which takes photos in intervals of 5/1000 s.[7] At a flow rate of 3 ml/min after a first deflection of the jet at the electrode a second deflection is observed nearby the vertical wall. Judging from the appearance of the interference by the wall it is possible to suggest an influence on the physical properties of the WJC (e.g. the α-value). A further hint is the result of Yamada and Matsuda,[9] who obtained $\alpha=0.73$ for their WJC. This seems to be very close to the optimum.

From this point of view a flat WJC of greater diameter should guarantee an undisturbed radial flow.

Calibration Plots

Calibration plots were drawn at a constant ratio of flow rates $\dot{V}_e$ / $\dot{V}_s$ and a constant $t_e=30$ s. The WJC type B was investigated in the range from 20 to 100 μg/l Pb^{2+} and from 100 to 1000 μg/l Pb^{2+}. The matrix is the same as described in previous experiments. For the ratio $\dot{V}_e/\dot{V}_s= 3$ ml min^{-1}/0.08 ml min^{-1} the cell shows a sixfold sensitivity enhancement compared with the non-modulated flow. This would have been expected from the arithmetical factor $(\dot{V}_e/\dot{V}_s)^{0.5} = (38)^{0.5} = 6.16$. In the case of the TLC the sensitivity increases by the factor 3 for $\dot{V}_e/\dot{V}_s= 4$ ml min^{-1}/0.08 ml min^{-1} at the same conditions. It must be emphasized, that the plots t_s vs. $c(Pb^{2+})$ are linear while there were obtained nonlinear diagrams for the stopped-flow technique ($\dot{V}_s \rightarrow 0$).[10] If the latter is due to an instationary concentration-profile of the oxidans there would exist one flow rate $\dot{V}_s$ which determines the transition between stationary and instationary states. Thus, in the modulated flow technique $\dot{V}_s$ has a lower limit. On the other hand, at higher flow rates (up to 20 ml min^{-1}) the signals decrease (see Fig.4). This is probably caused by a loss of amalgame on the electrode. Besides this, the hydrodynamic resistance increases at higher flow rates. Hence, it can be supposed an upper limit for $\dot{V}_e$. As conclusion it can be seen that the wall-jet cells equipped with an inlet of 0.5 mm are suitable in modulated flow technique because of the high α-values.

Acknowledgements

This work was supported by the Umweltbundesamt. We also thank W. Frenzel who made the TLC available and H. Rybczinsky for perfect realisation of the WJCs.

References

1. D. Jagner, Trends in Anal. Chem., **2** (1983) 53.
2. D. Jagner, Anal. Chem., **51** (1979) 342.
3. F. Wahdat and R. Neeb, Fresenius' Z. Anal. Chem., **316** (1983) 770.
4. A. Hu, R.E. Dessy and A. Graneli, Anal. Chem., **55** (1983) 320.
5. L. Anderson, D. Jagner and M. Josephson, Anal. Chem., **54** (1982) 1371.
6. G. Schulze, E. Han, O. Elsholz and W. Frenzel, Fresenius' Z. Anal. Chem., **327** (1987) 15.
7. G. Schulze, M. Koschany and O. Elsholz, Anal. Chim. Acta, **196** (1987) 153.
8. H.B. Hanekamp and H.J. van Nieuwkerk, Anal. Chim. Acta, **121** (1980) 13.
9. J. Yamada and H. Matsuda, J. Electroanal. Chem., **44** (1973) 189.
10. W. Frenzel, Thesis, Technische Universität Berlin, 1985.

ELECTROLUMINESCENCE DETECTOR FOR FLOW ANALYSIS

K. Haapakka, J. Kankare and K. Lipiäinen

Department of Chemistry, University of Turku
SF-20500 Turku, Finland

Introduction

Our previous studies have unambiguously pointed out that the cathodic electroluminescence generated at the rotating oxide-covered aluminum and tantalum electrodes in the presence of hydrogen peroxide and peroxodisulfate can be utilized for aqueous inorganic[1-3] and organic[4,5] analysis even at the nanomolar level.

Encouraged by these results, we have constructed for flow analysis an electroluminescence detector, which utilizes the oxide-covered aluminum and gold minigrid electrodes as the working (light generating) and counter electrodes, respectively. This communication presents a detailed structure of the detector and demonstrates the analytical applicability of the detector on the basis of the experiments carried out by using the 9,10-diphenylanthracene (9,10-DPA)-induced cathodic electroluminescence in the aqueous micellar solution[5] as the model electroluminescent system.

Experimental

Chemicals

Quartz-distilled water was used throughout the work. Sodium acetate (Suprapur) and potassium peroxodisulfate (pro analysi) were products of Merck and were used as received. Polyoxyethylene(23)dodecanol (Brij-35) and polycyclic aromatic hydrocarbons were purchased from Aldrich and used without further purification. A 0.010 M stock solution of 9,10-diphenyl-anthracene was prepared in ethanol and was daily diluted to an appropriate concentration in the eluent. Nitrogen was of the purest grade available.

Instrumentation

The main aim in designing the present flow cell was to construct a cell which measures the light emitted at the working electrode surface as efficiently as possible, and the design adapted is shown in detail in Fig.1. The working electrode imbedded in PTFE (C) is a 9.0-mm diameter aluminum rod (Alfa, 99.999 %) and its end was polished mechanically with 0.3 μm alumina suspension. The gold minigrid (Buckbee-Mears Company, 250 mesh, maximum transmittance 70 %) counter electrode (B) with the surface area of approximately 0.6 cm^2 was placed on top of the 3 mm thick and 25 mm diameter quartz window (A). The volume of the cell compartment could be easily changed by changing the distance of the aluminum electrode from the quartz window. The eluent is taken into the cell compartment through the stainless steel tubing (i.d. 0.25 mm, o.d. 1/16") and through a 0.5 mm diameter hole drilled in the center of the aluminum rod electrode. The eluent leaves the cell compartment via four symmetrically placed 0.3 mm diameter outlets. The cell body was made from PTFE and was painted black to prevent light leaks from reaching the photomultiplier. As shown in Fig.1, the cell body was placed onto the photomultiplier (Hamamatsu end-window type R-375) housing

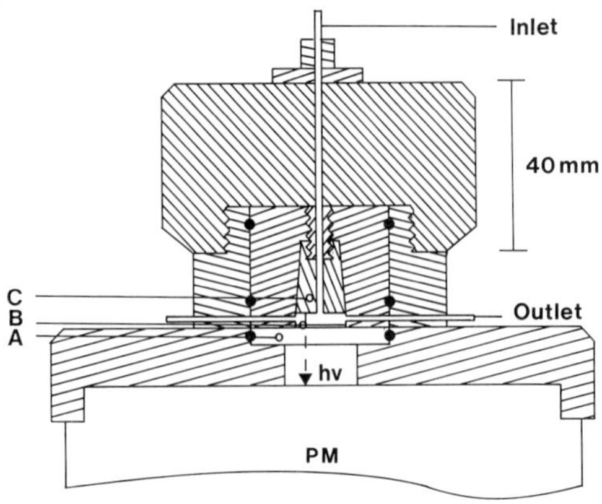

Fig.1. Electroluminescence detector for flow analysis. (A) quartz window, (B) gold minigrid electrode, (C) oxide-covered aluminum electrode.

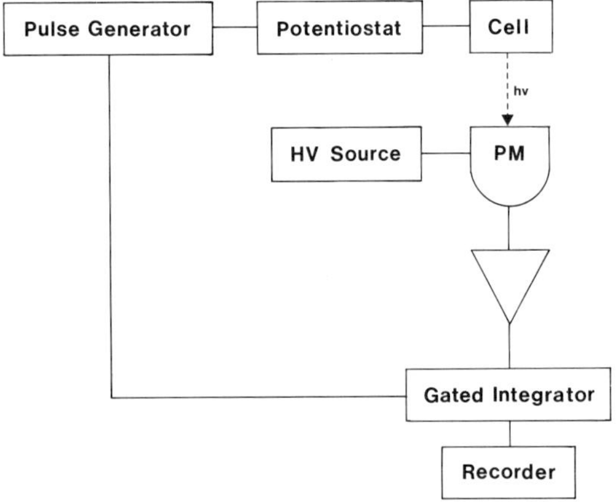

Fig.2. Circuitry for electroluminescence detector.

so that the distance between the working electrode surface and photomultiplier window is 10 mm, which enables efficient measurement of the light emitted at the electrode surface.

The block diagram of the electronic circuitry is presented in Fig.2: the cathodically pulsed potential (Fig.3a) for the electroluminescence generation was provided by a microprocessor-based pulse generator with a home-made potentiostat (single operational amplifier (741) with a transistor booster); photomultiplier output was amplified, demodulated by a simple gated integrator[6] and finally registered on a recorder.

The eluent flow was provided by an Altex pump (Model 110A) and the samples were introduced into the eluent flow by a Rheodyne valve (Model 7125) fitted with a 100 μl loop.

Results and Discussion

It is well known that polycyclic aromatic hydrocarbons (PAHs) can be solubilized in aqueous solutions by micelle forming surfactants.[7] As pointed out in our previous paper,[5] these aqueous micellar conditions can be utilized in generating PAH-specific light emissions by electrochemical means at the rotating oxide-covered aluminum electrode. Because the aqueous micellar solutions are also well suited for chromatographic analysis of PAHs,[8] the analytical feasibility experiments of the flow detector were carried out by utilizing the PAH-induced electroluminescence with Brij-35 and 9,10-diphenylanthracene (9,10-DPA) as the

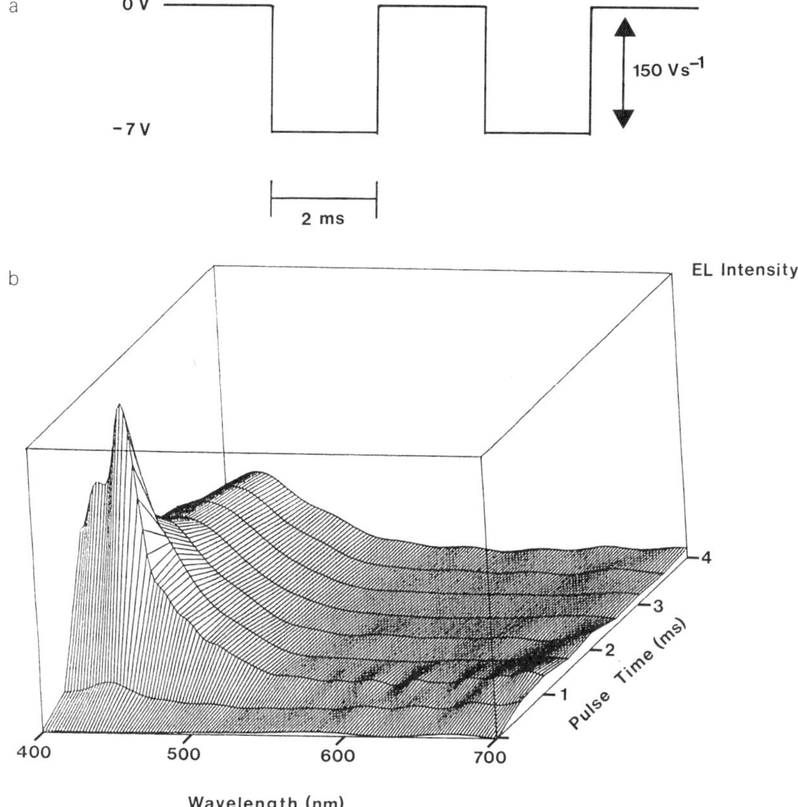

Fig.3. (a) Pulse potential for generation of cathodic electroluminescence, (b) 9,10-DPA-induced cathodic electroluminescence. Conditions:
sodium acetate 0.10 M adjusted to pH 8.0, Brij-35 1.0×10^{-2} M, peroxodisulfate 1.0×10^{-3} M, 9,10-DPA 1.0×10^{-5} M, solutions deaerated
with nitrogen, eluent flow rate 2.0 ml min^{-1}, excitation potential as in
Fig.3a, cell volume 5 μl.

representatives for micelle forming surfactants and PAHs, respectively.[5] Fig.3b is a time-
resolved display of this cathodic electroluminescence, the spectrum of which is quite similar
to the corresponding fluorescence spectrum.

On the basis of Fig.4 the high amplitude of the cathodically pulsed potential (Fig.3a)
is needed to initiate an intense 9,10-DPA-induced electroluminescence and because of the
resulting high current density, e.g., approximately 30 mA cm^{-2} at pulse amplitude -7.0 V,
the vigorous gas evolution was generated in the cell compartment and made the detector
response irreproducibly. This drawback could be partially solved by fast pulse amplitude
cycling between 0 V and -7.0 V, and as optimized in respect to the detection limit and
stability of the detector response the most beneficial cycling rate proved to be 150 V s^{-1}.
Besides, high eluent flow rates, e.g., 2 ml min^{-1}, removed gas bubbles more efficiently from
the cell compartment and thus to some extent improved the reproducibility of the detector
response.

On the ground of the experiments carried out to elucidate the effect of cell volume
on the detector response, the most intense electroluminescence signal is obtained when the
volume is as small as possible. Consequently, the further measurements were performed with
the cell volume of approximately 5 μl when the distance between the working and counter
electrodes was approximately 0.1 mm.

Optimization of the experimental conditions in respect to other variables gave the fol-

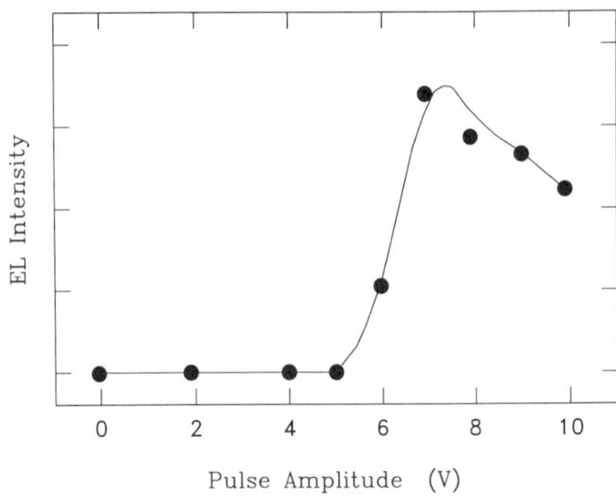

Fig.4. Effect of amplitude of pulse potential on the 9,10-DPA-induced electroluminescence. Conditions: as in Fig.3b.

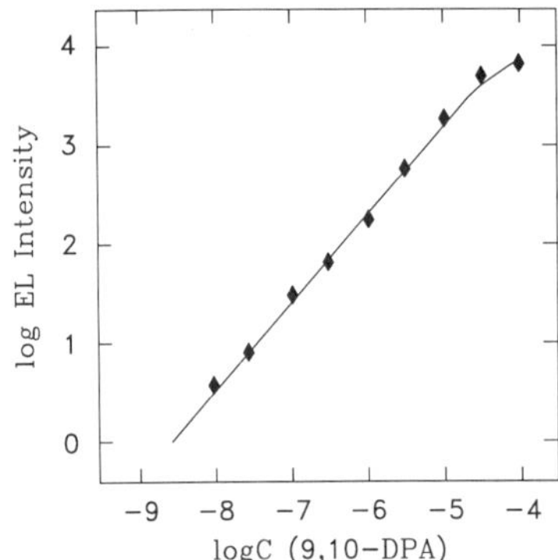

Fig.5. Calibration graph for the determination of 9,10-DPA. Conditions: as in Fig.3b.

lowing results. Firstly, the 9,10-DPA-induced electroluminescence was generated only in the aqueous acetate solutions where the concentration of Brij-35 exceeded its critical micelle concentration of 1×10^{-4} M[7] and the maximum electroluminescence intensity and the most reproducible detector response were obtained when the concentration of Brij-35 was 1×10^{-2} M. Secondly, as the case in the other cathodic electroluminescences at the oxide-covered aluminum electrode studied so far,[1,2,4] the present electroluminescence is also generated only at the uncharged electrode surface with the maximum intensity at the eluent pH of approximately 8. Finally, the peroxodisulfate concentration of 1×10^{-3} M produced the highest detector response and was used throughout the work.

As concluded from Fig.5 which presents the detector response in logarithmic scale as a function of 9,10-DPA concentration, this electroluminescence detector fulfills the demands of the usable detector: its detection limit is very low (approximately 10^{-8} M, 0.3 ng); it can be used over a wide concentration range (approximately from 10^{-8} M (0.3 ng) to 3×10^{-5} M (1000 ng)). Furthermore, it can be regarded as nearly universal for fluorescent compounds because out of twenty PAHs tested eighteen proved to be electroluminescent with their fluorescence-like spectra under the conditions stated in Fig.3.

Conclusions

The electroluminescence detector has been constructed and tested for use in flow analysis. The benefits of this flow detector can be summarized as follows: it is simple in construction and works linearly over the wide concentration range. One disadvantage is the inefficient removal of gas bubbles from the cell compartment makes the detector response somewhat irreproducible. Secondly, the oxide-covered aluminum electrode will be electrolytically dissolved to some extent during the long-term use making the cell volume larger and thus affecting unfavorably the precision of the detector response.

These results have, however, clearly indicated that the detector based on the electroluminescence principle is feasible in flow analysis, and the work to overcome the aforementioned disadvantages has already given promising results and will be published in the near future.

Acknowledgements

Financial support of the Academy of Finland is gratefully acknowledged.

References

1. K. Haapakka, J. Kankare and S. Kulmala, Anal. Chim. Acta, **171** (1985) 259.
2. K. Haapakka, J. Kankare and S. Kulmala, Anal. Chim. Acta, **211** (1988) 105.
3. K. Haapakka, J. Kankare and S. Kulmala, Anal. Chim. Acta, **208** (1988) 69.
4. K. Haapakka, J. Kankare and O. Puhakka, Anal. Chim. Acta, **207** (1988) 195.
5. K. Haapakka, J. Kankare and K. Lipiäinen, Anal. Chim. Acta, **215** (1988) 341.
6. K. Haapakka and J. Kankare, Anal. Chim. Acta, **138** (1982) 253.
7. L. Cline Love, J. Habarta and J. Dorsey, Anal. Chem., **56** (1984) 1132A.
8. R. Weinberger, P. Yarmchuk and L. Cline Love, Anal. Chem., **54** (1982) 1552.

DETERMINATION OF POLYAMINES (SPERMINE, SPERMIDINE, PUTRESCINE) IN BIOLOGICAL SAMPLES

R. Rips and C. Guette

Faculté de Pharmacie, Pharmacologia, INSERM
rue J.B. Clement
F 92296 Chatenay Malabry Cedex, France

Introduction

It has now been established that polyamines are involved in the process of cell growth and maturation.[1,2] The measurement of polyamines in pathotological tissue specimens can be useful in the diagnosis of cancers and in the assessment of anticancer treatments.[3,4] There is increasing evidence that polyamines, where they play an important role in the central nervous system, could be intracellular messengers in Ca^{2+} dependent release of neurotransmitters.[5] A "neurotransmitters" or "neuromodulators" role for spermine has also been discussed.[6] Relations are even more evident between polyamines and aminobutyric acid (GABA), and some agents which modify the metabolism of GABA also modify the concentration of putrescine.[7,8] This method, employing HPLC and electrochemical detection, has been developed to quantify polyamines in the different brain fractions. It will be useful initially, to determine physiological concentrations in small areas of mouse brain and later to correlate these concentrations with pharmacological effects of psychotropic drugs.

Material and Method

Chromatographic System

The HPLC system consisted of Merck-Hitachi L-6200 pump (Hitachi Ltd., Tokyo, Japan), a SEDEX 100 automatic autosampler (Touzart et Matignon, Vitry sur Seine, France), a Nucleosyl CN 5 μm column (15 × 0.4 cm), a BAS LC 4B (West Lafayette, Ind, U.S.A.) electrochemical detector fitted with a vitreous carbon working electrode and a Ag/AgCl reference electrode. The potential used was +0.7 V. The sensitivity was 1 to 50 nA and the output was 1 V. The mobile phase was 0.05 M KH_2PO_4 (Merck), 0.1 mM EDTA (Prolabo) in water (60 %) and acetonitrile (40 %). The pH was adjusted to 5.7 by addition of 2 N NaOH. The solution was filtered and degassed through 0.45 μm Millipore filter (San Francisco, Cal., U.S.A.) before use. Compouds used were Spermine tetrahydrochloride (Spm) (Sigma, St Louis, Mo., U.S.A.), Spermidine trihydrochloride (Spd) (Sigma, St Louis, Mo., U.S.A.), Putrescine dihydrochloride (Put) (Sigma, St Louis, Mo., U.S.A.), 3-3'diaminodipropylamine as internal standard (I.S.) (Aldrich, Strasbourg, France), Ortho-pthalaldehyde (OPA) (Janssen, Beerse, Belgium), Ethanethiol (ESH) (Merck-Schuchard, Darmstadt, Germany). Polyamines and internal standards were dissolved in water (2 mg/10 ml H_2O). Biological samples preparation was as follows. Male mice (26-30 g, 5 weeks old) were sacrified by decapitation between 9 and 10 a.m. to avoid circadian changes.[9,10] The brains were quickly removed from the skull and dissected on ice (cerebellum, midbrain, pons, medulla oblongata, frontal cortex, hypothalamus, thalamus, hippocampus, striatum and the remainder). All samples were stored at $-80°C$ until homogenization. Samples were homogenized in an Ultra-Turrax with 200 μl of 0.2 M $HClO_4$ and centrifuged at 8000 rpm ($4°C$) for 20 minutes. The protein content of the biological samples was measured by the Lowry method.[11]

Contemporary Electroanalytical Chemistry, Edited by A. Ivaska *et al.*
Plenum Press, New York, 1990

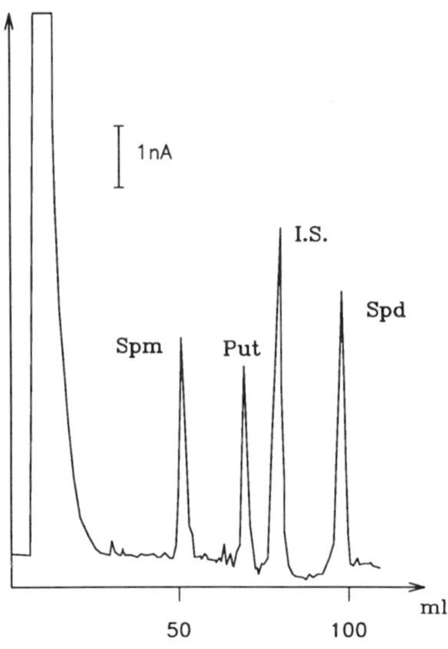

Fig.1. Isocratic separation of OPA-ethanethiol derivatives of polyamines with 40 % acetonitrile and 60 % buffer (KH$_2$PO$_4$ 0.05 M and EDTA 0.1 mM), pH 5.7. 30 pmole of Spm, 34 pmole of Put, 46 pmole of I.S. and 41 pmole of Spd; sensitivity 1nA.

Derivatization Reaction

The derivatization reagent was 10 mg OPA in 9 ml methanol and 1 ml of a 0.1 M potassium tetraborate buffer plus 50 μl of ethanethiol. 100 μl of 1/100 dilution of stock polyamine solution and 100 μl of OPA-thiol reagent were mixed together and 20 μl was injected after 5 minutes with a potential of +0.7 V and a sensitivity of 20nA. For biological samples, 50 μl of biological homogenates in 0.2 M HClO$_4$, 50 μl of internal standard (dilution 1/100), 100 μl of OPA-thiol reagent were mixed with 100μl of tetraborate buffer (0.1 M, pH = 9.5) for 5 minutes.

Results and Discussion

The aim of this study was to achieve separation of polyamines by isocratic reversed phase HPLC and electrochemical detection, and then to apply this technique to unpurified mouse brain samples. Polyamines are not electrochemically active, but the isoindole derivatives are and the optimal potential is +0.7 V.

Determination of the Chromatographic Conditions

The choice of the column and the composition of the mobile phase were studied to obtain the best compromise between resolution and analysis time. The Nucleosyl CN reversed phase column was used because C 18 and C 8 retention times were excessive. The eluant composed of acetonitrile (40 %) and a buffer (60 %) (KH$_2$PO$_4$ 0.05 M, EDTA 0.1 mM) gave the best separation of polyamines, when the pH was adjusted to 5.7. This pH is very important because changes in pH modify retention times of polyamines. Spermine and spermidine are the most affected by pH. In these conditions, the chromatograms showed satisfactory results and symmetrical peaks. The retention times were 40 minutes for spermine, 57 minutes for putrescine, 68 minutes for internal standard and 75 minutes for spermidine (Fig.1.). They were constant in repeated analyses. A good linear relationship (r = 0.99) existed between polyamine concentration and the peak height over the range 1 pmole to 10 nmole when the derivatization time was carefully controlled (5 min).

300

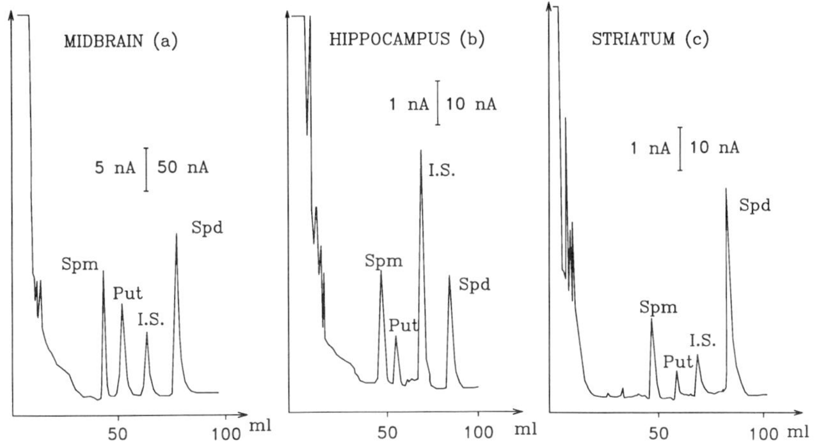

Fig.2. (a, b and c). Isocratic elution of polyamine derivatives in tissue homogenates. Eluant 40 % acetonitrile and 60 % buffer (KH_2PO_4 0.05 M and EDTA 0.1 mM), pH 5.7.

Estimation of Polyamines in Brain Samples

Samples were deproteinized with 0.2 M $HClO_4$.[12] Before reaction with OPA-thiol, a known amount of 3-3'-diaminodipropylamine was added to the tissue homogenates as an internal standard and perchloric acid extracts were adjusted to pH 9.5 by addition of borate buffer. For quantitation, polyamines were compared to the internal standard. The peak height ratios of polyamine to internal standard for a known concentration of polyamine was presented in Table 1.

Table 1. Peak height ratios of polyamine to internal standard.

Polyamines	Ratio
Spermine	0.9
Spermidine	0.94
Putrescine	0.81

In order to ascertain the non-interference of amino acids with peaks attributed to polyamines, brain samples were spiked with polyamines and amino acids.

Brain samples contain large amounts of amino acids which elute early in the chromatogram with the OPA-thiol reagent in excess. Fig.2 (a,b and c) illustrates the separation of polyamines in different brain samples.

Conclusions

A combination of HPLC on a reversed-phase CN column, prederivatized with OPA-ethanthiol and electrochemical detection has been used for direct assay of polyamines Spm, Spd and Put in deproteinated homogenates of mouse brain. This optimized system is rapid, simple and sensitive. It can be automated to provide a convenient assay for polyamines in, e.g., studies on the variations in polyamine levels in response to neuropharmacological agents.

References

1. S.I. Laitinen, P.H. Laitinen, O.A. Hietala, A.E. Pajunen and R.S. Piha, Neurochem. Res., **7** (12) (1982) 1477.
2. D. Chaudhury, I. Chadhury and M. Mukherjea, in: "Developmental Brain Research", **10** (1983) 143.

3. G. Scalabrino and M.E. Ferioli, Adv. Cancer Res., **36** (1982) 1.

4. J.M. Smit, H. Jurjens, J. Bouman, J. Bijzet, D.Th. Sleijfer, M.G. Woldring and N.H. Mulder, Eur. J. Nucl. Med., **10** (1981) 276.

5. R.J. Harman and G.C. Shaw, Br. J. Pharmacol., **73** (1) (1981) 165.

6. Z. Iqbal and H. Koenig, Biochem. Biophys. Res. Commun., **133** (2) (1985) 563.

7. N. Seiler and M.J. Al-Therib, Biochem. J., **29** (1974) 144.

8. N. Seiler, Physiol. Chem. Phy., **12** (5) (1980) 411.

9. L.D. Rodichok and A.H. Firedman, Life Sciences, **23** (1978) 2137.

10. J.P. Lapinjoki, O.A. Hietala, A.E.I. Pajunen and R.S. Piha, Neurochem. Res., **6** (1981) 377.

11. O.H. Lowry, N.J. Rosenburg, A.L. Farr and R.J. Randall, J. Biol. Chem., (1951) 193.

12. N. Seiler, in: "Liquid Chromatography Methods for Assaying Polyamides Using Prechromatographic Derivatization, Methods in Enzym", Vol. **94**, pp. 10-25, Acad. Press, 1983.

CLINICAL APPLICATIONS

OPPORTUNITIES AND LIMITATIONS IN THE USE OF ISEs IN CLINICAL CHEMISTRY: ASSAYS WITHOUT CALIBRATION?

Gudrun G. Rumpf, Lucas F.J. Dürselen, Hans W. Bühler and Wilhelm Simon

Department of Organic Chemistry
Swiss Federal Institute of Technology (ETH)
CH-8092 Zurich, Switzerland

Introduction

Among the ion-selective potentiometric sensors, the glass and the solvent polymeric membrane electrodes have become a routine tool in clinical chemistry and neurophysiological laboratories (for reviews see).[1-5] Up to now, carrier and ion-exchanger based ISEs for determing the activities of H_3O^+, Li^+, Na^+, K^+, Ca^{2+}, Mg^{2+}, Cl^-, HCO_3^- and CO_3^{2-} in extracellular environment (blood serum, whole blood, see[1-5]) are available. Membranes selective for Li^{+4} and $Mg^{2+6,7}$ suffer from interference due to Na^+ and Ca^{2+}, respectively, but chemometric procedures[8] allow an adequate assay of these ions by utilizing the Nicolsky – Eiseman equation:[9]

$$EMF \, [mV] = E_0 + s \, log \, [a_i + \sum_{j \neq i} K_{ij}^{Pot}(a_j)^{z_i/z_j}] \tag{1}$$

$$E_0 = E_D + E \tag{2}$$

$$s = \frac{2.303 \, RT}{z_i \times F} = 59.16 \, mV/z_i \, (25°C) \tag{3}$$

Where E, a constant potential difference, depends on the reference electrode and on the temperature (independent of sample solution); E_D is the liquid-junction potential difference generated between reference electrolyte and sample solution (by adequate choice of reference electrolyte often sufficiently independent of sample solution); R is the gas constant (8.314 $JK^{-1}mol^{-1}$); T the absolute temperature; F the Faraday equivalent (9.6487 $\times 10^4$ C mol^{-1}); z_i and z_j are the charge numbers of the primary ion i and the interfering ions j, respectively (in units of the proton chrage); a_i and a_j are the activities of the primary ion i and the interfering ions j, respectively, in the sample solution (mol/l); K_{ij}^{Pot} is the selectivity factor, a measure of the preference by the sensor for the interfering ion j relative to the ion i. For an ideally selective membrane electrode all K_{ij}^{Pot} values equal zero. Although the dimension of the selectivity factor is $(mol/l)^{1-z_i/z_j}$, K_{ij}^{Pot} is commonly referred to molarity concentrations (mol/l).

When using the ISEs mentioned above, a fundamental problem is the need for a very careful calibration, i.e., the determination of the sensor characteristics (E_o, s and K_{ij}^{Pot} in equation (1)) and the adaption of the electronic data processing system to changes in these parameters. Especially, the frequent recalibration of E_o is a must.[4,10-12] This is partly due to the small signal range corresponding to the variation of the activity of the ion to be sensed over the physiological range (Na^+ and Ca^{2+}, 2.4 mV each at 25°C).[4] A stability of about 0.12 mV

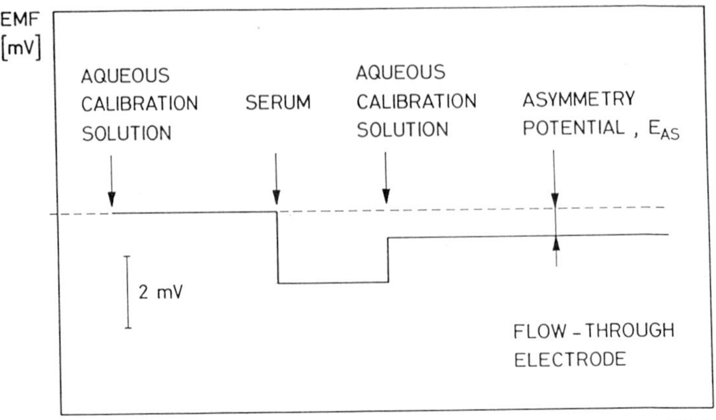

Fig.1. Schematic diagram of a potentiometric measurement in physiological samples. After serum contact the aqueous recalibration shows the asymmetry potential E_{AS}.

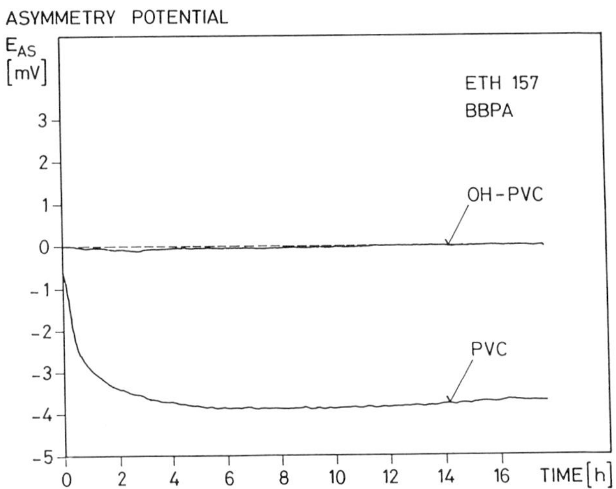

Fig.2. Asymmetry potentials of PVC and OH-PVC based Na^+-selective solvent polymeric membranes. One side of the membrane was exposed to 7 (wt)-% solutions of bovine serum albumin for 24 hours before EMF measurements in 0.1 M NaCl.

in the cell EMF between two subsequent recalibrations is therefore required for an adequate assay of Na^+ and Ca^{2+}.[4] The major reasons for the poor stability of ion-selective electrodes in clinical applications are shifts in the standard cell potential E_o. This effect is usually observed when the ion-selective membrane is brought into contact with protein-containing sample solutions after conditioning and calibrating with aqueous standard solutions.[10-12]

Recently,[10,13] we reported that these poorly reproducible E_o shifts were caused by proteins adsorbed on the membrane surface. The protein layer may induce a membrane asymmetry potential E_{AS} (see Fig.1) in inherently symmetric membranes.[10] The poor stability of E_{AS} is confirmed in Fig.2 for a conventional solvent polymeric membrane based on PVC. On the contrary, membranes prepared with vinyl chloride/vinyl alcohol copolymers (OH-PVC) do not suffer from such shifts[10] and are therefore attractive candidates to simplify the calibration of ISEs during clinical applications. Here we report on the possibility of ion assays without a need for sensor calibration.

Experimental

Reagents

All solutions for the potentiometric measurements were prepared with doubly quartz distilled water and salts of p.a. grade (E. Merck, Darmstadt; F.R.G.). The reference electrolyte solution consisted of 140 mM NaCl, 4.25 mM KCl, 0.625 mM $MgCl_2$ and 1.1 mM $CaCl_2$ and the sample solutions of varying NaCl concentrations (100, 120, 140, 160 and 180 mM, respectivley) with a constant background of 4.25 mM KCl, 0.625 mM $MgCl_2$ and 1.1 mM $CaCl_2$. For membrane preparation, poly(vinyl chloride) (PVC hochmolekular), the plasticizer BBPA (bis(1-butylpentyl) adipate), the ionophore ETH 2120 and tetrahydrofuran (THF, puriss, p.a., distilled before use) were obtained from Fluka AG, Buchs, Switzerland. The OH-PVC was prepared according to reference.[10]

Membrane Preparation

Conventional PVC membranes were prepare % according to reference.[14] The OH-PVC membranes were prepared by dissolving 1 % (by weight) ionophore, 33 % OH-PVC and 66 % plasticizer in THF by using ultrasonic bath. The solution was poured into a glass ring of 24 mm diameter resting on a glass plate covered with a fluoroethylene propylene foil which was then placed in a closed desiccator containing $\sim$250 ml of THF to give a saturated atmosphere. After 24 h the tap of the desiccator was slightly opened, and after further 24 h the lid of the desiccator was slightly opened ($\sim$0.5 cm). About 24 h later the glass plate with the membranes was removed and stored dust-free for another 24 h before use.

Membrane Conditioning

For conditioning, the membranes were mounted into the measuring cell (Fig.3), both half-cells being filled with the reference electrolyte solution described above. The conditioning time was 24 hours. Afterwards, both half-cells were washed with reference electrolyte solution before starting the EMF measurements.

Electrode System, Cell Assembly and EMF Measurements

All measurements were performed in the cell shown in Fig.3. Each half-cell contained about 15 ml of reference electrolyte and sample solution, respectively, which were not stirred. Two reference electrodes with free-flowing free-diffusion liquid junctions described earlier[14] were used, the bridge electrolyte being 3 M KCl. For the remote control, data storage and handling a personal computer Apple IIe was used with an IEEE 488 interface, an extended RAM memory (Apple Computer Inc., Cupertino, CA), a real-time clock (Thunderware Inc., Oakland, CA), a matrix printer RX 80 (Epson Corp., Nagano, Japan) and a Graphic Plotter Color Pro (Hewlett-Packard, San Diego, CA), managed by a high-level language program (UCSD Pascal 1.3). The measurements were performed at 20 $\pm$ 1 $^{\circ}$C. First the offset of the reference electrodes was obtained by immersing them into a glass beaker containing the reference electrolyte solution for 15 minutes. The measurements on the sample solutions were then performed utilizing a fresh, electrically symmetric membrane for each measurement. The EMF values of the 140 mM sample solution were taken at the beginning and end of each measuring series. The electrode response functions were calculated using a computer program correcting the raw EMF data for the contribution of the liquid-junction potential to the cell EMF.[14]

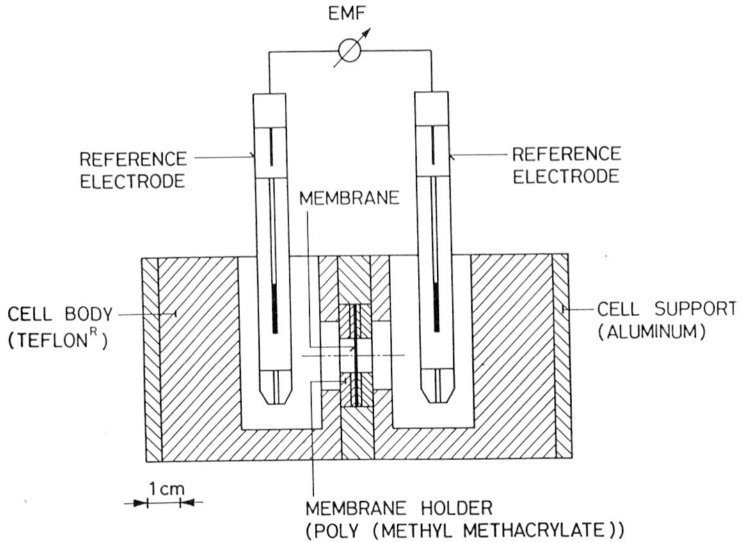

Fig.3. Cell used for the EMF measurements.

Results and Discussion

As described above, measurements with ISEs require adequately low interference from other ions (equation (1)), a constant standard cell potential E_o (equation (2)) and an acceptable slope s of the ion-selective electrode response (equation (3)).

For determing the activity a_i of various ions, membranes are available which exhibit sufficient selectivity to make corrections for interfering ions obsolete (equation (1)).[4]

On the other hand, a sufficiently constant E_o has to be guaranteed during the interval between two calibrations. Therefore, the membrane has preferably to fulfill two requirements. Firstly, the membrane preparation as well as membrane conditioning should be performed so that electrically symmetric membranes are obtained. Secondly, all asymmetry inducing effects (e.g. interference due to sample components) have to be avoided during measurements, the asymmetry thus being reduced to below 0.1 mV.[10]

By utilizing adequate membrane technology, ISEs with theoretical slopes of the electrode response (equation (3)) are obtained[14]. If fully symmetric cells with symmetric membranes ($E_{AS} = 0$) of the type shown in Fig.3 are used, the calibration procedure for measurement with ISEs could be drastically simplified.

With an electrolyte containing the ion i at an activity $a_i(r)$ (reference solution) in one compartment and the sample solution (activity of i: $a_i(s)$) in the other, the cell emf E_x is obtained from equations 1 to 3:

$$E_x\,[mV] = \Delta E_D + s \times log\frac{a_i(s)}{a_i(r)} \tag{4}$$

By adequately choosing the reference solution and bridge electrolyte the change in the liquid-junction potential ΔE_D due to a change from reference to sample solution can be kept sufficiently small.[14] If necessary, corrections can be applied.[14]

For aqueous electrolytes as sample (100 to 180 mM NaCl, 4.25 mM KCl, 0.625 mM $MgCl_2$, 1.10 mM $CaCl_2$) and an aqueous reference solution (140 mM NaCl, 4.25 mM KCl, 0.625 mM $MgCl_2$, 1.10 mM $CaCl_2$) the results in Table 1 were obtained by applying equation (4). In all the experiments, the slope of the electrode response is very close to the theoretical value (equation (3)). The standard deviation of the individual readings from the regression line is surprisingly small (average: 0.06 mV) and is well below the tolerated values of 0.12 mV for Na^+.[4] Since the induction of asymmetry in membranes due to proteins can be avoided or

Table 1. Assay of Na^+ with ISEs without cell calibration.

Membrane	Experimental slope of electrode response in % of s (eq (4)	Standard deviation of individual measurements from linear regression [mV]	Number of measurements
PVC	98.9	0.13	8
PVC	99.3	0.05	5
PVC	98.6	0.03	6
PVC	100.3	0.06	6
PVC	99.2	0.09	6
OH-PVC	99.7	0.02	6

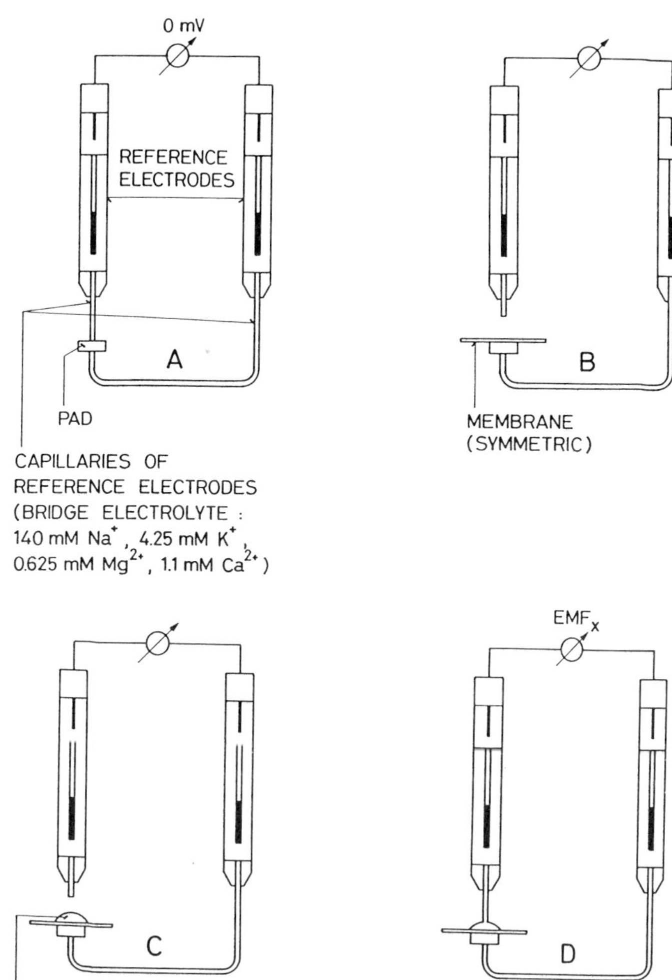

Fig.4. Design of an assay without sensor calibration: A: control of the reference electrode potentials, B: mounting of the electrically symmetric ion-selective membrane, C: application of the physiological sample, D: EMF measurement.

reduced when utilizing OH-PVC[10] instead of PVC (see Fig.2), we hope to successfully apply the same procedure to blood serum or whole blood samples, as outlined in Fig.4, A – D.

Acknowledgements

This work was partly supported by the Swiss National Science Foundation and by Eppendorf Gerätebau, Hamburg.

References

1. "Medical and Biological Applications of Electrochemical Devices", J. Koryta (Ed.), John Wiley & Sons, Chichester, New York, Brisbane, Toronto, 1980.
2. "Ion and Enzyme Electrodes in Biology and Medicine", M. Kessler, L.C. Clark, D.W. Lübber, I.A. Silver and W. Simon (Eds), Urban & Schawarzenberg, München, Berlin, Wien, 1976.
3. J.D. Czaban, Anal. Chem., **57** (1985) 345A.
4. U. Oesch, D. Ammann and W. Simon, Clin. Chem. (Winston – Salem N.C.), **32** (1986) 1448.
5. P. Nabet, Analusis, **15** (1987) 379.
6. M.V. Rouilly, M. Badertscher, E. Pretsch, G. Suter and W. Simon, Anal. Chem., **60** (1988) 2013.
7. M. Müller, M.V. Rouilly, B. Rusterholz, M. Maj-Zurawska, Z. Hu and W. Simon, Mikrochim. Acta, **III** (1988) 283.
8. G.J. Moody and J.D.R. Thomas, Ion-Selective Electrode Rev., **1** (1979) 3.
9. G.G. Guilbault, R.A. Durst, M.S. Frant, H. Freiser, E.H. Hansen, T.S. Light, E. Pungor, G. Rechnitz, N.M. Rice, T.J. Rohm, W. Simon and J.D.R. Thomas, Pure Appl. Chem., **48** (1976) 127.
10. L.F.J. Dürselen, D. Wegmann, K. May, U. Oesch and W. Simon, Anal. Chem., **60** (1988) 1455.
11. N. Fogh-Anderen, T.F. Christiansen, L. Komarmy, O. Siggaard-Andersen, Clin. Chem. (Winston – Salem, N.C.), **24** (1978) 1545.
12. U. Oesch, D. Ammann and W. Simon, in: "Methodology and Clinical Applications of Ion-Selective Electrodes", A.H.J. Maas, F.B.T.J. Boink, N-E.L. Saris, R. Sprokholt and P.D. Wimberley (Eds), Interprint A/S, Copenhagen, 1986, p. 273.
13. W. Morf, L.F.J. Dürselen and W. Simon, in: "Reviews on Analytical Chemistry", E. Roth (Ed.), les editions de physique, Les Ulis Cedex, France, 1988, p. 271.
14. R. Dohner, D. Wegmann, W.E. Morf and W. Simon, Anal. Chem., **58** (1986) 2585.

PROPOSED IFCC RECOMMENDATIONS FOR ELECTROLYTE MEASUREMENTS WITH ISEs IN CLINICAL CHEMISTRY

Anton H.J. Maas and Ron Sprokholt

Heart Lung Institute
University Hospital
Utrecht, The Netherlands

Problems of Clinical Chemical Application

Ion-selective electrodes (ISEs) make it possible to determine analytes, which cover such diverse substances as ions, dissolved gases, enzymes etc. in serum, plasma or whole blood.[1] Recently the direct potentiometric measurement of sodium, potassium and ionized calcium with ISEs became routinely available. Thermostated flow through electrodes and semi-automated calibration facilities have made the determination of sodium, potassium and ionized calcium as easy as modern blood gas analysis. ISEs for sodium and potassium ions may be used in two different ways, namely in undiluted and diluted samples. For this purpose the denomination <u>direct</u> and <u>indirect</u> potentiometry has been used although it in the fact is incorrect: in both cases the measurement is performed directly in different samples. When the sample has been <u>diluted</u> highly the total concentration expressed in mM serum or plasma should be measured at the end. This concentration is comparable with the concentrations measured for example with a flame photometer. The medium in which the ISEs are used is relatively constant concerning ionic strength, pH and further composition and thus indepenent of the sample. This means that the activity coefficient of the measuring ion and the mass concentration of the water in the dilute medium hardly differ from sample to sample (and calibration solution) resulting in that the measured active molality is proportional to the concentration and even in a given case equal to the concentration. Herewith it may be noticed, that some instrument should be calibrated not only with aqueous solutions but at the same time with a serum calibrator to compensate the possible matrix effects wich may appear in a (insufficiently) diluted sample. The problems of measurement in <u>undiluted</u> specimens with ISEs for sodium, potassium and also calcium ions in the clinical laboratories are:[2]

(a) What should be reported? History has largely determined that ions such as sodium and potassium are usually measured by methods on prediluted specimens using flame photometry. In principle the flame photometer measures the total substance concentration of the ion and the ISE measures the activity of the free ion in the water phase. Therefore differences between the results obtained with these two methods were observed by several authors. So we must decide whether we want to report activity or substance concentration in either the total volume of plasma (mol/l) or only in the water phase of plasma (mol/l); this means molality (mol/kg)?

(b) Differences in measuring technique. Different instruments on the market for the measurement of sodium, potassium and ionized calcium by ISEs on undiluted specimens give different results indicating a problem of measurement technique, i.e., calibration solutions, selectivity and sensitivity of the electrodes, measuring time, liquid junction potential and temperature.

(c) Differences in reference values. It may be clear, that above mentioned problems at the same time lead to differences in reference values, e.g., for ionized calcium, for which until now no reference method is available and that due to variable preanalytical factors sample handling should also be standardized.

European Working Group on Ion-Selective Electrodes

To consider the mentioned problems an European Working Group on Ion-Selective Electrodes (EWGISE) has been set up by The International Federation of Clinical Chemistry Scientific Division Committee on pH, Blood Gases and Electrolytes (IFCC/SD-CBGE) consisting of clinical chemists and manufacturers with a common interest in exchanging ideas and information on ISEs and trying to resolve some of the problems by the increasing use of this new technology. The group started in 1982 and has held meetings in Oslo (1983), Oxford (1984), Helsinki (1985), Graz (1986), Danvers (1987), in which European, American, Australian and Japanese invited scientists both from profession and industry participated. This year the EWGISE meeting will be held in Stresa (1988). This group proposes the following recommendations on sodium, potassium and ionized calcium measurements with ISEs in serum, plasma of whole blood in clinical practice.

Recommendation on the Expression of Results of Sodium and Potassium ISE Measurements[3]

Convention: Reporting Results of Ion-Selective Electrode Determinations of Sodium and Potassium

A convention is proposed, for routine clinical purposes:
1. Results of ion-selective electrode determinations of sodium and potassium in whole blood and undiluted plasma should be reported in terms of concentration (mmol/l).
2. Results of measurements on standard normal specimens should conform exactly with those obtained by flame photometer on the same specimens.
3. Standard plasma specimens are herein defined as having mass concentration of plasma water of 0.93 ± 0.005 kg/l, plasma bicarbonate concentration of 24 ± 2 mM, plasma pH of 7.40 ± 0.05 and concentrations of albumin, total protein, cholesterol and triglycerides within the reference range for healthy subjects.

By using this convention the results observed by ion-selective electrodes in normal plasma are equivalent to the substance concentration of the ion. However, in samples with <u>abnormal</u> plasma water mass concentration or with <u>abnormal</u> concentrations of complexes of the ion, results are numerically different from the true substance concentrations.

Standardisation: ISE versus Flame

For practical purposes, an <u>accurate</u> activity adjustment factor is not necessary as long as the main goal is achieved, i.e., values for normal plasma are identical with ISE and flame photometer. This may be achieved in three ways:

(a) Normal plasma sample method (single sample). The ISE system has to be calibrated on a concentration basis with at least two aqueous electrolyte solutions containing sodium chloride and potassium chloride (ionic strength of 0.16 M should be the same) spanning at least 5 mV thus obtaining a Nernstian slope. The advantage of achieving a Nernstian slope is that it enables us an easy assessment of the relative sensitivity which should generally not deviate more than plus two percent to minus five percent from the theoretical value in the physiological concentration ranges. The aqueous calibration solutions are then assigned an appropriate value to photometer result from normal plasma. This may be done by measuring the correlation for a number of normal plasma samples or by using a pooled specimen of normal plasma or serum.

(b) Pooled plasma sample method (few spiked samples). It has been suggested to calibrate the ISE with a standard serum pool and furthermore to obtain the slope with standard serum samples of which the concentration of sodium or potassium has been increased by adding known amounts of sodium chloride or potassium chloride and decreased by partial ultrafiltration.

(c) Patient plasma sample method (multi samples). It has been suggested that the slope should be obtained from a correlation study between ISE and flame photometry of 100 – 200 sera with electrolyte concentration covering the clinical range and fulfilling the normal conditions. Results are analysed by regression analysis after which the output of the ISE electrode analyser is adjusted giving the same results as the flame photometer.

It has been suggested that these methods of standardisation represent a short-term interim solution and that the use of a series of serum-based reference materials (generated by National Bureau of Standards of U.S.A., its European equivalent, or an internationally recognized Bureau) is a more practical and superior longer-term solution.

Recommendations on the Reference Method for the Determination of Ionized Calcium in Serum, Plasma or Whole Blood[4]

Convention: Reference Method

By the proposed reference method for the measurement of ionized calcium in serum, plasma or whole blood, the amount of substance concentration of ionized calcium in the water phase of plasma may be reliably determined on the basis of primary reference materials. These are aqueous solutions whose compositions are established by convention to contain known amount of substance concentrations of ionized calcium and which have a constant ionic strength of 0.160 mol/kg which value is commonly used for _normal_ plasma. The proposed IFCC reference method for ionized calcium measurement in plasma is based on the use of a cell consisting of an external reference electrode (R_1) with a concentrated potassium chloride liquid/liquid junction in combination with an ion-selective electrode with an inner reference electrode (R_2) according to the scheme:

$$\overbrace{R_1 \mid KCl\ (conc.) \parallel solution\ X}^{\text{Reference Electrode}} \mid \overbrace{M \mid inner\ reference\ sol. \mid R_2}^{\text{Ion Selective Electrode}}$$

Electrodes R_1 and R_2 are connected to a millivolt meter, e.g., ISE analyzer. Although the electrochemical cell responds to changes in the activity of calcium ions the cell is calibrated in terms of concentration by means of primary calibration soultions of which the composition is chosen such that the activity coefficient of the calcium ion is assumed to be identical both in the calibration solution and in normal plasma (see Table 1).

Convention: Reporting Results

Ideally ionized calcium measurements should be reported as molality (mol/kg) but in principal could also be reported as substance concentration (mol/l) or as activity. For normal plasma with substance concentrations of calcium of 2.5 mM we would find a substance concentration of ionized calcium of 1.25 mM and a relative millimolal calcium ion-activity of about 0.36. This indicates that we get entirely different values when we report activity instead of concentration and we should need new reference values. To avoid a proliferation of units, the convention is hereby adopted to report ionized calcium measurements as substance concentrations (mol/l). Reporting the results of ISE measurements in terms of substance concentration entails a bias when the ionic strength of the sample differs from that of the calibration solution, since ISEs response to activity rather than substance concentration. It is emphasized that with the present choice of calibration solutions the term ionized calcium concentration refers to ionized calcium in the plasma water and not in the entire volume of plasma. The concentration of ionized calcium in the plasma will be lowered by a factor of 0.933.

Instrumentation and Equipment

The sensor system comprises a calcium ion-selective electrode, a reference electrode and a liquid/liquid junction. The whole system is at 37°C. The associated measuring instrument provides the means of measuring the output from the sensor system. The calcium ion-selective electrode consists of an assembly in which a membrane, containing a calcium ion electroactive substance, encloses an inner reference solution held within a stem of plastic or glass. An internal reference electrode is immersed in the inner reference solution. As the electroactive substance either a charge carrier or a neutral carrier may be used. The requirements for the electrode, cell and high impedance mV-meter are specified in the same manner as described in the IFCC pH document.[5]

Calibration Solutions

Primary reference materials are issued by the NBS and the user is required to prepare the calibration solutions according to a prescribed procedure. All calibration solutions have an ionic strength of 160 mmol/kg, which actually refers to the ultrafiltered water phase of plasma, a pH between 7 and 8, and have the composition given in Table 1.

Table 1. Composition of the primary calibration solutions.

Solution No.	$c_{Ca^{2+}}$	$m_{Ca^{2+}}$	m_{Na^+}	m_{Cl^-}
1	1.250±0.006	1.266	156.25	158.78
2	0.250±0.001	0.253	159.25	159.76
3	2.500±0.01	2.526	152.50	157.55

c in mol/l; m in mol/kg; ionic strength 160.0±0.5 mmol/kg. pH should be adjusted between 7 and 8.

Solutions 1, 2 and 3, as given in Table 1, are required to establish the relative sensitivity of the calcium electrode and should be used to establish secondary calibration solutions. Solutions 2 and 3 differ by a decade. In principle the primary reference solutions should not contain buffer.

Specimen Collection

Whole blood is collected in a glass syringe (5 ml) containing a flat stainless steel plate for mixing. The syringe contains dry calcium titrated sodium heparin sufficient to achieve a final heparin concentration of 50 IU per ml of blood. The contents of the syringe has to be mixed after sampling and immediately before measuring. If it is certain that the final heparin concentration is below 15 IU/ml blood, the use of calcium titrated heparin is not obligatory.

Measurement Procedure

Concerning the measurement procedure, a typical protocol including the sample sequence of calibration and measurement has been prescribed.

Analytical Variability

The <u>inaccuracy</u> is affected by the following:

(a) The residual liquid/liquid junction potential included in the electric potential difference of the potential of the cell filled with sample minus the potential of the cell filled with the calibrator introduces an uncertainty in the activity of calcium ions. The magnitude of this uncertainty cannot be estimated using the Henderson equation in this system for plasma, for instance because this system does not fulfill the geometrical requirements of the equation and because of the high protein content.

(b) The ratio between the activity coefficient of the calcium ion in the calibrator and the activity coefficient of the calcium ion in the sample, the true value of which cannot be determined; the activity coefficient of the calcium ion in the calibrator depends on the ionic strength of the calibrator solution and the activity coefficient of the calicum ion of the sample varies with the ionic strength in the plasma sample. The effects of differences in ionic strength between samples and calibration solutions on the residual liquid/liquid junction potential (a) tends to be opposite in sign to their effects in (b).

An <u>imprecision</u> of 0.00 – 0.02 mmol/l should be obtained with the reference method when measuring the concentration of ionized calcium in the biological range.

The Use of the Reference Method

The reference method should be used to evaluate routine methods for measurement of ionized calcium in plasma. Routine methods are based on a cell similar to that of the reference method and are usually calibrated with secondary calibration solutions. These are aqueous buffer solutions to which values of ionized calcium concentration have been assigned by means of the routine instrument using the primary calibration solutions as reference material. Routine method often produce highly precise, but not necessary accurate ionized calcium data because of variations in electrode systems, liquid/liquid junctions, calibration and measurement procedures. In order to monitor the inaccuracy of routine ionized calcium measurements, it is necessary to employ <u>quality control solutions,</u> e.g., plasma samples for which the ionized calcium value have been established with the reference method.

Topics under Investigation by EWGISE

(a) A prototype measuring cell system for the reference method for the measurement of ionized calcium in serum, plasma and whole blood is under development and will be tested by an international group of scientists.

(b) Recommendations for optimal conditions for the collection and processing of blood specimens measurement both for routine purposes and to establish reference ranges are in preparation.[6]

(c) For users of ISEs who wish to express results of ion measurements in terms of activity, values of activity coefficients will be recommended in the near future providing calculated activity values of sodium, potassium and calcium in a number of specified aqueous solutions which are suitable for instrument calibration.

References

1. W. Oesch, D. Ammann and W. Simon, Clin. Chem., **32** (1986) 1448.
2. A.H.J. Maas, O. Siggaard-Andersen, H.F. Weisberg and W.G. Zijlstra, Clin. Chem., **31** (1985) 482.
3. A.B.T.J. Boink et al., Recommendations on the expression of results of ion-selective electrode measurement of sodium and potassium ion activities in undiluted serum, plasma or whole blood in clincal practice. In: "Methodology and Clinical Applications of Ion-Selective Electrodes", A.H.J. Maas et al., (Ed.), Vol. **8**, p. 137, Graz, 1987.
4. A.B.T.J. Boink et al., Reference method for the determination of the substance concentration of ionized calcium in serum, plasma or whole blood. In: "Methodology and Clinical Applications of Ion-Selective Electrodes", A.H.J. Maas et al., (Ed.), Vol. **8**, p. 281, Graz, 1987.
5. A.H.J. Maas et al., Clin Chim Acta, **165** (1987) 97 and J. Clin. Chem. Clln. Biochem., **25** (1987) 281.
6. A.B.T.J. Boink et al., Recommendations on blood sampling, transport and storage for the determination of the substance concentration of ionized calium. In: "Methodology and Clinical Applications of Ion-Selective Electrodes", A.H.J. Maas et al., (Ed.), Vol. **8**, p. 81, Graz, 1987.

INFLUENCE OF SOME DRUGS ON ISE MEASUREMENTS OF SERUM ELECTROLYTES

Ryszard Lewandowski, Tomasz Sokalski and Adam Hulanicki

Department of Chemistry
University of Warsaw
PL 02-093 Warsaw, Poland

Introduction

Since the early days of analytical applications of ion-selective electrodes they have been used in clinical analysis.[1-6] The progress in construction and miniaturisation of electrodes as well as contemporary development of computerised potentiometric apparatus have led to the production of automatic analyzers designed especially for clinical applications.[7-11] The high degree of response selectivity of the membrane sensors used today eliminates practically the mutual interaction of various blood, serum, plasma or urine components. Nevertheless when drugs or their metabolites are introduced into the body they in some cases may influence the electrode response and cause errors in estimation of the content of the ions present naturally, i.e., potassium, sodium, calcium and chloride. Such parasitic effects may be caused by the interaction of drugs with the electrode membrane. The aim of this study was to check whether some selected drugs can influence the determination of the above mentioned electrolyte ions in the serum.

Experimental

The measurements were performed using the commercial potentiometric analyzer "Microlyte" by KONE Corp. (Finland) and in batch conditions using the electrodes of the same membrane composition as those used in the "Microlyte" analyzer. In batch measurements with model aqueous solutions the potential was measured with PHM 64 research pH meter by Radiometer (Copenhagen). For measurements in serum the preparation "Seronor" produced by Nyegaard & Co, Diagnostic Division (Oslo) was used. All reagents used were of analytical grade and the drugs were products of Fluka or Sigma and used as received. Double distilled water was used throughout. Procedure of determination was the following: The drugs for experiments were dissolved in water, or in aqueous solution of 1 % ethanol. The added portions of drugs contained a 10 fold daily dose, or when an influence was observed a daily dose. Only in the case of lidocain a 7 fold daily dose was used. Liophylized serum was dissolved in such solution and pH was adjusted to 7.4 using small amounts of H_2SO_4 or Tris. The measurements were done within few hours in the presence of the added drugs.

Results

The mean concentration of electrolytes in the serum is indicated in Table 1, together with the repetetive results of determination by the "Microlyte" analyzer. The results of N sets of measurements were normalised using the Seronom specification values (C), and the bias for the total of measurements was calculated as recovery $\Delta = c/C \times 100$ %. For the total of normalized measurements the relative standard deviation was calculated and expressed in per cents, ncv. The data in Table 1 indicate that the values given by the producers of

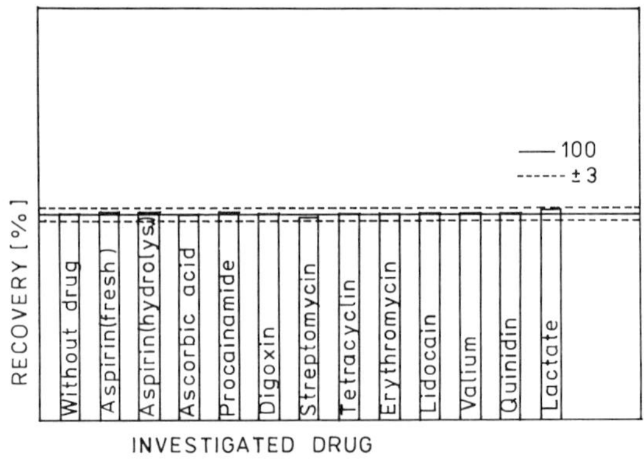

Fig.1. Influence of drugs upon K^+ electrode.

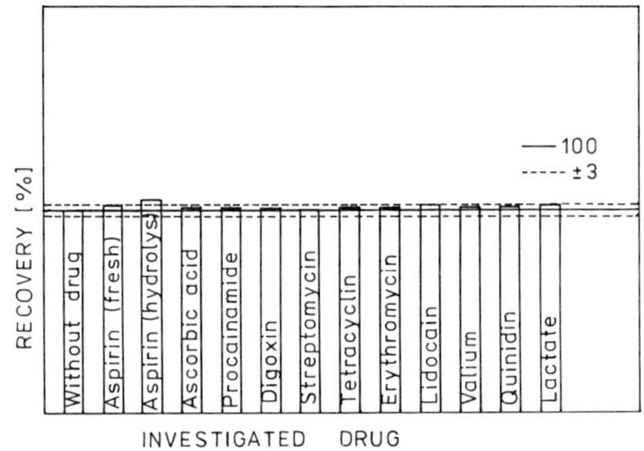

Fig.2. Influence of drugs upon Na^+ electrode.

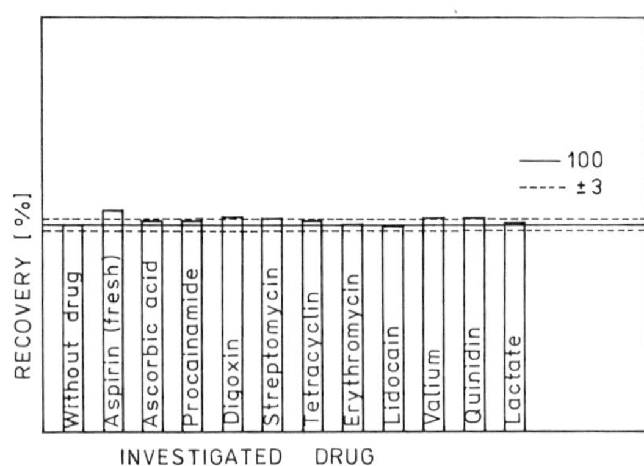

Fig.3. Influence of drugs upon Ca^{2+} electrode.

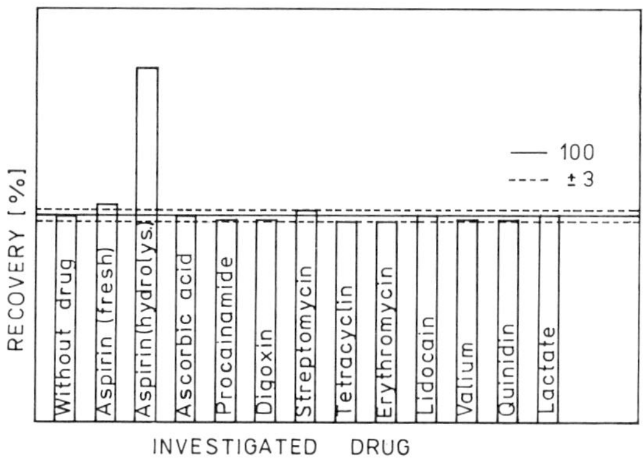

Fig.4. Influence of drugs upon Cl^- electrode.

Table 1. Mean concentration of K^+, Na^+, Ca^{2+} and Cl^- in Seronorm samples in absence of drugs at pH=7.4. n – number of samples, N – number of sets, c – mean concentration of given ion, ncv – relative standard deviation of the normalized content of the given ion in the Seronorm, $\Delta = c/C \times 100$, where C (recommended) concentration of given ion according to the Seronorm specification.

Ion	Certificate values		Results obtained with the analyzer				
	Range, mM	C, mM	n	N	c, mM	ncv	Δ
K^+	4.30 – 4.45	4.40	55	13	4.424	0.39	100.5
			47	10	4.507	0.25	102.4
Na^+	137.0 – 140.2	137.0	55	13	138.7	0.30	101.2
			47	10	139.1	0.22	101.5
Ca^{2+}	2.16 – 2.24	2.22	47	10	1.315	1.29	59.2
Cl^-	105.4 – 110.0	106.4	55	13	105.5	0.97	99.2

Table 2. Selectivity coefficient, $\log K^{Pot}_{Cl,Sal}$ for the electrode containing methyltridodecylammonium chloride.

Salicylate concentration, mM	Separate solution method	Mixed solution method
0.15	0.50	0.50
0.25	0.85	0.85
0.50	2.15	2.15
1.00	2.18	2.20
1.67	2.20	2.22
5.00	2.17	2.20
10.00	2.19	2.22

Seronorm differ from the results of measurements not more than 3 % for sodium, potassium and chlodide.

In the case of calcium such comparison was not possible as the certificate contains the total calcium content and not the ionized calcium. The effects of drugs on the potentiometric determination of serum electrolytes is shown on Figs 1 to 4. All measurements were carried

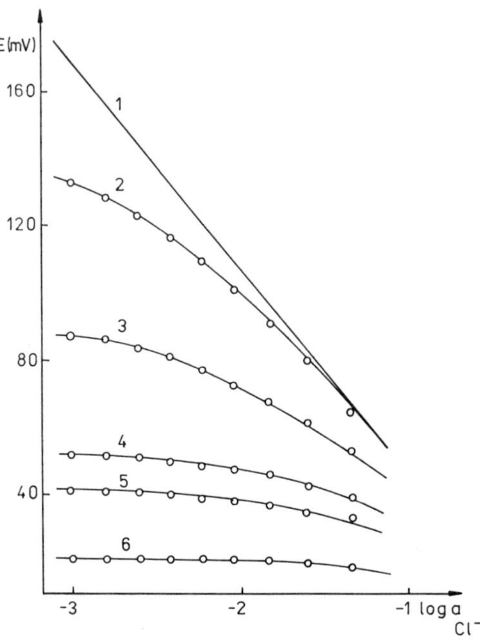

Fig.5. Calibration curves of chloride electrode in KCl solutions: without aspirin – 1, in KCl solutions containing 1.67 mM aspirin after 2 hours – 2, 1 day – 3, 3 days – 4, 1 week – 5, and 3 weeks – 6.

out in precisely the same conditions as for serum samples not containing additives. In determination of all cations the error was not larger than 3 %. Only in the case of chloride determination a large positive error was observed in the presence of aspirin. Its magnitude increased with time elapsing from the addition of aspirin solution. To get a clear picture of the aspirin interference , experiments were performed in model solutions. When the calibration curve for chloride electrode was obtained in the presence of 3.0 g/l of aspirin (1.67×10^{-2} M) at pH 7.4 depending of the time period between solution preparation and measurements the interference increases in a similar way as in the serum containing aspirin (Fig.5). Such behaviour results from hydrolysis of the acetylsalicylate ions at pH 7.4 with formation of salicylate and acetate ions. Similar effects were observed when the measurements were performed in solutions containing salicylate ions or equimolar solution of fully hydrolysed aspirin (Fig.6). Those measurements made it possible to evaluate the selectivity coefficient $K^{Pot}_{Cl,Sal}$ which in the Nernstian range equals $\log K^{Pot}_{Cl,Sal} = 2.2$ (Table 2). The selectivity coefficient for acetate ions ($K^{Pot}_{Cl,Ac}$) has been evaluated using the mixed solution method and equals 0.2 $\pm$ 0.06. This indicates that the influence of acetate on chloride determination in the physiological range in serum is not significant. Those experiments indicates that aspirin levels exceeding seriously the therapeutical doses may interfere the determination of chloride but this is not the case for amounts normally used.

Conclusions

The performed experiments indicate that ascorbic acid, procainamide, digoxine, streptomycin, tetracyclin, erythromycin, lidocain, valium, quinidin and lactate do not influence the determination of potassium, sodium, calcium and chloride in serum in the conditions existing in the KONE "Microlyte" analyzer. A positive error in chloride determination is observed when the dose of aspirine strongly exceeds the therapeutical dose.

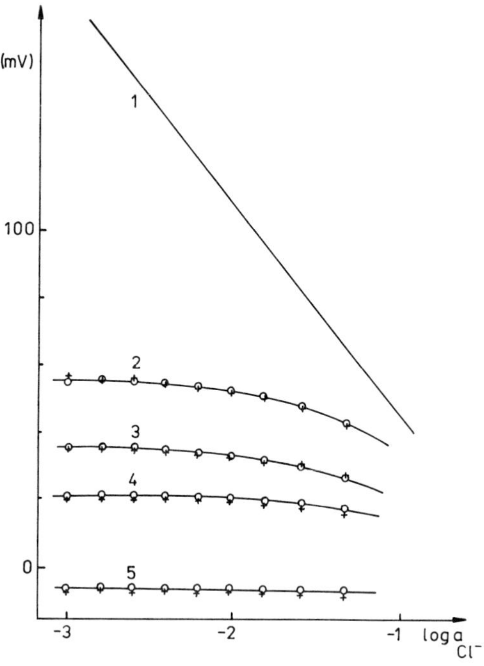

Fig.6. Calibration curves of chloride electrode in KCl solutions: without aspirin – 1, in KCl solutions containing respectively hydrolyzed aspirin (o) or salicytale ions (+) in concentrations: 0.5 mM – 2, 1.0 mM – 3, 1.67 mM – 4, and 5.0 mM – 5.

Acknowledgements

The authors are grateful to KONE Corp. for supplying analyzer "Microlyte" to the experimental work. The financial support of the project CPBP 01.17 is acknowledged.

References

1. J.H. Ladenson, Anal. Proc., **20** (1983) 554.
2. G.J. Moody and J.D.R. Thomas in: "Ion-Selective Electrodes in Analytical Chemistry", H. Freiser (Ed.), Vol. **1**, p. 339, Plenum Press, New York 1978.
3. M.A. Arnold and M.E. Meyerhoff, Anal. Chem., **56** (1984) 20R.
4. R.J. Cooke and R.L. Jensen, Clin. Chem., **29** (1983) 867.
5. M.E. Meyerhoff, Y.M. Fraticelli, J.A. Greenberg, J. Rosen, S.J. Parck and W.N. Opdycke, Clin. Chem., **29** (1982) 1973.
6. R.J. Cooke and R.L. Jensen, Clin. Chem., **29** (1983) 1563.
7. A. Sibbald, A.K. Covington and E.A. Cooper, Clin. Chem., **29** (1983) 405.
8. D.D. Koch, D. Parrish and J.H. Ladenson, Clin. Chem., **29** (1983) 1090.
9. A. Sibblad, A.K. Covington and E.A. Cooper, Clin. Chem., **30** (1984) 135.
10. N. Fogh-Andersen, P.D. Wimberley, J. Thode and O. Siggaard-Andersen, Clin. Chem., **30** (1984) 433.
11. J.H. Ladenson, J.J. Apple, J.J. Aguanno and D.D. Koch, Clin. Chem., **28** (1982) 1447.

BUFFER SYSTEM FOR THE SIMULTANEOUS STANDARDIZATION OF pH AND ELECTROLYTES BY ISE DETERMINATION IN WHOLE BLOOD

Angelo Manzoni and Mario Belluati

Research and Development Division
Instrumentation Laboratory S.p.A.
Viale Monza 338, 20128 Milano, Italy

Experimental, Results and Discussion

The traditional "buffer couple" (pH 7.384 and pH 6.840 at 37°C) for the standardization of the present pH/ISE analysers, obtained by making a suitable mixture of alkali metal phosphates, according to the NBS formula, is not suitable for the simultaneous standarization of sodium, potassium, calcium and chloride .

In this paper a study has been made of a series of buffers with a combined composition for electrolytes and pH, suitable for the simultaneous standardization of analyzers which use the potentiometric technique with ion-selective electrodes .

The selection criteria for the formulation of these standardization solutions were:
- buffer solutions whose pH was in the range of 6 – 8 units pH at 37°C.
- solutions with ionic strength as near as possible to that of whole blood (0.16 M).
- solutions which have a variation of the ions concerned, reflecting that detected in blood.
- solutions which, as far as possible, did not contain ions normally absent in blood.
- solutions prepared with substances having a high degree of purity and easily obtainable on the market.
- solutions whose difference in ionic concentration caused marked difference in terms of ΔE ($\Delta E > 8$ mV), still obliged to reflect the variation of same in blood.

The data furnished by papers was the starting point for their formulation (R.G. Bates, C. Vega "Standard for pH Measurement in Isotonic Saline Media of I.SE = 0.16", Analytical Chemistry, Vol. **50**, number 9, August, 1978).

Once prepared, the buffers were subjected to a series of tests regarding two aspects which interested us:
1) pH
2) electrolyte concentration (Na^+, K^+, etc.)

The pH of the buffers was measured at 37°C with a pH glass electrode as indicator electrode and a KCl saturated calomel reference electrode with open junction.

The titration was carried out with buffers obtained with NBS standard reference materials: pH 6.840 at 37°C, 0.025 molal in KH_2PO_4 and Na_2HPO_4, pH 7.384 at 37°C, 0.008695 molal KH_2PO_4 and 0.03043 Na_2HPO_4.

The titre of the sodium and potassium composition was carried out in emission flame photometry and for calcium in atomic absorption. By means of this method we determined the total quantity of sodium, potassium and calcium to be compared with the result obtained by the electrodes (ISE) both at 37°C and 25°C: the electrodes have been previously standardized with pure sodium, potassium and calcium chloride solutions.

Table 1 shows an example of the results obtained for pH 7.4 at 37°C, I.S. 0.16 with phosphate buffers according to the N.B.S formulation.

Table 2 shows an example of the results obtained for pH 6.8 at 37°C, I.S. 0.16 with a phosphate buffer according to the N.B.S. formulation.

Table 1

	Flame Photometry mM	ISE 25°C mM	ISE 37°C mM
Na^+	154	135	140
K^+	5.4	4.4	4.8

Table 2

	Flame Photometry mM	ISE 25°C mM	ISE 37°C mM
Na^+	100	85	90
K^+	8	6.1	6.3

Table 3

pH	$\Delta\%\frac{ISE}{F.F.}25°C$ Na^+	K^+	$\Delta\%\frac{ISE}{F.F.}37°C$ Na^+	K^+
7.4	88	82	91	88
6.8	85	76	90	79

Table 4. Data referring to the calcium electrode potential in solutions mentioned below at 25°C and at 37°C.

	Sol. 1	Sol. 2	Sol. 3
25°C	+42.3 mV	+42.5 mV	+42.6 mV
37°C	+42.8 mV	+42.8 mV	+42.8 mV

Sol. 1	1 mV TRIS/TRIS·HCl pH 7.2 at 37°C (7.5 at 25°C) 160 mM NaCl 1 mM $CaCl_2$
Sol. 2	10 mM TRIS/TRIS·HCl 152 mM $CaCl_2$ 1 mM $CaCl_2$
Sol. 3	100 mM TRIS/TRIS·HCl 80 mM CaCl 1 mM $CaCl_2$

The results obtained show that the calibration of the ion-selective electrodes for sodium and potassium with saline phosphate buffers at 0.16 ionic strength give a different measurement if compared with non-buffered solutions and provide inferior results in respect of the flame photometry or, anyway, of the stoichiometric quantity present in the solution.

Table 3 shows that this effect may be attributed to the contribution of the liquid junction potential deriving form the buffer obtained through the phosphate species but the differences denote that the sodium and potassium ions are affected by a phenomenon of ionic association with the phosphate buffer.

In any case, the entity of this effect is such as to require a detailed study of the question

HEPES $\quad$ HOC$_2$H$_4$N $\diagdown$ C$_2$H$_4$ / C$_2$H$_4$ $\diagup$ N—H$^+$ C$_2$H$_4$SO$_3^-$

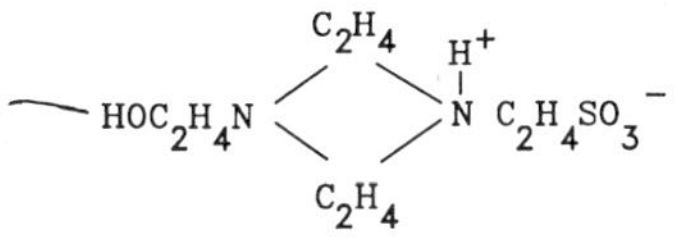

N (2 hydroxyethyl) piperazine – N ethanesulfonic acid $\qquad$ pKa 25°C = 7.565

TES $\qquad$ HOCH$_2$ / HOCH$_2$ / HOCH$_2$ $\Rrightarrow$ C —— N—H$^+$ C$_2$H$_4$SO$_3^-$

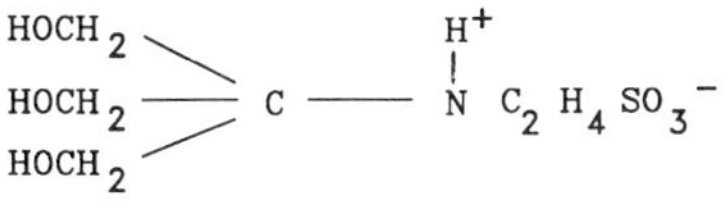

N (trishydroxymethyl) methylamino ethanesulfonic acid $\qquad$ pKa 25°C = 7.550

MOPS $\qquad$ O (morpholino ring) N—H$^+$ —CH$_2$–CH$_2$–CH$_2$ SO$_3^-$

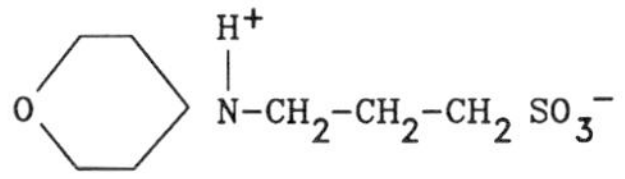

3 (N morpholino) propanesulfonic acid $\qquad$ pKa 20°C = 7.20

MOPSO $\qquad$ O (morpholino ring) N—H$^+$ – CH$_2$ – CH – CH$_2$SO$_3^-$ | OH

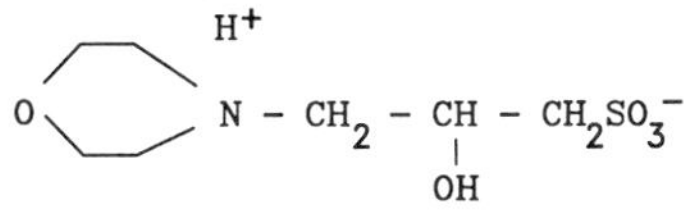

3 (N morpholino) 2 hydroxypropane sulfonic acid $\qquad$ pKa 20°C = 6.95

to arrive at a buffer couple which acts as a "primary" standardization solution towards the electrolytes, in the same way as an aqueous solution of NaCl, KCl and CaCl$_2$ whose pH is suitable for the standardization of an analytical system for the determination of pH, sodium, potassium, calcium and chloride in blood and non-diluted serum. For this reason, our attention was focused on certain substitute derivatives of ethan and propanesulfonic acids (i.e. TES, HEPES, MOPS and MOPSO), whose pK are such as to permit the preparation of buffers within the pH range set by us (6 – 8 pH).

To characterise each buffer system, particular attention was paid, in the initial phase, to the study of the compensation of the aspect connected to the incomplete dissociation. The analysis of this condition was carried out, using a "buffer" system as a reference, based on the trishydroxymethylaminomethane (TRIS)-TRIS·HCl pH 7.5 at 25°C, identified among those which, during the experiments, had shown poor binding properties both in respect of sodium and potassium and as compared to calcium.

As an example, Table 4 shows the variation of the electromotive force of the cell Ca^{2+} electrode vs SCE in solutions having the compositon, as given in Table 4.

Table 5 shows the variation of sodium and potassium ISE vs SCE at different concentrations of TRIS/TRIS·HCl.

By means of this preliminary analysis, we have studied the buffer pair (amongst those listed heretofore) which gave the best effect and the least complexing (binding) effect on the sodium, potassium and calcium ions.

The standardisation system used for sodium and potassium includes the use of a TRIS/TRIS·HCl 1 mM pH 7.4 buffer at 25°C, 140 mM Na$^+$ and 5 mM K$^+$, 100 mM Na$^+$ and 3 mM K$^+$.

Table 7 shows the effect of the concentration of the buffer pair HEPES/NaHEPES (pH 7.4 at 37°C) on the determination of sodium and potassium.

Up to 50 mM HEPES/NaHEPES we can exclude the effect due to the ion association by the buffer and attribute the differences (comparable) to the liquid junction potential which is established at the KCl saturated junction.

In confirmation of this hypothesis, we determined the concentration of calcium on the calcium electrode in a series of solutions with a constant concentration of sodium ion, using the sodium electrode as reference electrode. The calibration is carried out with 1 and 3 mM

Table 5. Data regarding the electrodes potential vs. SCE for Na^+ and K^+ at 25 and 37oC.

Sodium Electrode

Conc. in TRIS-TRIS·HCl, mM	1	50	100
Electrode pot. (mV) at 25°C	−11.1	−10.9	−10.9
Electrode pot. (mV) at 37°C	−13.3	−12.88	−12.7

Potassium Electrode

Conc. in TRIS-TRIS·HCl	1	50	100
Electrode Pot. at 25°C	+56.7	+56.9	+56.9
Electrode Pot. at 37°C	+47.7	+47.6	+47.6

Sol. 1   1 mM TRIS/TRIS·HCl pH 7.2 at 37°C
        140 mM NaCl
        5 mM KCl

Sol. 2   50 mM TRIS/TRIS·HCl
        140 mM NaCl
        5 mM KCl

Sol. 3   100 mM TRIS/TRIS·HCl
        140 mM NaCl
        5 mM KCl

Ca^{2+} in buffer TRIS/TRIS·HCl 0.01 M pH 7.5 at 25°C in 140 mM Na^+.

The results may be seen in Table 6. The measurement solutions differed from each other in the concentration of HEPES/NaHEPES.

This experiment, obtained by keeping the ionic strength constant, confirms that the contribution furnished by the liquid junction potential (vs ref. SCE) is decisive.

The largest difference in accuracy (roughly double) which derives from it for calcium as compared to that for sodium and potassium is to be attributed to the charge number of the first ion as compared to the others.

Table 6

HEPES/NaHEPES, mM	10	25	50	100
Observed calcium, 25°C vs Na^+ electrode as ref.	1.02	1.01	1.00	1.00
Calcium observed 25°C vs SCE	1.00	0.95	0.92	0.90

Table 8 shows a similar study made for sodium and potassium in buffer MOPS/NaMOPS at different concentrations.

Unfortunately, the situation is not so simple as it appears for HEPES/NaHEPES; in this case, too, if we deal with the concentration 50 mM of MOPS/NaMOPS, the effect on sodium and potassium as compared to the expected one, may be considered analogous. In any case, both for MOPS/NaMOPS and for HEPES/NaHEPES, our experiments have been continued and our attention has been focused on a concentration of 50 mM, putting into practice the compensations which the experimental results had shown to be necessary for the electrolytes in terms of ionic strength and liquid junction.

Table 7. Data regarding the concentration in mM of Na and K for various concentrations of HEPES/NaHEPES at the temperatures 25 and 37°C.

a) Sodium

Sol A 1	Sol B 10	Sol C 25	Sol D 50	Sol E 100	conc. HEPES/NaHEPES
140	140	138	137	133	conc. Na^+ mM at 25°C
139	138	136	135	130	conc. Na^+ mM at 37°C
-0.7	-1.4	-3	-3.6	-7.1	Δ C%/vs ref. 37°C

b) Potassium

Sol A 1	Sol B 10	Sol C 25	Sol D 50	Sol E 100	conc. HEPES/NaHEPES
5	5	5	4.9	4.8	conc. K^+ mM at 25°C
4.9	4.9	4.9	4.8	4.6	conc. K^+ mM at 37°C
-2	-2	-2	-4	-8	Δ C%/vs ref. 37°C

c) Calcium

Sol A	Sol B	Sol C	Sol D	Sol E	conc. HEPES/NaHEPES
	0.98	0.96	0.96	0.87	conc. Ca^{2+} mM at 25°C
	1	0.95	0.97	0.89	conc. Ca^{2+} mM at 37°C
	-	- 5	- 3	- 11	Δ C%/vs ref. 37°C

Table 8. Data referring to the concentration in mM of Na^+ and K^+ for various concentrations of MOPS/NaMOPS at the temperatures 25 and 37°C.

a) Sodium

Sol. 1	Sol. 2	Sol. 3	Sol. 4	Sol. 5	
1	10	25	50	100	Conc. MOPS/NaMOPS (mM)
101	99	98	99	92	Conc. Na^+ mM at 25°C
100	99	98	99	93	Conc. Na^+ mM at 37°C

b) Potassium

Sol. 1	Sol. 2	Sol. 3	Sol. 4	Sol. 5	
1	10	25	50	100	Conc. MOPS/NaMOPS (mM)
3	3.2	2.9	3	2.9	Conc. K^+ at 25°C (mM)
3	3.05	2.9	2.9	2.85	Conc. K^+ at 37°C (mM)

Conclusions

The standardization solutions adopted reduce the error deriving from the activity coefficient and the presence of the liquid junction potential supplying results on samples of normal blood (for content of total proteins, cholesterol, triglycerides and plasma water) or its serum or plasma in agreement with those determined for sodium and potassium through indirect methods (for example, flame photometry or ISE dil); otherwise through comparison methods for pH and ionized calcium (already existing commercial instrumentation).

The main object of the work described is to set up a standardization system intended

Table 9. Correction to be made in terms of stoichiometric concentration as compared to expected value.

	Na^+	K^+	Ca^{2+}
HEPES/NaHEPES 50 mM	+4%	+4%	+8%
MOPS/NaMOPS	+1%	+1%	+2%

to cope specifically with the simultaneous determination of pH and electrolytes in blood and its serum of plasma using a clinical analyzer which employs ion-selective electrodes for the discreet determination of the same quantities already described previously.

AN ANALYTICAL APPROACH TO THE DETERMINATION OF SOME MIXTURES OF SELECTED PTERIDINES BY ADSORPTIVE STRIPPING VOLTAMMETRY

P. Tuñón Blanco, J.M. Fernández Alvarez
and A. Costa Garcia

Department of Physical and Analytical Chemistry
Faculty of Chemistry
University of Oviedo
33071 Oviedo, Asturias, Spain

Introduction

As is well known, recent works in several laboratories have shown that controlled adsorptive accumulation of important molecules with biological significance can be used to enhance the sensitivity and selectivity of their voltammetric measurement. Accordingly, nowadays the continuous development of new adsorptive stripping voltammetric methods to analyse this kind of molecules is a reality. A great number of the molecules analysed in that way has already been reviewed.[1-3]

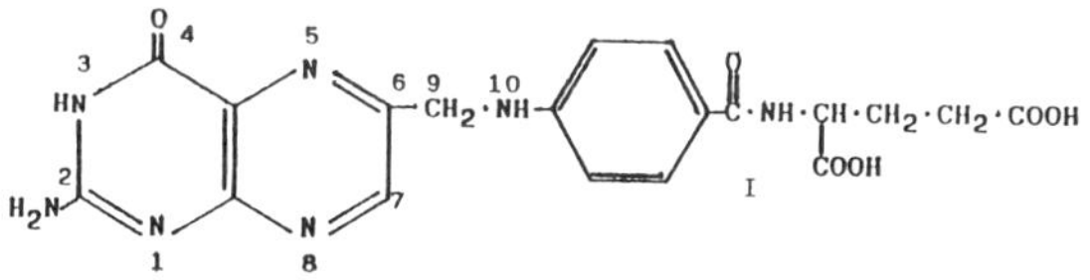

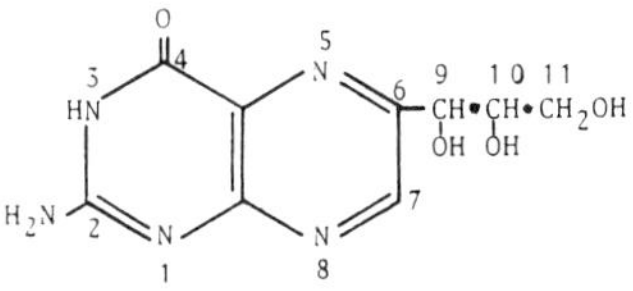

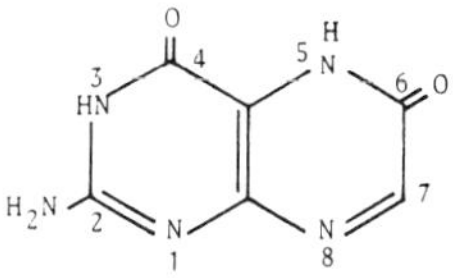

In spite of all, no attempt has been applied to the selective determination of mixtures of them by using only stripping voltammetry, without previous separation.

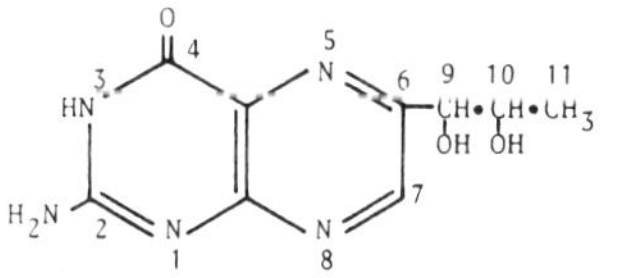

Contemporary Electroanalytical Chemistry, Edited by A. Ivaska *et al.*
Plenum Press, New York, 1990

The aim of this work is to prove the ability of this technique in the analytical resolution of structurally and biological related molecules in aqueous solutions. This is the first step for an analytical approach to the real problem of their analysis in biological fluids where they occur naturally.

Folic acid (I), biopterin (II), neopterin (III) and xanthopterin (IV) belong to the pteridine family. All of them are 2-amino-4-hydroxi-pteridines with different substituents at C-6.

Folic acid is part of the vitamin B complex,[4] and its electrochemical and biological reduction schemes have been reported by Dryhurst.[5] Its fully reduced form, tetrahydrofolic acid, is important because it acts as a carrier for a formate unit. Thus, formyl-N^{10}-tetrahydrofolic acid is involved in the biosynthesis of nucleic acid, primary constituents of living cells. Alternating current adsorptive stripping voltammetry has been applied to the determination of (I) in human seru.[6]

Biopterin, neopterin and xanthopterin have seen an increasing interest in their determination since a relation between their urinary level and certain malign proliferations has been found.[7]

Experimental

Apparatus

Linear and cyclic voltammetric experiments were carried out by using a Metrohm E-612 scanner coupled to a Metrohm E-611 detector. Current-potential curves were recorded by a Graphtec WX-4421 X-Y recorder. A Metrohm 663 static mercury drop electrode (SMDE) with a drop area of 0.47 mm^2 was used as working electrode for the stripping experiments. Potentials are referred to an Ag/AgCl/KCl 3 M electrode.

Reagents and Procedure

I, II, III, and IV were purchased from Sigma. Stock solutions (1.0×10^{-3} M) were prepared weekly and stored in the dark at 4 °C. Care was taken to protect them from direct light.

Britton-Robinson of constant ionic strength and 0.1 M sodium acetate buffers were used as background electrolytes. All reagents were of analytical grade (Carlo Erba RPE) and water was purified in a Milli-Q (Millipore) system.

The probe solution (20 ml) was transferred into the cell and deaerated with oxygen-free argon for 15 min (and for 60 s before each new experiment). Preconcentration (accumulation) was always done under a stirring of 3000 rpm unless otherwise stated. Afterwards, a quiescent period of 10 s was allowed before the potential scan was started. If the preconcentration step is carried out in open circuit, the electrolysis is established at the starting potential in the last 5 s of the rest period. A potential scan rate of 100 mV s^{-1} was used troughout.

Results and Discussion

All four pteridines adsorbed at a certain degree onto the mercury electrode. The adsorptive voltammetric behaviour of the folic acid has previously been described elsewhere.[8] Given its close structural relationship, biopterin and neopterin show a very similar behaviour, thus enabling a parallel study of both molecules. Voltammograms were recorded from a pH 5.0 acetate solution containing high (5.0×10^{-4} M), intermediate (5.0×10^{-5} M) and low (5.0×10^{-6} M) concentrations of biopterin, respectively. Even at high concentrations, where the process should be diffusion controlled, adsorption, takes place giving rise to a sharp peak superimposed to the diffusion plateau. When concentration decreases, the diffusion current becomes smaller, and at the lowest assayed concentration, a symmetric peak (II_{1c}) with 60 mV of width at its half height is developed (curve A, Fig.1).

The same study carried out for xanthopterin demonstrates its adsorption, although in a different fashion. Its adsorptive behaviour is much like that of the folic acid, showing a diffusion controlled peak at high concentrations (5.0×10^{-4} M) and an adsorption peak (IV_{2c}) at low concentrations (5.0×10^{-6} M) of the substance (curve B, Fig.1).

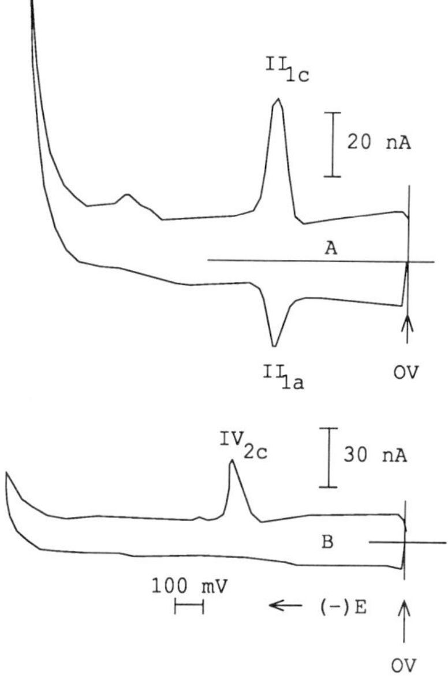

Fig.1. Cyclic voltammograms of (A) biopterin and (B) xanthopterin from a 0.1 M acetate buffer, pH 5.0. Assayed concentration: 5.0×10^{-6} M. Scanning rate: 100 mV s^{-1}.

A reverse potential scan gives rise to an anodic peak (II$_{1a}$) with the same peak potential and a lower peak intensity for I. This peak is due to the reoxidation of the 5,8-dihydro derivative to the oxidized I. The chemical reaction that takes place after the reduction process is completed (tautomerization of the 5,8-dihydro derivative to the respective 7,8-dihydro derivative), is to blame for the smaller current of the anodic peak. On the other side, IV is structurally unable to form the initial 5,8-dihydro derivative upon a 2e$^-$, 2H$^+$ reduction and hence is reduced in an irreversible process giving the 7,8-dihydro derivative.

The effective adsorption of the molecules onto the SMDE is further confirmed by carrying out medium exchange experiments. The electrode is kept in contact with a 5.0×10^{-6} M solution for 15 s and then it is removed, carefully cleaned and transferred to another cell containing only the background electrolyte. The voltammograms obtained in the new cell resemble those provided by the solutions containing the analytes.

Table 1. Peak potential, E_p, dependence on pH.

Compound	Equation $E_p - \mathrm{pH}$	Correlation coefficient	pH range
I (I$_{1c}$)	$E_p/V = -0.323 - 0.055\mathrm{pH}$	0.999	$2 - 10$
II (II$_{1c}$)	$E_p/V = -0.234 - 0.058\mathrm{pH}$	0.999	$2 - 7$
III (III$_{1c}$)	$E_p/V = -0.225 - 0.059\mathrm{pH}$	0.995	$2 - 7$
IV (IV$_{2c}$)	$E_p/V = -0.320 - 0.055\mathrm{pH}$	0.997	$2 - 7$

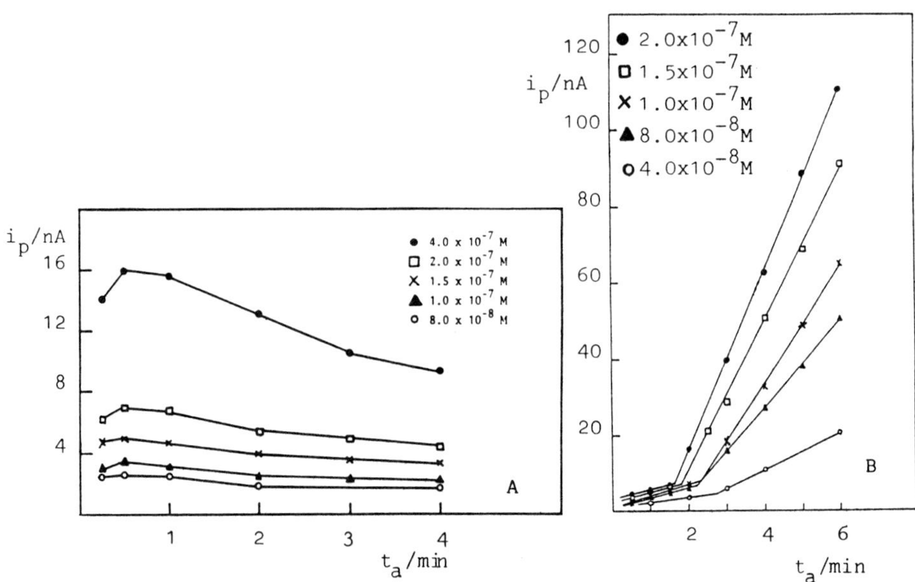

Fig.2. Accumulation curves for (A) biopterin and (B) xanthopterin under a stirring rate of 3000 rpm. Biopterin was accumulated at -0.3 V and xanthopterin was accumulated in open circuit. Other conditions as in Fig.1.

Table 2. Linear sweep voltammetry analytical data.

Compound	E_a/V	t_a/s	E_p/V	Linear range/M	Calibration plot i_p/nA vs C/M
I (I_{2c})	-0.700	600	-0.970	$8 - 400 \times 10^{-10}$	$-1.41 + 1.60 \times 10^9$
I (I_{2c})	-0.700	240	-0.970	$2 - 200 \times 10^{-9}$	4.94×10^8
II (II_{1c})	-0.300	15	-0.525	$8 - 800 \times 10^{-9}$	$-0.52 + 3.81 \times 10^7$
III (III_{1c})	-0.300	15	-0.520	$8 - 2000 \times 10^{-9}$	$-1.18 + 4.61 \times 10^7$
IV (IV_{2c})	-0.400	300	-0.600	$4 - 30 \times 10^{-9}$	$0.14 + 2.15 \times 10^8$
IV (IV_{1c})	$+0.300$	300	$+0.010$	$1.5 - 10 \times 10^{-8}$	$-14.3 + 1.19 \times 10^9$
IV (IV_{3c})	$+0.300$	300	-1.150	$8 - 200 \times 10^{-9}$	$-0.59 + 2.89 \times 10^9$

Adsorbed II, III and IV produce a symmetric, well defined reduction peak (II_{1c}, III_{1c} and IV_{2c}) in the pH range 2 to 7, this range being extended up to 10 in the case of compound I. The equations relating the peak potential, E_p, displacement with increasing pH values are summarized in Table 1. For every molecule the found slope is near to 59 mV/pH as expected for a $2e^-$, $2H^+$ process. The equations I and IV are almost identical and differ approximately by 90 mV from those of II and III.

The optimum stripping signal was always obtained at pH 5.0 and an ionic strength

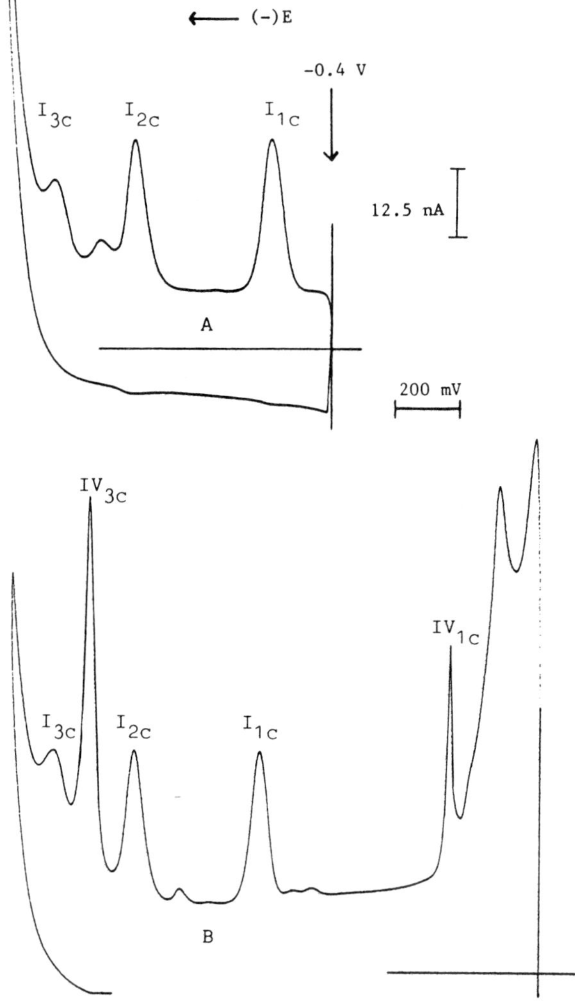

Fig.3. Equimolar $(2.0 \times 10^{-7}$ M) mixtures of I and IV. (A) Accumulation in open circuit.(B) Accumulation under electrolysis at +0.3 V. t_a = 60 s. Other conditions as in Fig.1.

above 0.01 seemed to be the most appropriate. An accumulation potential, E_a, of -0.300 V and a starting potential of -0.400 V (accumulation carried out in open circuit) have proved to be most suitable for biopterin or neopterin, and xanthopterin, respectively.

Accumulation Curves

The influence of the accumulation (preconcentration) time, t_a, on the peak current, i_p, has been studied by recording the voltammograms in the above mentioned optimized conditions for increasing concentration of every substance.

The data obtained for II are given in Fig.2 (A). It can be seen that adsorption equilibrium is reached within a short period of time, the maximun i_p being achieved for a t_a of 30 s. For larger t_a the i_p decreases showing a stabilizing trend towards accumulation times of 240 s. As bulk solution concentration gets higher, the value of i_p increases, showing a proportional increase with the concentration.

The evolution of i_p with t_a for IV is represented in Fig.2 (B). The peak shape and displacement of the peak potential (± 20 mV) when a certain electrode surface recovery is attained, agrees to a certain extent with that previously found for I.[8] However, unlike I, the first linear portions of the accumulation curves do not show a proportional slope to the solution concentration. This requirement is fulfilled by the second linear portions of the graph, which analytically can be used.

When preconcentration is done under electrolysis at $+0.300$ V, the subsequent voltammogram gives rise to two cathodic stripping peaks, with peak potentials of $+0.010$ V (IV_{1c}) and -1.150 V (IV_{3c}), due to the reduction of Hg(II)-xanthopterin salts.For several concentrations assayed, the two peaks show a linear increase of i_p with t_a, yielding straight lines whose slopes are proportional to the concentration. The second one of these peaks (IV_{3c}) gives better analytical slopes.

The second reduction peak (I_{2c}) of adsorbed I has also been studied since its peak potential, -0.970 V, does not merge with any other peak of the pteridines under consideration. This fact is important in analytical applications.

The analytical results obtained for all the mentioned processes of these pteridines are summarized in Table 2.

Mixture Assays

Folic Acid — Xanthopterin

Fig.3 (A) shows voltammograms corresponding to a 1:1 mixture of both I and IV after a preconcentration step in a stirred solution and open circuit. The adsorption stripping peak of xanthopterin (IV_{2c}) occurs at the same potential than the first one of folic acid (I_{1c}). Nonetheless, it does not interfere in the stripping signal of the I given the better adsorptive properties of this molecule, and external additions of I result in a linear increase of both its first and second stripping peak currents (I_{1c} and I_{2c}).

On the other hand, determination of IV in the presence of I requires the use of one of its mercury salt cathodic stripping peaks (IV_{1c} or IV_{3c}). Folic acid adsorptive stripping behaviour is not affected by the accumulation potential ($+0.300$ V) necesary for this purpose. Fig.3 (B) shows the possibility to carry out such a determination in 1:1 mixtures where the cathodic stripping peaks of IV are unaffected by the presence of I, and they show a linear response to external additions of IV. The second cathodic peak (IV_{3c}) yields a better linearity and keeps growing even in a four fold presence of I, whereas the first one (IV_{1c})is suppressed.

Folic Acid — Neopterine

The small difference between the peak potentials of the neopterin adsorption-reduction process (III_{1c}) and the first one of folic acid makes it impossible to distinguish them when they are in the presence of each other. In this circumstance the second process of folic acid (I_{2c}) is the alternative for its quantitation. In Fig.4 the voltammogram obtained for a mixture (2:1), after preconcentration at -0.300 V for 30 s under stirring is presented. As expected, the first reduction peak of I (I_{1c})is influenced by the presence of III, but the second one (I_{2c}) is unaltered and shows linearity with respect to the concentration of I.

Determination of III is based upon the different rate at which the coupled reaction to the first reduction step takes place for I and III (or II). Under the same experimental conditions, the voltammogram of I shows no anodic peak, while that of III still shows the reoxidation of remaining adsorbed 5,8-dihydroneopterin. This fact is exploited analytically, and this peak (III_{1a}) is used for the joint determination of III and II, under the assumption that they behave exactly the same way.

Folic Acid — Biopterin — Neopterin — Xanthopterin

Mixtures 1:1:1:1 of I, II, III and IV have been studied. The upper voltammogram in Fig.5 was obtained by accumulating at -0.300 V in a stirred solution for 60 s. All four compounds contribute to a certain degree to the current of the first distorted peak. The

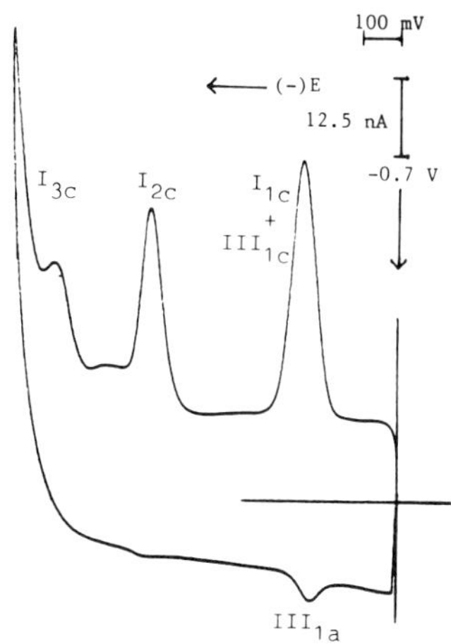

Fig.4. Cyclic voltammogram of a 1:2 mixture of I and III. $E_a = -0.3$ V. Other conditions as previous ones.

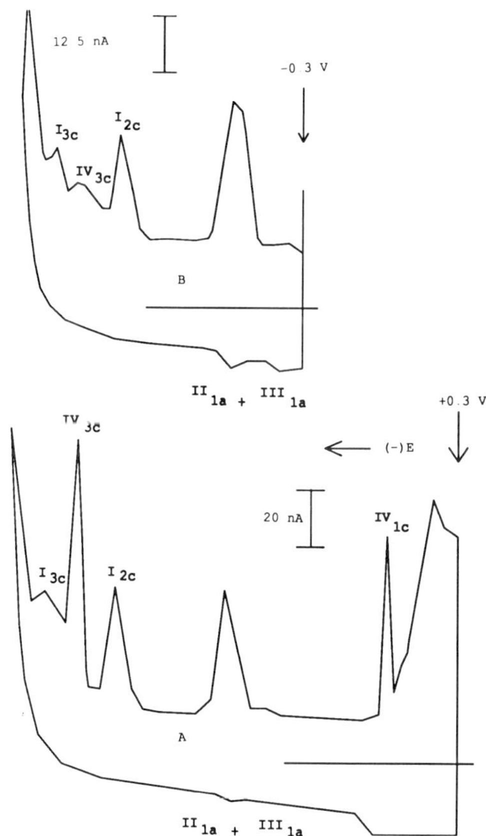

Fig.5. Cyclic voltammograms of a 1:1:1:1 mixture of I, II, III, and IV. Assayed concentration: 2.0×10^{-7} M. Other conditions the same as previous.

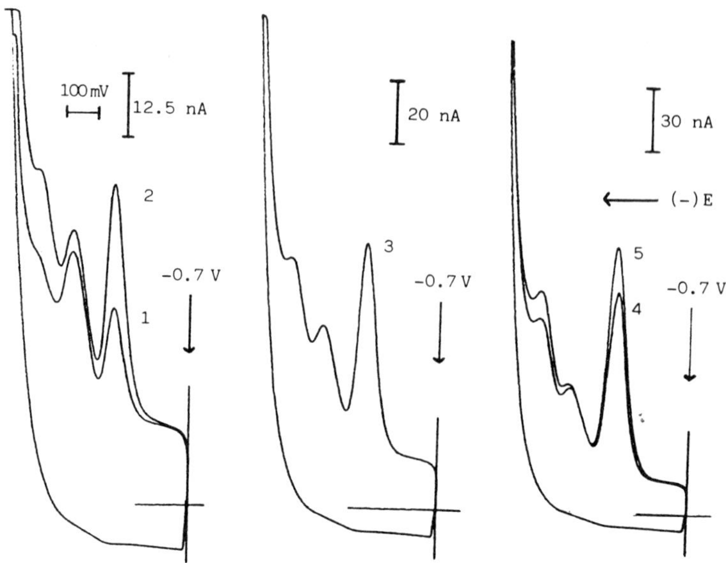

Fig.6. Adsorption-stripping cyclic voltammograms of I obtained in a 10 fold excess of II, III and IV. $E_a = -0.7$ V. $t_a = 30$ s. Increasing concentration of I: (1) 2.0×10^{-7} M; (2) 4.0×10^{-7} M; (3) 6.0×10^{-7} M; (4) 8.0×10^{-7} M; (5) 1.0×10^{-6} M.

second one belongs to the second reduction process of adsorbed I, which is followed by a hint of the second cathodic stripping peak of xanthopterin (IV_{3c}) and, finally, the third reduction process of I is observable. The anodic peak represents the contribution of II and III dihydro derivatives.

In these experimental conditions folic acid (through its second process, I_{2c}) and biopterin and neopterin (jointly, through their anodic peak, II_{1a} and III_{1a}) can be quantitatively determined.

The lower voltammogram was obtained for the same mixture in identical conditions, but using an accumulation potential of $+0.300$ V. In the cathodic sense, the two Hg(II)-xanthopterin cathodic stripping peaks, and the three adsorption-stripping peaks of I can be seen. The anodic side of the voltammogram is, again, the common contribution of 5,8-dihydro II and III.

Linear responses of one order of magnitude are obtained for the second process of I and for the second cathodic stripping peak of IV. Moreover, this linearity prevails for the folic acid response, no matter whether the preconcentration takes place at -0.300 V or at -0.700 V.

Finally, the especially favourable adsorption of I onto mercury electrodes has been tested, by increasing the concentration of the other three pteridines by ten fold. Thus, in a mixture 1:10:10:10 (2.0×10^{-7} M in I; 2.0×10^{-6} M in the rest) the second peak of I grew linearly with successive additions up to 1.0×10^{-6} M as can be observed in Fig. 6 (note the different sensitivity).

References

1. J. Wang, Int. Lab., **68** (1982) 15.
2. J. Wang, in: "Stripping Analysis", VCH Publishers Inc., Florida, 1985.
3. W.F. Smyth, in: "Electrochemistry, Sensors and Analysis", M.R. Smyth and J.G. Vos (Eds), Analytical Chemistry Symposia Series, Vol. **25**, Elsevier, Amsterdam, 1986, p. 29.
4. H.K. Mitchell, E.E. Snell and R.J. Williams, J. Am. Chem. Soc., **63** (1941) 2284.
5. G. Dryhurst, in: "Electrochemistry of Biological Molecules", Academic Press, New York, 1977, pp. 324-357.

6. J.M. Fernández Alvarez, A. Costa Garciaa, A.J. Miranda Ordieres and P. Tuñón Blanco, J. Pharm. Biomed. Anal., 6 (1988) 743.
7. H. Rokos, K. Rokos, H. Frisius and H.J. Kirstaedter, Clin. Chim. Acta, **105** (1980) 275.
8. J.M. Fernández Alvarez, A. Costa Garcia, A.J. Miranda Ordieres and P. Tuñón Blanco, J. Electroanal. Chem., **225** (1987) 241.

ELECTROCHEMICAL BEHAVIOUR OF METRONIDAZOLE

P. Siva Sankar and S.J. Reddy

Department of Chemistry
S.V. University
Tirupati 517 502, A.P., India

Introduction

Metronidazole (I) is an effective agent for a variety of protozoal diseases including trichominiasis, giardiasis, amoebiasis and balantidiasis.[1]

$$O_2N - \underset{\underset{CH_2CH_2OH}{|}}{N} \diagdown CH_2$$

(I)

The selectively toxic effect of metronidazole towards anaerobic bacteria and protozoal diseases depends on a number of factors. The killing of such drugs, as metronidazole, reguires the reduction of nitro group. The first studies on the mode of action of metronidazole indicated the output of hydrogen gas from T. Vaginalis before that of carbon dioxide, which parellels cell death. The polarographic behaviour of metronidazole in presence of DNA and interactions with nucleic acids have been studied extensively by Edwards et al.[3] and Nicholos et al.[4] HPLC,[5] spectrophotometry,[6] gas chromatography[7] and gravimetry[8] have been reported to be useful in the determination of metronidazole. In the present investigation, the detailed electrochemical reduction behaviour of metronidazole as well as polarographic determination procedure have been described.

Experimental

Metronidazole was supplied by unique pharmaceutical Labs Pvt. Ltd., Bombay. The purity of the sample was tested with thin layer chromatography. All chemicals used were of analar grade. The polarograms were recorded by Polarographic analyzer Model 364 coupled with BD 8 Kipp and Zonen recorder. Cyclic voltammograms, chronoamperograms, ac polarograms and differential pulse polaragrams were obtained with Metrohm unit Model E 506 polarecord coupled with VA-Scanner 612 and Model 2000 x-y/t recorder. Chronopotentiograms were obtained by Beckmann R Electroscan 30 analyzer from Beckman Instruments Inc., U.S.A. The dropping mercury electrode of flow rate 2.48055 mg s^{-1}, hanging mercury drop electrode of area 0.443868 cm^2 and dropping mercury electrode of area 0.02323 cm^2 were employed as working electrodes. Reference electrodes employed were SCE for dc polarography, Ag/Agl (s), Cl$^-$ for Metrohm unit and molybdenum electrode for chronopotentiograms. In all the cases platinum wire was used as auxiliary electrode. All the experiments were carried out at 25 $\pm$ 1°C.

Contemporary Electroanalytical Chemistry, Edited by A. Ivaska *et al.*
Plenum Press, New York, 1990

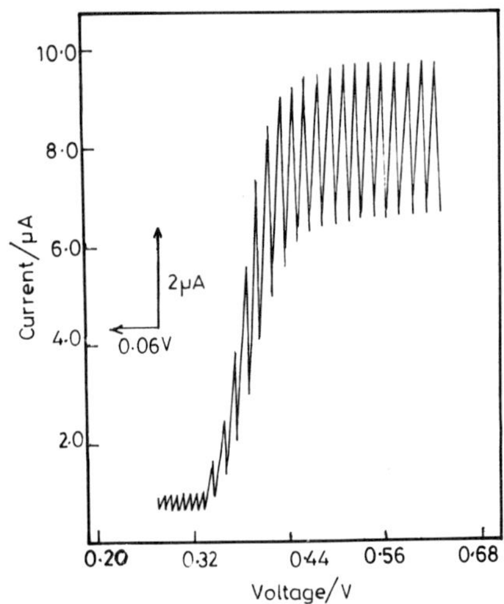

Fig.1. Typical dc polarogram of metronidazole in Clarks & Lubs buffer of pH 2.0. Concentration 0.5 mM; drop time 3 s.

An appropriate amount of metronidazole was dissolved in required quantity of triple distilled water and acurately diluted with the supporting electrolyte to 10 ml. The solution was deoxygenated by passage of nitrogen gas for 5 minutes before the polarogram was recorded. When tablets were concerned, about 100 mg or complete tablet of powdered sample was dissolved in 250 ml of water with repeated extractions. Supporting electrolyte was used to prepare various desired concentrations of metronidazole solutions from the standard solution.

Results and Discussion

A single polarographic wave (-0.638 V vs. SCE in acetate buffer of pH 4.00) / cyclic voltammetric peak (-0.8 V vs. Ag/Agl(s), Cl^- in the supporting electrolyte pH 9.25) of metronidazole has been observed in all the supporting electrolytes studied. (Figs 1 and 2). The wave/peak is attributed to the reduction of nitrogroup to the corresponding hydroxylamine. The wave height/peak height are found to depend on pH, solvent and on mercury column height (h)/sweep rate (v). The dependence of half-wave potential and peak potential are found to be linear with pH. Addition of methanol or ethanol or DMF shifts the half-wave potentials and peak potentials to more negative values. In dc polarographic techniques a maximum is observed in all the supporting electrolytes employed except in Clarks and Lubs buffer of pH 2.00. This maximum is suppressed by using 0.1 to 0.3 ml gelatin (0.05 %).

The linear plots of i_d vs. $h^{1/2}$ and i_p vs. $v^{1/2}$ passing through origin indicate the electrode process to be mainly diffusion controlled and free from adsorption complications in all the supporting electrolytes used except in pH 2.00. The plots of $i\tau^{1/2}$ vs. i in chronopotentiometry and ac polarogram in this media also confirm this. Conventional log-plot analysis, variation of $E_{1/2}$ and E_p values towards more negative potentials with increase of concentration of the depolariser and the absence of anodic peak for C_1 in the reverse scan in cyclic voltammetry and also the disobedience of Tomes criteria show that the electrode process is irreversible.

The colour of the experimental solution changes from colourless to thick yellow at higher pH value which can be explained on the basis of nitro group tautomerism. The nitro group

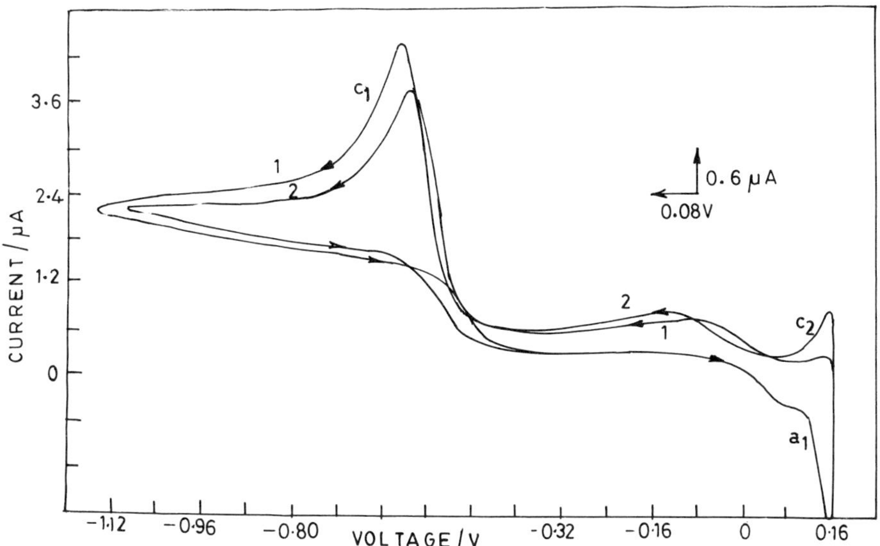

Fig.2. Typical cyclic voltammogram of metronidazole in phosphate buffer of pH 9.25. Concentration 0.5 mM; 1 – first scan, 2 – second scan.

in metronidazole exists in different forms depending on the pH of the supporting electrolyte.[9]

(I)

In acid   In neutral   In base

Millicoulometry is employed in Clarks and Lubs buffer of pH 2.00 to find out number of electrons in the electrode process. The results show that the number of electrons involved in the electrode process is four. The product of controlled potential electrolysis (carried out at -1.0V vs. SCE) in Clarks and Lubs buffer of pH 2.00 is identified as hydroxylamine.[10] The number of protons involved in the rate determining step is evaluated as one from $E_{1/2}$ vs. pH and E_p vs. pH plots. The electrochemical reaction order is shown to be one from log i vs. log C plots at constant E.

A small anodic peak (a_1) is observed in the reverse scan in cyclic voltammetry in acidic and neutral media (i.e., 2.00 < pH < 8.00). In the second scan, another small cathodic peak (C_2) is observed at more positive potential than C_1. The anodic peak may be due to the oxidation of reduced product and the peak C_2 may be due to the back reduction of oxidized nitroso to hydroxylamine as also observed by Morales et al.[11] in the case of nitrosubstituted furans. The redox couple is noticed to be irreversible as evidenced from the separation of the E_p values.

On the basis of the results of our own investigations as well as on the literature data, the following schemes A and B may be given for the electrochemical reduction of metronidazole in different pH ranges:

A) In 2.00 $<$ pH $<$ 8.00

$$O_2N \cdots \text{(I)} \xrightarrow{H^+} H^+O_2N \cdots \xrightarrow{e^-} HO_2N \cdots \cdots \xrightarrow[+e]{} HO\,HN \cdots \text{(II)} \xleftarrow{2\bar{e},2H^+} O\bar{N} \cdots \xleftarrow[-H_2O,\,+H^+]{}$$

B) In 10.00 $<$ pH $<$ 12.00

$$R-NO_2 \xrightarrow{2e^-} RNO_2^{2-} \xrightarrow{2H^+} RNO_2H_2$$

$$\downarrow \begin{array}{c} +2e^- \\ +2H^+ \\ -H_2O \end{array}$$

$$RNHOH$$

$$R = \text{(I)}$$

Though the electrochemical reduction mechanism consists of many intermediate steps probably, all are unstable, thus giving only one wave/peak. The intermediate (II) formed in cyclic voltammetry may be less stable finally being reduced to hydroxyl amine.

The variation of diffusion current and peak current with the pH of the supporting electrolyte, influences the diffusion coefficient values also to vary in the same manner. The reason for slight variations attributed in diffusion coefficient values with increase in pH may be due to the decrease in the availablity of protons with increase in pH. In general, forward rate constant values are noticed to decrease with increase in the pH of the solution. The values of heterogeneous forward rate constants obtained at DME are less when compared to those obtained at HMDE. This is probaply ascribed to the different electrolysis conditions existing at the two electrodes. The kinetic parameters are reported in Table 1 and 2.

Analysis

The analytical method described here is based on the results obtained with dc polarography and differential pulse polarography at dropping mercury electrode. The detection and determination of the drug metronidazole (I) continues to be of interest, particularly because of its status as a drug of abuse. The nitro group in metronidazole is useful in electrochemical analysis.

The polarograms are recorded over the range of applied potential from $+0.1$ V to -0.9 V. A series of polarograms for 5×10^{-4} M metronidazole in all buffers of pH ranging from 2.00 to 12.00 are recorded. Different concentrations of metronidazole (1.0×10^{-5} M to 1.0×10^{-3} M for dc polarograms and 1.0×10^{-8} M to 1.0×10^{-5} M for dpp) are examined in all the supporting electrolytes. The plots of i_d vs. C and i_m vs. C are also found to be linear passing trough origin. The calibration plot is seen to be linear in the range 1.5×10^{-4} M to 1×10^{-3} M in dc polarography and 2.5×10^{-7} M to 1.0×10^{-5} M in differential pulse polarography. Differential pulse polarographic technique is found to be more suitable at lower concentrations

Table 1. Typical voltammetric data of metronidazole. Concentration 0.5 mM.

Supporting Electrolyte	Dc Polarography t=3 s			Cyclic voltammetry, scan rate=40 mVs^{-1}		
	$\frac{-E_{1/2}}{V}$	$\frac{D\times10^6}{cm^2s^{-1}}$	$\frac{k^o_{f,h}}{cm\ s^{-1}}$	$\frac{-E_p}{V}$	$\frac{D\times10^6}{cm^2s^{-1}}$	$\frac{k^o_{f,h}}{cm\ s^{-1}}$
Clarks and Lubs buffer of pH 2.00	0.34	6.43	8.85×10^{-7}	0.23	4.16	3.32×10^{-8}
Acetate buffer of pH 4.00	0.40	6.08	3.18×10^{-8}	0.37	3.82	9.49×10^{-10}
Citrate buffer of pH 6.00	0.45	5.28	4.40×10^{-9}	0.45	2.84	6.31×10^{-12}
Phosphate buffer of pH 8.00	0.52	4.45	5.65×10^{-10}	0.63	2.42	4.79×10^{-13}
Carbonate buffer of pH 10.00	0.53	3.77	1.89×10^{-11}	0.69	1.76	7.05×10^{-16}
Bates and Bower buffer of pH 12.00	0.54	3.58	5.52×10^{-17}	0.89	1.82	1.24×10^{-17}

Table 2. Rate constants data for chronoamperometry and chrono potentiometry

Supporting Electrolyte	Chronoamperometry			Chronopotentiometry		
	$\frac{D\times10^6}{cm^2s^{-1}}$	$\frac{k^o_{f,h}}{cm\ s^{-1}}$	$\frac{Step\ potential}{V}$	$\frac{D\times10^6}{cm^2s^{-1}}$	$\frac{k^o_{f,h}}{cm\ s^{-1}}$	$\frac{Current}{\mu A}$
Clarks and Lubs buffer of pH 2.00	2.01	5.26×10^{-8}	0.21	3.80	4.48×10^{-8}	2.2
Acetate buffer of pH 4.00	2.18	9.14×10^{-10}	0.33	4.30	2.34×10^{-9}	2.6
Citrate buffer of pH 6.00	2.68	7.01×10^{-11}	0.40	4.80	9.34×10^{-12}	3.0
Phosphate buffer of pH 8.00	2.82	5.91×10^{-12}	0.59	4.80	1.14×10^{-12}	3.0
Carbonate buffer of pH 10.00	3.61	6.84×10^{-11}	0.65	5.80	5.43×10^{-15}	2.8
Bates and Bower buffer of pH 12.00	3.10	8.01×10^{-12}	0.85	6.10	8.36×10^{-16}	3.0

due to its high sensitivity and resolution.

Recommended Procedure: The solution was prepared by dissolving the required quantity of metronidazole in triple distilled water, and made up with the supporting electrolytes to get the 1.0×10^{-5} M concentrations of standard solution. All supporting electrolytes except Clarks and Lubs buffer of pH 2.00 are found to be suitable media for the determination of metronidazole because only slight adsorptions complications are involved. 1 ml of the standard solution was transfered to a polarographic cell and made up with 9 ml of the supporting electrolyte of 4.00 < pH < 12.00 to get the concentration of 1.0×10^{-6} M and also deareated with nitrogen gas before polarograms were obtained (Fig.3). In the present work, a modulation amplitude of 50 mV and drop time of 2 s are chosen as being suitable for analytical purposes. Using this procedure 20 polarograms were recorded for 20 standard additions. The relative standard deviation was found to be 2.1 % and the correlation coefficient as 0.9982 (from 20 replicants).

Metronidazole in tablets: Metronidazole can be successfully determined in different pharmaceutical formulations without any prior separation by using differential pulse polarog-

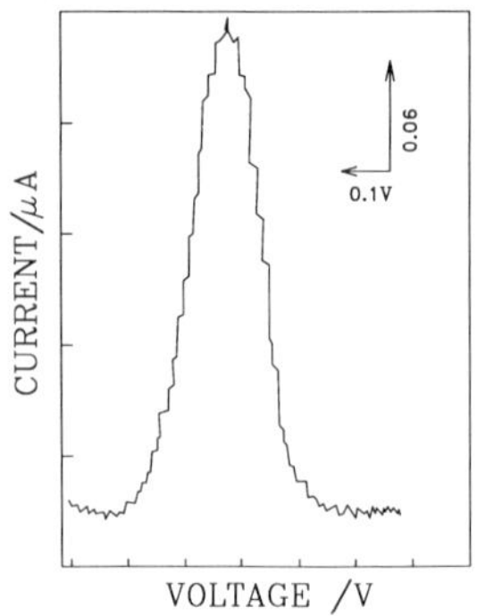

Fig.3. Typical differential pulse polarogram of metronidazole in phosphate buffer of pH 8.00. Concentration 5×10^{-5}; drop time 2 s; amplitude 50 mV.

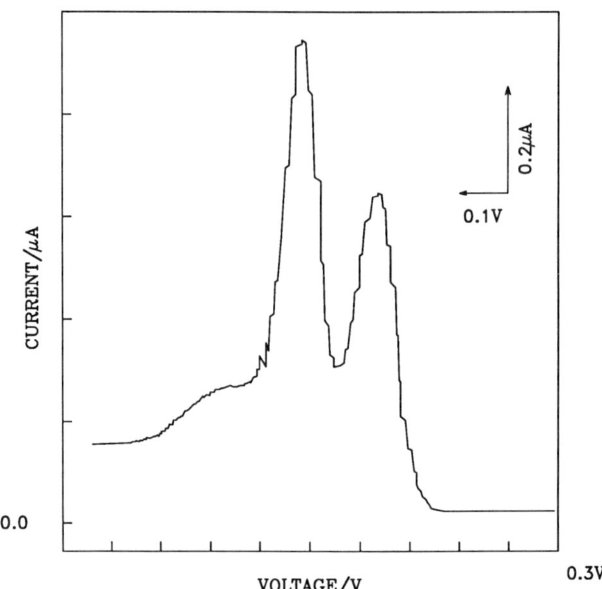

Fig.4. Typical differential pulse polarogram of metronidazole in presence of tinidazole in phosphate buffer of pH 8.00. Concentration of metronidazole 0.4×10^{-5} M; concentration of tinidazole 0.4×10^{-5} M; drop time 2 s; amplitude 50 mV.

raphy. The relative standard deviations are found to be 2.3 %. The correlation coefficient and coefficient of variation are found to be 0.9931 and ± 0.642 respectively. The results are shown in Table 3.

The peak obtained for metronidazole 2.5×10^{-7} M in phosphate buffer of pH 8.00 is not found to be disturbed by successive additions of other nitro group imidazoles like tinidazole, ornidazole and nimorazole, of concentrations 2.5×10^{-6} M (Fig.4).

Table 3

S. No.	Tablet	Composition	$i_m/\mu A$	Amount of the Compound	
				Observed mg	Prescribed mg
1	Dyrade-M	Metronidazole and Diloxanamide furoate	2.98	398.7	400
2	Metron	Metronidazole	3.71	198.7	200
3	Furoxone	Metronidazole and Furazolidone	3.73	197.1	200
4	Flagyl	Metronidazole	3.77	199.1	200
5	Metrogyl	Metronidazole	7.42	396.4	400

The procedure described can be employed for the successful estimation of metronidazole (I) without prior separation of diloxanamide furoate or furazolidone commonly available in certain pharmaceutical formulations and without interference from other nitroimidazoles such as tinidazole, ornidazole and nimarazole.

Acknowledgements

The authors are thankful to Depertment of Atomic Energy, Bombay for providing finacial assistance to carryout this work.

References

1. C. Cosar and L. Julon, Ann. Inst. Pasteur, **96** (1959) 238.
2. D.I. Edwards and G.E. Matheson, J. Gen. Microbiol., **63** (1970) 297.
3. D.I. Edwards, Br. J. Vener. Dis., **56** (1980) 285.
4. F.L. Nicholas, T. Maria, M. Muller and R. Lipman, Molecular Pharmacology, **13** (1977) 872.
5. Z. Chunfeng, Yu. Miaoyue, Flinlu and Le Jinque, Yaows Fenxi Zazhi, **6** (4) (1986) 229.
6. M.B. Devani, C.J. Shishoo, D. Kokila and A.K. Shash, Ind. J. Pharm. Sci., **43** (1981) (4) 151.
7. D.M. Singbal and G.B. Natekar, East. Pharm., **22** (266) (1979) 183.
8. C.R. Gutierrez and A. Garzon, Rev. Soc. Quim. Mex., **22** (6) (1978) 427.
9. Peter Smyth, in: "The Chemistry of Organic nitrogen compounds", Vol. **II**, 1966, W.A. Benjamin, Inc., New York.
10. Fritz Feigl, in: "Spot tests in organic Chemistry", Chapter 3, p. 284, Elsevier, Amsterdam.
11. A. Morales, R. Pablo and M.I. Toral, Analyst, **112** (1987) 965.

PHARMACEUTICAL APPLICATIONS

THE DETERMINATION OF SELECTED ANTIBIOTICS, ANTIBACTE-RIALS AND ANTICONVULSANTS BY VOLTAMMETRIC AND LIQUID CHROMATOGRAPHIC TECHNIQUES

W.F. Smyth*

Chemistry Department
University of Zambia
P.O. Box 32379, Lusaka, Zambia

Introduction

Modern voltammetric techniques such as differential pulse polarography and square wave voltammetry can determine electroactive drugs such as antibiotics and antibacterials over a relatively wide concentration range of approximately three orders of magnitude with detection limits of $10^{-7} - 10^{-8}$ M. Such drugs as raw materials and as they are contained in formulations can be determined rapidly with a minimum of sample treatment prior to electro-analysis. Biological matrices containing drug traces as found in therapeutic drug monitoring, pharmacokinetics and forensic toxicology frequently require solvent extraction and/or other sample treatment procedures prior to voltammetric determination of the drug actives. The selectivity of the these determinations of raw materials formulations and biological matrices towards structurally related process impurities, degradation products and metabolities respectively is generally admitted to be poor.

High performance liquid chromatography with a much superior selectivity is fast gaining acceptance as the most applicable instrumental technique for the determination of most of these antibiotics and antibacterials in raw materials, dosage forms and biological matrices. The rapid growth of this instrumental technique has been facilitated by development of reliable, moderately priced instrumentation and efficient columns.

This paper investigates the differential pulse polarographic, dc and square wave voltammetric determination of selected antibiotics, antibacterials and anticonvulsants at mercury and solid electrodes. The results of these determinations are compared and contrasted with rival analytical methods based on high performance liquid chromatography with ultraviolet detection.

Experimental

Apparatus and Techniques

Pulse polarographic measurements were performed using a Metrohm E-506 polarograph coupled with an E-505 polarographic stand. Rapid scan square wave polarography at single mercury drops was performed by an instrument designed and built at the Technion, Haifa. Rapid scan square wave voltammetry at the SMDE was carried out using an EG & G PAR 384B Polarographic Analyser which controlled operation of a Model 309 Automatic Voltammetric Electrode for rapid and automated sample handling. Dc linear sweep voltammetry was

---

* Visiting Professor of the Department of Pharmacy,
  The Queen's University of Belfast,
  Belfast BT97BL, N. Ireland

Contemporary Electroanalytical Chemistry, Edited by A. Ivaska *et al.*
Plenum Press, New York, 1990

stock solution of sulphadiazine (I) to known volumes of blood of the order of 10 ml. Typically standards are prepared in the range of 0.1 ppm – 10 ppm and they are treated in the same way as above prior to HPLC assay. The samples from the field can then be compared directly against standards in order to estimate the unknown concentration of sulphadiazine (I) present. A blank blood sample is also treated similarily in order to assure the analyst that the peak due to sulphadiazine (I) is due to it alone. Using the above HPLC conditions, sulphadiazine (I) elutes after 1.17 min and trimethoprim (III) after 4.12 min. This trimethoprim peak is only of use for determinations in blood samples at relatively high concentrations. Peak blood levels of (I) were found in three test animals two hours after injection with a mean value of 11.67 μg/ml being achieved. After 24 h, the mean value was 0.16 μg/ml. Levels of $\geq$ 1.50 μg/ml could then be maintained by injections repeated at daily intervals over a 5 day period. Peak blood levels in each day of ca 11-12 μg/ml were obtained approximately 2 h after each injection. The determination of trimethoprim (III) in biological matrices can also be achieved by HPLC, e.g., Weber et al.[6] have described such a rapid, precise and simple procedure for the quantitative determination of trimethoprim (III), sulphamethoxazole (II) and N-4-acetylsulphamethoxazole in body fluids, for use in therapeutic drug monitoring. The coefficient of variation was 3 %, the limit of detection was 0.5 μg/ml for all 3 compounds and recovery was > 97 %. Weinfeld et al.[7] have worked at lower concentrations and achieved a linear range of calibration of 20 – 200 ng/ml blood using the methylanalogue of trimethoprim as internal standard.

Penicillins

Penicillin G (VI) is widely used in veterinary medicine principally in controlled release penicillin G complexes such as procaine penicillin G, penicillin G benzathine and benethamine penicillin G. Penicillin G, in its complex with procaine (VII) has antibacterial activity in horses, cattle, sheep, cats and dogs and is manufactured by many drug firms, either as the complex on its own or mixed with other antibiotics such as streptomycins.

Penicillin G does not give polarographic waves/peaks in its intact form but can show electroactivity on derivatisation, e.g., alkaline hydrolysis, followed by acid hydrolysis, dilution with Brdicka's solution and recording of the catalytic wave due to the sulphydryl component of the penicillamine. Again, these electroanalytical methods do not possess adequate selectivity for differentiation of mixtures of penicillin G and degradation products.

Chromatographic methods do possess the necessary selectivity for resolution of different penicillins and their precursors and degradation products. Hendrickx et al.[8] have described a TLC scheme for 18 penicillins using silanised silica gel as the stationary phase with 35 different mobile phases. Classical column chromatography on Sephadex G-25 was found useful for the identification and determination of polymer degradation products formed from benzylpenicillin solutions stored on the dark (Ueno et al.).[9] Penicillin G salts are, however, not sufficiently volatile or thermally stable to be suitable for determination by GLC. High performance liquid chromatography is increasingly becoming the method of choice for the determination of penicillins and their degradation and metabolic products. This is illustrated by reference to the following examples in the author's experience. Firstly, a formulation which contains procaine penicillin G and benzathine penicillin G, together with methylhydroxybenzoate as preservative, can be determined by reversed phase HPLC on Sperisorb ODS-2 (5 μm) using a mobile phase of 0.018 M sodium hexane sulphonate in 0.016 M potassium dihydrogen orthophosphate-acetonitrile-dimethylformamide (70 + 23 + 7), pH 2.90, a flow rate of 1 ml/min and a detection wavelength of 254 nm. Procaine was eluted after 2.56 min, methyl hydroxybenzoate after 4.63 min, penicillin G after 8.94 min and benzathine after 10.14 min. Mean recoveries (n = 10) for each constituent were found to be – procaine penicillin G (99.97 %), benzathine penicillin G (100.66 %) and methylhydroxybenzoate (100.27 %). Reproducibility for six replicates was 0.982 % (procaine penicillin G), 0.903 % (benzathine penicillin G) and 1.062 % (methylhydroxybenzoate). Linearity of response was achieved over the following ranges – procaine penicillin G (0.438 – 1.283 mg/ml), benzathine penicillin G (0.372 – 1.114 mg/ml) and methylhydroxybenzoate (0.516 – 1.031 mg/ml $\times 10^{-2}$). Secondly, a formulation which contains procaine penicillin G and dihydrostreptomycine sulphate can be determined using a Sperisorb ODS column (5 μm), a mobile phase of 0.02 M sodium hexane

$$NH_2-\langle\bigcirc\rangle-SO_2.NH-\langle N \rangle \qquad (I)$$

$$NH_2-\langle\bigcirc\rangle-SO_2.NH-\underset{N-O}{\langle\ \rangle}CH_3 \qquad (II)$$

$$CH_3O,\ CH_3O,\ CH_3O-\langle\bigcirc\rangle-CH_2-\langle N \rangle-NH_2 \qquad (III)$$

carried out at C electrodes using a EG & G PAR 174A Analyser. All potentials were measured against the Ag/AgCl electrode. High performance liquid chromatography utilised a waters Model 510 pump, a Spherisorb ODS-2 (5 μm) column, a Waters Model 481 spectrophotometric detector and a Waters Model 730 Integrator for the sulphadiazine and penicillin determinations. HPLC of clobazepam and its N-desmethylmetabolite utilised a Waters H45 solvent pump coupled to a UK 6 injector with a 10 μm C_{18} Bondapak column (3.9 cm × 30 cm), a Waters 450 variable wavelength detector (set at 232 nm) and a Hewlett-Packard 3390A Integrator.

Reagents

Stock solutions of ca 10^{-3} M of the drugs were made up weekly in the appropriate solvent and stored in the dark and under refridgeration to minimise decomposition. Analar chemicals were used in the preparation of buffer solutions, mobile phases etc and L.C. grade solvents employed in the liquid chromatographic separations.

Results and Discussion

Sulphonamides and Trimethoprim Potentiated Sulphonamides

Sulphonamides such as sulphadiazine (I) can only be reduced at the dropping mercury electrode in non aqueous solvents such as CH_3CN with 0.1 M nBu_4NClO_4 supporting electrolyte. A pure solution of (I) will give two dc polarographic waves with $E_{1/2} = -1.31$ and -1.58 V (vs Ag/0.1 M nBu_4NI). The use of CH_3CN as a solvent and the negative potentials at which the waves are observed do not recommend this as a method of formulation analysis. Sulphonamides such as (I) can, however, be oxidised at the glassy carbon electrode through a reaction involving the primary amino group and this has been used by Momberg et al.[1] for quantitative analysis. The effect of trimethoprim (III) which also contains two aromatic amino groupings on the oxidative voltammetric behaviour of the sulphonamide is unknown but interference would be expected. Trimethoprim (III) gives well defined polarographic reduction waves/peaks suitable for its analysis in trimethoprim potentiated sulphonamide formulations. A well defined square wave voltammetric peak is given at the SMDE in $HClO_4$ supporting electrolytes, e.g., in 0.081 M $HClO_4$, a peak is given at -1.175 V and also further but badly defined cathodic phenomena at more negative potentials. The peak of -1.175 V can be enhanced by adsorptive accumulation at -0.8V when the concentration of (III) in the voltammetric cell is 2×10^{-6} M. 20 s accumulation at this potential gives an enhancement factor of 3.74. Detection of (III) at 10^{-7} M levels is possible using diluted supporting electrolytes such as 0.004 M $HClO_4$ and a frequency of 5 Hz. Tablets for human consumption containing 160 mg trimethoprim (III) and 800 mg sulphamethoxazole (II) can thus be conveniently and rapidly analysed for their trimethoprim content by the technique of automated cell – rapid scan square wave voltammetry as follows. A tablet is crushed and shaken vigorously with 25 ml Analar MeOH for 5 minutes, made up to a volume of 100 ml with the same solvent and then an appropriate aliquot diluted with $HClO_4$ supporting electrolytes (1.5, 0.08 and 0.004 M $HClO_4$ concentrations can be used with frequencies of 100, 25 and 5 Hz respectiv-

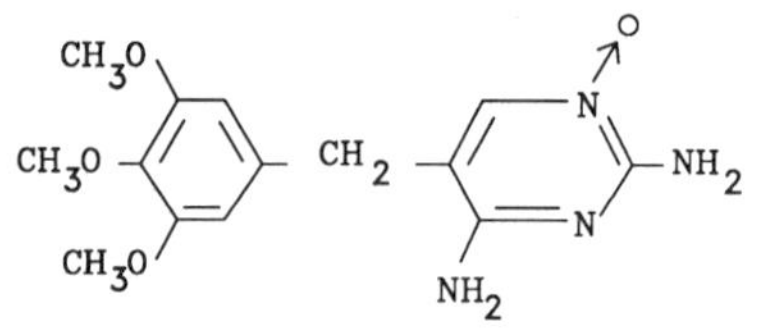

$$CH_3O-,\ CH_3O-,\ CH_3O- \quad C_6H_2 - CH_2 - \text{(pyrimidine N-oxide)} - NH_2,\ NH_2 \qquad (IV)$$

$$CH_3O-,\ CH_3O-,\ CH_3O- \quad C_6H_2 - CH_2 - \text{(pyrimidine N-oxide)} - NH_2,\ NH_2 \qquad (V)$$

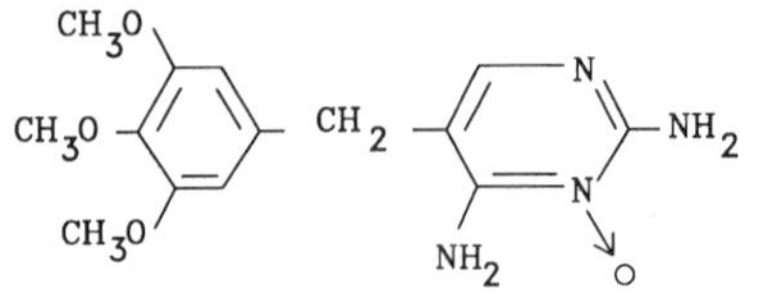

ley). This solution is then subjected to the technique of automated cell-rapid scan square wave polarogrphy as described by Peled et al.[2] Adsorptive preconcentration is not used due to possible interference of the sulphonamide and tablet excipients on the adsorption process. This technique can assay 24 samples of the trimethoprim-sulphonamide formulations per hour giving a coefficient of variation for replicate analysis of the same solution of 2 %.

Brooks et al.[3] have reported differential pulse polarographic analysis of trimethoprim and its N-oxide metabolites. To determine trimethoprim (III) the drug was extracted into $CHCl_3$ from blood and urine, buffered to pH 11.5. For blood, the drug was back extracted into 0.1 N H_2SO_4 and analysed by differential pulse polarography. For the drug in urine separation is done by TLC with an ethanol elution and dissolution of the residue in 0.1 N H_2SO_4.

The polarographic blood assay was found to be twice as fast as the fluorimetric assay, primarily due to the elimination of the time consuming conversion of trimethoprim to the trimethoxybenzoic acid. The differential pulse peak at -1.07 V (vs SCE) corresponds to azomethine reduction in the pyrimidine ring. To determine the N-1 and N-3 oxide metabolites of trimethoprim in urine, the procedure above was slightly modified to include a "salting out" with anhydrous Na_2CO_3 in order to facilitate extraction of the N-oxides into $CHCl_3$. Also the residue was extracted into 1 M phosphate buffer, pH 3 to provide optimal separation between the peaks for the N-1 oxide (IV, $E_p = -0.95$ V vs SCE) and the N-3 oxide (V, $E_p = -1.095$ V vs SCE). In the phosphate buffer, trimethoprim exhibits a peak at -1.190 V vs SCE. A TLC separation is still required for analysis of the N-oxide metabolites, since large quantities of trimethoprim in urine mask the N-oxide peaks.

High performance liquid chromatography is increasingly the method of choice for the determination of trimethoprim potentiated sulphonamide formulations and also for the determination of these two actives in biological matrices. The USP XXI has stated HPLC as its official method for sulphamethoxazole (II) and trimethoprim (III) oral suspensions and tablets. Using an octadecylsilane column and a mobile phase of dilute acetic acid (1 in 100) – acetonitrile (84:16) and detection wavelength 254 nm, (III) and (II) elute at 7 min and 13 min respectivley. Singleton[4] has achieved coefficients of variation as low as 0.36 % for (III) and 0.16 % for (II). Tammilehto[5] has used a normal phase HPLC procedure for the determination of these two actives in tablets, achieving linearity in the range 20 – 160 ng, a coefficient of variation of less than 1.6 % and recovery of 99.6 to 101.3 %.

A trial trimethoprim potentiated sulphadiazine formulation for veterinary usage has recently been evaluated in field trials and the sulphonamide was determined in the resulting blood samples as follows. Blood samples are stored in heparin coated tubes under refridgeration prior to sample treatment. This involves centrifugation at 3000 rpm for 5 minutes to remove cellular material and to isolate the plasma. 1 ml of acetonitrile is added to 1 ml plasma in order to precipitate proteins, which are then centrifuged off. The resulting supernatant is subjected to an HPLC assay using a reversed phase C_{18} column (Spherisorb 10 cm ODS), a mobile phase of 15 % acetonitrile and 85 % pH 6 KH_2PO_4, a flow rate of 1 ml/min, an injection volume of 20 μl and a detection wavelength of 258 nm.

Standard blood samples are prepared by adding small volumes of the order of μl of a

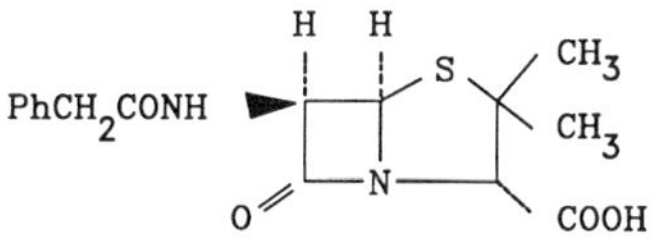

$PhCH_2CONH$ ... CH_3 / CH_3 / $COOH$ (VI)

NH_2 — C_6H_4 — $CO.OCH_2-CH_2.N(C_2H_5)_2$ (VII)

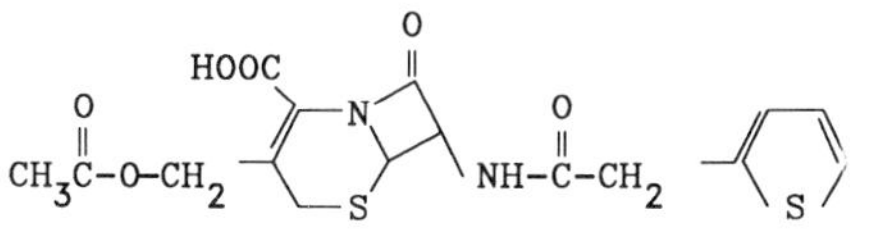

(VIII)

sulphonate, 0.025 M trisodium orthophosphate in water-acetonitrile (86:14), pH 6, a flow rate of 1 ml/min, a detection λ of 204 nm. Dihydrostreptomycin elutes after 1.95 min, penicillin G after 4.68 min and procaine after 10.45 min. A decomposition product of penicillin G is observed in aged formulations between the first and second peaks and this is useful in assessing the stability of the formulation. The recovery of the method was evaluated by adding known amounts of dihydrostreptomycin sulphate and procain penicillin G to the base, applying the analytical method quoted above and then calculating the % recoveries. The mean % recoveries (n = 5) were found to be 101.7 % and 101.9 % respectively. Reproducibility (n = 7 – replicate analysis on the same formulation) was calculated at 1.09 % and 1.22 % respectively. Linear calibration was found in the range 0.2 – 1.0 mg/ml for both actives.

Biological matrices such as milk, urine and plasma can be succesfully analysed by liquid chromatography. Moat's method for milk[10] utilised penicillin extraction from milk with acetonitrile, followed by clean-up by partitioning between buffers and organic solvents, first at acid pH (2.2) and then at neutral pH (7.0). Reversed phase high performance liquid chromatography on a C_{18} column with a mobile phase of phosphoric acid: acetonitrile (80 + 20 to 40 + 60) and a flow rate of 1 ml/min resulted in satisfactory resolution of the mixture in 20 minutes. A limit of detection of 0.005 ppm was reported for penicillin G and the method can distinguish one penicillin from another. Moats[10] also carried out a milk residue study using single cows given standard treatments for mastitis consisting of 3 succesive infusions of 100,000 units of procain penicillin G. Untreated quarters were used as controls and no cross over from treated to untreated quarters was observed. No detectable residues were present after 24 hours in milk from the treated quarters and one would have expected well under 0.01 ppm of the penicillin to have given a discernible peak, if present. The results of bioassays were consistent with these results except that the HPLC method was somewhat more rapid taking a total of 3 h for analysis.

Cephalosporins

In contrast to penicillins, cephalosporins can be electroactive at the DME, without derivatisation. Siegermann[11] has used differential pulse polarography to determine cephaloglycin in 1 M H_2SO_4. A linear calibration plot was observed in the range 1 – 5 ppm and the limit of detection was quoted as < 1 ppm. Peled, Yarnitzky and Smyth[2] have recently investigated the voltammetric behaviour of cephalothin (VIII) by rapid scan square wave voltammetry at the SMDE. They were able to use the optimised reduction signal for rapid, accurate and precise determination of cephalothin in a pharmaceutical preparation using an automated sample handling approach controlled by a Polarographic Analyser. Concentrations of cephalothin in the range $10^{-7} – 10^{-8}$ M can be determined using this technique after adsorptive accumulation at the SMDE.

The determination of (VIII) in biological matrices with and without adsorption preconcentration in urine and plasma was found impossible without prior solvent extraction and/or chromatographic separation.

Reversed phase HPLC has been applied with greater succes than polarography/voltam-

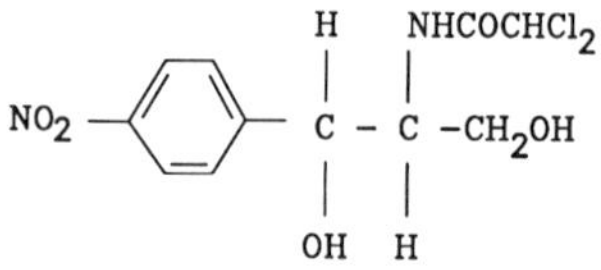

$$NO_2 - \langle \text{benzene ring} \rangle - \underset{\underset{OH}{|}}{\overset{\overset{H}{|}}{C}} - \underset{\underset{H}{|}}{\overset{\overset{NHCOCHCl_2}{|}}{C}} - CH_2OH \qquad \text{(IX)}$$

metry to the determination of (VIII) in plasma and urine by Nygård[12] – plasma proteins were precipitated with acetonitrile and 10 μl of the supernatant injected onto a reversed phase C_{18} column, using a mobile phase of 6 – 11 % acetonitrile in 0.01 M NaH_2PO_4 and UV detection at 254 nm. The limit of detection was < 1 $\mu g/ml$ and linearity was found in the concentration range 0.5 – 500 $\mu g/ml$ of plasma. This method was found to be particularly selective when the capacity factors for co-administered antibiotics and other drugs were calculated.

Chloramphenicol

Chloramphenicol (IX), at a concentration of 5×10^{-5} M in 0.1 M $HClO_4$ supporting electrolyte, was investigated by square wave polarography at single growing mercury drops. The peak current was found to be linearly realted to pulse amplitude for values in the range 5 – 20 mV with a gradient of 40 nA/mV. The peak current levelled off to a constant value of 1550 nA and peak broadening occured for pulse amplitudes of greater than 50 mV. Pulse amplitudes in the range 20 – 50 mV were then regarded as analytically optimal. Using a delay time of 10 s, peak height increased linearly with scan rate in the range 20 – 100 mV/s (gradient 11.5 nA/mVs^{-1}) and, again, levelled off with peak broadening at scan rates greater than 100 mV/s. Scan rates in the range 50 – 100 mV/s were then regarded as analytically optimal. The variation of peak height with pulse width in the range 20 – 100 ms gave rise to a single peak at -0.08 V using optimised scan speed of 50 mV/s and pulse amplitude of 50 mV. At pulse widths of less than 20 ms, a shoulder began to appear at -0.12 V which became a separate peak at -0.15 V using a pulse width of 2 ms. Application of a pulse width of 1 ms yielded a shoulder at -0.09 V and two distinct peaks at -0.16 V and -0.31 V. This phenomenum was also observed at a concentration of 5×10^{-6} M (IX) and could be used to identify this particular nitro-containing antibacterial. Other nitro-containing antibacterials such as metronidazole give different current-voltage patterns on application of square wave polarography at low values of pulse width, e.g., metronidazole at a concentration of 5×10^{-5} M in 0.1 M $HClO_4$ gives one main peak at ca 0 V and a second small peak at -0.32 V, using a pulse width of 2 ms.

The influence of adsorption on the reduction of 2×10^{-6} M chloramphenicol in 4×10^{-3} M $HClO_4$ was investigated by a study of the variation of i_p vs t_p, where t_p is the time during the drop life at which i_p is measured. The log – log plot gave a gradient of 0.8706, and intercept of 1.286 and a correlation coefficient of 0.9989. This gradient would suggest that chloramphenicol is not as strongly adsorbed at the mercury drop as, say 5×10^{-6} M cephalothin (VIII) in pH 2 supporting electrolyte[2] (gradient = 1.33). Square wave voltammetry at the SMDE was then applied to investigate whether adsorptive preconcentration could be used to lower the detection limit for chloramphenicol, bearing in mind the value of the gradient quoted above. Using a frequency of 25 Hz, a concentration of 2×10^{-6} M chloramphenicol in 0.08 M $HClO_4$, the well defined peak at -0.22 V did not undergo current enhancement using delay periods of up to 20 s at the potential of 0 V. This would suggest that gradients of at least the theoretical value of 1.16 for adsorption controlled currents[2] are required for adsorptive preconcentration to be effective prior to voltammetric determination at the SMDE. In the absence of adsorptive preconcentration, a detection limit of 3×10^{-7} M for chloramphenicol was found using square wave voltammetry at the SMDE in 0.081 M $HClO_4$ (frequency 25 Hz). This corresponds to 0.1 ppm identical to that quoted by Siegermann[11] for the differential pulse detection limit for chloramphenicol in 0.1 M acetate buffer.

The limit of detection can be lowered to 10^{-7} M (30 ppb) using a frequency of 5 Hz (scan rate 10 mV/s) and 4×10^{-3} M $HClO_4$ as supporting electrolyte. Again, adsorptive preconcentration at 0 V for 120 seconds had no effect.

While it is admitted that a technique such as rapid scan square wave voltammmetry

$$NH_2 - C_6H_4 - \underset{OH}{\underset{|}{\overset{H}{\overset{|}{C}}}} - \underset{H}{\underset{|}{\overset{NHCOCHCl_2}{\overset{|}{C}}}} - CH_2OH \qquad \text{(X)}$$

(XI)

(XII)

will yield accurate and precise results for the determination of the active in formulations, the determination of (IX) residues in milk, say and the determination (IX) metabolites is best carried out by chromatographic techniques. Wal et al.[13] have described an HPLC method for determination of chloramphenicol residues in milk and have modified the clean up procedure in this method to produce a new method for the determination of trace amounts of (X) in muscle, fat or liver.[14] Chloramphenicol was positively identified in the HPLC fractions by GLC-MS (selected ion monitoring) following formation of the trimethylsilyl derivatives. A simple ion pair HPLC method has bee used[15] as an improved analytical tool for chloramphenicol metabolic profiling. These HPLC separations combined with selective extraction allowed for qualitative and quantitative analysis of ^{3}H chloramphenicol metabolites in rat urine. Complete separation followed by identification of all the metabolites was obtained, especially for the toxicologically significant chloramphenicol arylamine (X):

Other -NO$_2$ containing metabolites such as the oxamic acid, chloramphenicol base and glucuronide could not be resolved by polarographic methods alone although it is probable that metabolite (X) could be determined by amperometric detection at a glassy carbon indicator electrode after HPLC separation.

Clobazepam

Clobazepam (XI) and its N-desmethyl metabolite (XII) are not reduced over the pH-range 2 – 12 at the dropping mercury electrode on application of differential pulse polarography. This is consistent with the absence in (XI) and (XII) of the electroreducible $>C=N<$ group. Amide carbonyl functional groups are not reducible under these conditions. (XI) is not oxidised at the glassy carbon electrode in BR buffer pH 4.0 on application of linear sweep voltammetry whereas (XII) gives a signal at approx +1.4 V (vs SCE). This is consistent with the findings of Smyth and Ivaska[16] who reported that some 1,4-benzodiazepines with an unsubstituted 1-N atom such as potassium chlorazepate, lorazepam, oxazepam could be similarly oxidised, probably by formation of radical cations on the 1-N atom with subsequent coupling reactions etc.

Table 1.

Molecule	Solvent	λ_{max}	$\varepsilon(\times 10^4)$
(XI)	EtOH	232	4.39
		254(S)	2.20
		290	0.6
(XII)	EtOH	229	3.65
		254(S)	2.02
		290	0.54

(S)=shoulder

Since both (XI) and (XII) exhibit relatively strong adsorption in the UV region of the electromagnetic spectrum (Table 1), it is possible by HPLC with UV detection to simultaneously determine (XI) and (XII) in mixtures. Even electrochemical detection on-line using a glassy carbon indicator electrode will only detect (XII). 0.5 ml acetonitrile containing the internal standard diazepam was added to 0.5 ml of plasma containing spiked amounts of (XI) and (XII) to give a solution which was 2000 ng/ml in (XI) and (XII) and 5 μg/ml in diazepam. This was vortex mixed briefly to precipitate proteins and then centrifuged at 3000 rpm for 10 minutes. A 20 μl aliquot was then injected onto the HPLC column using a mobile phase of phosphate buffer pH 6.5: acetonitrile:methanol (50 + 33.3 + 16.7). Three well defined and separated peaks occured at 11.88 min (XI), 8.56 min (XII) and 17.68 min (diazepam). A calibration curve of peak height ratio [(XI)]/[diazepam] vs concentration was linear in the range 60 – 2000 ng/ml as was the case of (XII) in the same concentration range. Limits of detection for (XI) and (XII) of 27 ng/ml and 20 ng/ml were calculated for (XI) and (XII) using the definition of that quantity 3σ where

$$\sigma = \sqrt{\sigma_S^2 + \sigma_B^2}$$

where

σ = standard deviation of the net signal

σ_S = standard deviation of the total signal

σ_B = standard deviation of the blank

In a blind spiking experiment, (XI) was added to plasma at concentrations 50, 200, 600 and 750 ng/ml and (XII) at concentrations 300, 500, 1500 and 2000 ng/ml and the analytical operator asked to estimate concentration in these samples according to preconstructed calibration graphs. The results are given in Table 2.

Table 2.

Clobazepam (XI)				N-desmethylclobazepam (XII)			
Spiked amount	Mean calculated amount (ng/ml)	%error	Coeff. of variation	Spiked amount	Mean calculated amount (ng/ml)	% error	Coeff. of variation
50	48.3	3.5	3.8	300	293.5	2.6	2.2
200	196.5	1.8	4.3	500	454.1	10.1	2.6
600	585.8	2.4	2.3	1500	1538.6	2.5	2.6
750	757.9	1.0	3.4	2000	2097.9	4.7	4.3

A sample taken from a patient on combined antiepileptic treatment of (XI), phenobarbitone and carbamazepine was then subjected to identical sample treatment as above followed by HPLC determination of (XI) and (XII). The retention time for (XI) was 11.79 min and the peak corresponded to a concentration of 274 ng/ml whereas (XII) eluted at 8.50 min giving a concentration of 3744 ng/ml.

References

1. A. Momberg, M.E. Carrera, D. von Bayer, C. Bruhn and M.R. Smyth, Anal. Chim. Acta, **159** (1984) 119.
2. D. Peled, Ch. Yarnitzky and W.F. Smyth, Analyst, **112** (1987) 959.
3. M.A. Brooks, J.A.F. de Silva and L.M. D'Arconte, Anal. Chem., **45** (1973) 263.
4. A. Singleton, J. Pharm. Sci., **69** (1980) 144.
5. S.A. Tammilehto, J. Chromatogr., **323** (1985) 456.
6. A. Weber, K.E. Opheim, G.R. Siber, J.F. Ericson and S.L. Smith, J. Chromatogr., **278** (1983) 337.
7. R.E. Weinfield and T.C. Macasieb, J. Chromatogr., **164** (1979) 73.
8. S. Hendrickx, E. Roets, J. Hoogmartens and H. van der Haeghe, J. Chromatogr., **291** (1984) 211.
9. M. Ueno, M. Nishikawa, S. Suzuki and M. Muranaka, J. Chromatogr., **288** (1984) 117.
10. W.A. Moats, J. Agric. Food. Chem., **31** (1983) 880.
11. H. Siegerman, in: "Differential Pulse Polarography of Antibiotics in Methods in Enzymology", J. Hash (Ed.), Academic Press, New York, 1975, Vol. **43**.
12. G. Nygård, J. Liquid Chromatogr., **7** (1984) 1461.
13. J.M. Wal, J.C. Peleran and F.G. Bories, J. Assoc. Off. Anal. Chem., **63** (1980) 1044.
14. F.G. Bories, J.C. Peleran and J.M. Wal, J. Assoc. Off. Anal. Chem., **66** (1983) 1521.
15. F.G. Bories, J.C. Peleran, J.M. Wal and D.E. Corpet, Drug Metabolism & Disposition, **II** (1983) 249.
16. W.F. Smyth and A. Ivaska, Analyst, **110** (1985) 1377.

APPLICATIONS AND POTENTIALITY OF ELECTROANALYTICAL METHODS FOR INORGANIC TRACE ANALYSIS IN THE PHARMACEUTICAL INDUSTRY

Juerg B. Reust

Analytical R+D, Sandoz Ltd.
CH-4002 Basle, Switzerland

Introduction

As the vast majority of the active principles of pharmaceuticals are organic compounds, inorganic analysis* has usually played a minor role in all the analyses performed in the pharmaceutical industry. But during the last few years its importance has steadily grown. The classical task of inorganic analysis was the simple analysis of residues, usually performed by relatively unselective methods. Examples are determination of sulphate ash, or "heavy metal" analysis by the method of sulphide precipitation. Methods such as complexometric titration played a certain role, e.g., in the case of assay determination of inorganic active principles, such as calcium formulations. The rise of the importance of inorganic analysis has various reasons. First of all come certainly safety considerations. These led to the pharmaceutical industry to use analyses which are more specific, selective, and sensitive than the methods prescribed by authorities. Nowadays other factors than safety considerations play an important role, as traces of metals may have an effect on the mechanism of the action of certain compounds (e.g. proteins which are prone to metal — protein interaction), or metals which strongly influence the production process (e.g. catalysis). The case of protein based innovative pharmaceuticals is also a good example to illustrate the change in inorganic analysis. Being up to now a macro trace analysis the amount of sample available for analysis was never an aspect to be considered. The restricted amount of synthesized proteins (milligrams only) necessitate micro scale trace determination.

Another reason for the rise of inorganic analysis within the pharmaceutical industry is the constantly expanding demand for determinations in the field of environmental analysis. Furthermore the expansion into fields other than pharmaceuticals makes inorganic analysis indispensable, as inorganic components are very often of higher interest in these areas, e.g. foods. These latter two developments extend inorganic analysis not only in a numerical sense, but also towards new classes of matrices, as well as to increased sensitivity.

Electroanalytical Methods

To understand the role of electroanalytical methods of inorganic analysis in the pharmaceutical industry one has to be aware of the main advantages and disadvantages of these methods. These points are listed in Table 1 and represent general characteristics which may not be universally true.

The use (qualitatively and quantitatively) of electroanalytical methods is based on an evaluation of the above mentioned advantages and disadvantages with respect to the

---

* The term inorganic analysis includes all elemental and ion analysis (whole concentration range from traces to major constituents), including organic ions, such as weak organic acids. This inclusion is made for historic reasons, as this type of analysis is generally performed in the trace elemental laboratories in the pharmaceutical industry.

Contemporary Electroanalytical Chemistry, Edited by A. Ivaska *et al.*
Plenum Press, New York, 1990

Table 1. General advantages and disadvantages of electroanalytical methods.

Item	Advantage	Disadvantage
multi-element	yes	
selectivity	high	
sensitivity	high	
speciation	available	
interferences		high
time consumption		high
maintenance		high
cost	low	

laboratory and its environment, e.g., classes of samples, size, quality control or research laboratory. For further discussion, electrochemical methods will be divided into three groups:
a) "stand-alone" electroanalytical methods (e.g. voltammetry, polarography)
b) combined methods where the electrochemical technique is usually the detection system (e.g. ion-chromatography with amperometric detection)
c) potentiometry.

The topic potentiometry will not be covered in this paper since it has been discussed by Simon.[1]

a) Stand-Alone Electrochemical Methods

1) Coulometric Titration

The coulometric titration (production of the iodide reagent) for the determination of water according to Karl Fischer is a well established technique. The method provides accurate, rapid, and practically interference free results. The instrumentation is easy to handle and the method can be regarded as rugged. Furthermore, it allows the determination of very low traces of water or the determination in very small samples which is not possible by the use of the classical titration reagents. All these aspects resulted in its wide spread use in pharmaceutical industry. The method is so well accepted that it is usually not regarded any more as an electrochemical method at all. As with any development, there are two sides; it is encouraging to see such a spread of an electroanalytical method, on the other hand it is somewhat disillusioning to realize that the fact "electroanalytical" is lost. Interestingly enough, the same development also happened in the area of electrochemical detectors (cf. later in this paper)!

2) Voltammetry

Determination of Heavy Metals. Voltammetry offers the possibility for determiniation of all the heavy metals of usual interest to the pharmaceutical industry (Zn, Cd, Pb, Cu, Ni, Co) within one experimental run. Sensitivity is high enough to permit the use of milligram amounts of samples. It was shown by Reust et al.[2] that this approach is feasible, even in the case of complex matrices, such as proteins or proteins in physiological buffer solutions if appropriate digestion conditions can be established. A DP-ASV voltammogram of the determination of cadmium, lead, and copper in bovine serum albumin is shown in Fig.1. The most critical step is the digestion, as minor amounts of undestructed organic matter influence the electroanalytical response significantly. This work showed for example that the conventional nitric/sulphuric acid digestion with addition of hydrogen peroxide is capable of destroying pharmaceutical samples with matrices such as ascorbic acid and gelatine, but for the complete oxidation of even small quantities of proteins a perchloric acid digestion step has to be used. Preliminary investigations published by Schramel and Knapp[3] indicate a possible improvement by the use of a high pressure / high temperature ashing system. Nevertheless one has to realize that the limiting step in electroanalysis of heavy metals is — and will be for some time — the pretreatment step required, which is usually a digestion procedure.

Digestion procedures do have some common characteristics, all of which are certainly far from ideal for a fast propagation into analytical laboratories. They are time consuming,

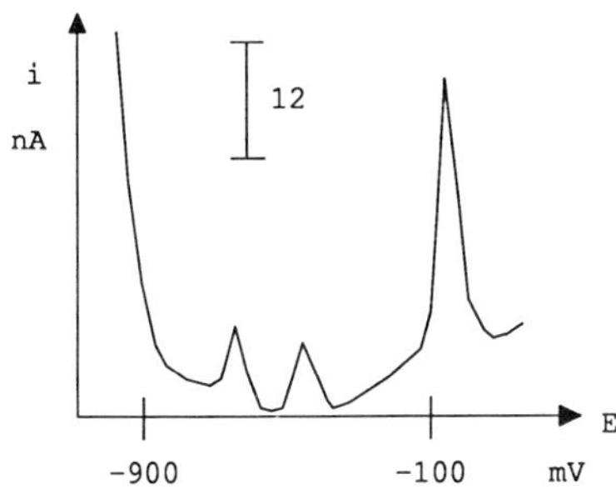

Fig.1. DP-ASV voltammogram of Cd, Pb, and Cu in bovine serum albumine. 150 ng of each metal added to 100 mg BSA in 50 ml phosphate buffered saline, perchloric acid digestion. Fig. from reference.[4]

complicated and require skilled personnel, difficult to automate, not generally applicable, and potentially dangerous. The fact that they are not generally applicable requires the use of different systems in a laboratory. We have seven systems available (not counting the fully manual digestion procedure in an erlenmeyer flask over the Bunsen flame). Considering that there are at least three different fundamental parameters influencing the digestion (chemicals, time/temperature profile, pressure), we end with more than two dozen different methods! Although our situation (being a research laboratory and having an extremely broad spectrum of different matrices and analytes) is certainly unusual, it does show the unsatisfying — even confusing — situation in this area of analytical chemistry, mostly mandatory prior to almost any electrochemical analysis. A fully automated, flow through digestion procedure resulting in adequate destruction of the sample matrix to allow electroanalysis would be a great step forward in inorganic trace analysis, especially for voltammetry. This would make the coupling with flow through DP-ASV systems as presented by Neeb et al.[4] possible. Another solution to the dilemma is the determination of extractable amounts of metals only e.g.,[5] and not to determine the total metal content. Disadvantages of such a procedure are evident, e.g., one cannot guarantee to find any metallic contaminant in a sample. A completely different approach would be, as in clinical chemistry, to define exactly a method and to use it worldwide. Results would then be reported only with respect to the reference values.

Strong acids are of great importance in analytical chemistry, e.g., they are used as the main reagent for the wet digestion. It is obvious that these acids have to be extremely pure. Purity control, specially with respect to heavy metals, is mandatory. Voltammetry is one of the few techniques, where a direct analysis is possible, contrary to e.g. graphite furnace atomic absorption spectrometry. Using a rotating mercury film glassy carbon electrode and a standard DP-ASV procedure the concentrated acid can be directly added to the cell containing the background electrolyte. Knowing the characteristics of the electrode, semi-quantitative evaluation of contamination by lead and cadmium can be established within a few minutes. At our laboratories a testing program will be started, specially to control the effect of the introduction of sub-boiling distillation units for further purification of high-purity acids.

Speciation Studies. One of the biggest advantages of electroanalytical techniques is the fact that they inherently allow the performance of *in situ* speciation studies. Other methods, e.g., atomic absorption spectrometry, require a chemical separation prior to the analysis to determine individual species of an element, a step adding great uncertainty to the result. An example from our laboratory is the determination of chromium(VI) in waste water according to a procedure with similarities to that described by Bersier[6] for such a determination in dyes. A determination limit as low as 0.02 μg/ml can be reached even where waste water contains thousandfolds excesses of chromium(III). Waste water is sucked into a syringe containing methanol and acitvated charcoal, thoroughly shaken and injected through a 0.45 μm filter to the background electrolyte containing diamine. Then chromium(VI) is determined by differential pulse polarography. Detailed experimental conditions are given elsewhere,[7] a polarogram of the determination of chromium(VI) in waste water spiked with chromate (ng/g range) is shown in Fig.2.

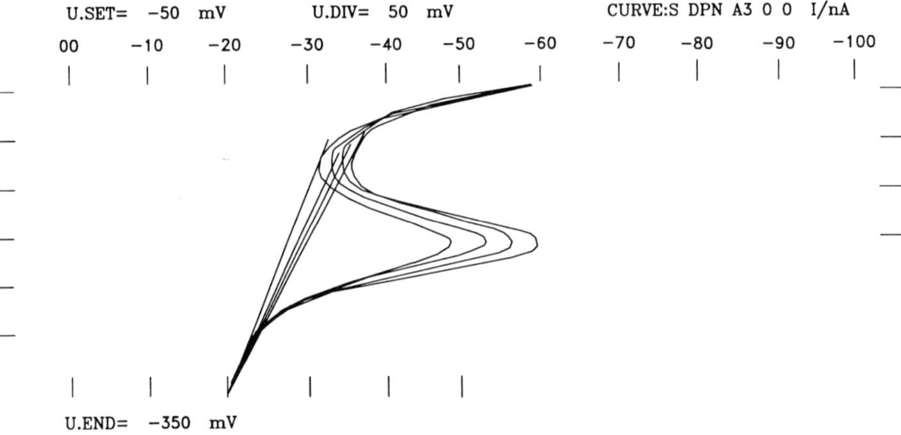

Fig.2. Dp polarogram of chromium(VI) determination in waste water. First signal corresponding to 135 μg/l.

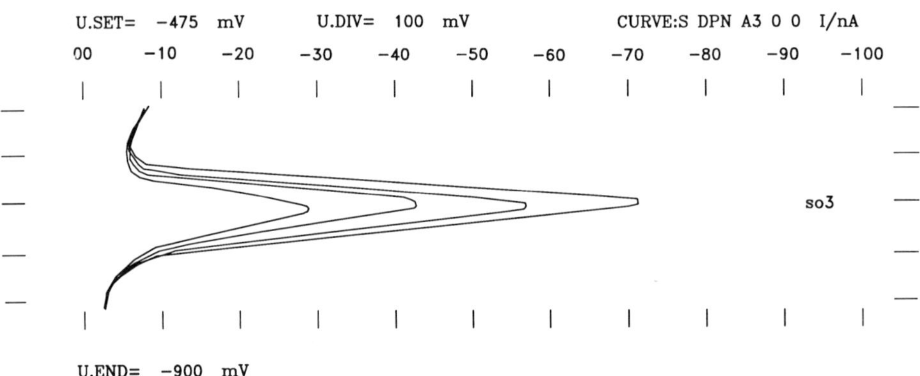

Fig.3. DP polarogram of sulphite determination in a research compound. First signal corresponding to 150 μg/ml sodium pyrosulfite.

Determination of Toxic Anions. An area of rising interest is the determination of (toxic) anions in active principles of pharmaceuticals, as well as in environmental matrices. In our laboratories we use ion-chromatography for such types of analyses as this procedure is a rapid multi-ion method. On the other hand there are certain cases, where the polarographic approach has advantages, e.g., as in the case of sulfite and sulphide determination in various matrices ranging from pharmaceuticals to residues from boiling systems. Dependent on the matrix the pretreatment step can range from simple suspension to direct distillation into the polarographic cell. As an illustration the polarogram for the sulfite determination in the μg/g range in a pharmaceutical research compound is shown in Fig.3.

3) Other Methods

At present other electroanalytical methods, e.g., tensammetry, are of no general interest in inorganic analysis in the pharmaceutical industry. Amperometric titrations are used occasionally for certain determinations, e.g., hydrogen peroxide, and of course for the end-point determination of the Karl Fischer method. The use of potentiometric stripping analysis as an alternative to the voltammetric approach is not generally used for pharmaceutical compounds.

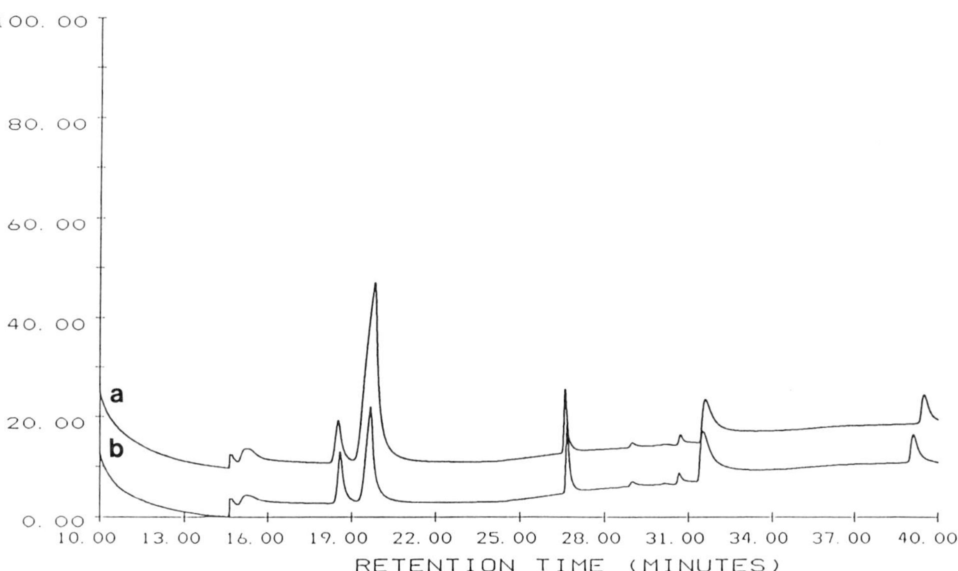

Fig.4. Gradient elution ion-chromatography of protein research compound (acetate form) spiked with anions, conductivity detection. a) standard solution (containing fluoride, acetate, chloride, oxalate, and citrate) b) protein plus standard addition of above solution.

b) Electrochemical Detectors

1) Electrochemical Detection in Ion-Chromatography

Conductivity Detection. The use of a conductivity detector is the basic configuration scheme of any ion-chromatography system. The sensitivity is high and it needs practically no maintenance, an important aspect in routine analysis. The detector is so simple to use that it is — cf. above the case of the Karl Fischer determination — not recognized as an electrochemical detector. Latest developments with membrane suppressor systems allow the use of the gradient elution technique. Using this procedure it is possible to analyse in the same run ions of extremely different characteristics (cf. Fig.4) as demonstrated for the analysis of a protein based active principle.

Amperometric Detection

Amperometric detectors, or specifically their electrodes, need much more care and maintenance and are therefore used only in cases where conductivity measurements are not possible or sensitive enough. Analytes range from sugars, metal-oxo-anions, to toxic anions, such as cyanide and sulphide. The determination of cyanide and sulphide in a drug substance is shown in Fig.5. The sample is analyzed after on-line matrix removal and accumulation of the analyte (a system similar to that described by Reust et al. in[8]) amperometrically on a silver electrode.

By combining the three most widely used detectors (conductivity, amperometric, and spectrophotometric) and using a gradient elution technique, the range of anions which can be determined in the same experimental procedure is extreme and offers a great flexibility.

2) Coulometric Detection in Halogen Analysis

The use of coulometry as a detection technique has become of interest as the legal requirements for very low TOX values (TOX = total organic halogen). Mitsubishi[9] has

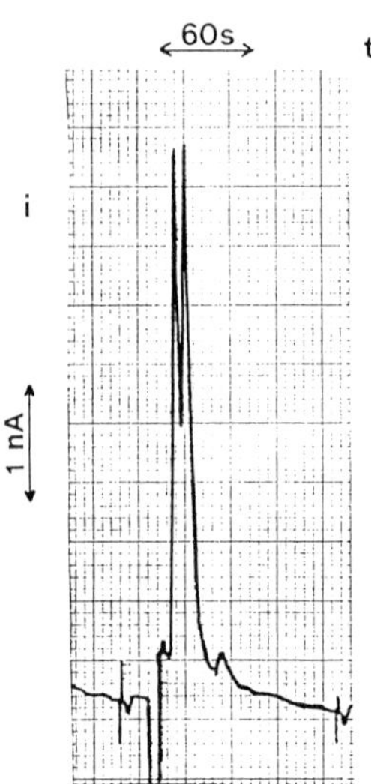

Fig.5. Amperometric determination of cyanide and sulphide in a pharmaceutical research compound in the acetate form. 7.72 mg sample containing approx. 1 mg acetate, addition of 70 ng cyanide and sulphide resp., in 7 ml 0.0147 M ethylenediamine, 500 μl injected (amount of analyte detected: 5 ng each).

shown that using a coulometric detector after combustion allows rapid determination of TOX values in waste water. Latest developments extend the use to solid samples.

Conclusions

Electrochemical methods offer a wide range of possible analyte determinations, covering both anion and cation analysis over a large concentration range. Advantages and disadvantages have to be considererd carefully before deciding on their use. Generally the larger a laboratory is, the less it will make use of the full spectrum offered by electroanalytical methods. They will be used only selectively (e.g. speciation studies, toxic anions at low concentrations), whereas analyses such as heavy metal determination will be done routinely by faster (e.g. graphite furnace atomic absorption spectrometry) or broad multi-element (e.g. X-ray fluorescence spectrometry) methods. An important use of electroanalytical methods in a large laboratory is for method validation aspects. On the other hand a smaller laboratory will make full use of the whole spectrum. In such cases the time required will be less important, as the number of analyses and the range of samples will be smaller. Furthermore the price of other instrumentation is prohibitive for small laboratories, e.g., a graphite furnace atomic absorption system costs approx. five times more than the comparable electroanalytical instrumentation and is restricted to elemental analysis only.

Acknowledgements

The author likes to express his thanks to R. Maurer and E. Grollimund for their contributions to this work and their participation in the experimental part.

References

1. W. Simon, Proceedings of the ElectroFinnAnalysis, Turku 1988.
2. N. Gopalsamy, V. Pfund, P. Rach and J.B. Reust, Analusis, **16** (1988) 55.
3. P. Schramel, S. Hasse and G. Knapp, Fresenius' Z. Anal. Chem., **326** (1987) 142.
4. R. Neeb, et al., Analusis, submitted for publication.
5. D. Frahne, J.V. Geil, K. Geng and U. Schrader, Metrohm Monografie, Metrohm AG, Januar 1982, Herisau (CH).

6. P.M. Bersier, personal communication.
7. R. Maurer and J.B. Reust, in preparation.
8. E. Grollimund and J.B. Reust, in: "Electrochemistry, Sensors and Analysis", M.R. Smyth and J.G. Vos (Eds), Anal. Chem. Symposia Series, **25** Elsevier, Amsterdam, 1987, 411.
9. Mitsubishi application manual, Mitsubishi Chemical Industries Ltd., Tokyo (J), 1988.

DIRECT ELECTROCHEMICAL IMMUNOASSAYS INVOLVING
ADSORBED OR IMMOBILISED SPECIES

Malcolm R. Smyth[1], Eileen Buckley[1], Juana Rodriguez
Flores[2], Richard John[3] and Gordon G. Wallace[3]

1. School of Chemical Sciences
   NIHE Dublin, Glasnevin, Dublin 9, Ireland
2. Department of Analytical Chemistry and Electrochemistry
   Universidad de Extremadura, Badajoz, Spain
3. Department of Chemistry, University of Wollongong
   Wollongong, N.S.W. 2500, Australia

Introduction

Since the early 1970's there has been much interest shown in the development of non-isotopic immunoassays. The main reasons for this stem from the perceived dangers of using radio-labelled substances (as required in radioimmunoassay), and the search for more selective, sensitive and precise methods of analysis. Much work has therefore been devoted to the development of fluorescent and enzyme-linked immunoassays, but it is only in recent years that there has been strong interest in the application of electrochemical techniques in this regard.[1,2]

To date, the analytical approaches that have been used in this regard have been largely based on either the development of immuno-selective electrodes,[2] or on labelling the antibody or antigen with either an enzyme (in which case the product of an enzyme-substrate reaction is monitored by HPLC or FIA with amperometric detection) or an electroactive molecule (in which case the current due to the oxidation or reduction of the label is usually decreased on binding of the antigen with the antibody).[1] Approaches using enzyme-labelled antigens or antibodies are inherently more sensitive than those involving electroactive labelling because of the catalytic nature of the enzymic reaction and the high sensitivity one can achieve with electrochemical detection systems in flowing streams.[3]

Direct Electrochemical Immunoassays

In addition to these approaches, there have been some attempts to monitor immunological reactions directly using both potentiometric and voltammetric methods of analysis. In the early 1950's, Breyer and Radcliff[4,5] reported their results on the use of polarography to study the interaction of an azoprotein with a specific antiserum raised in a rabbit. In the presence of non-specific antisera, no decrease (other than that due to a diluent effect) was seen in the polarographic wave due to the antigen. A sharp decrease in this signal was noted, however, when the specific antiserum was added to the cell.

In the late 1970's, Yamamoto et al.[6,7] reported on the use of chemically modified metal electrodes for potentiometric investigations of antibody-antigen complexes. A titanium wire was chemically activated in an aqueous CNBr solution and then treated with the specific antiserum to the antigen under study, i.e., human chorionic gonadotropin (hCG). The electrode was then soaked in urea to deactivate any remaining surface active sites and prevent any

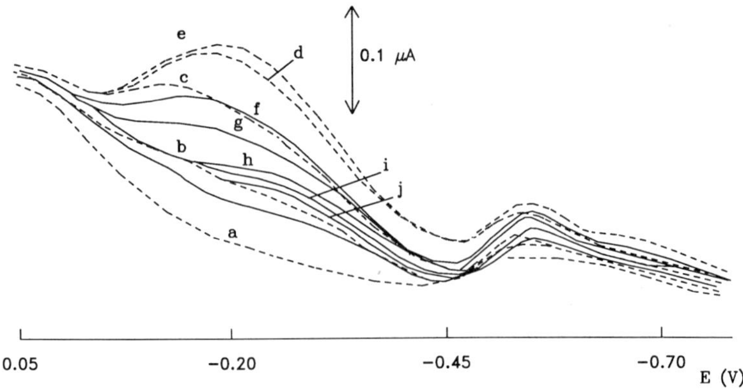

Fig.1. Differential pulse adsorptive stripping voltammograms of mouse IgG in the presence of anti-mouse IgG.   a) 0.1 M phosphate buffer   b) 2 mg/l IgG,   c) 4mg/l IgG,   d) 6 mg/l IgG,   e) 8 mg/l IgG,   f) e + 0.01 % anti-IgG,   g) e + 0.02 % anti-IgG,   h) e + 0.03 % anti-IgG, i) e + 0.04 % anti-IgG,   j) e + 0.06 % anti-IgG (conditions: scan rate 10 mV/s ; drop time 1 s ; t_{acc} 50 s).

excessive adsorption of protein. On addition of hCG to the electrolyte solution, the potential of the indicator electrode was found to shift to more positive potentials by about 3-5 mV, but returned to its original value when it was returned to a fresh electrolyte solution. This shift in potential was attributed to a specific reaction between the anti-hCG bound at the electrode surface and the hCG in the solution, and the electrode was tested for the analysis of hCG in the urine of pregnant women. However, the electrode was found to exhibit some non-specific responses, e.g., with albumin, and could therefore not be used in routine clinical practice.

More recently, Robblee et al.[8] have reported that the oxidation currents obtained for 5-hydroxytryptamine (5-HT) were higher at a glassy carbon electrode containing chemisorbed anti-5-HT than for an uncoated electrode. In addition the modified electrode exhibited greater selectivity for 5-HT over its metabolite 5-hydroxyindole acetic acid (5-HIAA). These findings were again attributed to formation of a surface antibody-antigen complex. Hertl[9] has suggested that the formation of such surface antibody-antigen complexes perturbs the electrical double layer resulting in a current change, and reported on dose-response curves obtained for immunoglobulin G (IgG), anti-IgG, anti-ferritin and *S. Aureus* cells.

Adsorptive Stripping Voltammetry

We have recently demonstrated similar findings for several antigen-antibody reactions at the hanging mercury drop electrode (HMDE) using the technique of adsorptive stripping voltammetry (AdSV). These have included the reaction of human serum albumin (HSA) with anti-HSA,[10] mouse IgG with anti-mouse IgG[11] and hCG with two anti-β-hCG antibodies.[12] In all cases, the current observed following accumulation of the antigen (at the HMDE) decreased on addition of the specific antibody. This is illustrated for the case of mouse IgG with anti-mouse IgG in Fig.1.

In all of these studies, we have demonstrated the importance that choice of accumulation potential (E_{acc}) has on the cyclic and adsorptive stripping voltammetric behaviour of the protein under study. By applying an initial potential of about −0.10 to −0.15 V (vs Ag/AgCl), it can be seen from cyclic voltammetry (cv) that the adsorbed protein species exhibits a reversible redox couple on repetitive scanning, indicating that the structure of the protein is not greatly affected under these conditions. Under conditions of adsorptive accumulation at this potential, two peaks can usually be seen using cv, the first due to the product adsorption and the second due to the main faradaic process. It has not yet been established what this faradaic process is due to, because most of the proteins we have studied give rise to their

368

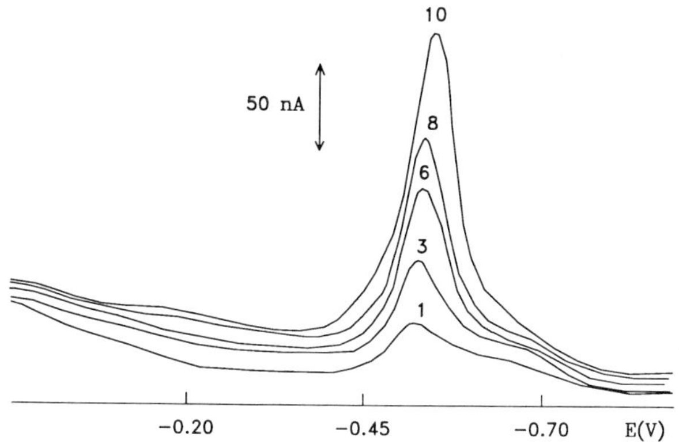

Fig.2. Effect of accumulation time (in min) on differential pulse adsorptive stripping voltammetric behaviour of rabbit anti-β-HCG (conditions: E_{acc} + 0.05 V ; scan rate 5mV/s ; drop time 1 s).

main peak around the same potential, regardless of their amino acid composition. It was originally thought that this process was due to reduction of cystine residues in the protein structure,[10] but a recent study[13] has suggested that it may be due to non-faradaic processes involving reorientation of the protein at the surface of the electrode, causing changes in the double layer structure. Our results would indicate, however, that the behaviour is likely to be both faradaic and non-faradaic in nature.

When the initial potential is changed to more positive potentials, i.e. +0.10 → +0.20V, then the protein appears to undergo structural change at the surface of the electrode causing loss of electrochemical behaviour after about 4-5 repetitive scans. Under conditions of adsorptive accumulation at these positive potentials, much higher current (and hence greater sensitivity) can be obtained using AdSV. Greater sensitivity in AdSV is also achieved by increasing the accumulation time (t_{acc}). This is illustrated for the case of rabbit anti-β-hCG in Fig.2.

The fact that these antigen-antibody reactions cause a decrease in the current of the main faradaic peak of the antigen means decreased sensitivity for detection of the antibody. We have, however, demonstrated in the case of the lectin concanavalin A, that enhancement can also occur. This protein exists predominantly as a tetramer, with every protomeric unit capable of selectively binding one mannose molecule. We have shown that the peak current due to concanavalin A actually increases linearly with respect to [mannose],[4] as would be expected from structural considerations.[14]

Mediation of Response by Electron Transfer Reagent

The response of a surface antigen-antibody reaction can also be mediated by an electron transfer reagent. This has been demonstrated by Robinson et al. in their studies on electrochemical immunoassays for hCG[15] and thyroxine.[16] In the analysis of hCG, a "two-site" amperometric immunoassay was developed in which monoclonal "capture" antibodies were immobilised on the surface of a glassy carbon electrode. A second antibody against hCG was labelled with glucose oxidase. The electrode was used both to separate "free" from "bound" enzyme-antibody conjugate and to assay the enzyme activity electrochemically by use of dimethylaminomethyl ferrocene as an electron transfer mediator. This method was found to correlate well with an immunoradiometric assay.[15] In the analysis of thyroxine, another ferrocene derivative, namely ferrocenemonocarboxylic acid, was used as the electron transfer mediator.[16]

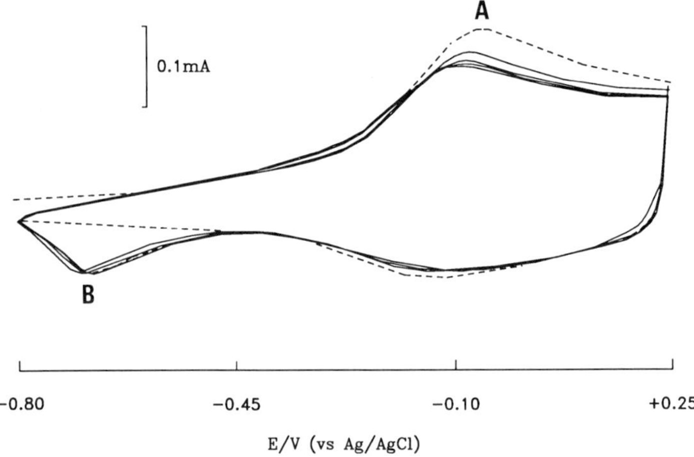

Fig.3. Cyclic voltammograms obtained for a polypyrrole-anti-HSA electrode generated galvanostatically at a platinum electrode from a solution of 0.5 M pyrrole that was 100 ppm in anti-HSA. Dotted line represents scan for polypyrrole-anti-HSA electrode cycled in 0.1 M NaNO$_3$, whereas solid lines represent subsequent scans in 0.1 M NaNO$_3$ that was 100 ppm in HSA (conditions: scan rate 100mV/s).

Polymer Modified Electrodes

An approach which is now becoming of more importance in the development of enzyme-based biosensors is that of immobilisation of the active protein in a polymer matrix.[17,18] We have now carried out a preliminary investigation into the incorporation of anti-HSA into an electrochemically-generated conducting polypyrrole film.[19] Conducting polymers such as polypyrrole (I)

are known to contain counterion loadings as high as 30-40 % w/w. The counterion (C$^-$) is incorporated from the supporting electrolyte during electrochemical synthesis. Based on this knowledge, polymers containing various counterions which may be used as chemical reagents or electrocatalysts have been synthesised.[20] We have demonstrated that polypyrrole-anti HSA electrodes can be prepared galvanostatically from solutions of pyrrole which are 10-500 ppm in the antibody. The scanning electron micrographs obtained for such electrodes at both platinum and gold foil electrodes have demonstrated that the morphologies of the resulting polymer films are quite different, but both are much more porous than corresponding galvanostatically generated polypyrrole films grown from solutions containing simple salts e.g. KCl. A typical cyclic voltammogram obtained for a polypyrrole-anti-HSA electrode is shown in Fig.3. When the electrode was cycled in 0.1 M NaNO$_3$ solution, a peak (peak A) was seen around -0.1 V, believed to be due to the influx of NO$_3^-$ ions into the polymer film. When the electrode was then cycled in 0.1 M NaNO$_3$ which was 100 ppm in HSA, a new peak (peak B) was seen on the reverse scan around -0.7 V which does not occur in the absence of the antigen or when a polypyrrole-chloride electrode was cycled in the presence of the antigen.[19] The nature of this peak is not yet understood, but it may be due to the effect that the binding of the antigen to the immobilised antibody has on the polymer structure. Work is underway to see if this polypyrrole-anti-HSA electrode can be used as a biosensor for HSA, based on these findings.

370

References

1. W.R. Heineman and H.B. Halsall, Anal. Chem., **57** (1985) 1321A.
2. D. Monroe, Immunoassay Technol., **2** (1986) 57.
3. D.C. Johnson, S.G. Weber, A.M. Bond, R.M. Wightman, R.E. Shoup and I.S. Krull, Anal. Chim. Acta, **180** (1986) 187.
4. B. Breyer and F.J. Radcliff, Nature, **167** (1951) 79.
5. B. Breyer and F.J. Radcliff, Aust. J. Expt. Biol., **31** (1953) 167.
6. N. Yamamoto, Y. Nagasawa, M. Sawai, T. Sudo and H. Tsubomura, J. Immunol. Methods, **22** (1978) 309.
7. N. Yamamoto, Y. Nagasawa, S. Shuto, H. Tsubomura, M. Sawai and H. Okumura, Clin. Chem., **26** (1980) 1569.
8. L.S. Robblee, M.J. Mangaudis, M.S. Blankship and R.B. Salmonsen, Proc. Electrochem. Soc., 1986, 86-14 (Electochem. Sens. Biomed. Appl.), 139.
9. W. Hertl, Bioelectrochem. Bioenerg., **17** (1987) 89.
10. J. Rodriguez Flores and M.R. Smyth, J. Electroanal. Chem., **235** (1987) 317.
11. M.R. Smyth, E. Buckley, J. Rodriguez Flores and R. O'Kennedy, Analyst, **113** (1988) 31.
12. J. Rodriguez Flores, R. O'Kennedy and M.R. Smyth, Biosensors, 4 (1988) 1.
13. H. Emons, G. Werner and W.R. Heineman, manuscript in preparation.
14. J. Rodriguez Flores, R. O'Kennedy and M.R. Smyth, Analyst, **113** (1988) 525.
15. G.A. Robinson, V.M. Cole, S.J. Rattle and G.C. Forrest, Biosensors, **2** (1986) 45.
16. G.A. Robinson, G. Martinazzo and G.C. Forrest, J. Immunoassay, **7** (1986) 1.
17. P.N. Bartlett and R.G. Whittaker, J. Electroanal. Chem., **224** (1987) 27, ibid, **224** (1987) 37.
18. C.D. Crawley, Diss. Abst. Int. B., **47** (1987) 4845.
19. R. John, G.G. Wallace and M.R. Smyth, Manuscript in preparation.
20. G.G. Wallace in: "Chemical Sensors", T.E. Edmonds (Ed.), Blackie, Glasgow and London (1988), 132-154.

ELECTROCHEMICAL BEHAVIOUR AT SOLID ELECTRODES AND METABOLIC FATE OF DRUGS

J-M. Kauffmann, J-C. Vire, O. Chastel and G.J. Patriarche

Université Libre de Bruxelles
Institut de Pharmacie
Campus Plaine 205/6, Boulevard du Triomphe
1050 Brussels, Belgium

Introduction

Electron-transfer and redox reactions play a central role in the metabolic fate of drugs. Number of compounds are intensivley metabolized in living organisms by enzymatic redox processes leading to the formation of by-products which may be active, inactive or toxic.[1] The mechanistic process underlaying these biotransformations are complex and not fully understood and their knowledge is capital for structure-activity studies and for synthesis of more active and less toxic products. With respect to the aforementioned needs, electrochemical techniques can be regarded as methods of choice allowing several interesting approaches:
1. sensitive and specific determinations
2. controlled drug electrosynthesis
3. *in vivo* monitoring
4. drug-membrane interaction modelling
5. precise insights into redox mechanisms

In this report we will illustrate the latter aspect by presenting some relevant results obtained in our investigations.

Experimental

Apparatus

Voltammetric measurements were carried out by using a PAR Model 175 and PAR Model 174 potentiostatic system connected to a PAR recorder and by using a BAS CV-27 voltammograph connected to a HP 7090 A recorder. The working electrodes are a Metrohm EA 267 carbon paste, a Metrohm AG glassy carbon, a Tacussel EDI platinum and a home made composite carbon-polyethene (30 % w/w) electrode.[2] Non aqueous measurements were performed in lithium perchlorate 0.1 M in acetonitrile media. Coulometry and exhaustive electrolysis were carried out at a platinum gauze in a PAR 9,600 cell using a PAR 173 and 179 potentiostatic system. Thin-layer chromatography (Merck Kieselgel 60F254), infra red (Beckman 4240), UV (Beckman DB-T) and mass spectrometry were applied when necessary for identification of the electrolyzed products.

Reagents

All reagents were of analytical grade (Merck p.a.). Pharmaceutical compounds were of pharmacopoeia quality and were used without further purification.

Contemporary Electroanalytical Chemistry, Edited by A. Ivaska *et al.*
Plenum Press, New York, 1990

I

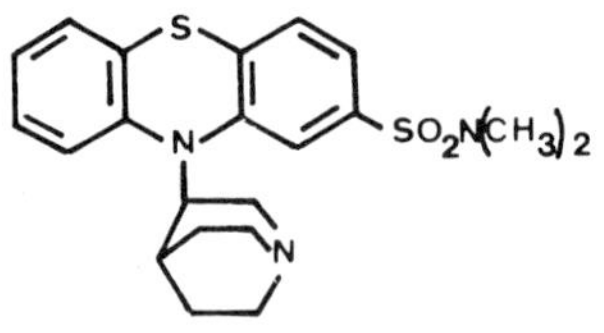

II

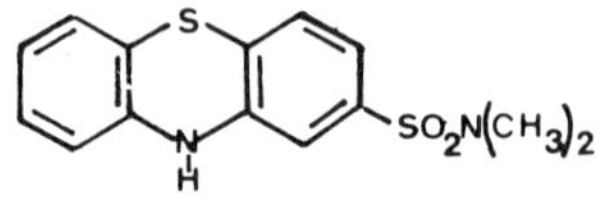

Fig.1. Schematic structure of quisultidine (I) and its metabolite 2-(N,N-dimethylsulphamoyl)phenothiazine (II).

Results and Discussion

Phenothiazine Nucleus

As we know, the first step in the metabolization of phenothiazines is a one electron oxidation mechanism giving a cation radical. This structure is supposed to be responsible for the psychotic properties but also for the toxic side effects of phenothiazines.[3] Cyclic voltammetric studies of a recent phenothiazine, quisultidine (Fig.1, structure I) possessing little psychotic activity have permitted to point out in a rapid and easy manner its unusual behaviour.[4] Phenothiazines are readily electrooxidized giving a cation radical which is rather stable in nonaqueous or high acidic aqueous media.

Quisultidine is surprisingly difficult to oxidize and gives no stable cation radical. However by reversing the scan direction, after the peak potential (E_{pa}), several cathodic peaks are detected at less positive potentials (Fig.2, curve a). By running the cyclic voltammogram of the metabolite of quisultidine (Fig.2, curve b) we observe a perfect matching of the curve with the curve shape of the less positive redox couples (curve a). These results suggest that a significant amount of the unsubstituted analogue (Fig.1, structure II) is formed during the oxidation of quisultidine. The unusual high oxidative potential and irreversible behaviour of quisultidine are probably related to the steric hindrance caused by the 3-quinuclidinyl side-chain which stabilizes the molecule by hindering the formation of the coplanar radical cation. Coplanarity is achieved by cleavage of the side-chain liberating the 2-(N,N-dimethylsulphamoyl)phenothiazine. The electrochemical observations can be correlated to the unique pharmacological properties of the molecule. The fact that quisultidine gives no stable cation radical is in agreement with its low psychotic activity. The biotransformation of the molecule gives rise to the appearance of the corresponding metabolite (structure II) which is responsible for the slight psychotic side effects.[5] Cyclic voltammetric curves have shown that structure II appears after oxidation of quisultidine. The former possess the phenothiazine nucleus and behaves classically giving a stable cation radical (peak a1).

Piperazine Ring

Nitrogen, after carbon and hydrogen, is one of the most abundant element present generally as amines in drugs. The first step in biotransformation of xenobiotic amines is usually an oxidation process enzymatically catalyzed and the nature of products formed as well as the biochemical bases for the reactions are of considerable interest.[1] N,N'-disubstituted piperazine is the basic part of a number of psychotic active drugs. The presence of this basic chain in the molecular structure seems to be a prerequisite for binding to the receptor. The biotransformation of the piperazine chain pass through several routes including N-oxygenation, N-dealkylation or ring opening reactions connected with the loss of two carbon atoms (Fig.3). The mechanisms underlaying these reactions are still not fully understood.[6,7]

We have realized the electrochemical study of a series of tricyclic neuroleptics possessing a piperazine moiety and suggested an oxidation mechanism at physiological pH. Previous experiments have been done in aqueous media[8] then after non aqueous investigations and spectroscopic characterization of the electrolyzed products[9] the mechanism of oxidation have

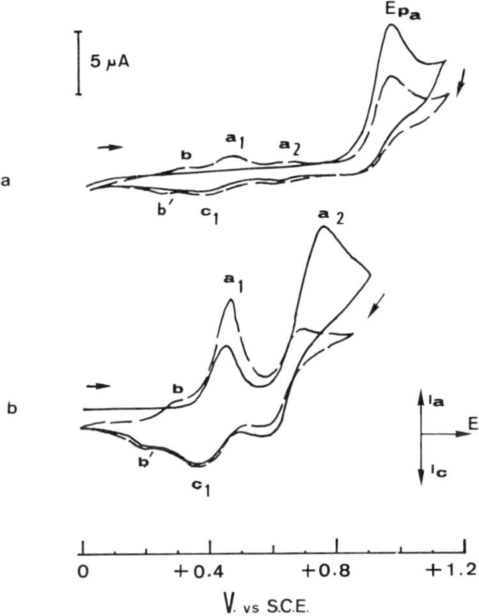

Fig.2. Cyclic voltammogram of 5×10^{-5} M quisultidine (a) and 5×10^{-5} M 2-(N,N-dimethyl sulphamoyl)phenothiazine (b) in 3 M H_2SO_4 + 20 % methanol at the carbon poyethylene electrode. Scan rate: 50 mV/s. Broken line = second scan.

been established. In non aqueous media the oxidation occurs preferentially on the piperazine group giving a imminium cation which can be trapped by nucleophilic agents like triphenylphosphine. The oxidation process is irreversible since the imminium cation is unstable; in the presence of water it hydrolyzes rapidly giving the N-dealkylated product which is further oxidized (Fig.4). A close parallelism was found between the electrochemical oxidation and pharmacological activity.[10] The integrity of the piperazine side-chain is of great importance for binding to receptors. For optimal neuroleptic activity, the distal nitrogen must be tertiary and the lone pair of electrons on the nitrogen must be unoccupied, otherwise binding would be prevented.[11] The electrochemical results suggest also the importances of the piperazine side-chain. When the lone pair of electrons is protonated, oxidation is more difficult. If both nitrogens are protonated, oxidation no longer occurs on this site. Moreover, the distal nitrogen must be tertiary otherwise no imminium cation and no addition product are obtained. These findings suggest that oxidation phenomena may occur in the brain, giving rise to reversible or irreversible bonds, and could play a key role in the activity of these drugs.

Trazodone is another example of a molecule possessing a piperazine ring in its structure and which is extensively metabolized (Fig.3). The electrooxidation of the molecule has been investigated by cyclic voltammetry as a function of pH (Fig.5).[12] From the comparative study of tradozone and its metabolite 1-m-chlorophenylpiperazine, we are able to assume that the first oxidation step of trazodone is located on the piperazine ring. We may postulate that when the aliphatic nitrogen of the piperazine ring, distal to the benzene ring, is protonated ($TRZH^+$), oxidation occurs on the proximal nitrogen with possible formation of benzene hydroxylated forms reversibly and more readily oxidized than the parent compound (Fig.3). Above pH 8.5, oxidation may exclusively occur at the most basic piperazine nitrogen (distal) following the well established aliphatic tertiary amine oxidation pathway.[9,13] The second oxidation step of trazodone occurs at the triazolopyridine moiety of the molecule.

375

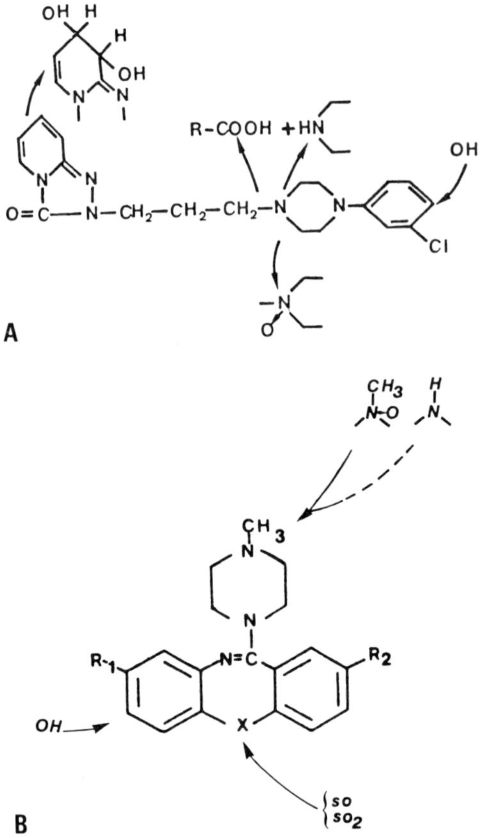

Fig.3. Literature reported metabolites of trazodone (A) and of dibenzothiazepines (B), R_1 = H, R_2 = Cl.

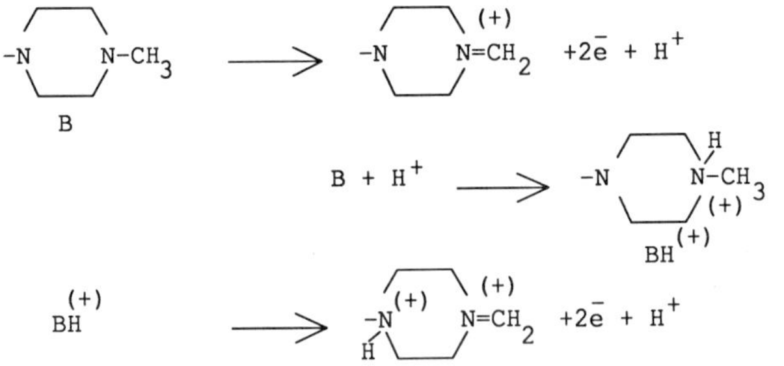

Fig.4. Oxidation mechanism of piperazine with formation of imminium cation.

Conclusions

The electrochemical behaviour of several drugs at solid electrodes closely parallels the biotransformation processes. Even if it is not possible to demonstrate a direct causal relation, such an approach could facilitate the interpretation of the mechanism by which drugs are degradated. Cyclic voltammetric experiments are simple techniques allowing mechanistic studies but could also provide an easy way to evaluate pharmacological activites.

376

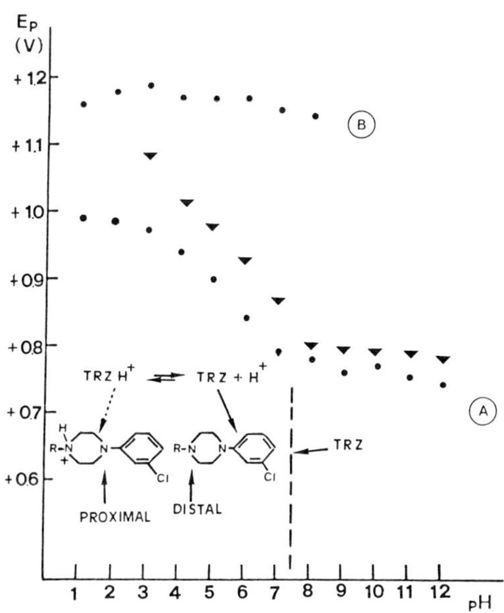

Fig.5. Peak potential evaluation as a function of pH. Trazodone = 1×10^{-4} M. methanol 20 %. Carbon polyethylene electrode: A = first peak, B = second peak. Trazodone metabolite 1×10^{-4} M, 20 % methanol.

Acknowledgements

Thanks are expressed to the Fonds National de la Recherche Scientifique (F.N.R.S. Belgium) for the help to one of us (G.J.P.) and to the SPPS (Belgium Politic Research, A.R.C.) contract number 86/91-89.

References

1. J.W. Gorrod and L.A. Damani, in: "Biological Oxidation of Nitrogen in Organic Molecules", J.W. Gorrod and L.A. Damani (Eds), Ellis Horwood series in health science (1985).
2. M.P. Prete, J-M. Kauffmann, J-C. Vire, G.J. Patriarche, B. Debye and G. Geuskens, Anal. Letters, **17** (1984) 1391.
3. J.S. Mahood, J.E. Packer, A.J.F. Seale, R.L. Willson and B.S. Wolfenden, in: "Phenothiazines and Structurally Related Drugs", E. Udsin, H. Eckert and I.S. Forrest (Eds), Elsevier-North Holland, Amsterdam, (1980) p. 103.
4. J-M. Kauffmann, G.J. Patriarche, J-C. Vire and W.R. Heineman, Analyst, **110** (1985) 349.
5. H. Koch, Pharm. Int., **1** (1984) 6.
6. S.H.Y. Wong and N. Marzouk, J. Liquid Chromatogr., **8** (1985) 1379.
7. H. Oelschläger and M. Al Shaik, in: "Biological Oxidation of Nitrogen in Organic Molecules", J.W. Gorrod and L.A. Damani (Eds), Ellis Horwood series in health science, (1985) p. 60.
8. J-M. Kauffmann, G.J. Patriarche and G.D. Christian, Anal. Letters, **12** (1972) 1217.
9. J-M. Kauffmann, G.J. Patriarche and M. Genies, Electrochimica Acta, **27** (1982) 721.
10. J-M. Kauffmann, J-C. Vire and G.J. Patriarche, Bioelectrochem. Bioenerg., **12** (1984) 413.

11. J. Schmutz, in: "Phenothiazines and Structurally Related Drugs", E. Usdin, H. Eckert and I.S. Forrest (Eds), Elsevier-North Holland, Amsterdam, (1980) p. 3.
12. J-M. Kauffmann, J-C. Vire, G.J. Patriarche, L.J. Nunez-Vergara and J.A. Squella, Electrochimica Acta, **32** (1987) 1159.
13. J.R. Lindsay Smyth and D. Masheder, J. Chem. Soc. Perk., **2** (1977) 1732.

ADSORPTIVE STRIPPING SQUARE WAVE VOLTAMMETRY OF PHARMACEUTICAL QUINONIC DERIVATIVES

J-C. Vire,[1] G.J. Patriarche,[1] H. Zhang,[2] B. Gallo[3] and R. Alonso[3]

1. *Free University of Brussels (ULB), Institute of Pharmacy*
   Campus Plaine C.P. 205/6, Bd du Triomphe, B-1050 Brussels, Belgium
2. *Free University of Brussels (ULB), Laboratory of Chemical Oceanography*
   Campus Plaine C.P. 208, Bd du Triomphe, B-1050 Brussels, Belgium
3. *Universidad del Pais Vasco, Facultad de Ciencias*
   Departamento de Quimica, Apdo 644, 48080 Bilbao, Spain

Introduction

It is well known that the polarographic reduction of quinone containing compounds gives rise to a well defined reversible two-electron wave.[1-4] However, this faradic reaction is frequently accompanied by an adsorption process, more or less pronounced depending on the morphology of the investigated molecule.[3] As illustrated by numerous examples,[5] these undesirable interfacial interactions may become useful to enhance by an adsorptive preconcentration step the performances of voltammetric techniques. This procedure, made very sensitive by applying a square-wave potential scan,[6] has been adapted to the determination of vitamins of the K group at the hanging mercury drop electrode. Vitamin K_1 (phylloquinone, 2-methyl-3-phytyl-1,4-naphthoquinone) and vitamin K_3 (menadione, 2-methyl-1,4-naphthoquinone) have long been recognized as essential constituants for blood coagulation and, owing to the reversible redox character of quinones, as electron mediators in the respiratory processes of the cells.[7] Marcellomycine, a new anthracyclinic antitumor antibiotic previously submitted to a polarographic investigation,[8] also exhibits strong interfacial phenomena.[8-10] The adsorptive stripping behaviour of this compound have been compared to those of vitamins K.

Experimental

Instrumentation

Cyclic and square wave voltammograms have been recorded using a PAR Model 384 B Polarographic Analyzer coupled with a PAR 303 A Static Mercury Drop Electrode (medium-size drop) and a Houston "HIPLOT" DMP 40 Digital Plotter. The potentials are refered to the $Ag/AgCl/KCl_s$ reference electrode and a platinum wire is used as counter electrode. A PAR 305 stirrer is connected to the 303 A SMDE . The peak heights are automatically measured using the "tangent fit" capability of the instrument.

Reagents

Marcellomycine, as tartrate salt (Bristol, Belgium), vitamin K_1 (Merck 501890) and vitamin K_3 (Merck 5793) have been used without further purification. Stock solutions were

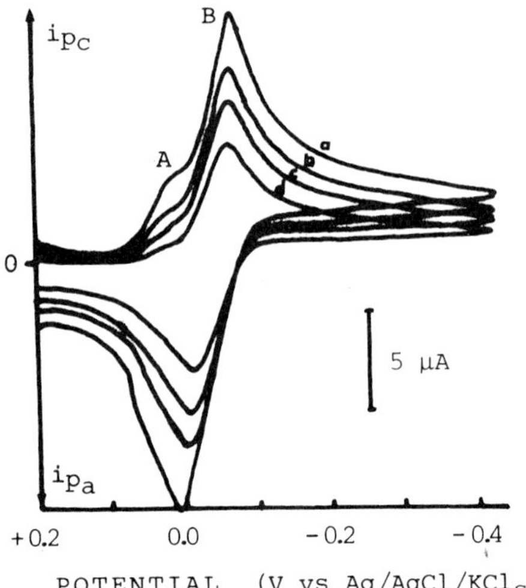

Fig.1. Vitamin K$_3$ 5×10^{-4} M in 0.3 M HClO$_4$. Cyclic voltammetry.
Starting potential: +0.20 V; scan rate: a) 500 mV/s, b) 300 mV/s, c)
200 mV/s, d) 100 mV/s.
A: capacitive shoulder, B: faradaic reduction peak.

prepared by dissolving the appropriate amount of the compound in water (marcellomycine),
methanol (vitamin K$_1$) or in water – methanol 80:20 mixture (vitamin K$_3$) and preserved
in the dark. These solutions were renewed daily. Supporting electrolytes, described in the
"Results and Discussion" part, were prepared from double distilled water and analytical grade
reagents.

Procedure

10 ml of the supporting electrolyte were deaerated during twelve minutes. The deposition potential, which is also the initial potential, is then applied to a new drop the solution
being stirred at 400 rpm. After an equilibration time, the potential is scanned using the
following parameters: scan rate: 200 mV/s, pulse amplitude: 20 mV, frequency: 100 Hz.
The same procedure is repeated after spiking increasing concentrations of the analyte.

Results and Discussion

Influence of the Operational Parameters

A preliminary investigation of the adsorptive stripping properties of vitamin K$_3$ has
showed that performing the deposition step at a potential less negative than that of the
reduction peak results in a very low signal, indicating a weak adsorption of the oxidized form
of the vitamin. Such a behaviour was expected from cyclic voltammetry where a shoulder
precedes the faradaic reduction peak (Fig.1). The intensity of this shoulder is proportional
to the scan rate, demonstrating its capacitive origin.

Taking advantage of the high reversibility of the quinone – hydroquinone system, a
stronger adsorption of the compound, and hereby a better sensitivity, should be obtained if
the preconcentration step is performed at a potential more negative than the reduction one,
followed by a positive going potential scan.

As a matter of fact, a cyclic scan performed after a 60 s deposition time at a potential corresponding to the oxidized form exhibits both small reduction and reoxidation peaks
(Fig.2A, peaks a and a') while the same deposition time imposed at a potential corresponding
to the reduced form provides a resulting oxidation peak enhanced by a factor of ten (Fig.2B,
peak b). The reduction peak (peak b') however has a lower intensity, indicating that the
strongly adsorbed reduced form is partly released after its oxidation.

380

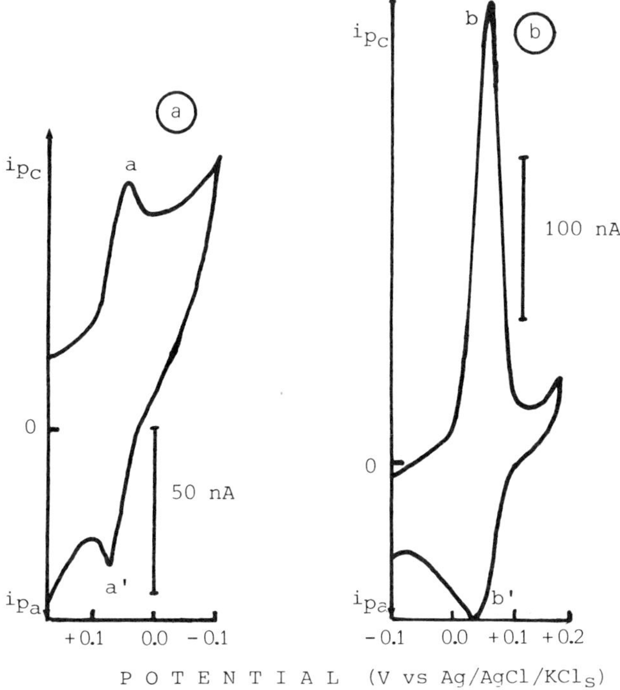

Fig.2. Vitamin K_3 5×10^{-7} M in 0.3 M $HClO_4$. Cyclic voltammetry. Deposition time before scan 60 s and scan rate 100mV/s.
A: deposition and initial potential: +0.175 V, negative going scan.
B: deposition and initial potential: −0.100 V, positive going scan.

The same procedure has been applied to vitamin K_1. Preconcentration of the oxidized form provides a higher signal, showing in this case a stronger adsorption of the initial compound. Moreover, the reduction peak is immediately followed by a sharp symmetrical peak corresponding to the desorption of the reduced form.

Marcellomycine behaves similarly and, as for vitamin K_1, these processes are attenuated in acidic media. They have been attributed to the long side chain fixed on the quinonic structure: the phytyl chain in the third position for vitamin K_1 and the daunosamine sugar moiety in the seventh position for marcellomycine.

Each compound has been submitted to a comparison of the electrochemical response given by the three following techniques: linear scan (lsv), differential pulse (dpv) and square wave (swv) stripping voltammetry. Results are collected in Table 1 and illustrated in Fig.3.

Table 1. Influence of the applied wave form on the recorded peak current (currents expressed in nA).

Compound	Vitamin K_3	Vitamin K_1	Marcellomycine
Concentration	5×10^{-8} M	4×10^{-7} M	3×10^{-6} M
Deposition time	60 s	300 s	60 s
lsv	5	-	245
dpv	25	10	57
swv	525	1920	6600

When lsv is able to give a signal, it is poorly defined and the background current is of comparable magnitude. A similar current is recorded using dpv but as the baseline is

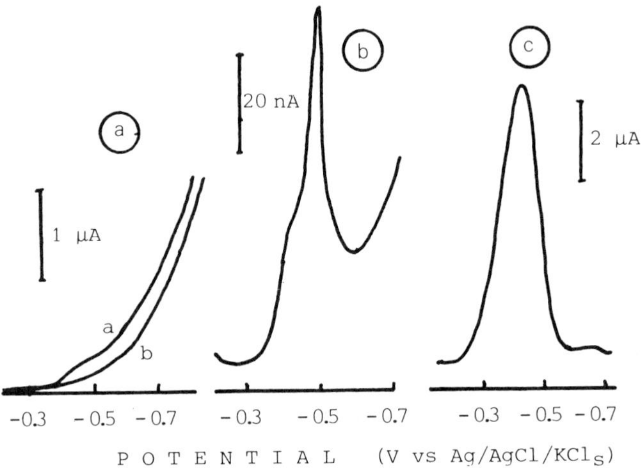

Fig.3. Marcellomycine 3×10^{-6} M; pH 4.0. Comparison of the voltammetric response of lsv, dpv and swv. Deposition time: 60 s; deposition potential: -0.20 V; scan rate: lsv: 100 mV/s, dpv: 4 mV/s, swv: 200 mV/s; pulse height: dpv and swv: 20 mV; frequency: swv: 100 Hz. A: lsv: a) linear scan, b) derivative scan; B: dpv, C: swv.

considerably improved, the peak can be easily determined. A marked enhancement of the response is obtained by applying swv. A 20-fold to 200-fold current increase is observed with respect to dpv, depending on the analyzed compound. This study clearly demonstrates the ability of swv voltammetry to determine low concentrations, owing to the applied wave form and the scan rates which can be used.

Table 2. Operational parameters used for the determination of marcellomycine, vitamin K_1 and vitamin K_3 and resulting peak potential.

	Vitamin K_3	Vitamin K_1	Marcellomycine
Supporting electrolyte	0.3 M $HClO_4$	0.03 M K_2SO_4 + 1.5 x 10^{-4}M H_2SO_4 pH = 4.2	0.03 M Acetate buffer pH = 4.0
Deposition potential	-0.10 V	+0.10 V	-0.20 V
Equilibration time	10 s	15 s	5 s
Peak potential	+0.08 V	-0.08 V	-0.45 V

The nature of the supporting electrolyte influences the voltammetric response. Among the five buffers used in the analysis, the acetic acid – sodium acetate mixture exhibits the higher signal. The response of vitamins K however is enhanced by using an unbuffered medium. As the ionic strenght and the pH of the solution also influence the peak height and with respect to the morphology and the definition of the peak which sometimes appears very close to the mercury oxidation, the supporting electrolytes have been selected as given in Table 2.

The effect of the deposition potential on the peak current is also to be considered as

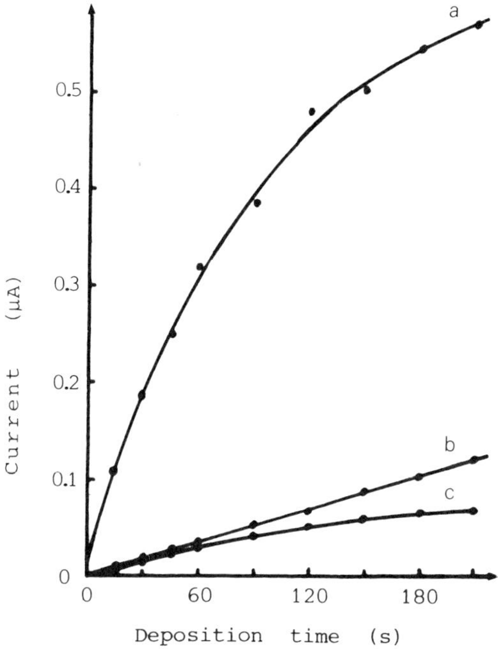

Fig.4. Influence of the deposition time on the peak current of vitamins K_3; 0.3 M $HClO_4$; swv; deposition potential: -0.10 V; scan rate: 200 mV/s; pulse height: 20 mV; frequency: 100Hz.
A: vitamin K_3: 2×10^{-8} M, deposition under stirring.
B: vitamin K_3: 2×10^{-8} M, deposition without stirring.
C: vitamin K_3: 4×10^{-9} M, deposition under stirring.

adsorption often depends on this parameter. The peak developed by vitamin K_3 remains unaffected as the accumulation potential is fixed between -0.10 and -0.30 V. It decreases markedly when more negative values are imposed. The potential range where vitamin K_1 can be deposited is more restricted as the compound is reduced at -0.08 V and then mercury oxidation starts at $+0.10$ V. This value has been chosen because the peak height increases when the deposition potential is made more positive. This choice is easier with marcellomycine as the peak current remains unchanged for deposition performed between 0.00 V and -0.30 V. The selected potentials are listed in Table 2.

Influence of the Deposition Time

Marcellomycine gives rise to a linear relationship between the peak height and the deposition time, indicating a constant adsorption rate of the compound. Moreover, the duration of the equilibration time imposed between the deposition step and the potential scan has no effect on the recorded current; this proves that no equilibrium occurs during this strong adsorption process.

Inversely, the current recorded in a 2×10^{-8} M vitamin K_3 solution does not increase linearly with increasing deposition time (Fig.4, curve A). This cannot be attributed to the saturation of the drop since the use of a 4×10^{-9} M concentration gives rise to a similarly shaped relationship (Fig.4, curve C). Such a behaviour indicates an equilibrium between the adsorbed and the solution species. A decrease of the current is also observed when the equilibration time exceedes twenty seconds, showing that adsorption is effective under stirring and became submitted to an equilibrium process when stirring is stopped. This phenomenon only depends on time, not on the vitamin concentration since working at a fixed accumulation time with increasing concentrations provides linear relationships. Similar properties have already been pointed out studying vitamin K_1, but also 9,10-phenanthrene

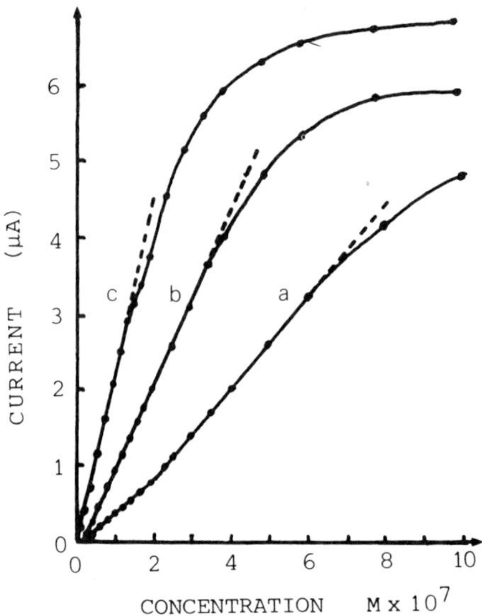

Fig.5. Marcellomycine; pH 4.0; calibration plots. Swv; deposition potential: -0.20 V; scan rate: 200 mV/s; pulse height: 20 mV; frequency: 100 Hz; deposition time: A: 30 s, B: 60 s and C: 120 s.

quinone.[11] However, curves A and C have been obtained after a deposition step performed under stirring. Curve B (Fig.4) shows that a linear relationship appears in a 2×10^{-8} M quiescent solution. As these currents, and hereby the amount of adsorbed compound, are higher than that of curve C, the equilibrium alone cannot be evoked but a rearrangement of the molecules after their adsorption at the electrode surface may occur, this final state being immediately established in quiescent solution.

It may be concluded that the adsorptive properties of the investigated compounds are to be attributed to the quinone containing moiety of the molecule, not to the presence of the side chain. The structure of this quinonic moiety is of particular importance since the anthracycline is strongly adsorbed without any equilibrium while adsorption of naphthoquinonic compounds is an equilibrium process between the adsorbed and solved species when the deposition step is performed under stirring. The side chain of vitamin K_1 may however play a role in the rearrangement of the molecules because it increases the area occupied by each molecule. This favours a rapid modification of their position, and can be seen in the immediate decrease of the current within ten seconds after the deposition step, due to stabilization of the structure of the adsorbed layer.

Influence of Concentration

Table 3 summarizes the characteristics of the calibration plots established for each compound using different deposition times. Fig.5 illustrates these plots for marcellomycine.

In spite of the adsorption equilibrium occuring with vitamin K_1 and K_3, linearity between the peak current and concentration can be obtained for a constant preconcentration time as indicated by the correlation coefficients of the calibration plots. The effect of this equilibrium appears in the case of vitamin K_3 by comparing the slopes of the relations which are not directly proportional to the deposition time, the deviation increasing with this latter. The proportionality is observed for the slopes of vitamin K_1 as the equilibration time (15 s) is long enough to reach the complete equilibrium at the electrode surface. This means that the equilibration rate is faster for vitamin K_1 than for vitamin K_3. The reproducibility of the total process (adsorption and equilibration) however is better for vitamin K_3 as indicated by

Table 3. Characteristics of the calibration plots of marcellomycine, vitamin K_1 and vitamin K_3 using different deposition times (d.l.=detection limit).

Compound	Deposition time (s)	Linearity range (M)	Equation (slope in $nA/10^{-8}$ M)	Correl coeff.
Marcellomycine	30	2.0×10^{-8} - 2.0×10^{-7}	$y = 45.0$ x - 83.3	0.9993
d.l. =	30	2.0×10^{-7} - 6.0×10^{-7}	$y = 59.9$ x - 395.0	0.9998
5.0×10^{-9} M	60	1.0×10^{-8} - 3.5×10^{-7}	$y = 108.7$ x - 160.0	0.9997
	120	1.0×10^{-8} - 1.4×10^{-7}	$y = 212.9$ x - 56.2	0.9984
Vitamin K_1	30	8.0×10^{-8} - 1.0×10^{-6}	$y = 4.3$ x + 1.0	0.9967
d.l. =	60	4.0×10^{-8} - 1.0×10^{-6}	$y = 7.9$ x + 0.9	0.9991
1.0×10^{-9} M	120	1.0×10^{-8} - 8.0×10^{-7}	$y = 15.3$ x + 3.5	0.9992
	300	1.0×10^{-8} - 6.0×10^{-7}	$y = 34.6$ x - 7.3	0.9982
	300*	1.0×10^{-9} - 1.0×10^{-7}	$y = 33.9$ x - 1.5	0.9989
Vitamin K_3	30	1.0×10^{-9} - 4.8×10^{-7}	$y = 60.7$ x - 3.5	0.9999
d.l. =	60	5.0×10^{-10} - 2.5×10^{-7}	$y = 105.2$ x - 13.3	0.9999
1.3×10^{-10} M	120	3.0×10^{-10} - 1.1×10^{-7}	$y = 178.6$ x + 13.0	0.9998
	300*	2.0×10^{-10} - 2.6×10^{-8}	$y = 234.2$ x - 7.3	0.9995

*using blank subtraction

the correlation coefficients as well as by the relative standard deviations of ten measurements being 2.3 % for vitamin K_3 and 3.9 % for vitamin K_1. Proportionality of the slopes is also respected for marcellomycine as this compound is not submitted to any equilibrium, but interactions of the adsorbed material may occur since the 30 s deposition time plot exhibits two slopes (Fig.5, plot A).

Comparing the slopes of both vitamins K_1 and K_3, it appears that the adsorption method is about ten times more sensitive for the latter compound. This could be attributed to the phytyl chain in vitamin K_1 which increases the size of the molecule. However, marcellomycine exhibits sensitivities comparable to those of vitamin K_3. This may be explained by a favourable orientation of the daunosamine moiety of marcellomycine while the longer phytyl chain of vitamin K_1 leads to a steric hindrance.

In each case, the upper limit of the linearity range indicates the complete coverage of the electrode surface and is determined by the deviation from the linearity of the current – concentration realtionship.

The detection limits given in Table 3, based on a signal-to-noise ratio of 3, have been calculated for the longest deposition time.

Conclusions

The investigated quinone containing derivatives exhibits sufficiently strong adsorption properties to allow their preconcentration at the electrode surface. Owing to the reversibility of the quinonic structure, the compound in its reduced form can also be deposited and this form is even more strongly adsorbed. The adsorption behaviour has been attributed to the quinonic moiety of which the structure more than the side chain, may influence the adsorption process. Vitamins K are found to have an equilibrium between the adsorbed and the solution species. This equilibrium is immediately achieved in a quiescent solution after the deposition step is performed. Determinations of the compounds studied can easily be done down to 10^{-9} – 10^{-10} M using a squarewave voltammetric technique.

Acknowledgements

Authors are indebted to Bristol S.A. (Brussels, Belgium) who generously provided the marcellomycine tartrate. Thanks are expressed to the Fonds National de la Recherche Scientifique (F.N.R.S. Belgium) for help to one of us (G.J.P.) and to the SPPS (Belgium Politic Research, ARC) contract number 86/91-89.

References

1. S. Patai, in: "The Chemistry of the Quinonoid Compunds", J. Wiley, London, Vol. **1**, chapter 9, 1974.
2. J-C. Vire and G.J. Patriarche, Analusis, **6** (1978) 155.
3. J-C. Vire and G-J. Patriarche, Analusis, **6** (1978) 395.
4. G.J. Patriarche, M. Chateau-Gosselin, J.L. Vandenbalck and P. Zuman, in: "Polarography and Related Electroanalytical Techniques in Pharmacy and Pharmacology", A.J. Bard (Ed.), Electroanalytical Chemistry, Marcel Dekker, New York, Vol. **11**, p. 141, 1979.
5. J. Wang, in: "Stripping Analysis, Principles, Instrumentation and Applications", Verlag Chemie, Deerfiel Beach, Fl., 1985.
6. J. Osteryoung and J.J. O'Dea, in: "Square-Wave Voltammetry", A.J. Bard (Ed.), Electroanalytical Chemistry, Marcel Dekker, New York, Vol. **14**, p. 209, 1986.
7. W.H. Sebrell Jr. and R.S. Harris, in: "The Vitamins", Academic Press, New York, 2nd ed., Vol. **3**, 1971.
8. F. Mebsout, J-C. Vire and G.J. Patriarche, Anal. Letters, **18** (1985) 1431.
9. M. Chateau-Gosselin, J-C. Vire and G.J. Patriarche, Microchim. Acta, **III** (1983) 457.
10. F. Mebsout, J-C. Vire and G.J. Patriarche, Anal. Letters, **17** (1984) 805.
11. H.Y. Cheng, L. Falat and R.L. Li, Anal. Chem., **54** (1982) 1384.

THE DETERMINATION OF TIMOLOL IN BIOLOGICAL FLUIDS BY ADSORPTIVE STRIPPING VOLTAMMETRY

R.J. Barrio Diez-Caballero and J.F. Arranz Valentin

Department of Analytical Chemistry
Colegio Universitario de Alava
Aptdo. 450, Vitoria, Spain

Introduction

Timolol (1-(1,1-dimethylethyl)amino-3-[[4-morpholinyl-1,2,5-thidiazol-3-yl]oxy]-2-propanol) is a β-adrenergic blocker used in the treatment of arterial hipertension and above all is used very efficiently as an anti-glaucoma agent. The margin between the toxic and therapeutic effects of the drug is so slight that very sensitive methods are required for its determination at trace levels. The most recent of these methods is chromatography, g.l.c. or HPLC, with different detection systems. G.l.c. is used for the determination of timolol in urine,[1] and more recently g.l.c.-mass spectrometric assay with selected ion monitoring, has been proposed for the determination of timolol in human plasma[2] with a detection limit of 0.5 ng/ml. Timolol has also been determined in milk and serum by HPLC with electrochemical detection with detection limits of 2 ng/ml.[3]

In this study low concentration of timolol are determined by adsorptive stripping voltammetry. The preconcentration of the analyte is done by adsorption of the drug on a surface of the electrode and the adsorbed species are determined by application of a voltammetric scan on the electrode. This technique has recently been employed by different authors[4,5,6] for the determination of a number of drugs.

As the present study shows, submicromolar levels of timolol can be measured using this technique, controlling their accumulation on a hanging mercury drop electrode and optimizing the instrumental parameters. As a specific application, timolol was determined in aqueous humor of rabbits and human serum.

Experimental

Apparatus

A Metrohm polarecord E-506 coupled with a Metrohm 663 VA Stand was used. A multimode mercury drop electrode (Metrohm 6.1246.000) which contains a hanging mercury drop electrode (HMDE) was used, which served as the working electrode. The average drop size was 0.52 mm^2. A glassy carbon was used as the auxiliary electrode and the SCE as a reference electrode. A Metrohm VA 612 Scanner and a rapid X-Y register, Linseis LY 1800, were used to obtain the cyclic voltammograms. All measurements were performed at room temperature.

Reagents and Solutions

Stock solutions (1×10^{-3} M) of timolol (Merck) were prepared daily by dissolution in de-ionised water. The supporting electrolyte was Britton-Robinson buffer at an ionic strength of 0.25 M in NaClO$_4$ (pH = 4). The aqueous humor samples were from five different rabbits and the human serum were pools of seven subjects.

Contemporary Electroanalytical Chemistry, Edited by A. Ivaska *et al.*
Plenum Press, New York, 1990

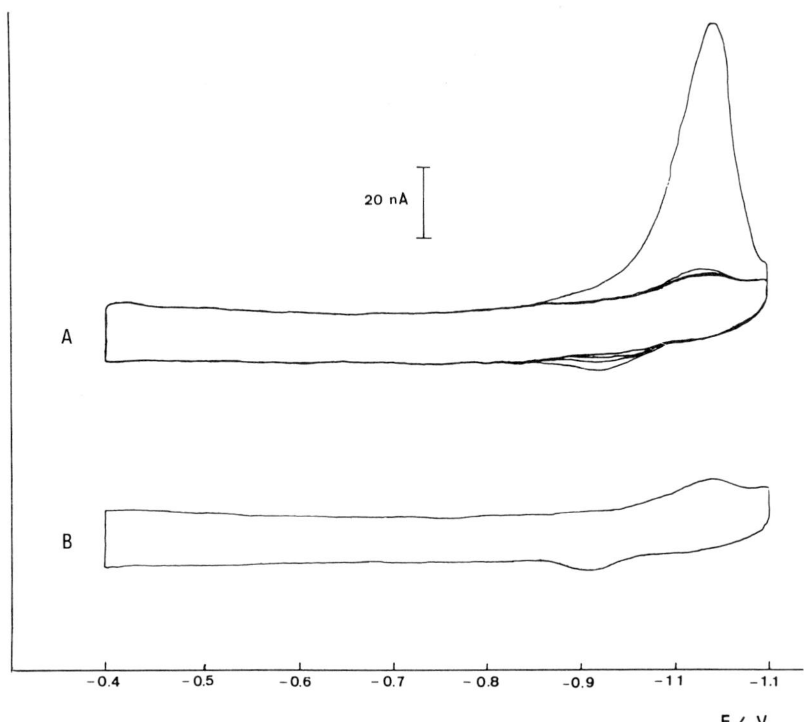

Fig.1. A: Repetitive cyclic voltammograms for 5×10^{-7} M of timolol at Britton-Robinson buffer (pH 4), after stirring for 60 s at -0.4 V. B: An analogous voltammogram without accumulation. Scan rate 100 mV/s.

Adsorptive Stripping Voltammetry

The supporting electrolyte solution (20 ml) was added to the cell and deoxigenated with nitrogen for 10 minutes. The pre-concentration potential (usually -0.40 V vs. SCE) was applied to the electrode for a given time period while stirring the solution at 1500 rpm. After 15 seconds rest period, a scan to negative direction was initiated and the resulting voltammograms were recorded with different operational parameters.

Timolol Assay in Biological Fluids

A Waters Sep-Pak C_{18} extraction cartridge was pre-wet with 2 ml of methanol and 5 ml of water. For the timolol extraction, 1 ml of human serum (or aqueous humor) was passed through the cartridge, the drugs being adsorbed on the Sep-Pak matrix. The matrix was then rinsed with 6 ml of Britton-Robinson buffer (pH = 1.8). For elution of timolol 3 ml of acetone was used and the first ml discarded. The eluent was dried under a stream of nitrogen. The residue was diluted to 20 ml with Britton-Robinson buffer (pH = 4.0). The voltammograms were recorded under the optimum instrumental conditions.

Results and Discussion

The study by cyclic voltammetry confirms the spontaneous accumulation of timolol on the hanging mercury drop electrode after the solution was stirred for 60 seconds. Fig.1 shows the cyclic voltammogram of a solution of 5×10^{-7} M of timolol at pH = 4.0. A big, well-defined cathodic peak can be seen at -1.03 V, which in succesive scans decreases rapidly assuring a fast desorption from the electrode surface. Voltammogram B shows the same pattern, but without accumulation. The resulting peak is much smaller than that obtained by accumulation.

The confirmation that we are dealing with an adsorption process on the electrode surface

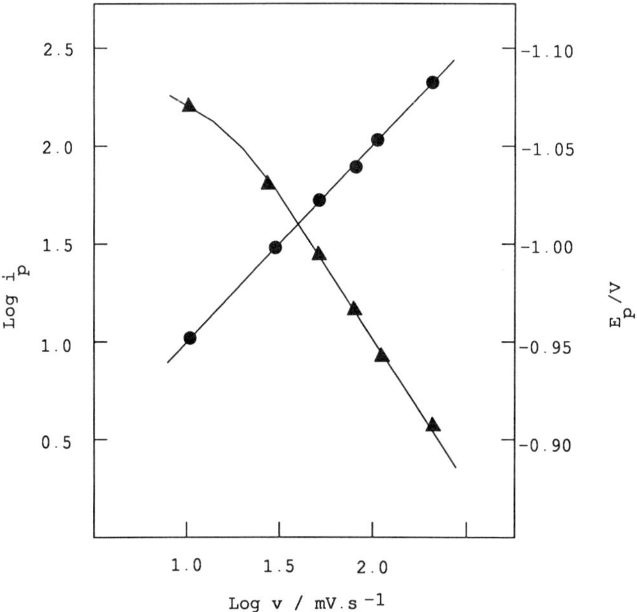

Fig.2. Dependence of the logarithm of the peak current ($\circ$) and of peak potential ($\triangle$) on the logarithm of the potential scan rate. Timolol concentration: 5×10^{-7} M.

can be obtained by plotting log i_p versus log v. In such a case a slope of 1.0 is to be expected. In Fig.2 a straight line can be seen whose slope is 0.9702 and correlation coefficient 0.9973.

The electrode area occupied by each molecule of timolol can be calculated by a cyclic voltammetric study. The surface saturation of the HMDE, using a concentration of 5×10^{-7} M of timolol, is reached after 60 seconds. The cyclic voltammogram obtained under these conditions can be used to determine the surface coverage: So, the value of the charge transferred, obtained by integration of the peak resulting from the reduction process, is 5.47×10^{-8} C. This means a value of 2.73×10^{-10} mol/cm^2 which can be calculated by division of the charge by the nFA factor (A being the electrode area, 0.52 mm^2 and n=4 for reduction of the two azometine groups). Thus each timolol molecule adsorbed occupies an area of 6.10 nm^2.

The spontaneous adsorption of timolol can be used for its voltammetric determination. In Fig.3 the variation of peak current with the preconcentration time is shown for timolol solutions of 10^{-7} and 10^{-8} M. A linear dependence for the 10^{-8} M concentration, up to 120 s, can be seen, with a slope of 0.193 nA/s. Thus the choice of optimum accumulation time depends of the range of concentration studied.

The influence of the accumulation potential on the peak stripping current was studied in the range of 0.0 to -0.6 V. The results are shown in Fig.4. The highest peak current was obtained with accumulation at -0.40 V.

The effect of pH on the height of the peak was also studied. In Fig.5 a the voltammogram is shown in the pH range 1.5 to 6.0. From pH 5.5 onwards the adsorption wave begins to fold, until it disappears completely at pH 6.0. The ideal value is thus pH 4.0 in Britton-Robinson buffer at an ionic strength of 0.25 M in NaClO$_4$. With the i_p variation, a displacement of the peak potential can be seen from -0.73 to -1.04 V, coincident with the rise in pH.

Different instrumental parameters affect the voltammetric response and especially the shape and resolution of the waves. The peak height was found to increase linearly with pulse amplitude between 20 and 80 mV. The value of 60 mV was chosen for this variable because with a larger value the wave becomes wider, decreasing the resolution. The height of the peak was also found to vary with the size of the drop, between 0.25 and 0.52 mm^2; the latter value was chosen as the optimum value. The stirring speed during the accumulation was not

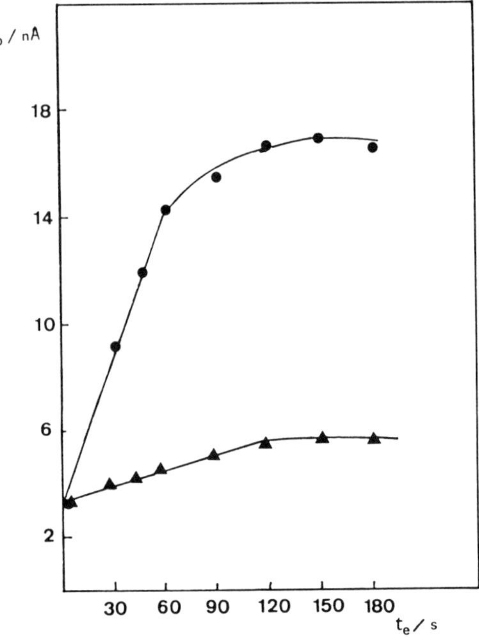

Fig.3. Effect of pre-concentration time on the stripping voltammogram;
o: 10^{-7} M, $\triangle$: 10^{-8} M. Scan rate: 8.3 mV/s. Britton-Robinson buffer
(pH 4) in $NaClO_4$ 0.25 M.

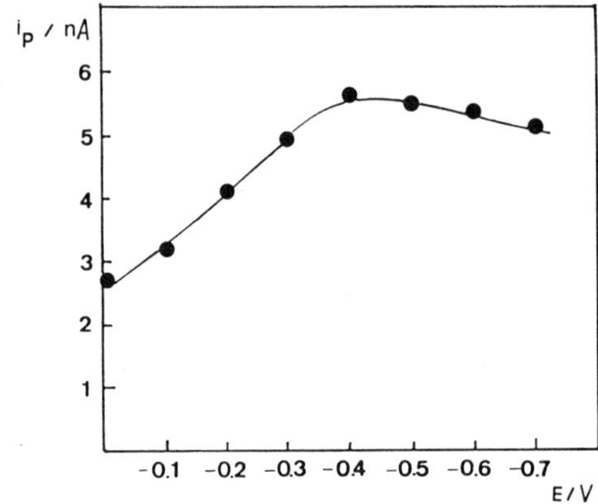

Fig.4. Influence of accumulation potential on the peak stripping current. Timolol concentration is 5×10^{-8} M and the other conditions as
in Fig.3.

found to have any effect on the peak height. The rotation speed was varied between 500 and
3000 rpm and 1500 rpm was chosen in this study.

Quantitative determination of timolol is based on the linear dependence of the peak
current on the concentration. Voltammograms of different timolol concentrations varying
from 3×10^{-9} to 10^{-8} M and using 60 seconds preconcentration are shown in Fig.6. A linear
calibration graph is obtained with a slope of 0.63 nA per 10^{-8} M and a correlation coefficient
of 0.9990.

The reproducibility of the method is determined by succesive measurements of ten
solutions of 10^{-8} M timolol. The average peak current is 7.7 nA in a range of 7.6 – 7.8 nA
and with a relative standard deviation of 1.9 %.

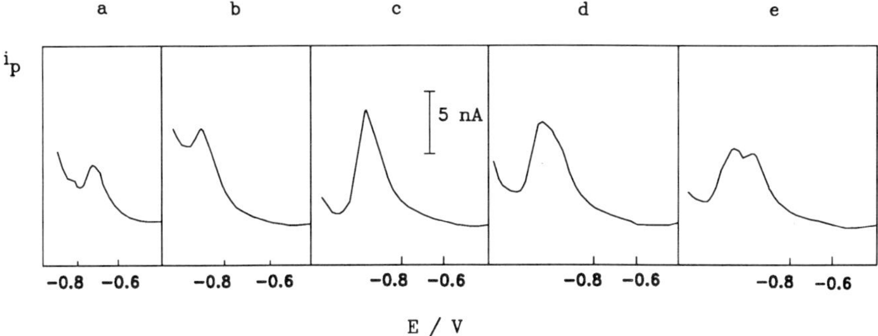

Fig.5. Influence of pH on the voltammogram peak current in Britton-Robinson buffer and in 0.25 M NaClO$_4$. Timolol concentration is 10^{-8} M. a) pH = 1.6, b) 3.5, c) 4.3, d) 5.1, e) 6.0.

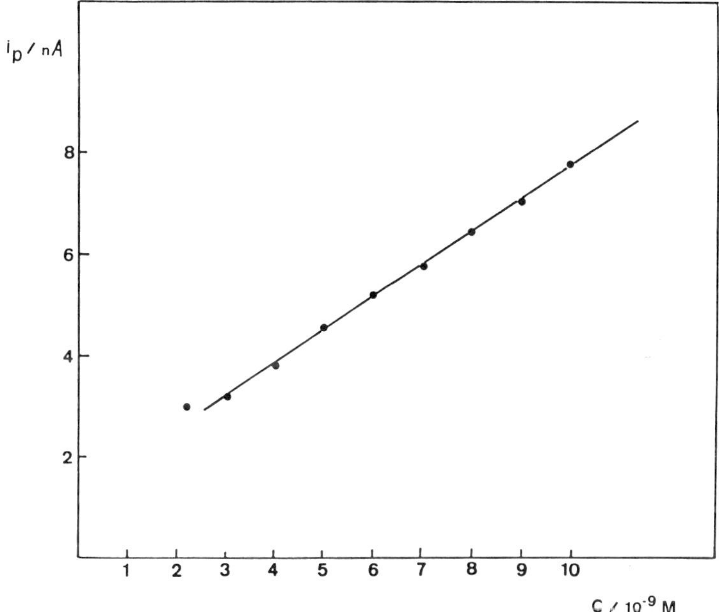

Fig.6. Calibration graph with 60 s accumulation time. Accumulation potential is -0.4 V. Scan rate: 8.3 mV/s.

The presence of other surface-active compounds which might affect the adsorptive re-disolution response was investigated, particularly those which might be present in biological samples. The effects of chlorides, albumin, gelatine and dodecylbenzene sodium sulfonate have been studied in a timolol solution of 5×10^{-7} M. Addition of 2 ppm of albumin and 2 ppm of gelatine produces a timolol peak depression of 39.0 and 29.9 % respectivley. The peak disappears completely in presence of 5 ppm of gelatine and 9 ppm of albumin. The addition of up to 200 ppm of chlorides does not have any effect on the adsorption of timolol. Also other electroactive surfactants such as dodecylbenzensulphonate were tested. It causes a depression of 32 % at a concentration of 2 ppm the response remaining unchanged after 5 ppm.

Using the results of the studies mentioned above, timolol in human serum samples was determined. A standard additions method, was used as shown in Fig.7. The detection limit of the method is 4.5 ng/ml and the mean recovery is 95.4 %.

In some types of sera it was necessary to change the preconcentration conditions in

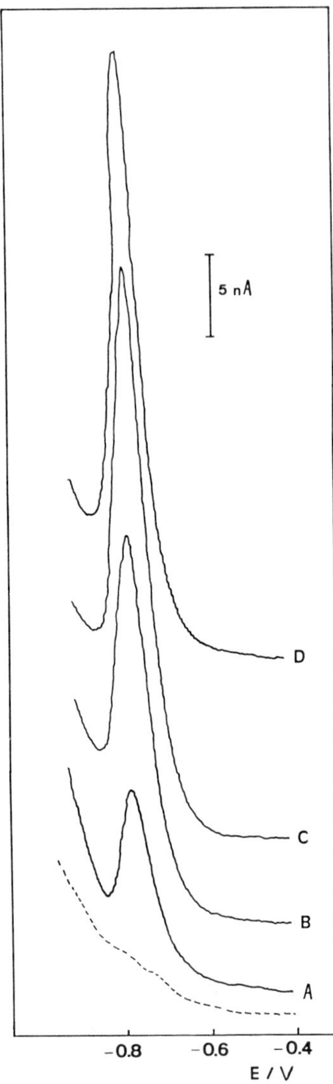

Fig.7. Determination of timolol in human serum after solid-liquid extraction. A: sample solution. B, C and D: standard additions of 100 μl of 10^{-4} M timolol solution. Dotted line: blank. For conditions see text.

order to avoid the interferences previously mentioned. Good results were obtained in cases where the wave, under standard conditions nearly disappears, changing the electrodeposition potential to -0.6 V and/or reducing the accumulation time to 20 s.

The determination in aqueous humor was performed in the same way. The detection limit in this case is 2.5 ng/ml with a mean recovery of 94.0 %.

It can be concluded that the determination of timolol in human serum and aqueous humor of rabbits is practicable by adsorptive stripping voltammetry on the hanging mercury drop electrode. In order to avoid interferences it is necessary to use a solid-liquid extraction method with Sep-Pak C_{18} cartridges.

Acknowledgements

The authors thank Mr. John N. Warner for revision of the text.

References

1. L. von Meyer, G. Drasch, G. Kanert and A. Riedl, Beitz. Gerichtl. Med., **37** (1973) 363.
2. J.R. Carlin, R.W. Walker, R.O. Davies, R.T. Ferguson and W.J.A. Vadenheuvel, J. Pharm. Sci., **69** (1980) 1111.
3. M.R. Gregg and D.B. Jack, Chromatog. Biomed. Appl., **30** (1984) 244249.
4. M.R. Smyth, E. Buckley, J. Rodriguez and R. O'Kennedy, Analyst, **133** (1988) 31.
5. J. Wang, M.S. Lin and V. Villa, Analyst, **112** (1987) 1303.
6. R. Alonso, R. Jimenez and A.G. Fogg, Analyst, **113** (1988) 27.

PULSE VOLTAMMETRIC DETERMINATION OF SULPHUR CONTAINING ORGANIC COMPOUNDS

Zenon J. Karpinski

Department of Chemistry
University of Warsaw
PL – 02093 Warsaw, Poland

Introduction

Biological and medical importance of sulphur containing organic compounds (SCOC) prompted development of various methods for their determination. Several classes of SCOC, particularly containing sulfhydryl group, exhibit anodic polarographic waves due to their reactions with mercury ions.[1,2] Similar anodic behaviour has been observed recently for thiazolidine-4-carboxylic acid (thioproline, Thz) and its 2-substituted derivatives.[3] Practical applications of anodic polarographic currents for SCOC determinations have been limited, however, due to difficulties related to autoinhibiting effects of the oxidation products accumulating on the electrode surface. Short electrolysis times of pulse voltammetric techniques may help to overcome these problems.

In the present communication the anodic voltammetric behaviour of cysteine (Cys), Thz and its 2-substituted derivatives: 2-propyl-thiazolidine-4-carboxylic acid (PrThz) and 2-tetrahydroxybutyl-thiazolidine-4-carboxylic acid (AThz) at mercury electrode is reported and applications of pulse voltammetric techniques for SCOC determinations are discussed.

Results and Discussion

Distortions of dc polarographic waves for SCOC appeared at concentrations close to 1×10^{-4} M. For 2×10^{-4} M Cys the distinct distortions were evident for drop times longer than 2 seconds. Similar distortions for PrThz appeared already at 1 s drop time (Fig.1). On the other hand, at the same Cys concentration, well developed curves were obtained in all pulse voltammetric techniques (Fig.2). The techniques studied were normal pulse, differential pulse and the fast square wave voltammetry (oswv) according to Osteryoungs.[4] Pulse width dependence of normal pulse voltammetric waves confirmed their diffusional character and forward and reverse current curves in square wave voltammetry indicated reversibility of the oxidation process aiding in the increased sensitivity of this technique.

Anodic reaction of thiols at mercury electrodes are relatively simple comparing to anodic behaviours of thioproline derivatives. The latter compounds form initially unstable oxidation products, observed in fast scan rate cyclic voltammetry (Fig.3, curve 1). The primary oxidation products undergo transformations into other mercury compounds, reducing at more negative potentials, evident at slower sweep rates (Fig.3, curve 2). New species formed in the cathodic processes are oxidized during the second potential scan showing new anodic peaks. These transformations probably involve the thiazolidine ring-opening reactions leading to the formation of sulfhydryl group interacting with mercury ions. The oxidation product transformations were also evident in square wave voltammograms. Even at high frequences (100 Hz) cathodic reverse current was observed only for PrThz, but concerning the other thioproline derivatives the initial products have been transformed during the 5 ms forward step (Fig.4).

Contemporary Electroanalytical Chemistry, Edited by A. Ivaska *et al.*
Plenum Press, New York, 1990

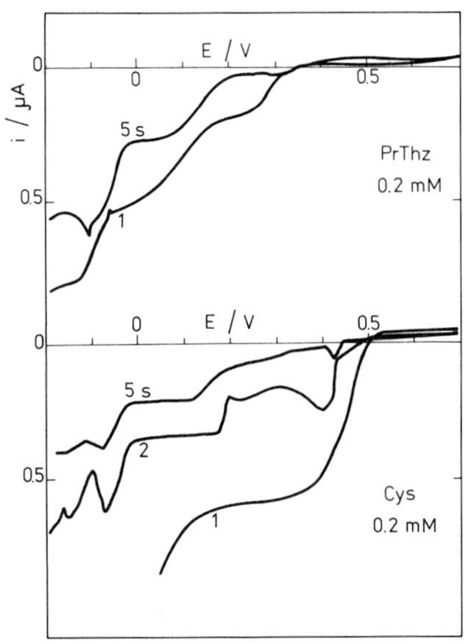

Fig.1. Sampled currents dc polarograms for 0.2 mM PrThz and Cys. Drop times indicated at the curves.

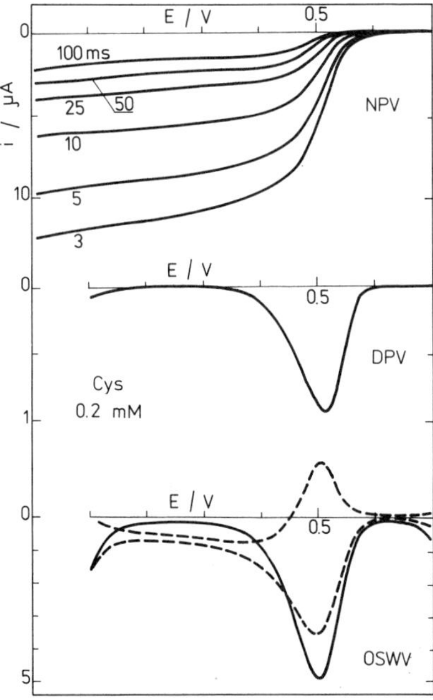

Fig.2. Normal pulse, differential pulse and square wave voltammograms for 0.2 mM Cys.

Np voltammograms are least affected by transformation of the oxidation products and exhibited similar distortions for the thioproline derivatives and for Cys (Fig.5). Such, however, distortions appear at different concentrations of SCOC,. For PrThz, distorted normal pulse waves are obtained with concentrations over 2.5×10^{-4} M and at pulse widths larger than 25 ms, while for Cys they appear only at concentrations exceeding 1×10^{-3} M and with the pulse width 50 ms.

Autoinhibiting effects of the SCOC oxidation products resulted in characteristic shapes

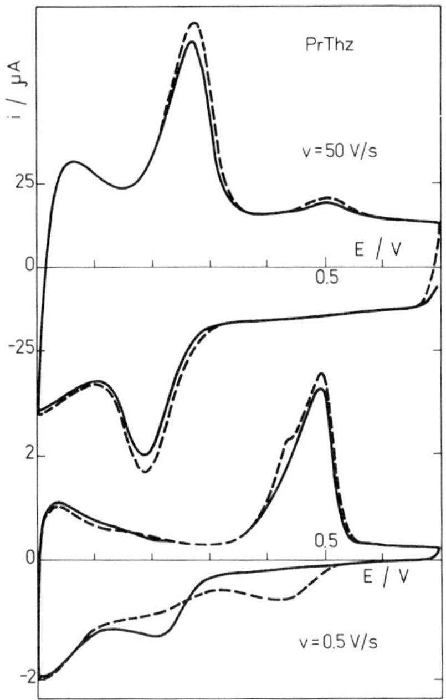

Fig.3. Effect of potential scan rate on cyclic voltammograms of 0.25 mM PrThz.

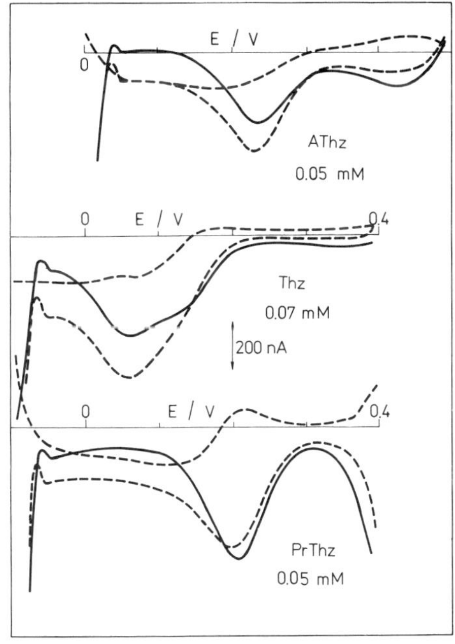

Fig.4. Square wave voltammograms for Thz and its derivatives.

of their calibration plots in all voltammetric techniques. They exhibited linear segments at low concentrations followed by gradual leveling off of the i vs. C plots at increasing concentrations. Linear segments of the calibration plots are very short in dc polarography, with their upper limits close to 5×10^{-5} M for Thz derivatives and equal 2×10^{-4} M for Cys. Longer linear segments of the calibration plots, however, are obtained in pulse voltammetric tech-

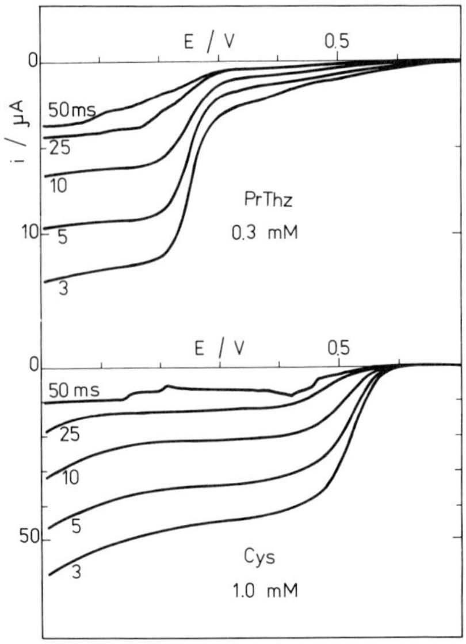

Fig.5. Normal pulse voltammograms for PrThz and Cys.

Table 1. Parameters of linear segments of calibration plots.

Compound	Technique	Concentration range (μM)	Sensitivity ($\mu A/\mu M$)
	npv	1 to 500	7.3×10^{-3}
Thz	dpv	1 to 150	3.6×10^{-3}
	oswv	2 to 500	4.6×10^{-3}
	npv	1 to 250	5.6×10^{-3}
PrThz	dpv	1 to 100	3.1×10^{-3}
	oswv	2 to 300	10.2×10^{-3}
AThz	dpv	1 to 70	3.2×10^{-3}
	oswv	2 to 150	5.1×10^{-3}
	npv	1 to 800	8.3×10^{-3}
Cys	dpv	0.5 to 200	5.3×10^{-3}
	oswv	1 to 1000	20.3×10^{-3}

niques (Table 1). The upper limits of the concentration ranges in various pulse voltammetric techniques clearly correlate with the amounts of the oxidation products accumulatied during the potential pulse sequence. In np voltammetry the limits are ca 5 times higher than in dc polarography, as expected for an electrolysis charge proportional to square root of the electrolysis time. The lower limits in dp voltammetry resulted from additional amounts of the products formed before the pulse application. In osw voltammetry the entire potential scan is done on one mercury drop and, for systems with products not reduced in the reverse pulse, the product accumulation is similar to that during the linear potential scan. High frequences and scan rates used in the former technique limited the product formation, however, and permitted fast determinations of Thz derivatives. Over 50 osw voltammograms could be obtained in a period of time needed for one np or dp experiment. Osw voltammetry exhibited also good sensitivity, especially for reversible systems. Thus, for Cys and PrThz the sensitivity of osw voltammetry is higher than that of np voltammetry (Table 1). High speed of osw voltammetry permits its application in flow systems.[5] Preliminary experiments indicate that this approach is very promising in extending applications of anodic reactions of SCOC for their determination.

Conclusions

1. Fast thiazolidine ring opening resulting from interactions with mercury ions enables voltammetric determinations of Thz and its derivatives.
2. Np voltammograms are least affected by transformations of the oxidation products of Thz derivatives.
3. Short electrolysis times in pulse voltammetric techniques limit autoinhibiting effects of SCOC oxidation products.
4. Square wave voltammetry permits very fast and sensitive determinations of SCOC.

References

1. I.M. Kolthoff and C. Barnum, J. Am. Chem. Soc., **62** (1940) 3061.
2. M. Brezina and P. Zuman, in: "Polarography in Medicine, Biochemistry and Pharmacy", Interscience, New York, 1958.
3. Z.J. Karpinski and J. Radomski, Anal. Letters, submitted.
4. J.G. Osteryoung and R.G. Osteryoung, Anal. Chem., **52** (1980) 101 A.
5. R. Samuelsson, J.J. O'Dea and J. Osteryoung, Anal. Chem., **52** (1980) 2215.

GENERAL

ADSORPTIVE STRIPPING VOLTAMMETRY IN TRACE ANALYSIS

Robert Kalvoda

*UNESCO Laboratory of Environmental Electrochemistry at the J. Heyrovský
Institute of Physical Chemistry and Electrochemistry
Czechoslovak Academy of Sciences, Prague, Czechoslovakia*

In trace analysis the voltammetric stripping methods are popular because of their sensitivity – ranging down to sub ppb concentration. These methods are accurate and precise and the instrumentation is of low cost. The stripping methods are based on previous accumulation of the ion or compound to be determined on the working electrode. In addition to the electrolytical accumulation – used in anodic or cathodic stripping voltammetry (ASV or CSV resp.) also adsorptive accumulation of the species on the working electrode can be exploited, as many organic compounds exhibit surface active properties that are manifested by their adsorption from solution onto the surface of the solid phase.

The first description of this method was published many years ago in connection with the observation that the faradaic signal is increased after adsorptive acccumulation of sulphur,[1] poorly soluble inorganic compounds, alkaloids[2] and some benzophenones[3] on a mercury electrode. The adsorptive accumulation of the reduction products of methylene blue at a HMDE leading to an increase in the height of the anodic polarographic peak in dependence of the accumulation time has also been described.[4] Further development of this method, primarily for practical applications, was closely connected with the developement of a suitable sensor, such as various types of carbon electrodes and static mercury electrodes.[5]

Adsorptive stripping (AdSV) can be employed in the trace analysis of a wide variety of organic compounds exhibiting surface active properties. When the given compound contains an electrochemically reducible or oxidizable group, the peak current on the voltammetric curve recorded after completion of the accumulation period then corresponds practically only to the reduction or oxidation of the whole amount of the adsorbed electroactive species. Only tensammetric adsorption/desorption peaks are obtained for electroinactive compounds.

In AdSV a very rapid diffusion controlled adsorption is assumed. When the amount of substance adsorbed on the electrode surface is determined by the actual adsorption rate, which is smaller than the diffusion rate, it can be assumed that the concentration of surface active substance at the surface of the electrode equals that in the bulk of the solution. This is also true of low adsorptivity of the surface active substance. These last two cases are not useful in AdSV.

In AdSV the measured current signal depends on similar parameters as in ASV: concentration of the compound in the solution, duration of the accumulation, transport velocity of the compound to the electrode (intensity of the convection of the solution during the accumulation step). The peak height in this kind of analysis is also dependent on the surface active properties of the particular compound.

AdSV can be carried out at practically all types of electrodes employed on voltammetry for which a completely reproducible constant surface area can be ensured over the whole measuring period or during a series of measurements.

AdSV enables frequently the determination of adsorbable organic compounds in the concentration range from 1×10^{-6} M to 1×10^{-9} M . The lowest detection limit (until now) was obtained with riboflavine[6] – 2.5×10^{-11} M (HMDE), $t_{acc} = 30$ minutes and with the pesticide DNOK[7] (2-methyl-4,6-dinitrophenol) 5×10^{-10} M ($t_{acc} = 3$ minutes), using differential pulse voltammetry to record the curve. The sensitivity achieved at the determination of organic compounds is similar to the sensitivity of ASV in the determination of metal ions. Until now examples of the determination of more than 150 organic compounds by AdSV are described.[6,8] These compounds include alkaloides, hormones, glycosides, vitamines, antibiotics, drugs, carcinogenes, pesticides, dyes, surfactants, macromolecular compounds, etc.

Ions of many metals form with complexing agents complexes which are frequently adsorbed on the surface of the working electrode. This property can be utilized in the adsorptive accumulation of these metal chelates. AdSV thus permits sensitive determination mainly of ions of metals that are hard or impossible to determine by ASV. The most extencively used AdSV method in practice is the determination of nickel at a mercury electrode as Ni-dimethylglyoximate.[9] From other metals U, V, Pd, Cr, Al, Ca, Mg, Sr, Ba, La, Mo, W, etc. can be determined by AdSV.[6,8]

The high sensitivity of adsorptive stripping methods is obviously their greatest advantage. On the other hand, a serious drawback is interference from other surface-active substances that may be present in the solution. In this case, competitive adsorption usually occures and leads to a decrease in the measured current or, at very high surface-active substance concentrations, to significant suppression of the signal. In such cases it is then necessary to employ suitable separation of interfering compounds, e.g., the application of LC or gel chromatography,[10] TLC separation,[11] ultrafiltration[12] or application of membrane covered electrodes.[13]

If the sample contains interfering compounds that are electrochemically active but are not adsorbed on the electrode surface, then classical separation procedures are not necessary – good results can be obtained when accumulation from the sample solution is followed by exchange of this solution to the pure supporting electrolyte solution. If adsorptive accumulation is carried out using a carbon paste electrode, the medium exchange is very simple: the paste electrode is simply transfered from the sample solution after completion of the accumulation period into the pure supporting electrolyte solution, after brief rinsing with water. Chlorpromazine, phenothiazine and similar substances can be determined in blood serum after accumulation on a carbon paste electrode.[14] The transfer of the electrode after accumulation to the pure base electrolyte permits the determination of these substances in samples containing compounds that are oxidized at the same potential (e.g. ascorbic acid.[15]) Using this techniques adriamycine in urine[16] and uric acid in blood serum and urine[17] can be determined.

The transfer of the HMDE after completion of accumulation to a different base electrolyte solution in which the actual voltammetric measurement is carried out[18] permits the extraction of biomolecules (nucleic acids, some proteins, polysaccharides, lipids) from a medium that is not suitable for the polarographic determination (nonaqueous media, or solution containing various interferents – such as ascorbic acid). The transfer method also permits the study of interactions at the electrode surface of immobilized biomolecules with substances from the solution, without interactions in the solution that would affect the measurement.

The combination of the effect of spontaneous adsorption of the analyte with the medium exchange principle led to the application of AdSV in flow through systems. Here, accumulation (at a given potential) is carried out during the interval when the carrier solution with the injected sample flows through the detector. The accumulation period begins when the injected sample plug comes into contact with the working electrode inside the detector and terminates when it has passed completely through the detector. Afterward fresh carrier solution is allowed to pass through the detector ("washing period") and a voltage scan starting from the accumulation potential is applied. The passage of fresh carrier solution through the detector during the reduction or oxidation step ensures electrochemical stripping of the adsorbed analyte into pure electrolyte with no electrochemically interfering compounds. The amount of accumulated analyte generally depends, e.g., on the analyte concentration, electrode surface area, volume of sample injected and the flow rate. The application of AdSV in flow through systems improves the selectivity and sensitivity of the determination, simplifies the analytical procedure and increases the sample throughput.

As example of the application of AdSV in flow-through systems the determination of Cyadox, one of the quinoxaline-N-dioxide derivatives used as growth promoter in animal breading can be mentioned.[19] A simple detector slipped on to the capillary of a static mercury drop electrode immersed in an electrolyte solution together with the reference and auxiliary electrode were used for this determination. The optimized conditions for analysis of Cyadox in blood plasma are as follows: plasma diluted with carrier solution (0.05 M $NaClO_4$ with 5 vol-% DMF) in the ratio 1:3, flow rate 0.2 ml min^{-1}, accumulation potential -0.10 V, scan rate 10 mV s^{-1}, washing period 5 min, sample loop 1000 μl. These conditions enable the determination of Cyadox from the amount corresponding to 10 ng ml^{-1}. As other examples there can be mentioned the determination of chlorpromazine in body fluids[20] and adriamycine in urine[21] using detectors equipped with carbon paste electrodes. Nickel and cobalt in materia with complex composition were determined as dimethylglyoximates using a mercury film electrode.[22] The determination of nickel in steel samples using HMDE is described[23] and the determination of uranium(VI) (as a catechol complex) in sea water using a mercury film-coated fibre electrode and the constant current stripping method is also described in the litterature.[24] Further progress in this field can be expected in the near future.

References

1. R. Kalvoda, Coll. Czechoslov. Chem. Commun., **21** (1956) 852.
2. R. Kalvoda, Chem. Listy, **54** (1960) 1265.
3. R. Kalvoda and G.K. Budnikov, Coll. Czechoslov. Chem. Commun., **28** (1963) 838.
4. V. Čermák, Chem. Listy, **52** (1958) 413.
5. R. Kalvoda, Anal. Chim. Acta, **138** (1982) 11.
6. J. Wang, American Laboratory, May 1985, 41.
7. H. Beňadiková and R. Kalvoda, Anal. Letters, **17** (A13) (1984) 1519.
8. R. Kalvoda and M. Kopanica, Pure and Appl. Chem., **61** (1989) 97.
9. B. Pihlar, P Valenta and H.W. Nürnberg, Fresenius' Z. Anal. Chem., **307** (1981) 337.
10. R. Kalvoda, Anal. Chim. Acta, **162** (1984) 197.
11. Z. Tocksteinová and M. Kopanica, Anal. Chim. Acta, **199** (1987) 77.
12. Z. Tocksteinová and R. Kalvoda, Chem. Listy, **82** (1988) 1209.
13. F.B. Jarbawi and W.R. Heineman, Anal. Chim. Acta, **186** (1986) 11.
14. J. Wang, B.A. Freiha and B.K. Desmuth, Bioelectrochem. and Bioenergetics, **14** (1985) 457.
15. J. Wang and B.A. Freiha, Anal. Chim. Acta, **148** (1983) 79.
16. E.N. Chaney and R.P. Baldwin, Anal. Chem., **54** (1982) 2556.
17. J. Wang and B.A. Freiha, Bioelectrochem. and Bioenergetics, **12** (1984) 225.
18. E. Paleček and I. Postbieglová, J. Electroanal. Chem., **214** (1986) 359.
19. M. Kopanica and V. Stará, J. Electroanal. Chem., **214** (1986) 115.
20. J. Wang and B.A. Freiha, Anal. Chem., **55** (1983) 1285.
21. E.N. Chaney and R.P. Baldwin, Anal. Chim. Acta, **176** (1985) 105.
22. F. Wahdat and R. Neeb, Fresenius' Z. Anal. Chem., **320** (1985) 334.
23. M. Kopanica, J. Vorlíček and M. Chudačíková, Rudy, **34** (1986) 368.
24. C. Hua, D. Jagner and L. Renman, Anal. Chim. Acta, **197** (1987) 265.

ADSORPTION EFFECTS USED IN ELECTROANALYSIS

Hendrik Emons and Gerhard Werner

Department of Chemistry
Karl-Marx-University of Leipzig
Talstr. 35, Leipzig 7010, G. D. R.

Introduction

Adsorption phenomena at electrodes are often blamed to cause negative or non-understandable effects during electroanalytical measurements. But now there is an increasing interest in the utilization of such interfacial processes for the determination of bulk concentrations of analytes. Therefore a better understanding of the phenomena is necessary. In voltammetric experiments one has to distinguish between the measurement of faradaic and capacity currents. The non-faradaic signals are useful for the analysis of electroinactive surfactants.[1] Mainly ac voltammetry, differential pulse polarography or chronocoulometry are applied to record capacity currents after the adjustment of the adsorption equilibrium, but also non-equilibrium data can be used for analytical purposes. The correlation between the experimental data and the concentration of individual species is still empirical because of the lack of a general adsorption isotherm, which could quantitatively describe the correlation between the surface coverages of certain compounds at different electrode materials and the bulk concentrations for each species also in multicomponent mixtures. Therefore calibration curves are always necessary.

Influence of Adsorption on Capacitive Current

Differential capacity (C_d)-potential (E) curves allow a relatively fast overview about the adsorption of soluble compounds at electrodes. An example for this is shown in Fig.1 by C_d-E curves of (R)-3-((R)-3-(O^2-α-L-rhamnopyranosyl-α-L-rhamnopyranosyloxy) -decanoyloxy) decanoic acid (abbreviated rhamnolipid RL), measured with phase-selective ac voltammetry at the hanging mercury drop electrode. The C_d-values in the potential region of maximum biosurfactant adsorption are not strongly dependent on the concentration. But well-pronounced ad-/desorption peaks around -1.3 V (vs. SCE) are formed. The dependence of the capacity increase (with respect to the C_d-value of the supporting electrolyte) on the bulk concentration of the rhamnolipid can be described by two straight lines (Fig.2). The intersection point appears at the same rhamnolipid concentration, which causes the maximum surface tension decrease in electrocapillary curves. It corresponds to the critical micelle concentration (c.m.c.). A remarkable pecularity of Fig.2 consists in the change of the capacity peak above the c.m.c. By using these non-equilibrium signals it is possible to determine biosurfactant concentrations also above the c.m.c.

In recent years non-faradaic processes which occur within adsorption layers at electrodes are gaining increasing attention.[2-5] Experimentally, additional capacity currents were observed due to phase transitions between different immissible adsorption layer structures. An example is illustrated in Fig.3. The potential dependence of the non-faradaic charge density (q_c) at the interface: static mercury drop electrode/0.9 M $NaNO_3$ and 3-methylisoquinoline (3 MiQ) with a surfactant concentration of 0.545 times the saturation value (c_s) were measured with the integrated dc polarography. The q_c — steps at certain potentials are caused

Contemporary Electroanalytical Chemistry, Edited by A. Ivaska *et al.*
Plenum Press, New York, 1990

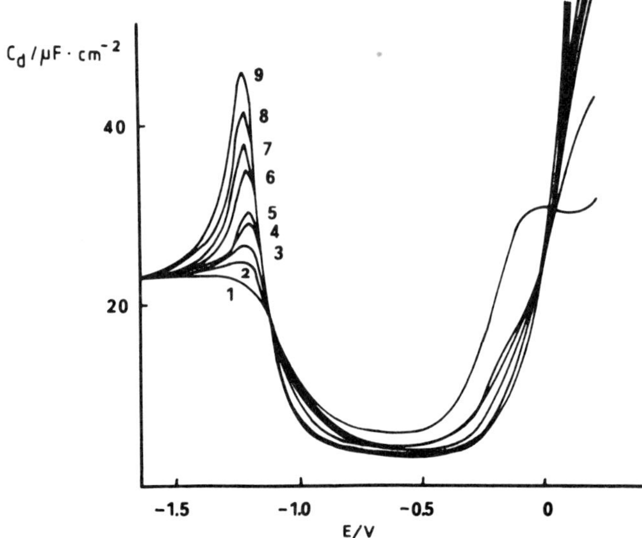

Fig.1. Differential capacity-potential curves at the HMDE in 0.5 M LiClO$_4$ (20 % ethanol) with rhamnolipid-concentrations: (1) 0; (2) 1.5×10^{-5} M; (3) 3×10^{-5} M; (4) 6×10^{-5} M; (5) 9×10^{-5} M; (6) 1.5×10^{-4} M; (7) 2.1×10^{-4} M; (8) 3.6×10^{-4} M; (9) 5×10^{-4} M.

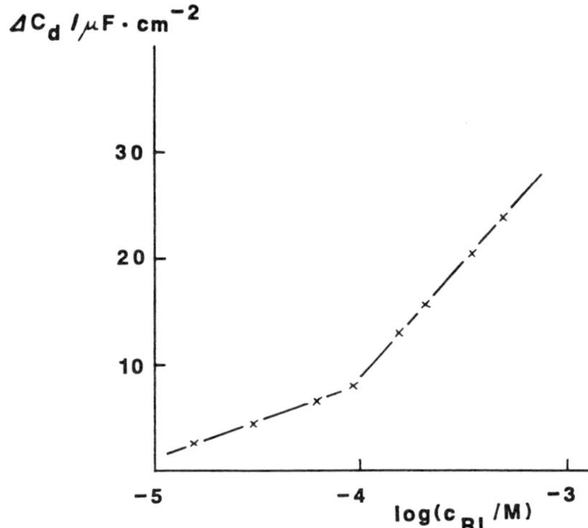

Fig.2. Differential capacity peak height in the presence of different rhamnolipid concentrations with respect to the supporting electrolyte.

by non-faradaic phase transitions. More detailed investigations of these phenomena showed the existence of three different immissible adsorption layer structures.[6] Phase A represents a solid-like film of closely packed 3 MiQ-molecules, phase B a liquid-like structure consisting of 3 MiQ and water molecules and phase C a solid-like coadsorption layer of 3 MiQ and specifically adsorbed nitrate ions. The possible formation of coadsorption layers is often not considered in electroanalytical measurements, but the coadsorption of electrolyte anions can alter the interfacial structure significantly. The potential range, where a definite adsorption film is thermodynamically stable, depends on the surfactant bulk concentration. Fig.4 shows the influence of the 3 MiQ-concentration on the phase transition potentials E$_T$. The E$_T$-values were determined by a double potential step method[7] to avoid kinetic effects. On principle

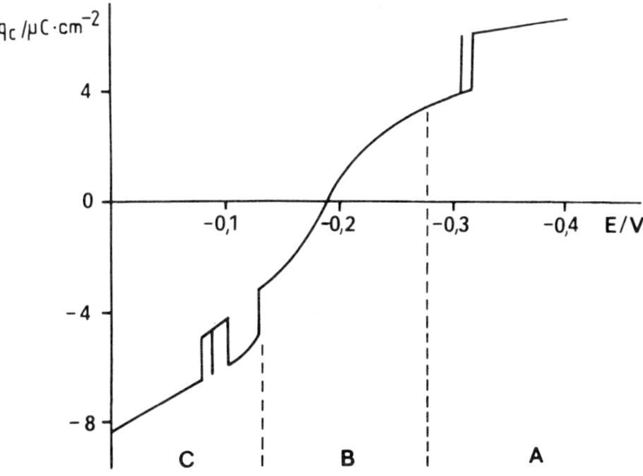

Fig.3. Potential dependence of the capacity charge density of 3 MiQ
(c/cs = 0.545) in 0.9 M NaNO₃ at the SMDE.

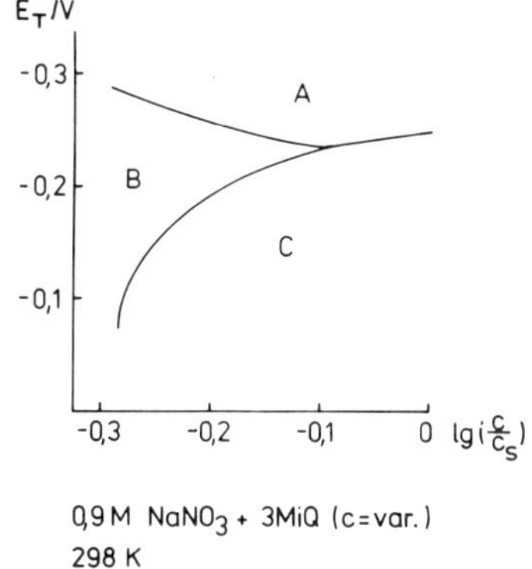

Fig.4. Phase transition potentials in dependence on the
3 MiQ-concentration in 0.9 M NaNO₃.

the E_T — log c plots can be used for determination of surfactant bulk concentration. But the phase transition potentials are also strongly influenced by other parameters, like temperature and supporting electrolyte.

An interesting application of well-defined adsorption layer structures is the *in situ* modification of the electrode/solution interface. The surfactant layers can influence faradaic processes differently.[8] This is demonstrated in Fig.5 by the reduction of Tl^+ and p-benzoquinone, respectively, in a solution of 0.9 M NaNO₃ with various 3 MiQ-concentrations. The liquid-like adsorption layer formed by 10^{-3} M 3-methylisoquinoline at the static mercury drop electrode is not able to inhibit the fast reduction processes and the differential pulse polarographic peaks remain unchanged with respect to measurements without 3 MiQ (curve a). But in the presence of highly condensed 3 MiQ - or coadsorption films (curve b), the electrode reactions are inhibited in different ways: The ion-transfer reaction with amalgam formation $Tl^+/Tl(Hg)$ is totally depressed (peak 2). The electron-transfer reaction of benzoquinone (peak 1) is only partially retarded, which becomes evident by a shift of the peak potential to more negative values and a decreasing peak height (peak H 1'). In the current-potential curve (b) an additional dpp-peak (3) appears. This signal is caused by the non-faradaic phase transition between the adsorption structures C and A. It was also observed in the pure electrolyte solution with saturated 3-methylisoquinoline. Consequently the interpretation of voltammetric experiments can be complicated by non-faradaic processes occuring in the time window of the current measurement. But it is possible to discriminate between electron- and

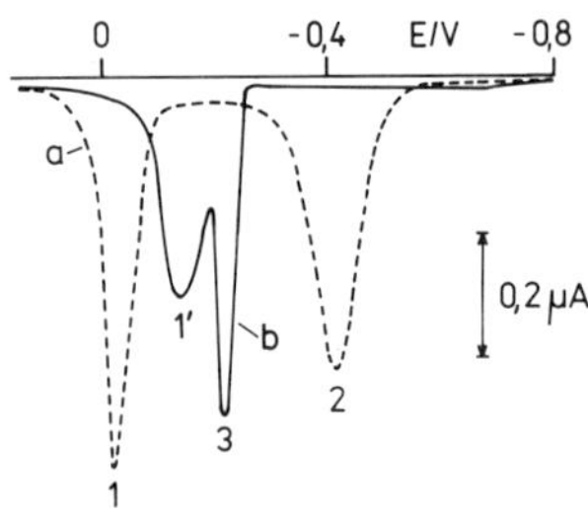

Fig.5. Dpp curves of 10^{-4} M TlNO$_3$ and 10^{-4} M benzoquinone in 0.9 M NaNO$_3$; 3 MiQ-additions: (a) $c/c_s = 0.001$; (b) $c/c_s = 1$.

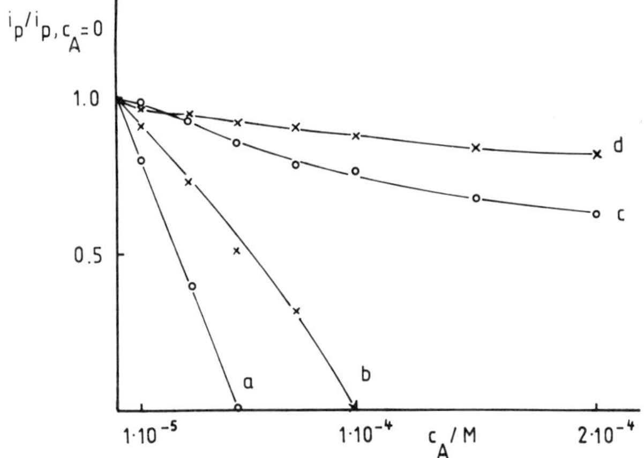

Fig.6. Surfactant influence on the dpp-peak of 5×10^{-4} M Zn^{2+} in 0.1 M NH$_4$Cl + 0.1 M citrate + 0.01 M DMG (10 % methanol); (a) tetrabutylammonium fluoride; (b) triphenyl-phosphine oxide; (c) isoquinoline; (d) BME.

ion-transfer reactions with the help of well-defined adsorptive modified electrode interfaces.

Influence of Adsorption on Faradaic Current

The influence of adsorption phenomena on faradaic processes can be used in different ways for analytical purposes. Faradaic detector reactions, mostly electrode processes of simple inorganic ions, can serve as a probe for the composition of the interface. Therefore bulk concentrations of electroinactive surfactants are determined by the measurement of the inhibition action of the surfactant. Experimental parameters would be the shift of the half-wave potential or the decrease of the diffusion limited current of the detector reaction. Further details are described in the literature about inhibition (for a review see.[9]) In recent years an increasing number of electroactive species is determined by adsorptive stripping techniques.[10,11] The method is used for the trace analysis of surface-active compounds, but also non-surface-active substances can be accumulated at the electrode surface after the formation of complexes with surface-active ligands. A general problem during adsorptive stripping experiments are the interfering processes from other solution constituents. Relating to this it is especially difficult to determine trace components in an electroactive matrix. An application example represents the determination of Co^{2+} in zinc electrolyte plating baths.

It is known, that the polarographic determination of Co^{2+} can be improved significantly by adding dimethyl glyoxime (DMG) to the solution.[12] The adsorption of the Co^{2+}-DMG complex at the mercury electrode enhances the analytical signal. But the reduction of the complex in 0.1 M NH$_4$Cl occurs at nearly the same potential as the Zn^{2+}-reduction. An

410

Table 1. Surfactant influence on the dpp-peak of 5×10^{-6} M Co^{2+} in 0.1 M NH_4Cl + 0.01 M DMG (10 % methanol) with or without 0.1 M citrate

Surfactant (3×10^{-4} M)	i_p (without citrate)	i_p (with citrate)
without surfactant	100 %	100 %
tetrabutylammonium fluoride	41 %	30 %
sodium dodecylsulphate	124 %	50 %
isoquinoline	33 %	60 %
1-(benzylsulfonyl)-2-(N-morpholino)ethane	7 %	122 %

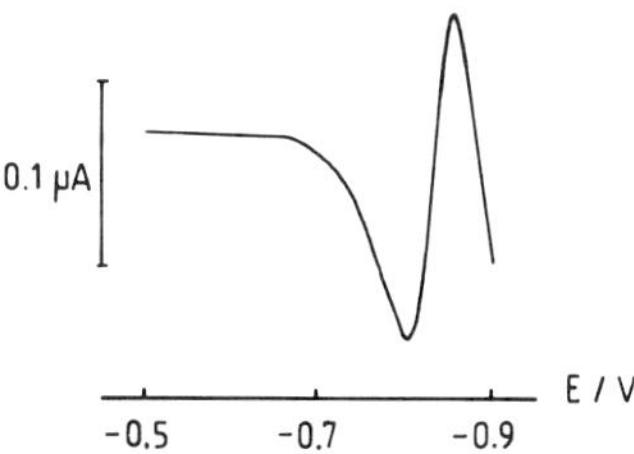

Fig.7. Dpp curve of 10^{-6} M Co^{2+} and 0.4 M Zn^{2+} after adding 10^{-4} M tetrabutylammonium fluoride.

addition of citrate shifts the Zn^{2+}-signal to more negative potentials due to the formation of Zn^{2+}-citrate complexes.[13] But the resulting dpp-peak potential difference of about 200 mV is not sufficient to separate these two faradaic signals in the presence of a very large excess of Zn^{2+}. Therefore the possibility for a selective inhibition of the Zn^{2+}-reduction by an *in situ* modification of the mercury electrode with adsorption layers was investigated. The retardation effect of different surfactants on the Zn^{2+}-reduction is illustrated in Fig.6 by the peak current ratio in the presence and absence of the surfactant. Low bulk concentrations of tetrabutylammonium fluoride are sufficient to depress this electrode reaction totally. Also 1-(benzylsulfonyl)-2-(N-morpholino) ethane (abbreviated BME) shows a significant inhibition effect at surfactant concentrations above 5.10^{-4} M. The influence of the modification layers on the reduction of the Co^{2+}-DMG complex is strongly dependent on the chemical structure of the surfactant, but also on the composition of the supporting electrolyte solution. Table 1 summarizes the relative changes of the dpp-peak currents due to different surfactants before and after the addition of citrate. The most striking phenomenon is the compensation of the strong retardation effect of BME in the presence of citrate. This results from the coadsorption of citrate, which causes also a change in the mechanism of the rate determining step. The surfactants tetrabutylammonium fluoride and BME are the most favourable compounds for a selective inhibition of the Zn^{2+}-reduction without any big changes in the Co^{2+}-DMG-signal. The Fig.7 shows the reduction peak of 10^{-6} M Co^{2+} at -0.8 V in presence of a 400 000 fold excess of Zn^{2+}. The Co^{2+}-signal is not detectable without additions of 10^{-4} M tetrabutylammonium fluoride. It was possible to determine Co^{2+} in the μM-range even in an one million fold excess of Zn^{2+} after modifying the mercury electrode with an adsorption layer of BME.[14] As summarizing conclusion one may say that adsorption phenomena are often useful for the determination of bulk concentrations. But to utilize all the potentials such an effect can offer, a certain knowledge about interfacial processes and structures is necessary.

References

1. P.M. Bersier and J. Bersier, Analyst, **113** (1988) 3.
2. Cl. Buess-Herman, J. Electroanal. Chem., **186** (1985) 27.
3. Cl. Buess-Herman in: "Trends in Interfacial Electrochemistry", A.F. Silva (Ed.), Reidel, Dordrecht, 1986.
4. L. Pospisil, E. Müller, H. Emons and H.-D. Dörfler, J. Electroanal. Chem., **170** (1984) 319.
5. R. Sridharan and R. de Levie, J. Electroanal. Chem., **230** (1987) 241.
6. Cl. Buess-Herman, H. Emons and L. Gierst, J. Electroanal. Chem., (submitted).
7. Cl. Buess-Herman, J. Electroanal. Chem., **186** (1985) 41.
8. J. Lipkowski, Cl. Buess-Herman, J.P. Lambert and L. Gierst, J. Electroanal. Chem., **202** (1986) 169.
9. H.-D. Dörfler and H. Emons, Z. Chem., **27** (1987) 273.
10. R. Kalvoda, Anal. Chim. Acta, **138** (1982) 11.
11. J. Wang, Am. Lab., **17** (1985) 41.
12. P. Nangiot, J. Electroanal. Chem., **14** (1967) 197.
13. A. Meyer and R. Neeb, Fresenius' Z. Anal. Chem., **315** (1983) 118.
14. Th. Schmidt, M. Geißler, G. Werner and H. Emons, Fresenius' Z. Anal. Chem., **330** (1988) 712.

MICROANALYTICAL APPLICATION OF A KISSINGER TYPE THIN LAYER CELL POLARIZED BY LSV, NPV AND DPV EXCITATION

G. Farsang and T. Dankházi

Institute of Inorganic and Analytical Chemistry
L. Eötvös University
1088 Budapest, VIII. Múzeum krt 4/B, Hungary

Introduction

The Kissinger type twin electrode thin layer cell is a widely used tool in the everyday analytical practice especially in the field of flow injection techniques and as an amperometric detector in liquid chromatography.[1-3] Less attention is paid to the possibilities offered by this cell as a microanalytical tool when it is filled with a quiescent solution sample. In this way the determination of electroactive components in a volume of about 50-100 μl can be carried out by applying a proper excitation potential program.

The attention was drawn to this opportunity by W.R. Heinemann and his coworkers when they used a thin layer cell for the determination of chloropromazine and diazepam in urine and plasma samples. They used differential pulse voltammetric (dpv) excitation and reported a lower detection limit of 5×10^{-8} M whe using sample volumes of 50-100 μl.[4] The Kissinger type twin electrode cell was found very promising for the analysis of very small sample volumes (50-100 μl) containing reversible redox systems when it was polarized with a slow dc scan.[5] When comparing the behaviour of the same cell filled with quiescent solution and a flowing stream of sample solution the sensitivity for the two cases was found to be nearly equal. The extremely low sample volume needed ($\leq$ 50 μl) and the favourable shape of the signal (stationary limiting current) makes it interesting to extend this type of measurements by applying the three most generally used excitation waveforms, namely the linear sweep voltammetry (lsv), normal pulse voltammetry (npv) and the differential pulse voltammetry (dpv). This gives a chance to compare the analytical parameters of these powerful techniques for the thin layer cells filled with quiescent analyte solutions. The aim of this paper is to present the most significant results for these methods.

Experimental

Instruments

A Bruker E-100 type Universal dc-Pulse Polarograph was used for polarization. A home made Kissinger type twin electrode thin layer cell with carbon paste working and counter electrodes of 3.14 mm^2 geometric area each was used. The reference electrode was a standard Ag/AgCl/1 M KCl cell saturated with AgCl. 40 and 80 μm thick Teflon spacers with 4×13 mm channel cuts were used. The cross section of the cell is shown in Fig.1. The approximate volume of the cell used is 2-5 μl depending on the spacer thickness. A Hamilton type syringe (manufactured by Scientific Engineering PTY.LTD. Australia) of 100 μl volume was used to flush and fill the cell with the solutions to be measured.

Chemicals and Solutions

10 - (3' - dimethylaminopropyl) - 2 - chlorophenothiazine hydrocloride (further on chloropromazine CPZ) prepared by EGIS Pharmaceutical Factory, (Pharmacopeia

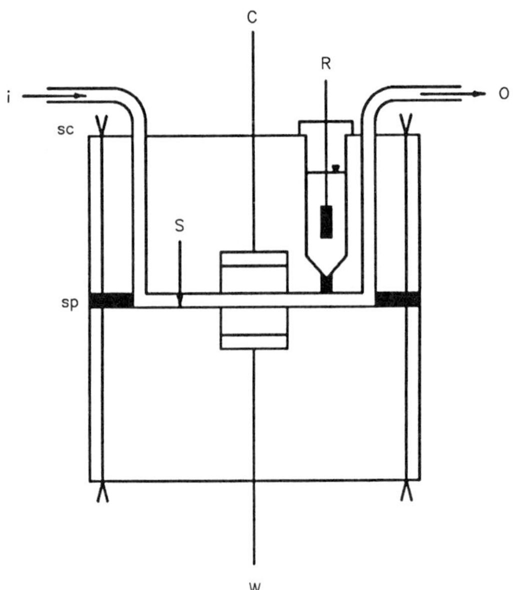

Fig.1. The cross section of the home built Kissinger type twin electrode thin layer cell. R: Ag/AgCl wire, in a closed salt bridge compartment. The salt bridge is connected to the thin layer by a piece of "Vycor" disc. The reference compartment is filled with 1 M KCl saturated with AgCl.
sp: Teflon spacer
i: Inlet tube
o: Outlet tube both made from Teflon
s: Solution layer
sc: screws at the four corners of the cell body to provide leakproof assembling of the two halves of the cell.

Hungarica VII.Eds. Vol.II. quality), was dissolved in 0.01 M hydrochloric acid. It was diluted with distilled a.lt. hydrochloric acid prepared by Reanal Fine Chemical Factories. All dilutions of the solutions used were made of double distilled water immediately before use and the oxygen was removed before the dilution by bubbling high purity nitrogen gas through the base electrolyte.

Results and Discussion

As a model compound for the measurements a major tranquilizer, the well known chloropromazine was chosen. The electrode reaction of this compound is well estalished.[6-8] In the potential range and time scale used in the measurements with thin layer voltammetry one needs to take into consideration only one fast electron exchange reaction of chloropromazine:

$$CPZ \rightarrow CPZ^{+\cdot} + e^- \tag{1}$$

because the radical cation is stable in the medium used (0.01 M HCl). Three methods were used for polarization of the thin layer cell:
a. linear dc scan (lsv)
b. differential pulse voltammetric excitation (dpv)
c. normal pulse voltammetric excitation (npv)

In routine polarographs only the following three excitation program parameters can be set:
− the rate of the dc voltage scan (v: mV/s)
− the amplitude of the potential pulses (ΔE: mV)

– the clock time controlling the synchronizing pulses (T: s)

In the case of pulse methods, the pulse length and the sampling time are fixed to given values, except for computerized instruments. The Bruker E-100 type polarograph used in this work has a pulse time of 55 ms and a current sampling time of 15 ms. The clock time can be varied in the range of 0.1-99.9 s in steps of 0.1 s. The dc scan rate can be varied from 0.1 to 500 mV/s. In order to achieve the best analytical performance (sensitivity, selectivity, accuracy and low detection limit) for the three methods with a thin layer cell one has to find the optimal settings for recording the voltammetric curves.

Linear Sweep Voltammetry

In this case only the rate of polarization giving stationary diffusion profiles in the thin solution layer needs to be found. This results in a stationary limiting current instead of a current peak as for semi-infinite conditions. The rate was experimentally found to be 2 mV/s at a space thickness of less than 100 μm. The slow rate of polarization does not affect the limiting current, but the charging current of the signal decreases which results in a better lower detection limit. The stationary limiting current can be calculated with the equation first derived by Anderson and Reilley:[9]

$$i_L = \frac{nFAD_{SS}C_T}{d} \tag{2}$$

where A is the area of the working electrode, C_T the sum of concentrations of oxidized and reduced forms in the diffusion layer, d the thickness of the solution film. D_{SS} can be taken to be $2D_R$ if D_R is equal to D_O. F and n have their usual meanings. An i_L vs. C calibration curve was determined for the range $1\times10^{-6} - 1\times10^{-4}$ M CPZ using a 25 μm cell spacer. The following values were calculated for the calibration curve:
– slope: 2.24×10^{-3} A/dm$^3\times$ mm$^2\times$ M
– intercept: 2.35×10^{-9} A/mm^2
– correlation coefficient: 0.9979

The lower detection limit is improved when a thinner spacer is used, and 1×10^{-7} M can be achieved easily. The method is very simple and even simple and inexpensive dc polarographs can be used. The selectivity of the method is the same as for normal dc polarography. A high excess of an easily oxidized compound prevents the measurement of another compound oxidized at a higher positive potential.

Differential Pulse Voltammetry

DPV was first tried with twin electrode thin layer cell. From the point of optimization its behaviour is quite similar to the cell used by Jarbawi, Heinemann and Patriarche.[1] In Table 1 the effect of clock time is shown for a 10^{-4} M CPZ solution. The pulse amplitude was 20 mV and the dc rate of polarization 2 mV/s. Two different spacer thicknesses, 40 and 80 μm, were used.

Table 1. The effect of clock time (T) on the DPV curves recorded at two spacer.

T(s)	i_p (nA) $d = 40\ \mu$m	i_p (nA) $d = 80\ \mu$m	E_p (V) $d = 80\ \mu$m
0.1	272	283	+0.569
0.2	265	252	+0.575
0.5	265	250	+0.575
1.0	266	259	+0.574
2.0	263	255	+0.574
5.0	264	245	+0.580
9.9	260	224	+0.580

The increase of clock time at a properly chosen pulse amplitude and scan rate does not effect the peak current or the peak potential seriously. The curves are smoothed as T decreases (v and ΔE fixed) and can consequently be evaluated more precisely. In spite of this a time T = 1.0 is recommended, because the complete decay of the reverse current spike due to the potential step back to the actual dc potential disturbes less the current sample before the application of the next potential pulse, as can be seen in Table 1. The thickness of the spacer does not have any greater effect on the measured dpv current.

Due to the dc sweep a stationary concentration profile is built up in the thin layer cell. The effect of this profile is eliminated in the dpv technique by the double sampling of the current and therefore only the current sample taken at the end of the pulse is actually measured. There is no time for redox cycling during the pulse duration, as was proved for the npv-polarization. The dpv excitation has several analytical advantages:
i) Sensitivity can be increased by increasing the pulse amplitude.
ii) Peak type curves offers very good selectivity.
iii) Significantly lower detection limits can be achieved as will be shown in the comparison of the three methods.

By varying different pulse amplitudes and scan rates at different clock times the following were found to be the most useful:
v$\leq$10 mV/s, T = 0.5–1.0 s and pulse amplitude ΔE $\leq$50 mV.

Normal Pulse Voltammetry

It is known that npv excitation gives a larger analytical signal than dpv or dc for a given concentration of the analyte. Because pulses with large amplitudes are needed to reach the limiting current region of the npv curve, the charging current cannot be eliminated properly using a fixed pulse duration and a fixed width of a current sample. This means that the sensitivity of npv is highest at the detection limit. The detection limit, however, is quite poor. In order to optimize the parameter settings at npv excitation at a Kissinger type cell the optimal scan rate and clock time had to be measured first. The results are given in Table 2:

Table 2. Effect of scan rate and clock time on the limiting current of npv curves at a given concentration: 10^{-4} M CPZ in 0.01 M HCl.

i_L = limiting current of npv (μA)				
T (s)	\multicolumn{4}{c}{v (mV/s)}
	10	5	2	1
0.1	1.21	1.21	1.00	0.97
0.2	1.71	1.76	1.57	1.48
0.5	2.16	2.09	1.90	1.83
1.0	2.50	2.19	2.02	1.98
2.0	2.35	2.09	2.07	2.04

From Table 2 and the dpv curves obtained the following optimized conditions can be selected:
Decreasing the scan rate at a given clock time causes only a slight change in the current. The curves, however, become smoother when v is decreased. The sensitivity can be improved by increasing the clock time at a low scan rate. A clock time of 1 s and a scan rate of 2 mV/s give an acceptable recording time, well evaluable curves and a high sensitivity. The explanation of this behaviour is quite clear. By increasing the clock time a longer relaxation periode is allowed for the perturbed diffusion layer. It is quite obvious, that with a 40 μm spacer thickness the diffusion layer behaves like under semiinfinite conditions. No recycling effect is expected during the 55 ms pulse duration. The less exhausted diffusion layer in the neighbourhood of the anode at longer T values provides larger faradaic signals.

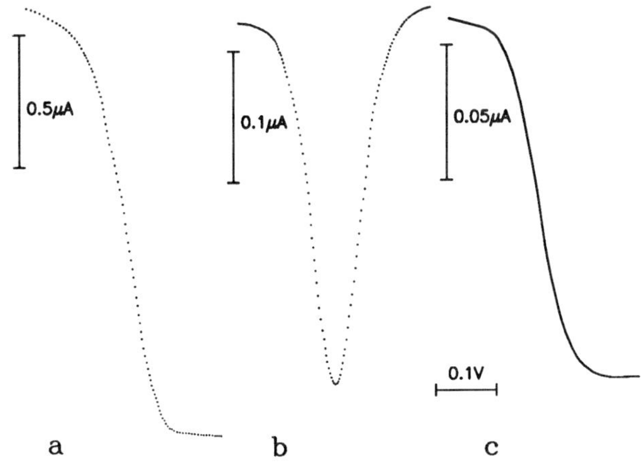

a   b   c

Fig.2. The voltammetric curves recorded with the three excitation signals used.

Curve a: Normal pulse voltammetric curve

Curve b: Differential pulse voltammetric curve

Curve c: Linear sweep voltammetric curve

CPZ concentration: 10^{-4} M

Base electrolyte: 0.01 M HCl.

Linear scan rate applied: 2 mV/s; clock time: 1.0 s;
d = 40 μm; ΔE = 20 mV when recording the dpv curve.

Evaluation of the Comparative Results Obtained by lsv, dpv and npv Excitation in a Thin Layer Cell

Each of the three methods tried can be used for analytical measurements of samples of extremely small volume (about 50 μl is needed to carry out each measurements). In Fig.2 the three different signals are shown when recorded at optimal conditions.

It can be seen from Fig.2 that from the analytical point of view the three methods have different properties. The npv method is much more sensitive compared to the other two techniques. The use of npv is, however, almost impossible at concentrations lower than 10^{-6} M. Apparently the lsv technique has the lowest sensitivity.

The dpv excitation signal can be increased by increasing the pulse amplitude. It was found that even in the case of such a fast electron exchange reaction as the radical formation from CPZ both techniques have about the same detection limits: 10^{-7} M. The pulse techniques are insensitive to the spacer thickness while the lsv signal can be doubled by halving the spacer thickness. The lsv technique has a definite advantage, when compounds of slower electron transfer kinetics are measured.

It is quite informative to study the summary of the calibration curves for the three different techniques in the concentration range of 5×10^{-7} to 2×10^{-5} M. The experimental parameters were:

lsv: v = 2 mV/s

dpv: v = 2 mV/s, T = 1.0 s, ΔE = 20 mV

npv: v = 2 mV/s, T = 1.0 s

spacer thickness: 40 μm

Conclusions

All the three methods studied are useful for polarization of electrodes in a thin layer cell. The main advantage of Kissinger type cells filled with quiescent solution is that small sample volumes, less than 50 μl, can be used. The sensitivity is highest for npv, but this technique is not recommended because of the relatively high charging current. Dpv is an

Table 3. Data from the calibration curves measured by lsv, dpv and npv techniques:

Concentration M	lsv i_L (nA)	dpv i_p (nA)	npv i_L (nA)
5×10^{-7}	1.9	2.4	5.2
1×10^{-6}	2.4	3.9	10.0
2×10^{-6}	4.4	6.3	28.5
5×10^{-6}	9.4	15.6	79.0
1×10^{-5}	16.9	32.1	162.0
2×10^{-5}	34.2	64.7	320.0
slope (nA/M)	1.653×10^{6}	3.208×10^{6}	1.626×10^{7}
intercept (nA)	0.927	0.247	1.00
corr. coeff. (r)	0.9997	0.9998	0.9998

excellent method due to its sensitivity (at high ΔE it is superior) and very good selectivity.

The lsv method is proved to be an excellent and simple method when reversible systems are to be measured. The steady state limiting current gives a possibility to carry out measurements with high sensitivity and a lower detection limit of 10^{-7} M was found for this technique.

The optimization procedure could have been further refined when computerized instruments are used. The sigmoid type current curves in the lsv technique would certainly give a better selectivity if the first derivative of the current would be used.

References

1. D.C. Johnson, S.G. Weber, A.M. Bond, R.M. Wightman, R.E. Shoup and I.S. Krull, Anal. Chim. Acta, **180** (1986) 187.
2. P.T. Kissinger, in: "Laboratory Techniques in Electroanalytical Chemistry". P.T. Kissinger and W.R. Heinemann (Eds), Marcel Dekker, New York (1984), p. 611.
3. K. Stulik and V. Pacakova, C.R.C. Crit. Rev. Anal. Chem., **14** (1984) 297.
4. T.W. Jarbawi, W.R. Heinemann and G.J. Patriarche, Anal. Chim. Acta, **126** (1981) 57.
5. G. Farsang, T. Dankházi, J. Loranth and L. Daruhazi, Talanta (to be published).
6. P. Kasbakalian and J. Mc Glotten, Anal. Chem., **31** (1959) 431.
7. M.D. Hawley, S.W. Tatwawadi, S. Piekarski and R.N. Adams, J. Am. Chem. Soc., **89** (1967) 447.
8. F.H. Merkle and C.A. Dishen, Anal. Chem., **36** (1964) 1639.
9. L.B. Anderson and C.N. Reilley, J. Electroanal. Chem., **10** (1965) 295.
10. G.Farsang and T.Dankházi, Anal. Lett., **22**(5) (1989) 1305

DIFFERENTIAL PULSE POLAROGRAPHIC DETERMINATION OF TeO_4^{2-} AND VO_3^- IN PRESENCE OF SOME OXYHALIDE ANIONS

M.M. Kamal, Z.A. Ahmed and Y.M. Temerk

Chemistry Department, Faculty of Science
Assuit University
Assuit, Egypt

Introduction

Resolution and trace determination of inorganic oxyanions are important for assay purposes in commercial samples. The dc polarographic behaviour of these oxyanions and the resulting application have been described earlier.[1-5] The detection limit of dc polarography is usually about 10^{-4} M but may be higher under non-optimum conditons. Since differential pulse polarography (dpp) and voltammetry should improve sensitivity at trace level concentrations, the use of dpp and voltammetry for determination of some oxyanions has been studied.[6-12] In the present work applicability of dpp for quantitative trace determination and resolution of TeO_4^{2-} and VO_3^- in presence of some oxyhalide anions such as BrO_3^-, IO_3^- and IO_4^- is investigated.

Experimental

Materials

All chemicals used were of analytical grade. 0.01 M stock solution of TeO_4^{2-}, VO_3^-, BrO_3^-, IO_3^- and IO_4^- were prepared by dissolving the accurate weight of the solid potassium or ammonium salt in twice distilled deionized water. The modified universal buffer series of Britton and Robinson (pH 2 – 12) was used as supporting electrolyte and prepared as given by Britton.[13]

Instrumentation

All polarograms were recorded with a PAR Model 174A polarographic analyzer equipped with a PAR Model 303A hanging mercury drop electrode. Polarograms were recorded on an advanced Hewlett-Packard 7045A X-Y recorder. The electrode system was constructed with a platinum wire auxiliary electrode and Ag/AgCl reference electrode together with a dropping mercury electrode. Oxygen free nitrogen was used for de-gassing and a digital Radiometer pH-meter Model PH M64 was used for pH measurements. All measurements were performed at $22 \pm 0.5°C$.

Results and Discussion

Differential Pulse Polarographic Behaviour of TeO_4^{2-} and VO_3^- in Presence of Some Oxyhalide Anions

TeO_4^{2-} gives a well defined peak in 0.04 M universal buffer soluitons (pH 2.2 – 12.0) for a concentration of 10^{-5} M. In presence of an equimolar concentration of BrO_3^- oxyanion, this peak was found to be affected and interferred with the background discharge at 2.2 – 4.0 pH range. At pH 4.5, the large difference in E_p's for TeO_4^{2-} (E_p = 1.23 V) and BrO_3^- (E_p =

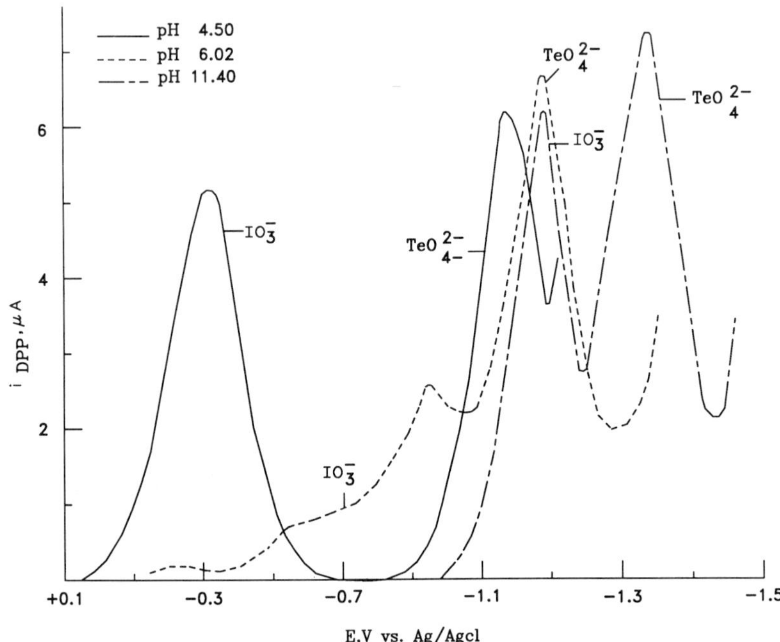

Fig.1. Effect of pH on the dpp behaviour of TeO_4^{2-} (5×10^{-5} M) in presence of an equimolar concentration of IO_3^-. Scan rate 5 mV/s, drop time 2 s and pulse amplitude 100 mV.

−0.93 V) reduction peaks permits the determination of TeO_4^{2-} in presence of an equimolar concentration of BrO_3^- with fair accuracy (the degree of recovery is 96 %). In neutral and alkaline solutions, the differential pulse peak of TeO_4^{2-} was recorded with recovery of 100 % in presence of BrO_3^- which does not give a cathodic signal in these media.

The differential pulse polarograms of 5×10^{-5} M of TeO_4^{2-} were investigated at various pH values in presence of an equimolar concentration of IO_3^- or IO_4^- (Fig.1). The response of TeO_4^{2-} reduction peak and its resolution from the cathodic peak of IO_3^- or IO_4^- are mainly dependent of pH. The maximum response and a better separation of TeO_4^{2-} reduction peak from the reduction peak of IO_3^- or IO_4^- were found at pH's 4.5 and 11.4. At pH < 4.0 TeO_4^{2-} reduction peak interfered with the background discharge whereas at 5.0 < pH < 9.5 the reduction peaks for TeO_4^{2-} and IO_3^- or IO_4^- overlap. Therefore, dpp determination of TeO_4^{2-} in presence of IO_3^- or IO_4^- was recorded at pH 4.5 and 11.4 in acidic and alkaline media, respectively with higher degree of recovery (96 − 106 %).

VO_3^- gives a well defined peak in the range pH 2.2 − 7.0 for a concentration of 10^{-4} M. This peak was found to be unaffected by equimolar concentrations of IO_3^- or IO_4^- oxyanions with a recovery of 100 % on the peak signal. The maximum response and better separation of VO_3^- reduction peak from the cathodic peaks IO_3^- or IO_4^- was obtained at pH 2.2. At higher pH values (pH > 8.0) the reduction peak of VO_3^- interferes with the background discharge. In presence of an equimolar concentration of BrO_3^- the VO_3^- reduction peak is completely overlapped with the cathodic response of BrO_3^- at pH 2.2 - 5.0. However, an equimolar concentration of BrO_3^- reduces the signal of VO_3^- to 72 % at pH 6.0 − 8.0.

At pH 4.5 and in presence of an equimolar concentration of IO_3^-, the differential pulse polarograms of 5×10^{-5} M TeO_4^{2-} at various pulse amplitudes (25 − 100 mV) indicate that the peak height increases as the pulse amplitude is increased. However, at 100 mV pulse amplitude the cathodic peak of TeO_4^{2-} give the highest sensitivity and is most sharply resolved from the IO_3^- peak.

Drop time 1 s was found to give the maximum response and a high degree of resolution for the investigated oxyanions compared with the drop times of 2 s or 5 s. The drop time 1 s was therefore used in the present study.

420

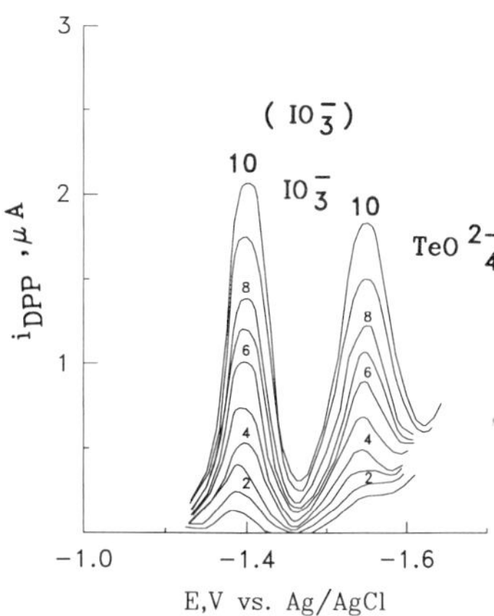

Fig.2. Differential pulse polarograms of an equimolar concentrations of TeO$_4^{2-}$ and IO$_3^-$ at pH 11.4, 1) 0.495; 2) 0.980; 3) 1.456; 4) 1.923; 5) 2.830; 6) 3.704; 7) 4.545; 8) 5.357; 9) 6.897 and 10) 8.333 μM. Curves recorded under the optimum conditions.

The resolution and the maximum response of the reduction peak of the selected oxyanions was markedly affected by the scan rate. Scan rate of 5 mV/s reduces the sensitivity of the reduction peak of the oxyanion by 18 – 28 %. The degree of resolution and the reduction peak response at 2 mV/s and 1 mV/s are similar, therefore, scan rate of 2 mV/s was used in the present investigation.

Table 1. Results of determination of TeO$_4^{2-}$ in presence of IO$_3^-$ and IO$_4^-$.

Conc. taken μM	Conc. observed (μM) in presence of		Recovery (%) in presence of	
	IO$_3^-$	IO$_4^-$	IO$_3^-$	IO$_4^-$
0.495	0.475	0.525	96.16	106.08
0.980	0.965	1.020	98.47	104.11
1.456	1.470	1.540	100.96	105.76
1.923	1.980	1.900	102.96	98.80
2.830	2.760	2.835	97.53	100.17
3.704	3.710	3.595	100.16	97.06
4.545	4.545	4.410	100.00	97.03
5.357	5.355	5.145	100.18	96.04
6.897	6.910	7.050	100.19	102.22
8.333	8.510	8.400	102.12	100.80

Determination of TeO$_4^{2-}$ and VO$_3^-$ in Presence of Some Oxyhalide Anions

The optimum conditions for the analytical determination of TeO$_4^{2-}$ in presence of BrO$_3^-$, IO$_3^-$ and IO$_4^-$ were found to be a drop time of 1 s, scan rate of 2 mV/s and a pulse amplitude of 100 mV. Response of the differential pulse peak height (i$_p$, μA) with the concentrations of TeO$_4^{2-}$ is linear at the optimum pH in presence and in absence of oxyhalide anions. The results of determination and resolution of TeO$_4^{2-}$ in presence of oxyhalide anions are based on the calibration curves of TeO$_4^{2-}$ alone by this technique. Table 1 shows the results of determination of TeO$_4^{2-}$ in presence of IO$_3^-$ and IO$_4^-$ at pH 4.5. The observed concentrations are based on the comparison of measured i$_p$ with the calibration curve for TeO$_4^{2-}$.

421

Table 2. Results of determination of VO_3^- in presence of IO_3^- and IO_4^-.

| Conc. taken | Conc. observed (μM) in presence of | | Recovery (%) in presence of | |
μM	IO_3^-	IO_4^-	IO_3^-	IO_4^-
0.996	1.050	1.100	105.42	110.44
2.964	2.975	2.990	100.37	100.88
4.901	4.910	4.900	100.18	99.98
9.615	9.615	9.620	100.00	100.05
14.151	14.150	14.165	99.99	100.01
22.727	22.725	22.735	99.99	100.03
34.483	34.480	34.495	99.99	100.03

Fig.2. shows the differential pulse polarograms of an equimolar concentration of TeO_4^{2-} and IO_3^- at pH 11.4. The variations of the i_p of TeO_4^{2-} reduction peak with concentration is a straight line with the slope of 0.235 ± 0.016 μA/μM, intercept -0.05 ± 0.003 μA and the regression coefficient of the fit was 0.999. The values of i_p/C, μA/μM was 0.252 with standard deviation of 0.0063; the detection limit was estimated to be 0.495 μM.

Results of analysis of VO_3^- in presence of IO_3^- or IO_4^- at pH 2.2 and under the optimum conditions are summerized in Table 2. The observed concentrations are based on the calibration curve constructed for VO_3^- alone in the solution.

References

1. I.M. Kolthoff and J.J. Lingane, J. Am. Chem. Soc., **61** (1939) 825.
2. I.M. Kolthoff and E.E. Orleamnn, J. Am. Chem. Soc., **64** (1942) 1044.
3. P. Souchay, Anal. Chim. Acta, **2** (1948) 17.
4. R.M. Issa and B.A. Abd-El-Naby, Fresenius' Z. Anal. Chem., **237** (1967) 339.
5. J.P. Arnold and R.M. Johnson, Talanta, **16** (1969) 1191.
6. D.J. Myers and J. Osteryoung, Anal. Chem., **45** (1973) 267.
7. Y. Foiseberg, J.W. O'Langhlin and R.E. Megargle, Anal. Chem., **47** (1975) 1586.
8. H.P. Davis, G.R. Dulude, R.M. Griffin, Z.R. Matson and E.W. Zink, Anal. Chem., **50** (1978) 137.
9. Y.M. Temerk, M.E. Ahmed and M.M. Kamal, Fresenius' Z. Anal. Chem., **301** (1980) 414.
10. W. Holak, Anal. Chem., **52** (1982) 2189.
11. D. Chakraborti, R.L. Nicholsand and K.J. Irgolic, Fresenius' Z. Anal. Chem., **319** (1984) 248.
12. M. Rausl Jan and W.F. Smyth, Analyst, **109** (1984) 1187.
13. H.T.S. Britton, in: "Hydrogen Ions", 4th edition, Chapman and Hall, London (1952), p. 313.

DIFFERENTIAL PULSE POLAROGRAPHIC DETERMINATION OF SOME SUBSTITUTED BENZYLIDENE ACETHYDRAZONES

M.M. Kamal, Y.M. Temerk, Z.A. Ahmed and
G. Abd-El-Wahab

Chemistry Department, Faculty of Science
Assiut University
Assiut, Egypt

Introduction

The electrochemistry of hydrazides and their derivatives is of potential interest due to their biological importance as antituberculosis compounds. In this context the polarographic behaviour and reduction mechanism of various arylidene acid hydrazone has been studied in the past over a wide range of pH.[1-6] The polarographic methods which has been applied previously for the determination of the hydrazide derivatives are insufficient for trace analysis.[4,5] The detection limit for dc polarography is usually ca. 10^{-4} M but be higher under non-optimum conditions.

In continuation of the applicability of dpp for quantitative trace determination of hydrazide derivatives,[7] the present investigation is focused on the analytical determination of some substituted benzylidene acethydrazone under the optimum conditions.

Experimental

Chemicals and Solutions

Substituted benzylidene acethydrazones were prepared and purified by a conventional method.[8] The hydrazones investigated have the general structure:

$$CH_3-\overset{\overset{\displaystyle O}{\|}}{C}-NH=N=CH-C_6H_4-X$$

Where X=H (Benzylidene acethydrazone, BAH); 2-OH(2-hydroxy-BAH);
4-OH(4-hydroxy-BAH); 4-N-$(CH_3)_2$ (4-N,N-dimethyl-BAH) and
4-NO_2 (4-nitro-BAH).

The purity of the compounds was checked by elemental analysis and m.p. determination. Stock 0.01 M solutions were prepared in absolute ethanol. The modified universal buffer series of Britton and Robinson[9] was used as supporting electrolyte. The pH was checked using a digital Radiometer pH-meter, Model PH M64, accurate to ±0.05 unit.

Instrumentation

Differential pulse polarographic measurements was carried out with a PAR Model 174A Polarographic Analyzer equipped with a PAR Model 303 static mercury electrode system. This system contained a dropping mercury electrode (dme) as the working electrode, a Ag/AgCl reference electrode and a platinum wire counter electrode. Polarograms were recorded on an advanced X-Y recorder (Hewlett-Packard 7045A). All measurements were performed at $20 \pm 0.5°C$.

Contemporary Electroanalytical Chemistry, Edited by A. Ivaska *et al.*
Plenum Press, New York, 1990

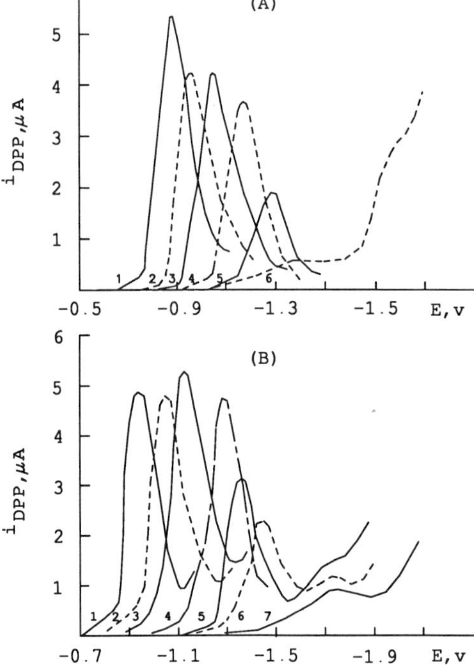

Fig.1. Effect of pH on 5×10^{-5} M of BAH (A) and 4-N,N-dimethyl-BAH (B). (1) pH 3.10; (2) pH 4.02; (3) pH 5.08; (4) pH 6.02; (5) pH 7.00; (6) pH 8.02 and (7) pH 9.06, Scan rate 5 mV/s, drop time 2 s and pulse amplitude 100 mV.

Working Procedure

10 ml of the buffer solution was placed in the electrolysis cell and was deaerated by passing a slow stream of pure nitrogen through it for 30 min. The nitrogen was previously humidified by passing it through a wash bottle filled with a blank solution. Sequential additions of millimolar stock solutions of the investigated compounds were used to obtain the calibration curves. For a total addition of less than 100 ml a Unimetric Model 1010 10 ml syringe was used. Statistical analysis of calibration curve parameters (slopes, intercepts, standart deviation of the fit and regression coefficients) is also given. The variation of i_p (μA) with the concentration (μM) of each compound is based on the straight line equation $i_p = aC + b$ where a and b are the slope and intercept of straight line, respectively.

Results and Discussion

Effect of pH on the Faradaic Response

Differential pulse currents of the various substituted benzylidene acethydrazones, recorded as a function of the potential in universal buffer solutions containing 40 % ethanol are given in Fig.1. Differential pulse polarographic current voltage curves of the investigated compounds in different buffer solutions reflects the reduction of the hydrazone centre CH=N-NH- for the compounds under investigation. According to the general mechanism for the polarographic reduction of some acethydrazone derivatives[6] the CH=N-NH- linkage undergoes a totally irreversible reduction with an overall uptake of $4e^-/4H^+$ and $2e^-/2H^+$ per molecule in acid and alkaline media, respectively.

The dpp of benzylidene acethydrazone derivatives in acidic (pH < 4.8) and in alkaline (pH > 7.2) buffer solutions is of particular interest. The reduction current decreases thermendously with time, while a second peak appears at more negative potentials (Fig.2).

Closer inspection of the second peak shows the conincidence of E_p with that corresponding to the reduction of aldehyde at the same pH. This indicates that the second peak must be attributed to the reduction of aldehyde formed in the hydrolysis reaction of the substituted benzylidene acethydrazones into their main constituents. The hydrolysis is caused by the localization of electrons on the N atom of the azomethine centre. In order to confirm the result, the variation of peak height (μA) with time (min.) was plotted as log i_p vs time. A good linear correlation was obtained for all substituted benzylidene acethydrazones which indicates that the hydrolysis process follows strictly first-order kinetics. The aformentioned result indicates that the range of stability of substituted benzylidene acethydrazones lies at the pH range of 4.8 to 7.2. The maximum response peak for the investigated compounds was found at pH = 5.08 (Table 1).

The optimum pH for the analytical determination of some substituted benzylidene acethydrazones in presence of some transition metal ions was also investigated. It is found that the high stability constants of the formed metal complexes decrease the rate of hydrolysis. The percentage of hydrolysis decreases with increasing pH up to pH 3.70 and at pH 4.02 the process of hydrolysis was not observed any more. In this case the mode of bonding the substituted benzylidene acethydrazones with transition metal ions was investigated carefully by ir spectra which showed that the chelate formation takes place through the C=O and C=N groups. Therefore the process of hydrolysis is decreased or eliminated as the bond order of C=N is increased due to the delocalization of electrons on the N atom of the azomethine centre by the chelation. This result indicates that the optimum pH for the determination of the investigated compounds in presence of metal ions was found at pH 4.02 without any effect of hydrolysis. At pH 4.02 the maximum response was observed.

Dependence of Differential Pulse Peak (dpp) on Pulse Amplitude, Drop Time and Scan Rate

Dependence of the dpp peak current on the pulse amplitude (ΔE) of the investigated compounds is given in Fig.3. A well-defined peak was obtained in the range of pulse amplitude of 25–100 mV. However, at a pulse amplitude of 100 mV the investigated compounds gave the highest sensitivity and complete resolution from the background discharge, hence a pulse amplitude of 100 mV was used in the present study.

The peak half-width of the various substituted benzylidene acethydrazones was found to be independent on pulse amplitude (170 ± 10 mV for all compounds at pH 5.08). This is in agreement with the observations of Parry and Oldham on the irreversible reduction system.[10]

Dependence of the dpp signal on the drop time is represented in Fig.3. It was found that the height of the reduction peak increases as the drop time (t_d) is decreased from 5 s to 0.5 s. The electronic noise and stirring, however,[11,12] result in a decreased signal to noise ratio when the drop time is decreased. The latter fact becomes more significant for drop time 0.5 s therefore a drop time of 1 s was used in the present investigation.

The differential pulse peak is scan rate dependent in that an increased scan rate decreases resolution and lowers the peak height. It was found that the peak height of benzylidene acethydrazone at 5 mV s^{-1} is only 76 % and 72 % of that at 2 mV s^{-1} and 1 mV s^{-1} respectively. Because response for a 2 mV s^{-1} was very similar to that with 1 mV s^{-1} the former was used for the optimum determination of the investigated compounds.

Quantitative Trace Determination of Some Substituted Benzylidene Acethydrazone

The optimum conditions for the analytical determination of some substituted benzylidene acethydrazones were found to be the following: pH 5.08, pulse amplitude 100 mV, drop time 1 s and scan rate 2 mV s^{-1}. The investigated compounds display a maximum peak height at the optimum pH value. Response of differential pulse peak height with concentration of each compound is a straight line (Fig.4). The validity of the method is supported by the constancy of the i_{dpp}/C (Table 2). The variation of i_p, μA with concentration of each compound is also represented by straight line equation $i_p = aC + b$. The data for three to five replicated measurements were subject to a least square refinement and the values of the regression coefficient were calculated and are given together with the straight line constants in Table 3.

Table 1. Maximum peak current (i_p), peak potential (E_p) and the percentage of hydrolysis (D%) at different pH values for some substituted benzylidene acethydrazones.

pH	BAH i_p, μA	$-E_p$, V	D%*	2-OH-BAH i_p, μA	$-E_p$, V	D%*	4-OH-BAH i_p, μA	$-E_p$, V	D%*	4-N-(CH$_3$)$_2$-BAH i_p, μA	$-E_p$, V	D%*	4-NO$_2$-BAH i_p, μA	$-E_p$, V	D%*
2.30	6.31	0.77	38.11	9.10	0.81	42.10	4.65	0.87	46.13	5.30	0.79	56.30	12.6, 6.60	0.19, 0.73	31.12
3.31	5.50	0.86	11.90	9.40	0.90	12.76	4.80	0.94	14.14	5.10	0.82	41.20	10.4, 6.80	0.25, 0.84	20.58
4.02	4.25	0.96	5.88	3.90, 4.20	0.82, 0.89	4.76	4.80	1.05	14.60	6.40	0.96	6.25	9.8, 8.80	0.33, 1.02	11.36
5.08	4.20	1.05	0.00	4.15, 4.40	1.0, 1.10	0.01	5.30	1.14	0.00	5.45	1.06	0.00	8.9, 9.15	0.39, 1.11	0.00
6.01	3.75	1.16	0.00	3.50, 3.65	1.11, 1.22	4.28	4.90	1.30	11.22	4.10	1.20	0.00	6.4, 8.00	0.45, 1.22	0.00
7.00	2.10	1.28	1.06	2.75	1.20	9.10	3.25	1.36	50.76	3.15	1.28	4.76	5.1, 5.80	0.56, 1.30	0.00
8.10	0.07, 2.90	1.31, 1.52	90.00	4.00	1.26	32.50	5.00	1.46	64.00	2.10, 7.70	1.25, 1.35	7.14	4.85, 2.2	0.64, 1.37	11.36
8.98	-	-	-	3.75	1.36	33.00	3.00	1.57	96.00	2.00, 2.75	1.25, 1.35	9.56	4.80, 1.8	0.66, 1.37	24.31
10.02	-	-	-	0.75	1.45	86.00	-	-	-	1.40	1.80	13.89	-	-	-

* The percentage hydrolysis after 30 min.

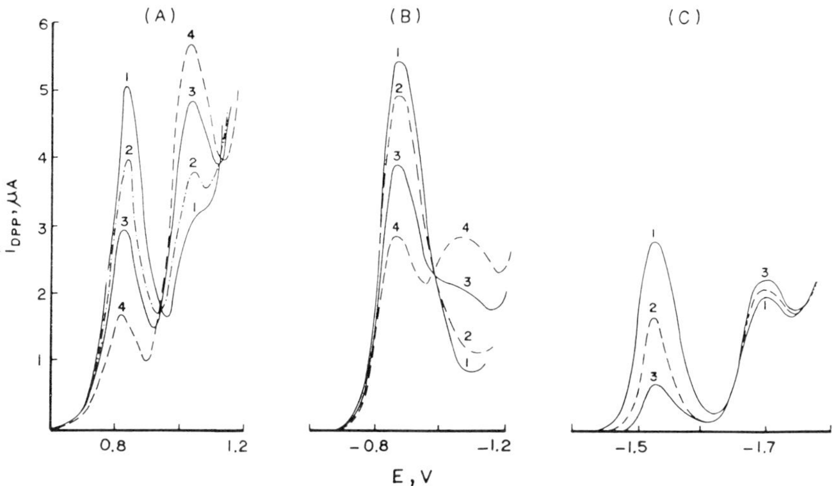

Fig.2. Time dependence of the differential pulse peak of 5×10^{-5} M of (A) 4-N,N-dimethyl-BAH (pH 3.1), (B) BAH (pH 3.1) and (C) 4-hydroxy-BAH (pH 8.02). 2 min (1), 15 min (2), 30 min (3) and 60 min (4). Conditions as in Fig.1.

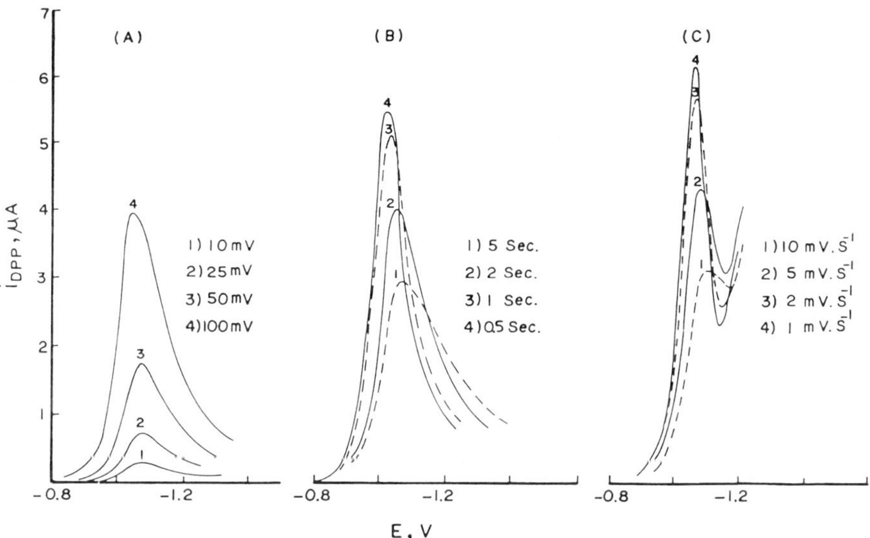

Fig.3. (A) Dependence of the dpp of 5×10^{-5} M of BAH on pulse amplitude at pH 5.08. (B) Dependence of dpp of 5×10^{-5} M of BAH on drop time at pH 5.08. (C) Dependence of dpp of 5×10^{-5} M of 4-N,N-dimethyl-BAH on scan rate at pH 5.08.

Quantitative Trace Determination of Some Substituted Benzylidene Acethydrazone in Presence of Some Metal Ions

Some substituted benzylidene acethydrazones and some transition metal ions form complexes of high stability. The investigated compounds can be determined with fair accuracy without any effect of hydrolysis. The height of differential pulse peak vs. concentration of each compound is linear at pH 4.02 (Fig.5). The detection limit is down to 0.099 μM at pH 4.02 whereas at pH 5.08 and in absence of metal ions it was estimated to be 0.199 μM at the same pH. The calibration curve (Fig.5), e.g., 2-hydroxy-BAH had a slope of $0.667 + 0.048\mu$A$/\mu$M with the intercept of $0.0041 \pm 0.0006\mu$A; the regression coefficient of the fit was 0.999. The

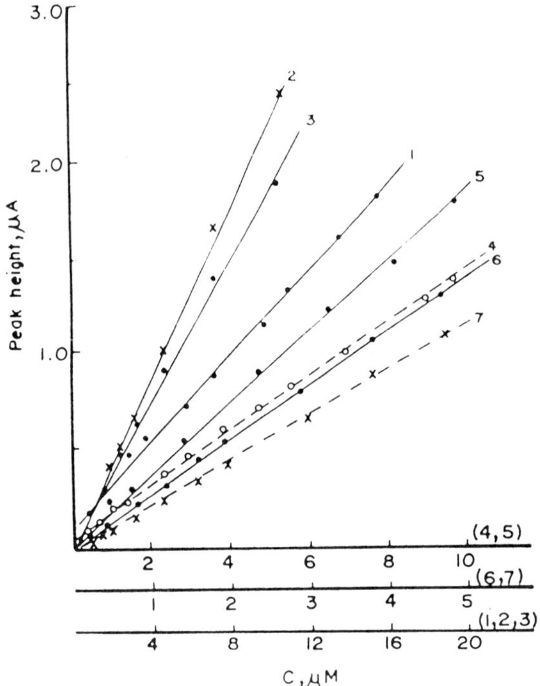

Fig.4. Calibration curves for BAH(1); 4-NO$_2$-BAH 1st peak (2) 2nd peak (3); 4-OH-BAH(4); 4-N,N-(CH$_3$)2-BAH (5); 2-OH-BAH 1st peak (6) and 2nd peak (7). Under the optimum conditions.

value of i_p/C was 0.728 with the standard deviation of 0.042.

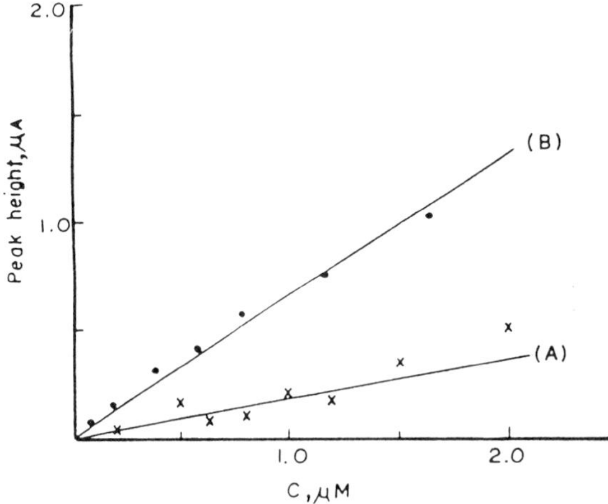

Fig.5. Calibration curves for 2-OH-BAH at pH 4.02 drop time 1 s, scan rate 2 mV/s and pulse amplitude 100 mV in absence (A) and in presence of Cu^{2+} (B).

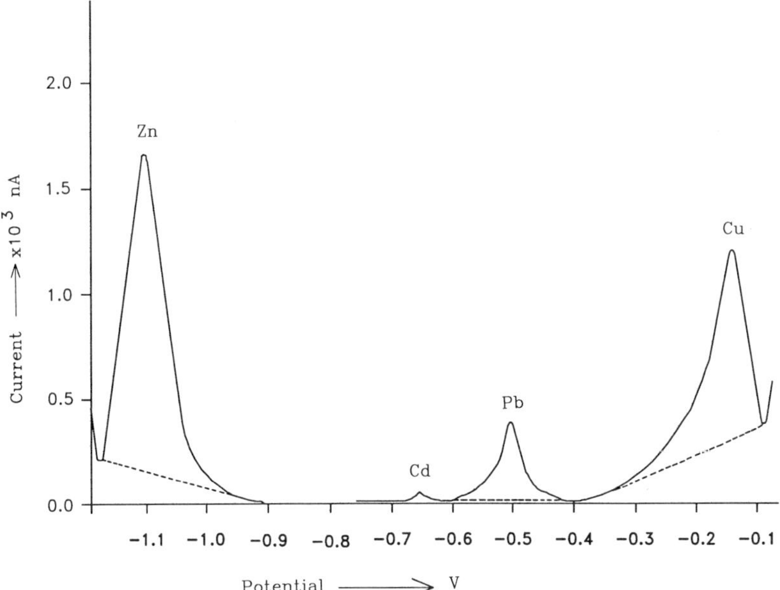

Fig.1. A typical dp anodic stripping voltammogram of a sample containing Zn^{2+}, Cd^{2+}, Pb^{2+} and Cu^{2+} in 1 M citrate buffer pH=4.5. Zinc was determined in a separate experiment by using -1.2 V and the other metal ions by using -0.8 V as the electrolysis potential.

Table 1. Concentrations of Cu, Cd, Pb and Zn in Indian Snuff Samples.

Sample No.	Pb (ppb)*	Cu (ppb)*	Zn (ppb)*	Cd (ppb)*
1	560	450	2200	< 2
2	505	90	1600	25
3	540	310	1820	< 2
4	495	400	975	10
5	550	650	2300	5
6	410	670	2000	< 2
7	530	250	1150	< 2
8	15	550	1800	15
9	470	220	980	< 2
10	140	300	1200	20
11	565	240	1080	< 2
12	525	480	1120	< 2
13	480	350	2100	< 2
14	490	380	1950	< 2
15	510	680	2320	5

* ppb = ng/ml of sample solution.

The extent of danger due to polycyclic aromatic hydrocarbons[13-17] and nitrosoamines in causing cancer cannot be ascertained as these substances were not determined in the present study.

Clinical Enquiry

All types cancer cases are monitored by National Cancer Registry through its centres at Bombay, Bangalore, Madras, Chandigarh, Dibrugarh and Trivandrum. Table 3 depicts number of cases of nasal pharyngeal carcinoma and nasal cavity cancer and total number of

Table 2. Concentration of Cr and Ni in Indian Snuff Samples.

Sample No.	Cr (ppb)*	Ni (ppb)*
1	660	190
2	260	190
3	430	180
4	430	160
5	640	430
6	580	210
7	430	190
8	110	190
9	100	160
10	100	120
11	240	230
12	480	240
13	250	230
14	300	120
15	440	120

* ppb = ng/ml of sample solution.

Table 3. Different cancer cases registered at National Cancer Registry.

Centre		Bangalore				Bombay			
Year		P 1982	P 1983	P 1984	H 1984	P 1982	P 1983	P 1984	H 1984
Total no. of	M	969	979	966	2260	3128	3125	3244	5744
cases registered	F	1168	1149	1178	2733	2526	2496	2715	4871
Cancer of Naso-	M	11	6	7	35	25	29	18	55
phyranx	%	1.14	0.61	0.72	1.55	0.80	0.93	0.55	0.96
	F	2	6	3	13	14	7	7	18
	%	0.17	0.52	0.25	0.48	0.55	0.28	0.26	0.37
Cancer of Nasal	M	3	2	6	19	28	24	28	87
Cavity	%	0.31	0.20	0.62	0.84	0.90	0.77	0.86	1.51
	F	6	4	1	13	20	12	24	58
	%	0.51	0.35	0.08	0.48	0.79	0.48	0.88	1.19

Centre		Madras				Chandigarh		
Year		P 1982	P 1983	P 1984	H 1984	H 1982	H 1983	H 1984
Total no. of	M	982	1050	962	1807	1127	1070	1206
cases registered	F	1315	1392	1281	2166	1328	1227	1220
Cancer of Naso-	M	11	10	13	13	4	18	12
phyranx	%	1.12	0.95	1.35	0.72	0.35	1.68	1.00
	F	4	5	5	5	4	7	6
	%	0.30	0.36	0.39	0.23	0.30	0.57	0.49
Cancer of Nasal	M	6	13	9	20	22	18	20
Cavity	%	0.61	1.24	0.94	1.11	1.95	1.68	1.66
	F	3	5	9	10	17	16	12
	%	0.23	0.36	0.70	0.46	1.28	1.30	0.98

Table 3. Continues

Centre		Dibrugarh			Trivandrum		
Year		H 1982	H 1983	H 1984	H 1982	H 1983	H 1984
Total no. of	M	902	888	863	1890	2024	2097
cases registered	F	324	379	335	1603	1691	1755
Cancer of Naso-	M	21	18	18	22	24	30
phyranx	%	2.33	2.03	2.09	1.16	1.19	1.43
	F	7	5	1	10	8	13
	%	2.16	1.32	0.30	0.62	0.47	0.74
Cancer of Nasal	M	3	1	2	23	28	27
Cavity	%	0.33	0.11	0.23	1.22	1.38	1.29
	F	2	2	2	20	25	12
	%	0.62	0.53	0.60	1.25	1.48	0.68

P = Population based cancer registy, H = Hospital based cancer registry,
M = Male, F = Female.

Table 4. Total Percentage of cases of Nasal Cavity Cancer and Nasophyrangal Cancer in Males and Females.

	Year	Male	Female
Bangalore	P 1982	1.45	0.68
	P 1983	0.81	0.87
	P 1984	1.34	0.33
	H 1984	1.39	0.96
Bombay	P 1982	1.70	1.34
	P 1983	1.70	0.76
	P 1984	1.41	1.14
	H 1984	2.47	1.56
Madras	P 1982	1.73	0.53
	P 1983	2.19	0.72
	P 1984	2.29	1.09
	H 1984	1.83	0.69
Candigarh	H 1982	2.30	1.58
	H 1983	3.36	1.87
	H 1984	2.66	1.47
Dibrugarh	H 1982	2.66	2.78
	H 1983	2.14	1.85
	H 1984	2.31	0.90
Trivandrum	H 1982	2.38	1.87
	H 1983	2.57	1.55
	H 1984	2.72	1.42

P = Population Based Cancer Registry
H = Hospital Based Cancer Registry

all types of cancer cases registered at different cancer registry centres in India. It is evident from Table 3 that, generally more cancer cases are reported in case of males than in females (Table 4). These data can be explained on the basis of the fact that the social custom of snuffing is more prevalent in male population than in female. It is reported that in Bantu the cancer of the nasal and accessory sinuses comprised 45.5 % of all the respiratory tract cancer.[2,18] In comparison with this value, the total percentage of nasal pharyngeal carcinoma and nasal cavity cancer in males and females together in India is much less. However, the number of all cases, particularly from rural India, not being reported may be higher.

Conclusions

All tobacco products, including snuff, contain toxic heavy metals and many carcinogenic substances. The concentrations of heavy metals in Indian snuff samples is significantly much less compared to those in snuff samples from South Africa and U.S.A.[12]

References

1. D.E. Redmond Jr., N. Engl. J. Med., 1970, Jan. 1, 282, (1) 18.
2. M.P. Shapiro, et al., Brit. J. Cancer, 7 (1953) 45.
3. R.L. Cooper, J.M. Campbell, Brit. J. Cancer, 9 (1955) 528.
4. S. Danday, et al., Western J. Med., 145 (1) July, (1986) 111.
5. W.H.O. Oral Surgery, 45 (1978) 518.
6. S. Silverman et al., Cancer, 53 (1984) 563.
7. Council Report, JAMA, 255, (8) Feb. 28, (1986) 1038.
8. M.A.H. Russell et al., Brit. Med. J. (Clinical Research), 283 (6295) Sep. 26, (1981) 814.
9. M.P. Shapiro, et al., S. Afr. Med. J., 29 (1955) 95.
10. P. Keen, et al., Brit. J. Cancer, 9 (1955) 528.
11. S.V. Bhide, U.S. Murdia and J. Nair, Cancer Letters, 24 (1984) 89.
12. Archives of Environmental Health, 23 (1) July, (1971) 1.
13. F. Haber, in: "Zur Geschichte des Gaskrieges". In Fünf Vortrage aus den Jahren 1920-23, 76, Springer, Berlin.
14. E.L. Wynder, Med. Clin. N. Amer., 40 (1956) 629.
15. H.V. Gelboin, Physiol. Rev., 60 (1980) 1107.
16. G. Grimmer, in: "Environmental Carcinogems: Selected Methods of Analysis", Vol. 3, pp. 55-61, M. Castegnaro, P. Bogovski, H. Kunte and E.A. Walker (Eds), IARC Scientific Publications No. 29, Lyon.
17. National Academy of Science Reports, U.S.A., (1972) Particulate Polycyclic Organic matter committee on biological effects of atmospheric pollutants, p. 30, Washington DC.
18. J.K. Selikth, in: "Carcinogenesis, Vol. 5, Modifiers of Chemical Carcinogenesis", T.J. Slaga (Ed.), Raven Press, New York, 1980, p. 1.
19. Y. Woo, and J.C. Arcos, in: "Environmental Chemicals in Carcinogens in Industry and the Environment", J.M. Sontag (Ed.), Marcel Dekker, New York, 1981.
20. P. Keen, UICC Monogr. Ser., 1, 15, (1964) 95.

ELECTROCHEMICAL BEHAVIOUR OF EPN

T.N. Reddy and S. Jayarama Reddy
Department of Chemistry
S.V. University
Tirupati-517 502, A.P., India

Introduction

Polarographic methods can be used to advantage for the determination of agrochemicals in food, water, body fluids etc., particularly if the pollutant contains nitro group in their molecules. The nitro group present in the pesticides generally produces well defined waves in a variety of solvents.[1] Many pesticides containing a nitro group such as parathion, methylparathion[2,3] and phenetrothion[4] have been subjected to polarographic study. Some workers[5,6] have reported slight differences in the behaviour of parathion, methylparathion and EPN which can be used for their determination in mixtures.

The aim of this work is to carry out detailed studies on the reduction of EPN in the solvent DMF over a wide pH range (2.0 to 12.0), to postulate the possible reaction mechanism for the electrochemical reduction of the compound employing different techniques such as cyclic voltammetry, dc polarography ac polarography and differential pulse polarography and to describe a procedure for the analytical determination of the pesticide in trace levels.

Experimental

Apparatus

Princeton Applied Research Corporation (PARC) model 364 polarographic analyser coupled with BD 8 Kipp and Zonen recorder was used to recorde direct-current polarograms. A three electrode cell was employed containing a dropping mercury eledtrode of flow rate 2.73 mg/second as working electrode, saturated calomel electrode as the reference electrode and a platinum wire as the auxiliary electrode. Differential pulse polarography and ac polarography were performed with Metrohm E 506 polarecord connected to E 612 VA-scanner. The electrode assembly, in this case, consisted of a dropping mercury electrode of area 0.0223 cm^2 as the working electrode, a saturated Ag/AgCl(s), Cl^- as the reference electrode and a platinum wire as the auxiliary electrode. A digital electronics 2000 X-Y/t recorder coupled with Metrohm E 506 was used to recorde the cyclic voltammograms. The hanging mercury drop electrode of the area 0.0443 cm^2 was used as working electrode in cyclic voltammetry.

All experiments were carried out at 26 $\pm$ 1 °C. Dissolved air was removed from the solutions by degassing with oxygen-free nitrogen for 10 minutes. pH measurements were made by using Elico digital pH meter.

Reagents

The technical compound EPN was supplied by Promochem, West Germany. The purity of the sample was tested by thin layer chromatography. Stock solution was prepared by dissolving the required amount of EPN in dimethylformamide (DMF) as a solvent and making up with the supporting electrolytes to get the desired concentration. All the supporting electrolytes used in this work were of analytical grade.

Contemporary Electroanalytical Chemistry, Edited by A. Ivaska *et al.*
Plenum Press, New York, 1990

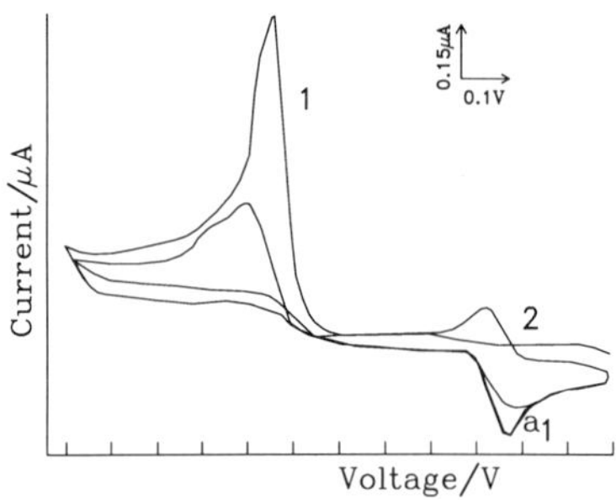

Fig.1. Typical cyclic voltammogram of EPN in McIlvaine buffer at pH 8.00. Concentration 0.5 mM, sweep rate: 40 mVs^{-1}, 1: first scan, 2: second scan.

Results and Discussion

Two well defined waves/peaks have been observed ($E_{1/2}$ values at -0.41 V and -0.83 V; E_p values at -0.53 V and -0.90 V) for the reduction of EPN in acetate buffer of pH 4.0 at 25 % DMF. With increasing pH and alcohol concentration, the half-wave potential/peak potential values are found to be shifted to more negative potentials. This behaviour can be discussed on the basis of the presence of an adsorbed film of solvent molecules at the electrode surface which increases the activation energy of the reaction and retards the reaction rate.[7,8] These observed waves/peaks correspond to the reduction of nitro group present in EPN to yield hydroxylamine in the first stage and form hydroxylamine to amine in the next stage. Only one wave/peak has been observed in the alkaline medium which may be attributed to the reduction of nitro group to hydroxylamine because the observed current is equal to that observed in the first wave in acidic media.

A small anodic peak (a_1, at -0.24 V) has been observed in the reverse scan of cyclic voltammogram at higher pHs as shown in Fig.1. It is quite likely that nitroxide ion[9] is formed, whose movement at the electrode surface may be responsible for the anodic peak. In the next scan, another small cathodic peak is observed which may be due to the reduction nitroxide formed at a_1.

Conventional log-plot analysis, variation of $E_{1/2}$ and E_p values towards more negative potential with increase of concentration of depolariser, disobedience of Tomes' criterion, the non-linearity between i_m vs $\frac{1-\sigma}{1+\sigma}$ where $\sigma = \frac{nF}{RT} \frac{\Delta E}{2}$ and the deviation of ac polarographic summit potential (E_s) from the dc polarographic half-wave potential ($E_{1/2}$) show that the electrode process related to the reduction of EPN is irreversible. The irreversibility of the electrode process is also evidenced from the absence of anodic peak in the cyclic voltammogram and the variation of peak potential with the scan rate.

The diffusion controlled nature of the electrode process is evidenced from the linear plots of i_d vs. $h^{1/2}$, i_p vs. $v^{1/2}$, i_m vs. $T^{2/3}$ and passing through origin in alkaline media indicating the absence of adsorption complications. But ac polarographic measurements have indicated weak adsorption of reactant in the acidic medium as seen from the depression of the base current before the ac peak (Fig.2):

The adsorption behaviour is attributed to the presence of $-\overset{.}{\underset{.}{P}} = S$ moiety in EPN. In alkaline conditions, this moiety may probably be removed from the electroactive species due to hydrolysis[10] thus making no adsorption in this media.

The number of electrons involved in the electrode process are found to be six and four in acetate buffer (pH = 4.0) and carbonate buffer (pH = 10.0) respectively from the results of microcoulometry. The products of controlled potential electrolysis (carried out at -1.50 V and -1.30 V vs. SCE) in the above mentioned supporting electrolytes are identified[11] as amine and hydroxylamine respectively. The nitro group in EPN exists in different forms depending on the supporting electrolyte.[12] The change in colour of depolariser to yellow at

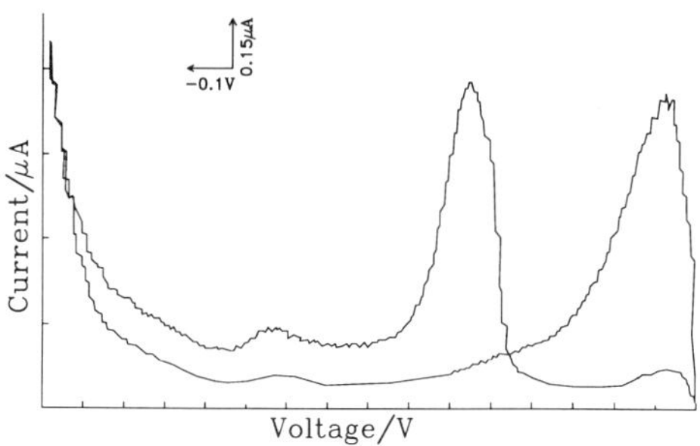

Fig.2. A typical ac polarogram of EPN in acetate buffer at pH 4.0. Concentration: 0.5 mM, drop time: 3 s, solvent: 40 % DMF and 60 % water.

higher pH can be explained as due to the anion formation shown:

$$R \;=\; \text{—} \langle C_6H_4 \rangle \text{—} O\text{-}\underset{\underset{C_6H_5}{|}}{\overset{\overset{S}{\|}}{P}}\text{-}OC_2H_5$$

$$R\text{-}N\Big\langle \begin{matrix} OH \\ O^- \end{matrix} \qquad R\text{-}N\Big\langle \begin{matrix} O \\ O \end{matrix} \qquad R\text{-}N\Big\langle \begin{matrix} O^- \\ O^- \end{matrix}$$

(In acid)   (In neutral)   (In alkaline)

Various mechanisms can be proposed which explain the electrochemical behaviour of EPN in the different pH ranges studied.

In acidic medium:

$$R - NO_2 + 4e^- + 2H^+ \longrightarrow R - NHOH + H^+ \rightleftharpoons R - N\overset{+}{H}\overset{}{O}H_2$$

$$R - NH\overset{+}{O}H_2 + 2e^- + 2H^+ \longrightarrow R - NH_2 + H_3O^+$$

In alkaline medium:

$$R - NO_2 + e^- \longrightarrow R - NO_2^-$$

$$R - NO_2^- \longrightarrow R - NHOH + 4OH^-$$

The relevant kinetic data is calculated and reported in Table 1. The heterogenous forward rate constant values are in general found to decrease with increase in pH indicating that the electrode reaction tends to become more and more irreversible. The rate constant values are found to be high in acidic medium when compared to neutral and alkaline medium indicating that the rate of reaction is fast in acidic medium since in this medium the protonated form is reduced. The standard rate constants obtained in ac polarography is found to be high when compared to other techniques since the rate constants are evaluated at standard potential in ac polarography. The rate constants have been calculated at potential E = 0 in other techniques. Dc polarographic diffusion coefficient values may be considered as more reliable in the case of EPN since it is a slow technique and the renewal of the surface of mercury drops under dc polarographic conditions might also be responsible for this situation.

Analysis

In the present investigation, dc polarography and differential pulse polarography have been used for the quantitative estimation of EPN. Detection limit for EPN in the present investigation is 0.5×10^{-4} M with dc polarography when compared to differential pulse polarography where it is observed that the determination of EPN could be carried out even down to 1.0×10^{-8} M levels, since differential pulse polarography should improve sensitivity

Table 1. Typical electrochemical kinetic data for EPN. Solvent : 35% DMF, †sweep rate : 40 mV s^{-1}, concentration : 0.5 mM.

Name of the supporting electrolyte pH	D.C. Polarography			A.C. Polarography			†Cyclic voltammetry			Differential pulse polarography conc. 0.05 mM		
	$\dfrac{-E_{1/2}}{V}$	$\dfrac{D \times 10^6}{cm^2 s^{-1}}$	$\dfrac{K^o_{f,h}}{cm\ s^{-1}}$	$\dfrac{-E_s}{V}$	$\dfrac{D \times 10^6}{cm^2 s^{-1}}$	$\dfrac{K_s}{cm\ s^{-1}}$	$\dfrac{-E_p}{V}$	$\dfrac{D \times 10^6}{cm^2 s^{-1}}$	$\dfrac{K^o_{f,h}}{cm\ s^{-1}}$	$\dfrac{-E_m}{V}$	$\dfrac{D \times 10^4}{cm^2 s^{-1}}$	$\dfrac{K^o_{f,h}}{cm\ s^{-1}}$
Clarks & Lubs buffer pH 2.0	0.28	2.36	4.51×10^{-5}	0.43	0.46	1.08×10^{-5}	0.50	0.34	3.34×10^{-18}	0.29	3.38	9.71×10^{-10}
Acetate buffer pH 4.0	0.40	2.09	5.19×10^{-8}	0.56	0.10	1.36×10^{-4}	0.53	0.26	3.26×10^{-18}	0.39	0.54	9.61×10^{-14}
Phosphate buffer pH 6.0	0.51	1.31	2.40×10^{-9}	0.75	0.33	2.47×10^{-4}	0.68	0.41	4.90×10^{-19}	0.58	0.72	1.20×10^{-16}
McIlvaine buffer pH 8.0	0.65	0.71	3.52×10^{-13}	0.81	0.24	6.95×10^{-5}	0.74	0.28	2.46×10^{-20}	0.71	0.35	1.79×10^{-19}
Carbonate buffer pH 10.0	0.78	0.48	3.74×10^{-13}	0.87	0.27	2.23×10^{-4}	0.84	0.01	1.24×10^{-22}	0.78	0.02	2.58×10^{-21}
Bates & Bower buffer pH 12.0	0.79	0.40	3.69×10^{-13}	0.87	0.17	1.82×10^{-4}	0.84	0.02	2.80×10^{-22}	0.79	0.02	1.49×10^{-21}

and resolution at still lower concentrations. In the present investigation, EPN is analyzed by standard addition method in the unknown samples.

Recommended procedure (dpp): The solution was prepared by dissolving the required quantity of the technical grade EPN in the solvent DMF, and made up with the supporting electrolyte to get the 1.0×10^{-5} M concentration of standard solution. A phosphate buffer of pH 6.0 in 10 % DMF was found to be suitable medium for the determination of EPN. 1 ml of the standard solution was then placed in a polarographic cell and made up with 9 ml phosphate buffer of pH 6.0 to get the concentration of 1.0×10^{-6} M, and then deareated with nitrogen. After recording blank polarogram, small increments (i.e. 1 ml of standard solution by means of micropipettes) were added and polarograms were taken. The optimum conditions for the analytical determination of EPN was found to be a drop time of 2 seconds and a pulse amplitude of 60 mV. The relative standard deviation was found to be 3.23 % (15 replicants). The correlation coefficient was 0.984 in this method.

Acknowledgements

The authors are thankful to the Department of Atomic Energy, Bombay for providing financial assistance to carry out this work.

References

1. H. Benadikova and V. Jakubickova, Collect. Czech. Chem. Commun., 48 (9) (1983) 2636.
2. M.R. Smyth and J.G. Osteryoung, Anal. Chim. Acta, **96** (1979) 335.
3. C.V. Bowen and F.I. Edwards, Anal. Chem., **4** (1950) 232.
4. M. Gruca and K. Mosinska, Pr. Inst. Przen. Org., **2** (1970) 89.
5. K. Hasegawa, Kagaku Keisatsu Kenkyusho Hokoku, **15** (1962) 62.
6. H. Yamano, T. Matsumara and K. Fukado, Kagaku Keisatsu Kenkyusho Hokoku, **15** (1962) 65.
7. M. Suzuki and P.J. Elving, Collect. Czech. Chem. Commun., **25** (1960) 3202.
8. H. Sadek, R.M. Izsa and B.A. Add-El-Nabey, Electrochim. Acta, **16** (1971) 401.
9. P.H. Reiger and G.K. Frainkel, J. Chem. Physic., **39** (1963) 609.
10. P. Ramakrishna and B.V. Ramachandran, Analyst, **101** (1976) 528.
11. F. Feigel, Spot Tests in Organic Chemistry, ch 3, p. 284, Elsvier, Amsterdam.
12. P. Smith, The Chemistry of Open-Chain Organic Nitrogen Compounds, Vol. II, 1966, published by W.A. Benjamin, INC, New York.

ELECTROANALYSIS OF THE ACRYLIC MONOMER/ POLYMER SYSTEM

Lothar Dunsch

Institute of Polymer Technology
Academy of Sciences of the G.D.R.
Dresden, G.D.R.

Introduction

Acrylonitrile is the starting material for the different acrylic compounds formed by various types of reactions.[1] Fig.1 shows the general scheme of the derivation of acrylonitrile and the polymerisation of the different monomers. The system within the dotted lines (Fig.1, right) is present in solutions for the modification techniques of polymers surfaces by radiation grafting. The use of this technique in technical application causes the need of a detailed analysis of the components of the system. The grafting by irradiation leads to a consumption of the monomer as well as to the formation of the so called homopolymers which are non-desired by-products of the graft copolymer. For the control of a grafting solution instrumental methods are needed which preferentially are based on electroanalytical techniques.

The same is true for other polymer modification reactions with acrylamide derivatives especially cationic structures for the formation of cationic surface layers. Furthermore acrylamide derivates like methylenediacrylamide are used in various polymerisation reactions and have to be determined in polymer solutions. The polycations formed by a radical polymerization and their reaction products with polyanions (symplexes) have to be characterized when they are used in polymer modification or other fields.

This broad spectrum of analytical problems can be solved by electroanalytical techniques as will be shown with the following results. The use of electroanalytical methods in polymer research is often restricted to direct current polarography and this paper lays emphasis on the demonstration of the application of other methods. Showing the applicability of electroanalytical methods in polymer research might initiate further work in this field to get a widespread use of these methods in polymer science.

Experimental

All electrochemical experiments were done with the electrochemical system GWP 673 (Zentrum für wissenschaftlichen Gerätebau, Berlin) using the three electrode configuration with a saturated silver/silverchloride reference electrode and the mercury pool counter electrode. With the exception of the maximum suppression method the drop time controller of the DME was used in the range of 1 to 8 s intervall. In ac polarography the frequency was generally 80 Hz and the amplitude 5 mV. Tast mode was used in both methods. In dpp the pulse height was 20 mV and the sampling time 20 ms. The solution was deaerated with nitrogen throughout.

All inorganic salts used as supporting electrolytes were of analytical grade. The monomers acrylamide (Feinchemie Sebnitz), acrylic acid (Fluka), methylenediacrylamide (Merck), methacrylamide (Roehm) and its cationic form trimethylammoniumpropylmethacrylamide-chloride (Roehm) as well as methacrylic acid (Roehm) were of pure grade.

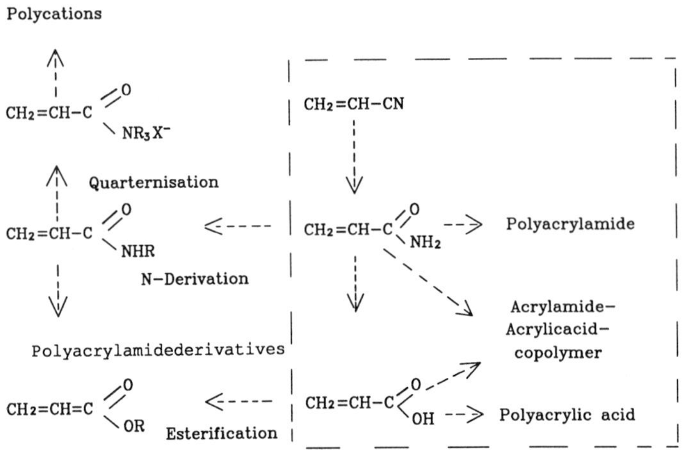

Fig.1. The chemistry of acrylic compounds.

The polymers were prepared by chemical or radiationchemical methods. The purification was done either by dialysis (polyacrylic acid) or by precipitation with ethanol.

Results

The Monomers

Technological conditions[2] as well as environmental health criteria[3] have their impact to the control or monitoring of acrylamide concentrations in aqueous solution. The first proposal to apply differential pulse polarography for determination of acrylamide was made in[4] but the use of methanolic solution, extraction procedures and some erroneous results with acrylic acid (see below) required the development of a new procedure. As can be seen in Fig.2 the methanol content in an electrolyte solution lowers the acrylamide peak height but the peak potential is nearly uneffected by the composition of the solvent system. Contrary to[4] we found well defined dpp peaks in 0.1 M aqueous tetramethylammonium hydroxide (TMAOH) solutions. The exchange of TMAOH by LiOH as supporting electrolyte causes a lower sensitivity of the signal. At low polyacrylamide concentrations (with at least a tenfold excess of acrylamide) the polarographic signal is not distorted by the polymer because the polymer is already desorbed from the electrode surface at that negative potential as will be shown below. But nevertheless there exist at least two sources of interference: reduction of the sodium ion (not the potassium) and reduction of the nitrogen derivatives of acrylamides. The peak potentials of some acrylamide derivatives and methacrylamide are listed in Table 1. It is obvious that all the substituents listed at the amide group do not have any significant effect on the electron density of the carbon-carbon double bond and no effect of any steric hinderanc occures in reduction at the mercury electrode. Therefore no real differences in the peak potentials do exist and the peak separation for different acrylic compounds is impossible. But the simultaneous determination of various acrylamides can be done in a single polarographic measurement if only the sum of the double bonds in all acrylamide derivatives is needed. This can be applied to determine low molecular substituents with an error of 10 to 15 % when a single calibration curve is used.

Acrylamide can be determined in pure aqueous solutions in absence of its derivatives with TMAOH as supporting electrolyte with a precision of +1.55 % in a single measurement and of 5.6 % for a sample. The use of a calibration curve is preferable if the DME parameters are constant within the entire test series.

It must be clearly pointed out that the determination of acrylic or methacrylic acid is impossible by a direct polarographic reduction of their double bond. The polarographic peak in the potential range of -1.6 to -1.75 V in the solution of TMABr is due to reduction of the

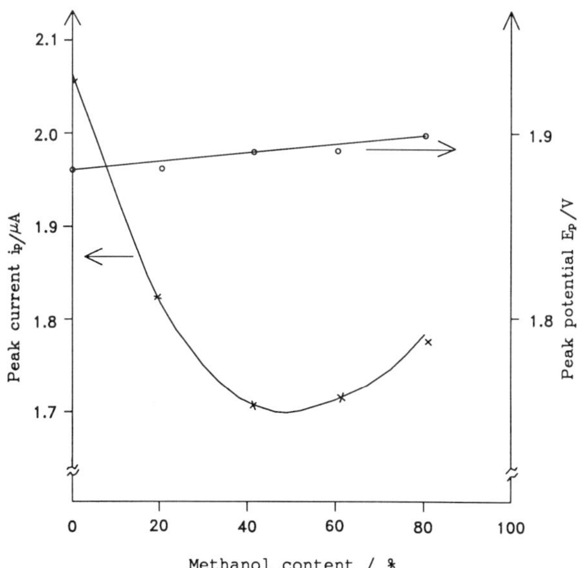

Fig.2. Dependence of the peak current I_p and peak potential E_p of acrylamide on the methanol content in the electrolyte solution. Method:
dpp, pulse height 20 mV, sampling time 20 ms Supporting electrolyte:
0.1 M $(CH_3)_4NOH$.

Table 1. Peak potentials of different monomeric acrylamides,
$(1.5 \times 10^{-3}$ M).

Substance	Peakpotential E/V
acrylamide	-1.88
acrylonitrile	-1.94
methacrylamide	-1.93
methylenediacrylamide	-1.88
hydroxymethylacrylamide	-1.85
hydroxymethylacrylamide-ethylether	-1.87
N-trimethylammoniumpropylmeth-acrylamide-chloride	-1.95
acrylic acid	(-1.67)
methacrylic acid	(-1.71)

dissociated proton. Addition of a mineral acid causes the amplification of the "acrylic acid
peak" as can be seen in Fig.3. The shape of the polarographic proton peak is dependent on
the presence of carboxylate anion. In this case the peak is better defined than in pure mineral
acid solutions of the same concentration. It can generally be said that the polarographic wave
of saturated aliphatic or aromatic organic acids in aqueous solutions occuring in the potential
range of -1.5 to -1.8 V[5] is due to the reduction of the proton of the carboxyl group. The
dependence of the half wave potential on the organic structure is not due to the different
electron density of the molecules but rather due to the different adsorption behaviour of the
molecule and the proton transfer reaction.

When studying the protonated form of the acrylamide the following facts have to be
considered. Adding mineral acids to the acrylamide solution (Fig.4) the peak at -1.88 V
disappears and a new peak at -1.64 V appears which is due to the reduction of the protonated

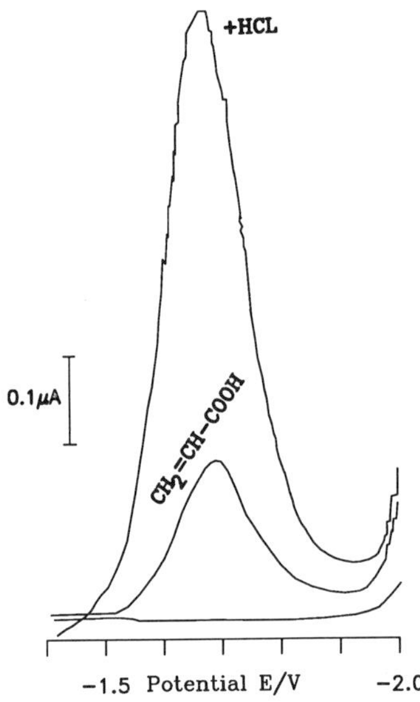

Fig.3. Pulse polarogram of a 10^{-3} M acrylic acid in 10^{-1} M tetramethylammoniumbromide.

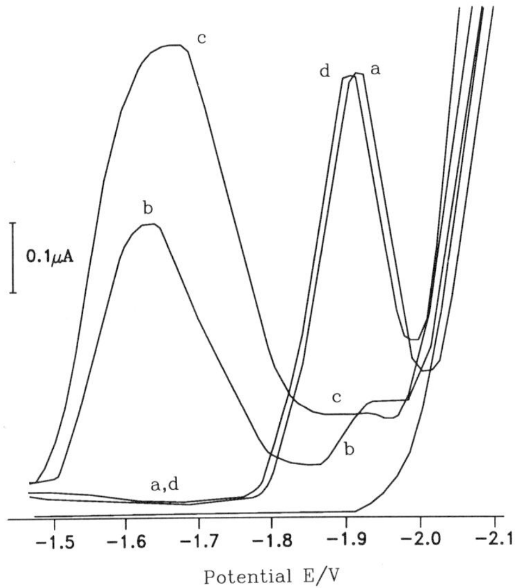

Fig.4. Dpp of acrylamide (1×10^{-3} M) in 0.1 M tetramethylammoniumbromide. Curve a) initial condition, b) after the addition of an equimolar amount of HCl, c) after the addition of acrylic acid, d) after the addition of LiOH (2×10^{-3} M).

form of acrylamide. By further addition of mineral acid this peak increases because reduction of the proton takes place at the same potential. By neutralization with LiOH to the initial condition a single acrylamide peak at −1.88 V with nearly the original height can be found. Only a few percent of the acryleamide might be converted into the acrylic acid during this procedure.

Reduction of the double bond of the acrylic acid in aqueous solutions at the mercury electrode does not take place. The reduction of acrylic acid at the platinum electrode[6] is

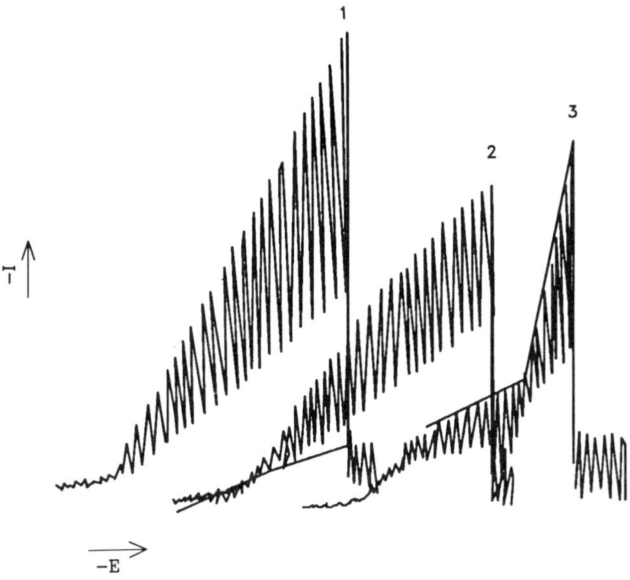

Fig.5. Polarographic oxygen maximum suppression of polyacrylamide (curve 2) and polyacrylic acid (curve 3) in 10^{-3} M Na_2SO_4-solution (curve 1). Polymer concentration 4 mg/l, drop time 2.8 s.

of interest for the preparative scale, but not for analytical applications especially not in industrial monomer solutions.

The Polymers

Adsorption of polymers, in general, on the electrode surface is the main property of these (most) electroinactive organic compounds that has been used in their polarographic determination. The polarographic adsorption analysis, the state of which was quite recently reviewed[7] embraces two different methods: dc polarographic maximum suppression and ac polarography.

The first method has proofed to be very useful in the determination of polyacrylic acid, polyacrylamide and the acrylic acid-acrylamide copolymer in industrial solutions.[8]

The suppression of a polarographic maximum, the theory of which is not yet totally understood,[9] appears to be very simple in use but some restrictions have to be taken into account.

The presence of surfactants or other interferring adsorbable compounds at the glassware should be avoided in studying the polarographic maximum suppression of the polymers. The concentration of the supporting electrolyte may not be higher than 10^{-2} M. The effect of the polymers on the maximum depends on the functional group in the polymer chain. When 10^{-3} M Na_2SO_4 is used as the electrolyte solution the oxygen maximum suppression is irregular in the case of polyacrylic acid as can be seen in Fig.5. Therefore the concentration dependence of the maximum suppression height Δh is quite different for these two polymers (Fig.6). Interaction of the polymers with divalent cations equalizes the different behaviour of the two polymers studied by the maximum suppression. Therefore the Ca^{2+}-ion was choosen because it is electroinactive in the potential range between $+0.2$ and -0.6 V. Thus the suppression activity of polyacrylic acid and polyacrylamide are identical within the concentration range of 1 to 4 mg l^{-1} in the test solution (Fig.6). In this way it is possible to determine the polymer content of solutions containing polyacrylic acid, polyacrylamide and the acrylic acid-acrylamide copolymer with a relative precission of 4.2 %.

If the aim of the analytical work is to distinguish between the different acrylic polymers, the non-specific maximum suppression method must be replaced by the less sensitive but more specific ac polarography.

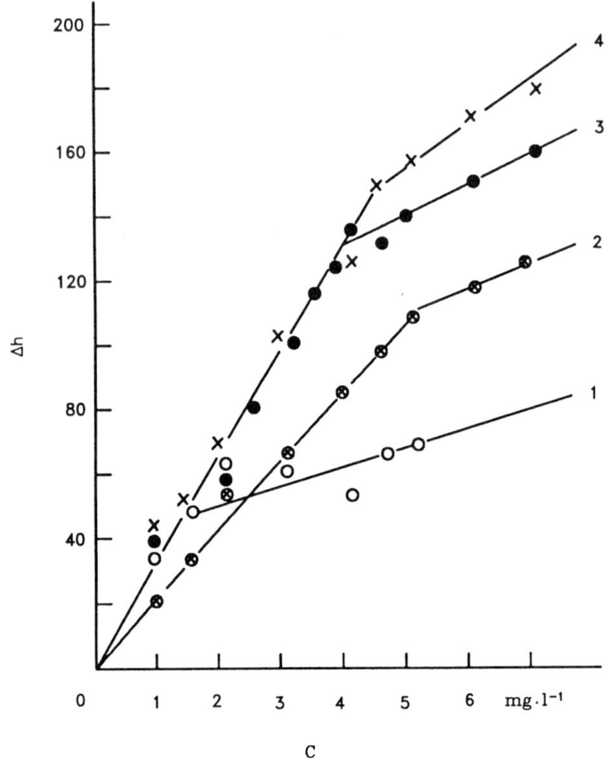

Fig.6. Concentration dependence of polarographic maximum suppression. 1,3-polyacrylic acid in 10^{-3} M Na_2SO_4 and 10^{-3} M $Na_2SO_4/10^{-3}$ M $CaSO_4$ resp.; 2,4-polyacrylamide in 10^{-3} M Na_2SO_4 and 10^{-3} M $Na_2SO_4/10^{-3}$ M $CaSO_4$ resp.

Until now ac polarography has scarcely been used for determination of polyelectrolytes.[10] Under standard conditions (1 M KCl-solution) the neutral form of polyacrylate or polyacrylamide gives no distinct change in the double layer capacity nor it does give any adsorption-desorption peak. As it is known from the polarographic maximum suppression the polymers are indeed adsorbed at the mercury electrode at these potentials. To explain this phenomenon it is supposed that the primary hydration shell of the polyacrylate and the polyacrylamide chain is preserved also in the adsorbed state of the polymer. Therefore the double layer capacity of the mercury electrode is only slightly changed by the polymer adsorption because the solvent water molecules in the inner double layer are replaced by the water molecules of the polymer hydration shell. The situation can be improved by changing the pH-values of the test solution to more acidic conditions. It has been found that the citrate buffer solution of pH = 3.3 with 0.1 M Na_2SO_4 offers the optimum condition where the effect of polymer adsorption is pronounced and the available potential range is large enough to study the adsorption or desorption phenomena of the polymers under study. The suitability of the citrate buffer system is demonstrated in Fig.7 where the ac polarograms of a few polymers are given.

The adsorption behavior of polyacrylamide is somewhat irregular with respect to the double layer capacity changes. A broad hump at -0.9 V is found with some double layer changes at its negative and positive sides. Therefore the tensammetric behavior of polyacrylamide is not suitable for analytical purposes.

The most important aspect of the polyacrylamide ac polarogram is the fact that in the region of the polyacrylic acid desorption $(-1.3$ V$)$ polyacrylamide does not interfere. Thus

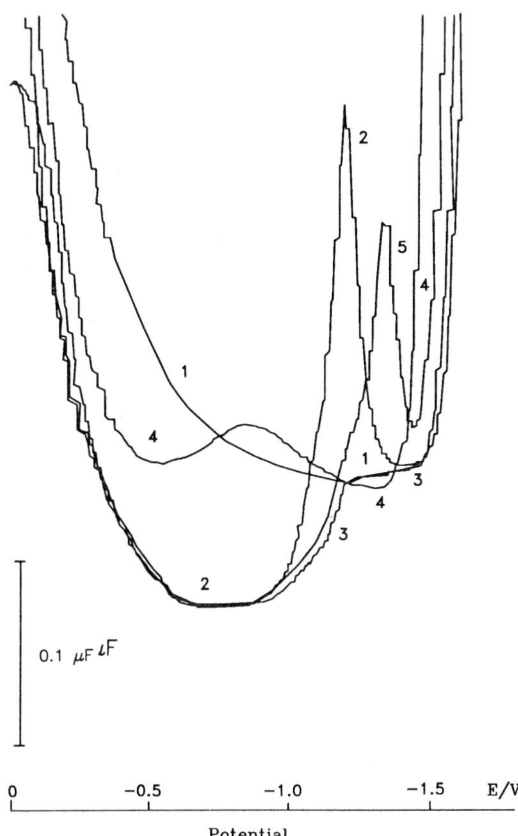

Fig.7. Ac polarogram of different polymers in citrate buffer solution
pH 3.3 (0.1 M Na$_2$SO$_4$). (1) supporting electrolyte, (2) propane-maleic
acid-copolymer, (3) styren-maleic acid-copolymer, (4) polyacrylamide,
(5) polyacrylic acid; concentration: 300 mg/l, frequency 80 Hz, amplitude 5 mV.

the desorption peak of polyacrylic acid, the concentration dependence of which is shown in
Fig.8, can be used for determination of polyacrylic acid in the presence of polyacrylamide.

The citrate buffer has also been applied successfully to polycations. This enables the
electroanalyst to use ac polarography for polycation determination. The choice of the supporting electrolyte is indeed of great importance because earlier results with 1 M LiCl[11] and
polydimethyldiallyl ammonium chloride gave small double layer effects and no adsorption
desorption phenomena. The main advantage of the citrate buffer solution in comparison to
alkali halids is the availability of more positive potentials.[12] Using the potential range up to
+0.2 V the adsorption region of polycations can be studied (Fig.9). Because of coulombic
interactions the polycations are adsorbed at negatively charged electrode surfaces. Therefore
the adsorption peak at −0.05 V can be investigated, but no desorption peak will be found at
negative potentials (Fig.9). The ac polarographic study was done with both polydimethyldiallyl ammonium chloride and poly(trimethylammonium propylmethacrylamide chloride).
For the first product the ac polarogram is shown in Fig.9.

Estimating the adsorption peak height, a linear correlation with the polymer concentration is found in the range of 2.5–50 mg/l. The relative standard deviation of the method
is +6.5 %. The change of the double layer capacity can also be used for analytical work but
the sensitivity is not as high as with the adsorption peak.

As the determination of polyanions and polycations could preferably be done in citrate
buffer solution it should be possible to study the symplex formation polarographically. Ad-

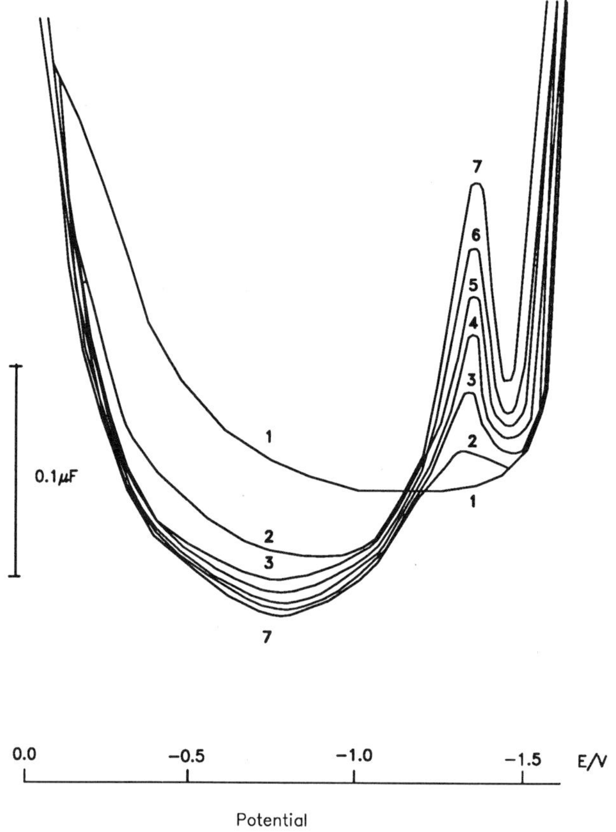

Fig.8. Electrosorption of polyacrylic acid in citrate buffer solution pH 3.3 Polymer content: (1) 0, (2) 60, (3) 120, (4) 250, (5) 500, (6) 1000, (7) 2000 mg/l; frequency 80 Hz, amplitude 5 mV, drop time 1.5 s.

dition of a polycation to a polyanion solution and vice versa could indeed be studied in such solutions. The symplex formed and applied for instance in polymer modified electrodes[13] does not cause a new adsorption characteristics at the electrode even in its soluble initial state but ac polarography could follow the diminuation of either the polycation or polyanion. Therefore the excess component of a symplex is polarographically detectable.

Conclusions

The characterization of acrylic monomer/polymer system can be done with polarographic methods in a rapid and sensitive manner. The monomers and polymers can be determined by different polarographic techniques using special supporting electrolyte composition. In this way electroanalytical methods can contribute to the study of the state and the interactions of acrylic monomer/polymer systems needed in research, industrial applications and pollution control.

Acknowledgements

The financial support of the Academy of Sciences and the technical assistance of Mrs. U. Feist are duly acknowledged.

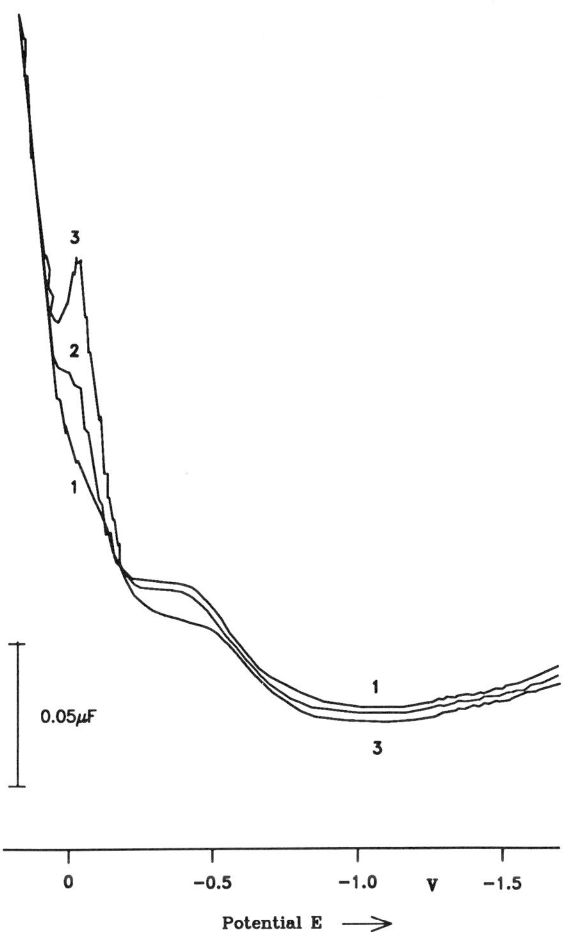

Fig.9. Ac polarogram of polydimethyldiallylammonium chloride in citrate buffer solution pH 6.0; polymer content (1) 0, (2) 20, (3) 30 mg/l, frequency 80 Hz, amplitude 5 mV, drop time 1.5 s.

References

1. For a review see:
   D.C. Mac Williams, Acrylamide and other alpha, beta unsaturated amides in: "Functional monomers", R.H. Yo and E.B. Nyquist (Eds), Vol. 1, New York, Marcel Dekker 1973, pp. 1-187,
   N.M. Bikales, Acrylamide and related amides, in: "Vinyl and diene polymers", E.C. Leonhard (Ed.), Part 1, New York, Interscience 1970, pp. 81-104,
   L.S. Luskin, Acrylic acid, methacrylic acid and the related esters, ibid, pp. 105-203,
   H. Rauch-Puntigam and Th. Voelker, in: "Acryl-und Methacrylverbindungen", Chemie, Physik und Technologie der Kunststoffe in Einzeldarstellungen, Bd. 9 Berlin-Heidelberg-New York, Springer 1967.
2. M. Raetzsch, H. Paessler and A. Heger, Radiat. Phys. Chem., 25 (1985) 461,
   A. Heger, H. Paessler, L. Dunsch and B. Ihme, Papers of the 10 th. Münchener Klebstoffseminar, Munich 1985, pp. 176-184.
3. Environmental Health Criteria 49, Acrylamide, Geneva, WHO, 1985.
4. S.R. Betso and J.D. McLean, Anal. Chem., 48 (1976) 766.
5. K. Schwabe, in: "Polarographie und chemische Konstitution organischer Verbindungen", (Scientia chimica 8) Berlin, Akademie-Verlag 1957, pp. 144-187.
6. M. Byrne and A.T. Kuhn, J. Electroannal. Chem., 60 (1975) 75.

7. P.M. Bersier and J. Bersier, Analyst, **113** (1988) 3.

8. L. Dunsch, U. Feist and J. Morgenstern, Acta Polymerica, **34** (1983) 73.

9. H.H. Bauer, Streaming Maxima in Polarography, in: "Electroanalytical Chemistry", A.J. Bard (Ed.), Vol. **8**, Marcel Dekker, New York, 1976, pp. 169-279.

10. H. Jehring, in: "Electrosorptionsanalyse mit der Wechselstrompolarographie", Berlin, Akademie-Verlag 1974.

11. B. Philipp, B. Reiche and H. Jehring, Faserforschung und Textiltechnik, **29** (1978) 431.

12. L. Dunsch, Z. Chem., **26** (1986) 337.

13. L. Dunsch, in: "Polymermodifizierte Elektroden", L. Dunsch (Ed.), Dresden, ITP 1987, pp. 51-75.

Abd-El-Wahab G.,423
Ahmed Z.A.,419,423
Alonso R.,379
Antila M.,237
Arranz Valentin J.F.,387
Asplund J.,85
Bühler H.W.,305
Barrio Diez-Cabarello R.J.,387
Belluati M.,323
Bersier J.,109
Bersier M.,109
Bobacka J.,21,37
Buckley E.,367
Chastel O.,373
Costa Garsia A.,329
Cox A.,267
Dürselen F.J.,305
Dankházi T.,413
Dhaktode S.S.,431
Diamond D.,51
Dunsch L.,59,443
Emons H.,407
Engblom S.O.,21
Farsang G.,413
Faulkner R.,5
Fernandez Alvarez J.M.F.,329,
Flores J.R.,367
Gallo B.,379
Ghali E.,213
Goldring S.,191
Gorton L.,183
Gray J.,267
Guette C.,299
Haapakka K.,293
Han E.,289
Hansen E.,245

Hernandez L.,205
Hernandez P.,205
Hulanicki A.,145,213,317
Ivaska A.,1,21,37
Jahngen E.G.E.,275
Janata J.,159
Janata P.,199
Johansson G.,183
John R.,367
Josowicz M.,199
Kalvoda R.,403
Kamal M.M.,419,423
Kankare J.J.,15,31,293
Karpinski Z.J.,395
Kauffmann J-M.,373
Kellner R.,223
Kowalski Z.,149
Laitinen H.,237
Lehtonen P.O.,237
Lewandowski R.,317
Lewenstam A.,145,213
Lindner E.,223
Lipiäinen K.,293
Lorenzo E.,205
Lukkari J.,31
Lundström I.,173
Maas A.H.J.,311
Manzoni A.,323
Meaney M.,283
Migdalski J.,149
Nayak M.S.,431
Neshkova M.,231
Niegreisz Zs.,71
Nijenhuis te B.,139
Patriarche G.J.,373,379
Persson P.,183

Polasek M.,183
Polos L.,71
Potje-Kamloth K.,199
Pungor E.,71,223
Reddy S.J.,339,437
Reddy T.N.,437
Reust J.B.,359
Rips R.,299
Rumpf G.G.,305
Schulze G.,289
Simon W.,305
Siva Sankar P.,339
Smyth M.R.,283,367
Smyth W.F.,349
Sokalski T.,145,317
Sprokholt R.,311
Stojek Z.,47
Swartz M.,275
Sänger R.,289
Temerk Y.M.,419,423
Toth K.,223
Tougas T.P.,275
Trojanowicz M.,255
Tuñon Blanco P.,329
Tupasela T.,237
Vire J-C.,373,379
Wallace G.G.,283,367
Walsh M.R.,5
Wasberg M.,21
Werner G.,407
Wingvist F.,173
Xu C.,5
Zhang H.,379
Zippel E.,223

INDEX

Acrylamide, 443
Acrylic acid, 443, 444
Acrylic, 443
Acrylonitrile, 443
Admittance, 21, 38
Ammonia sensor, 178, 243, 174
Amyloglucosidase, 247
Anthraquinone, 8, 123
Antibody, 81, 367
Antigen, 367
Asymmetry potential, 225, 306

Benzoquinone, 66, 409
Benzylidene acethydrazones, 423
Biotechnology, 139, 250

Calcium
    in biotechnology, 235
    in clinical samples, 311, 317, 323
Carbohydrate oxidation, 267, 275
Carbon dioxide electrode, 191, 236
Carbon fibre electrode, 64, 197
Cephalosporin, 353
CHEMFET, 81, 141, 173, 243
Chemical interference, 74, 118, 162,
        248, 289, 301, 319, 344, 350,
        360, 391, 404, 410, 444
Chemically modified electrode, 183,
        203, 247, 267, 283, 368
Chloride
    in clinical samples, 317, 323
    interference, 145, 211, 225
Chloropromazine, 413

Chronoamperometry, 14, 35, 59, 342
Chronocoulometry, 8, 14, 275, 407
Chronopotentiometry, 59, 111, 243,
        339, 342
Clay modified electrode, 203
Clobazepam, 355
Coated wire electrode, 159
Coenzyme, 247
Conducting polymer, 14, 16, 27, 33,
        54, 370
Coulometric
    detector, 363
    titration, 360
Coulometry, 59, 74, 126, 243, 267,
        363, 373
Creatinine, 178
Critical micelle concentration, 407
Current-to-photon conversion, 4
Cysteine, 270, 271, 368, 395

Dehydrogenase, 286
Dimethyl glyoxime, 115, 410
DNA, 278, 339

Encapsulation, 162, 198
Environmental analysis, 72, 109,
        250, 253, 289, 359
Enzyme, 139, 183, 311, 373
    FET, 168
    electrode, 80, 247, 285
    immobilized, 168, 173
    immunoassay, 367
    sensor, 164, 247, 248, 370

ESR spectroscopy, 59
Explosives, 85

Fermentation, 80, 139, 235
Fermenter, 140, 174, 236
FIA, 141, 243, 254, 275, 283, 289
    amperometry, 267, 270
    electrochemical detection, 111,
        132, 243, 367
    gradient, 243, 244
    impulse-response, 250
    potentiometry, 186, 243, 258, 260
    voltammetry, 244
    technique, 413
Flow technique, 77, 289
Flow through
    cell, 74, 225, 243, 244, 253, 270,
        271, 286, 289
    electrode, 244, 311
    system, 61, 173, 186, 229, 243,
        267, 273, 289, 360, 396, 404
Folic acid, 330
FT-FAM, 21, 39
FTIR, 221

Gas sensor, 51, 173, 196, 243
Glucose, 142, 147, 168, 238, 247,
        247, 275, 277, 278, 285, 286
    dehydrogenase, 247
    oxidase, 168, 248, 249, 285, 369
    sensor, 141, 168, 248
Glutathione, 271
Gran method, 145

Heavy metals, 109, 145, 179, 229,
        289, 360, 431
HPLC, 141, 351
    electrochemical detection, 104,
        111, 267, 275, 283, 299, 367,
        387

Immobilized enzyme, 247, 247, 285
Immunoassay, 17, 111, 367
Impedance, 14, 25, 37, 72, 238
Interference, 163, 168
interfering ion, 53, 146, 148, 211,
        219, 225, 286, 305, 308
Inverse stripping, 116

Ion-chromatography, 227, 253, 361
    amperometric detection, 360
    conductivity detection, 363
    potentiometric detection, 253
Ion-selective electrode, 53, 74, 141,
        145, 159, 192, 211, 221, 229,
        237, 253, 305, 311, 317, 323
    calcium, 237, 244, 305, 311, 314,
        317, 323, 324
    chloride, 52, 305, 317, 319, 370
    copper, 211, 229
    potassium, 221, 305, 312, 317,
        324, 325
    sodium, 52, 55, 305, 311, 312,
        317, 320, 323
Ionophore, 52, 163, 227, 258, 307
ISFET, 52, 81, 141, 159, 222, 243,

Karl Fischer titration, 363

Lactate oxidase, 249
Lactic acid, 235, 249
Luminescence, 4, 17, 247, 293

Marcellomycine, 379
Maximum suppression, 127, 443,
        447
Mediator, 184, 247, 267, 271, 275,
        277, 286, 369, 379
Mercury sensors, 176
Metronidazole, 61, 339, 354
Micro-organism, 139, 235
Microcell, 176
Microcomputer, 31, 74
Microconduit, 244
Microelectrode, 3, 47, 197, 253, 270
Microlitography, 55
Microprocessor, 28, 31, 51, 293
Microwave, 53, 118
Modified electrode, 285
Modulated flow, 289
Monomer, 14, 113, 192, 443, 444

NADH, 183, 247, 285
NADPH, 247
Nitrobenzene, 62, 203

Oxidase, 247

Oxide electrode, 275
Oxyanion, 419
Oxygen electrode, 74, 141, 191, 236,
    285
Oxygen maximum, 119, 447

PAH, 294
Penicillin, 168, 352
Pesticides, 116, 403, 437
Phenothiazine, 374, 404, 413
Phenoxazine, 183
Phenyl urea, 86, 87, 113, 127
Photoimpedance, 16
Piperazine, 374
Polarographic detector, 149, 244
Polarography, 88, 109, 149, 360, 379,
    395, 403
    ac, 37, 111, 339, 340, 437, 439,
        443, 448
    dc, 91, 111, 339, 407, 437, 443
    differential pulse, 89, 111, 342,
        349, 361, 407, 419, 423, 437,
        444
    normal pulse, 111
    reverse pulse, 111
    square wave, 111, 349
Pollutants, 110, 179, 437
Polyamine, 299
Polyelectrolyte, 443, 448
Polymer, 443, 447
    coated electrode, 27, 161, 169
    electrode, 54, 192, 268, 305
    electrolyte, 191, 194, 195
    membrane, 162, 198, 227, 236,
        306, 370, 443
    modified electrode, 59, 65, 283,
        286, 369, 373, 449
Polypyrrole, 169, 283, 370
Polythiophene, 14, 27, 35
Potassium in clinical samples, 311,
    317
Potentiometric stripping, 111, 243,
    289, 362
Process analytical chemistry, 71, 139
Process control, 71, 85
Process monitoring, 115, 149, 247
Product control, 85
Pteridine, 330

Quinone, 122, 379

RNA, 278

Selectivity coefficient, 146, 230, 319,
    305
Semiconductor sensor, 75, 169, 173
Signal processing, 21
Smoothing, 37
Sodium in clinical samples, 55, 317,
    323
Speciation, 115, 361
Spectroelectrochemistry, 14, 222
Static mercury drop electrode, 37,
    149, 330, 349, 379, 431

Tensammetric
    peak, 403
    techniques, 112, 119, 126
    waves, 114
Tensammetry, 362, 448
Thiazolidine, 395
Thin layer cell, 413
Timolol, 285, 387
Trace analysis, 111, 120, 203, 359,
    403, 410, 423

Urea, 173, 177, 249, 367
Urease, 173, 177

Vitamin B, 330
Vitamin K, 379
Voltammetry, 74, 85, 109, 203, 214,
    349, 360
    ac, 330, 407
    adsorptive stripping, 111, 116,
        123, 329, 368, 379, 387, 403,
        410
    anodic stripping, 111, 116, 121,
        431
    cathodic stripping, 111, 129, 334
    cyclic, 3, 59, 187, 200, 204, 245,
        271, 277, 283, 330, 340, 341,
        368, 374, 380, 388, 395, 438
    differential pulse, 102, 111, 396,
        413
    hydrodynamic, 271, 277
    inverse, 132

Voltammetry (continued)
   linear scan, 129, 267, 270, 331,
       381, 413, 417
   normal pulse, 111, 395, 413
   pulse, 395
   reverse pulse, 47
   square wave, 47, 111, 350, 379,
       395

Waste water, 76, 361